Thu 6:45 - 9:45

Boo?
M.E.

Data General

Systems Analysis, Design, and Development with Structured Concepts

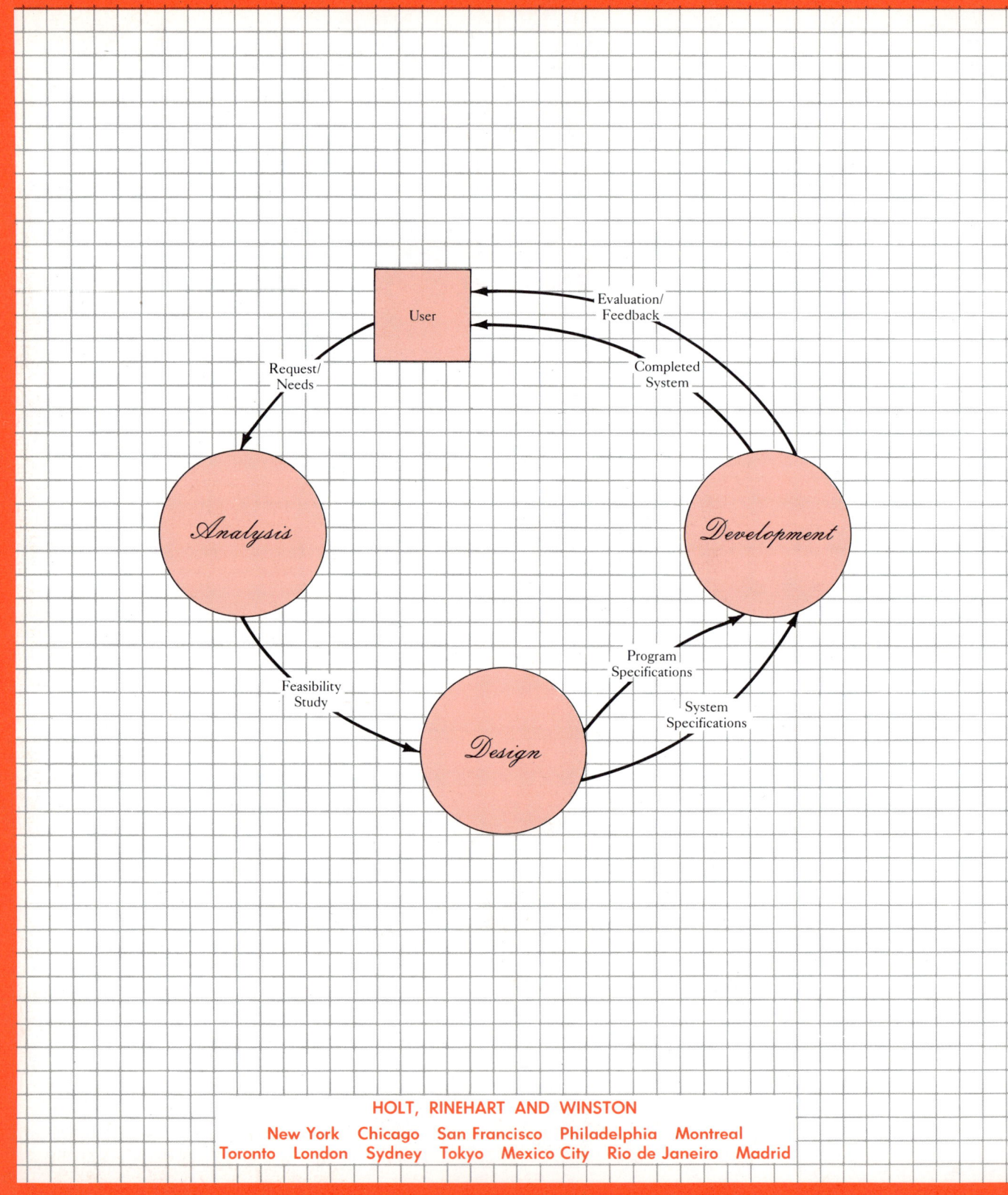

Systems Analysis, Design, and Development with Structured Concepts

PERRY EDWARDS

Computer Systems — Office Automation
Computer World

How to conduct a system study

Copyright © 1985 CBS College Publishing
All rights reserved.
Address correspondence to:
383 Madison Avenue, New York, NY 10017

Library of Congress Cataloging in Publication Data
Edwards, Perry.
  Systems analysis, design, and development with structured concepts.

  Includes bibliographies and index.
  1. System design.   2. System analysis.   I. Title.
QA76.9.S88E32   1985   003   84-28979
ISBN 0-03-000142-0

Printed in the United States of America

Published simultaneously in Canada

5 6 7 8   032   9 8 7 6 5 4 3 2 1

CBS COLLEGE PUBLISHING
Holt, Rinehart and Winston
The Dryden Press
Saunders College Publishing

Brief Contents

Preface		xvii
PART I	SYSTEMS ANALYSIS	1
1	The Systems Cycle and Business Organizations	2
2	Structured Systems Analysis	34
3	Detailed Analysis and the Feasibility Study	66
4	Tools for the Systems Designer	106
PART II	SYSTEMS DESIGN	147
5	Systems Design—A Broad Perspective	148
6	Output Design	170
7	Organizing and Designing Files	234
8	Designing a Data Base	264
9	Input Design	288
10	Process Design and Acquisition of Hardware and Software	324
11	Program Definition, Module Design, and the Design Review	350
PART III	SYSTEM DEVELOPMENT	383
12	Programming and Conversion	384
13	Testing, Training, Documentation, and System Maintenance	416
14	Management of the Systems Process	444
Appendix	Business Systems Today	464
Glossary		495
Index		503

Detailed Contents

Preface — xvii

PART I SYSTEMS ANALYSIS — 1

Chapter 1 The Systems Cycle and Business Organizations — 2

Goals and Preview — 3
Systems — 4
    Computer-Based Business Systems — 4
    Systems Analysis, Design, and Development — 6
Organizational Structures — 6
    Line and Staff Organizations — 7
    How Data Processing Fits Into an Organization — 8
The Roles of the Analyst and the User in the Organization — 9
Technical Developments and Information Systems — 11
    Hardware — 11
    Software — 16
    Structured Methodology — 16
Dijkstra, Bohm, and Jacopini: The Developers of Structured Programming — 18–19
The Systems Life Cycle — 20
    Step 1. Systems Analysis — 21
    Step 2. Systems Design — 22
    Step 3. Systems Development — 22
    The Systems Life Cycle Illustrated — 23
Case Study: Fleet Feet Athletic Shoes: An Ongoing Example of the Systems Life Cycle at Work — 25
Working with People: Communication Skills — 30
Summary — 31
New Terms — 32
Questions for Review and Discussion — 32
References — 33

Chapter 2 | Structured Systems Analysis — 34

Goals and Preview	35
Data-Flow Diagrams	36
An Overview of Analysis	37
Preliminary Analysis	39
Evaluation of a User's Request	39
Analysis of the Request	42
The Interview	43
Preparation	45
Warm-up	45
Questioning	45
Closure	47
Follow-up	47
The Preliminary Report and Management Action	47
The Politics of Systems 51–52	
Case Study: Camp Beaverbrook's Camper System	52
Working with People: Conducting a Meeting	61
Summary	62
New Terms	63
Questions for Review and Discussion	63
References	64

Chapter 3 | Detailed Analysis and the Feasibility Study — 66

Goals and Preview	67
Review and Assignment	69
Preliminary Report	69
Scheduling the Study with a Gantt Chart	69
Authorization and Notification	70
Fact Finding	71
Interviewing	72
Questionnaires	72
Observing the System	74
Fifteen Key Questions to Ask Any Software Supplier 76–77	
The Feasibility Study	77
Diagramming the Logical System	77
Sample Documents	81
Alternatives, Costs, and Benefits	83
Reporting Findings	86
Presentation of Findings to Management	86
Case Study: Detailed Analysis of Camp Beaverbrook's Camper System	90
Working with People: The Need for Accounting in an Analyst's Training	103
Summary	104
New Terms	104
Questions for Review and Discussion	104
References	105

Chapter 4	**Tools for the Systems Designer**	**106**
	Goals and Preview	107
	The Need for a Structured Approach	108
	The Top-Down Approach	108
	Modules	109
	Flowcharts	111
	System Flowcharts	112
	Program, or Detail, Flowcharts	115
	HIPO—Hierarchy Plus Input Process Output	115
	Constructing a VTOC	117
	IPO—Input Process Output	119
	Pseudocode or Structured English	120
	The Phenomenon of the Spreadsheet 123–124	
	Data-Flow Diagrams	125
	Data Dictionary	128
	Decision Tables and Decision Trees	130
	Warnier-Orr Diagrams	135
	Nassi-Shneidermann Chart	137
	Structured Walkthroughs	138
	Case Study: Data-Flow Diagrams, the Data Dictionary, and	
	Pseudocode for Fleet Feet's Accounts Payable System	139
	Working with People: New Tools Versus Old	143
	Summary	144
	New Terms	145
	Questions for Review and Discussion	145
	References	146
PART II	**SYSTEMS DESIGN**	**147**
Chapter 5	**Systems Design—A Broad Perspective**	**148**
	Goals and Preview	149
	Design Overview	151
	Review and Assignment of Tasks	152
	Output Design	153
	Data-Base, or File, Design	154
	Input Design	155
	Process Design	156
	Program Definition	157
	Module Design	157
	Package Design	157
	Design Review	158
	PERT Charts 160–162	
	Review and Assignment of Tasks	163
	Case Study: Review and Assignment of Tasks for Fleet Feet's	
	Accounts Payable System	164

Working with People: How Important Is Keeping Current with the State of the Art? 167
Summary 168
New Terms 168
Questions for Review and Discussion 169
References 169

Chapter 6 Output Design 170

Goals and Preview 171
Output Design: General Guidelines 173
Selecting the Best Output Medium 174
   Printers 176
   Paper 179
   CRTs 181
Formatting Reports: Headings, Footings, Details, and Totals 183
Types of Reports: Query, Detail, Summary, Exception, and Periodic 185
Designing Reports for Printers 187
Designing Reports for CRTs 191
Guidelines for CRT Screen Designs   190–191
Control of System Outputs 195
   An Overview of Control Systems 195
   Totaling and Cross-footing 196
   Auditing 196
   Security Checks 198
Building a Data Dictionary 199
Case Study: Designing the Reports for Fleet Feet's Accounts Payable System 200
Working with People: Involving the User from the Start 230
Summary 231
New Terms 232
Questions for Review and Discussion 232
References 233

Chapter 7 Organizing and Designing Files 234

Goals and Preview 235
An Introduction to File Design 237
   Storage Capabilities 237
   Types of Files 238
Data-Storage Methods: ASCII, EBCDIC, Numeric Methods 238
   Storing Alphanumeric Data 239
   Storing Numeric Data 239
Media and Data-Storage Techniques 241
   Tapes and Sequential Files 242
   Disks and Direct-Access Files 244
   VSAM and Direct-Access Files 248

DETAILED CONTENTS xi

Designing Disk or Tape Files	251
Guidelines for Designing Files 253	
File and Processing Controls	254
Record Counts	254
Backup	255
Case Study: Fleet Feet's Accounts Payable File Design	256
Working with People: Professional or Personal Computer Considerations	260
Summary	261
New Terms	261
Questions for Review and Discussion	262
References	263

Chapter 8 — Designing a Data Base   264

Goals and Preview	265
Data-Base Management Systems	266
Types of Data Bases	269
Hierarchical Model	269
Network Model	270
Relational Model	270
Defining Data Bases	272
Data Manipulation Languages	273
Relational DBMS: What's in a Name? 274–275	
Query Languages	276
Utilities	277
Data-Base Controls	278
Transaction Logging	278
Access Security	278
Case Study: Fleet Feet's Accounts Payable Data-Base Design	279
Working with People: Retraining for the Data-Base Environment	284
Summary	285
New Terms	285
Questions for Review and Discussion	285
References	286

Chapter 9 — Input Design   288

Goals and Preview	289
Methods of Data Entry	291
Controlling Data Entry	294
Verification	295
Validation	295
Batch, or Control, Totals	296
Check Digits, Transpositions, and Slides	296
Visual Verification and Computer-Assisted Validation	297
Data-Entry Hardware	298
Terminals: Dumb and Intelligent	298

xii DETAILED CONTENTS

Guidelines for Designing Data-Entry Screens 299	
Key-to-Tape and Key-to-Disk Devices	300
Optical Readers	304
Designing Data Entry for Terminals	305
Designing Data Entry for Key-Entry Devices	308
Case Study: Fleet Feet's Input Design	312
Working with People: Ergonomics and Design	320
Summary	321
New Terms	322
Questions for Review and Discussion	322
References	323

Chapter 10 Process Design and Acquisition of Hardware and Software 324

Goals and Preview	325
Modes of Processing Data	327
Batch Processing	327
Online Processing	329
Real-Time Systems	330
Distributed Systems: Completely Connected, Star, Ring, and Bus Networks	331
LANs and Distributing the Work Load 338–339	
Hardware and Software Acquisition: RFP and RFQ	340
Case Study: Fleet Feet's Online Accounts Payable System and an RFQ for New Terminals	344
Working with People: Too Many Controls	346
Summary	347
New Terms	348
Questions for Review and Discussion	348
References	349

Chapter 11 Program Definition, Module Design, and the Design Review 350

Goals and Preview	351
Program Definition	354
Module Design	356
Modules	356
Control Structures	358
Decomposition and Refinement	360
Coupling	363
Guidelines for Planning Modules 366–367	
Language Considerations	368
Package Design and Program Specifications	369
Design Walkthrough	371
Design Review	372

	Case Study: Design Specifications for Fleet Feet's Accounts Payable Voucher-Check Printing Program	373
	Working with People: A Program Specifications Walkthrough	380
	Summary	381
	New Terms	381
	Questions for Review and Discussion	382
	References	382

PART III — SYSTEMS DOCUMENTATION — 383

Chapter 12 — Programming and Conversion — 384

Goals and Preview	385
Overview of Development	386
Scheduling and Assignment of Tasks	388
Programming a Structured System	393
Standards	393
Stubs	393
Program Walkthroughs	395
Testing: Modules, Module Integration, and Programs	399
Guidelines for Writing IF and GO TO Statements **399**	
Conversion	403
Parallel	403
Phased	405
Direct	407
Programs, Facilities, and Procedures	408
Case Study: Developing Fleet Feet's Accounts Payable Voucher-Check System	410
Working with People: Coping with Change	413
Summary	414
New Terms	414
Questions for Review and Discussion	415
References	415

Chapter 13 — Testing, Training, Documentation, and System Maintenance — 416

Goals and Preview	417
Testing	418
Program Integration	418
System Test	418
Acceptance Tests	422
Training	423
Tools for Training	423
Management Training	424
Users and Operations Staff	424
Guidelines for Terminal Users **425**	
Documentation	426

	Management Documentation	427
	User Documentation	427
	Program Documentation	427
	Operations Documentation	430
	System Maintenance	433
	Case Study: Testing, Training, and Documentation for Fleet Feet's Accounts Payable Voucher-Check System	434
	Working with People: Writing with Style	440
	Summary	441
	New Terms	442
	Questions for Review and Discussion	442
	References	443
Chapter 14	**Management of the Systems Process**	**444**
	Goals and Preview	445
	Management Activities: An Overview	446
	Schedule Overruns	447
	People Problems	447
	Theories X, Y, and Z 448–449	
	Cost Overruns	449
	System Operations	450
	System Audit or Review	451
	System Documentation	453
	Keeping Current	454
	Professional Societies	457
	Journals and Periodicals	457
	Books	458
	Training Groups	459
	Case Study: Acceptance of Fleet Feet's Accounts Payable System	460
	Working with People: Analysts as Instant Experts	462
	Summary	462
	New Terms	463
	Questions for Review and Discussion	463
	References	463
Appendix	**Business Systems Today**	**464**
	Goals and Preview	465
	Information and Decision Making	466
	The General Ledger System	467
	The Accounts Receivable System	474
	Balancing Methods	474
	Aging	478
	The Accounts Payable System	479
	The Payroll System	482
	The Electronic Spreadsheet and VisiCalc	486

Other Systems	489
Order Entry	490
Fixed Assets	492
Summary	493
New Terms	493
Questions for Review and Discussion	494
References	494
Glossary	495
Index	503

Preface

The issues confronting analysts today—communications with users, exploding technology, and resistance to change—inspired this book. It offers a comprehensive and contemporary approach to the subject. Today's analysts, whether they receive their training in the undergraduate or graduate programs at 4-year schools or in 2-year programs at community colleges, not only need a solid foundation in systems principles and practice but also must have a sense of the people problems they will encounter in their work.

Employers who hire graduates with 4-year or graduate school degrees often complain that they lack real-world knowledge. For instance, recruits may have mastered "textbook" accounting but cannot implement what they know on a computer system. Other employers discover that computer science majors with weak accounting and management backgrounds have trouble understanding the information needs of a business. To cope with ever-changing computer systems, new analysts must master the general principles they can apply to evolving technology. *Systems Analysis, Design, and Development with Structured Concepts* presents perennial principles, discusses their practice within contemporary computing, and indicates an ongoing concern with people and organizations. It illustrates every principle with case studies, applications, and examples, and concludes each chapter with a unique "Working with People" story.

Since integrated structured techniques provide powerful new tools for professionals, they are presented in the first chapter (data-flow diagrams) and used wherever appropriate throughout the remainder of the book.

The text also stresses terminals and data-base systems available for mainframes, minicomputers, and personal computers. Without neglecting traditional subjects such as sequential files, magnetic tape, indexed sequential files, and direct files, the book presents and illustrates state-of-the-art topics such as screen menus; data-base schemas; hierarchical, network, and relational data structures; and query languages.

To bring the reader into contact with real information systems, an appendix examines the computerized business systems predominant in modern firms. Here students see how all of the systems (accounts receivable, accounts payable, payroll, general ledger, etc.) work together in a management information system. In addition to the usual accounting business systems, the book also covers electronic spreadsheets and order-entry and fixed asset systems.

Because programming is vital to any system, this book examines such topics as language selection, programming practices, and system security. This material is not concerned with language details, but rather concentrates on the procedures involved in programming a system, including walkthroughs, modularity, decomposition, coupling, span of control, cohesion, and stepwise refinement.

Each of the text's 14 chapters begins with a brief outline and a set of goals followed by a preview that sets the stage for the chapter. Whenever a new principle appears, an example, or "minicase," immediately reinforces it, and numerous illustrations help bring the principles to life. To marry principles and practice further, each chapter concludes with two case studies, one that applies principles to the evolving needs of a hypothetical business, and one ("Working with People"), that dramatizes the human dimension of computing and the systems process.

The ongoing case study involves a hypothetical business, a chain of athletic-shoe stores called Fleet Feet, whose growing information needs demand the increased use of computers. We follow Fleet Feet's analyst from the initial request for a new accounts payable system to its final installation and audit.

As catalysts of change, analysts must quickly master "people problems": person-to-machine, person-to-system, and person-to-person relationships. The "Working with People" sections dramatize such topics as user involvement, written and oral communications, effective meetings, interviewing techniques, accounting, and the effect of computerization on jobs.

After the case studies, readers will find a chapter summary, a list of new terms introduced in the chapter, questions for review and discussion, and a list of pertinent books and journal articles. The questions involve three levels of activity: review, application of material, and research.

Of special interest is the description of some exciting new development that appears at a key point in each chapter. These sections feature essays that bring readers up to date with some of the issues that will determine the future environment of systems analysis, design, and development: structured methodology; local area networks; guidelines for designing screens, files, and modules; relational data bases; questions to ask software suppliers; spreadsheets; and management theories X, Y, and Z.

The book concludes with a comprehensive glossary and index. The glossary defines words or terms in short, easy-to-read format. The multilevel index permits the reader to find quickly any topic in the text.

The text's teaching system includes a practical *Study Guide*, software, and a detailed *Instructor's Manual*. The *Study Guide* contains a second lengthy case study that allows students to apply what they have learned in the text. The software system drills the student in the new words and concepts, and the *Instructor's Manual* presents solutions to end-of-chapter questions, as well as suggestions for class projects, overhead transparency masters, and creative teaching ideas.

Systems Analysis, Design, and Development with Structured Concepts fulfills the requirements of the first systems class recommended by both the Data Processing Management Association (DPMA) and the Association for Computing Machinery (ACM) in their model curricula for undergraduate computer information systems education. Both curricula prescribe a course in systems analysis and methods as the fourth or fifth class in their respective programs.

ACKNOWLEDGMENTS

Writing any book is a creative effort on the part of many individuals. The author works with a team of people in bringing the manuscript from concept to final bound copies.

Early in the writing of the book, Kathleen Edwards (my wife) and Mike Snell (a long-time friend) joined me in writing and editing the various drafts of the book. They provided me with invaluable assistance through the entire 3 years of writing and revising the book. I am very grateful for their patience, helpful advice, and constant prodding to achieve the best book possible.

Thanks also to Mr. Clifford Burns of Sierra College for class testing the book. For three semesters Cliff and his students used the text and provided invaluable feedback to me.

The fine student *Study Guide* is the work of Mr. Jerry Ralya. His work enhances the text and helps students practice and master the materials presented in the text.

At least twenty people from all over the country, representing all types of colleges and universities reviewed the manuscript prior to its publication. My special thanks to: Mr. Brian Michaelson, Palomar College; Mr. William A. Jones, Modesto Junior College; Mr. Joseph Waters, Santa Rosa Junior College; Mr. Paul Ross, Millersville State University; and Mr. Richard Brightman, Golden West College.

A fine group of people produced the book at Holt, Rinehart and Winston, including Deborah L. Moore, Acquisitions Editor; Paul Nardi, Production Manager; and Lila M. Gardner, Project Editor at Cobb/Dunlop Publisher Services, Inc. My thanks go to each and every one of them.

Finally, normal family relations can resume. My two daughters and son will not have to fight with me over the Apple computer and can now continue to expand their knowledge of computers.

Perry Edwards

PART I

Systems Analysis

CHAPTER 1
The Systems Cycle and Business Organizations

- Goals and Preview
- Systems
- Organizational Structures
- The Roles of the Analyst and the User in the Organization
- Technical Developments and Information Systems
- Dijkstra, Bohm, and Jacopini: The Developers of Structured Programming
- The Systems Life Cycle
- Case Study: Fleet Feet Athletic Shoes: An Ongoing Example of the Systems Life Cycle at Work
- Working with People: Communication Skills
- Summary
- New Terms
- Questions for Review and Discussion
- References

GOALS AND PREVIEW

After reading this chapter you should be able to:

1. Define the word "system," and give at least three examples of common systems.
2. Draw a diagram of an organization chart.
3. Explain a systems analyst's role in a business.
4. ~~Differentiate between hardware and software.~~
5. List at least four benefits of structured design.
6. ~~State the steps in the systems cycle.~~
7. Describe effective communications skills.
8. Identify the components of an information system.

The use of computers in our society is accelerating so rapidly we find ourselves in the midst of changes as dramatic as those generated by the Industrial Revolution in the nineteenth century. Computers no longer benefit just large corporations and government agencies; the so-called personal computers are invading our classrooms, homes, and local small businesses.

What does the revolution mean? In a word, it means change. And change brings with it new fields of study and new ways of getting work done. Prior to the computer, organizations processed data manually or mechanically. As a result, information often arrived too late to use it to enhance productivity and make profitable decisions. Even worse, manually generated information often contained the sort of errors that cause costly mistakes. Modern computers not only process data faster, but they do it with extreme accuracy, giving businesses more time to concentrate on sales or personnel problems.

This text will teach you to analyze the ways organizations process data so that you can better understand and improve those ways. In this first chapter, you will learn the basic concepts of systems analysis, design, and development, beginning with a definition of systems.

SYSTEMS

System Combination of resources working together in converting inputs to outputs.

System. We see, hear, or read this word every day. We have all learned to read, write, and calculate within an educational system. The human body is a biological system. Nightly news reporters analyze our political system, often examining the way it interacts with the political systems of our allies and adversaries. As consumers, employees, or employers, we struggle with an economic system, operating under the laws of supply and demand, which affect interest rates and personal or business bargaining power. Businesses define their activities and needs as systems: accounting or finance systems, sales systems, personnel systems, and order-entry systems.

All these systems exhibit certain characteristics. First, they consist of interrelated and interdependent elements. Our own human biological system, for example, contains blood and nerve cells, muscle tissue, and so on. Similarly, business systems involve people, equipment (from paper and pencil to computers), programs, and procedures (rules). Within any system, the component elements work together to accomplish a specific task, job, or function. All the accountants, bookkeepers, and clerks in a company, for example, follow procedures to convert data about transactions into useful information.

To study and understand systems better, we can subdivide them into smaller systems, each of which contains its own interacting elements. For example, one can subdivide the U.S. political system into national, state, and local government subsystems, which consist of various agencies that are themselves subsystems. Similarly, within an organization's information system resides an accounting system, and this, in turn, contains certain subsystems—general ledger, accounts receivable, accounts payable, and payroll.

Computer-Based Business Systems

We will concentrate here on a particular category of systems—business systems—and how their operations may be improved. Despite the fact that not all business systems need a computer, we will emphasize computer-based systems throughout our discussion. Bear in mind that the result of the analysis of an existing manual system may be to retain that system (perhaps keeping sales contacts on 3-by-5 cards) rather than to automate it.

A computer-based system involves four interdependent elements—hardware (machines), software (instructions or programs), people (programmers, managers, users), and procedures (rules)—that interact to process or convert data into information. The word "data" means accumulated but unorganized facts (a list of the type of shoes worn by each American at a given time). Information implies usefully organized and reported facts (the most popular type of running shoe worn by men and women aged 25 to 35).

Figure 1.1 illustrates the conversion process. Data enter the system for processing, where they are organized and tabulated, and then leave the system as information. Organizing and reporting vast amounts of information by hand can be costly and time consuming. Computers can convert data to information so easily that management is becoming increasingly dependent upon them when making decisions.

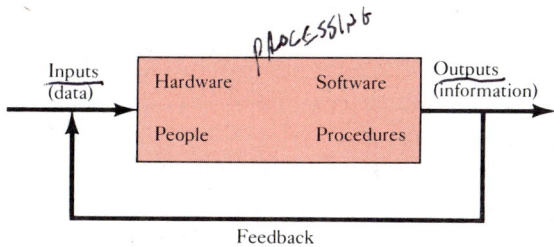

(a) Systems consist of seven components: inputs, hardware, people, software, procedures, outputs, and feedback.

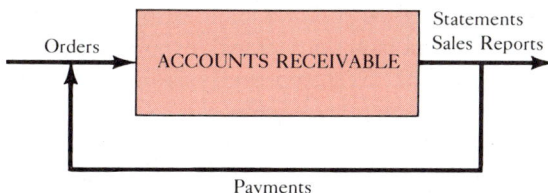

(b) In an accounts receivable system the data inputs are customer numbers, items ordered, and date of order. Outputs are statements sent to customers requesting payment, reports of items sold indicating where sales efforts might be stepped up or abandoned.

FIGURE 1.1. *An information system is a converter that receives data, acts on it, and releases it in a different form.*

As Figure 1.1 also shows, the computer-based system allows for adjustment. A manager will study the information produced by the system, comparing it with expected results. If the results seem to require an adjustment of some sort, the manager may decide to change the original data. We call this comparison and use of information in adjusting the system **feedback**. For example, if a computer-based inventory system shows a particular style of running shoe to be a poor seller, the manager may decide to lower the price. By entering the new price back into the system, the company can see how it might affect sales. Note that feedback can be negative, indicating a necessary adjustment, or positive, confirming that the system is producing the desired output. Feedback informs the system of the need for some type of remedy.

To see how computer-based systems have improved data processing, consider the following situation. Mountain Motors, an automobile-parts chain with five stores, 55 employees, and an appointments secretary, prides itself on its preventive auto-care practice. Customers periodically receive reminders that their cars are due for regular battery, tire, oil, alignment, and air-conditioning checkups. The appointments secretary uses a computer to record the date of each customer's last visit and the service performed. The computer scans customer files each month, alerting the secretary as to which customers have not been in for 6 months and should receive reminders. The computer's printer then produces reminder notices to be sent to those customers. Mountain Motors' method of reminding customers exhibits all the traits of a system. In this example, the input is the record of the customer's last visit. The outputs are the reminder notices. This system includes

Feedback Comparing actual output with desired output for system adjustment.

four components: people (the appointments secretary), hardware (the computer), software (programs extracting the customers' names and addresses), and procedures (6-month rule for recall). We see feedback in the system when customers fail to respond to their reminders and need a telephone call from the appointments secretary.

Systems Analysis, Design, and Development

The system described is fairly straightforward. Other systems are much more complex, involving large amounts and different kinds of data, intricate processing, and various outputs. No matter how simple or how complex the system, however, its successful use depends on careful and thoughtful preparation. This is the job of the **systems analyst**, the person (or persons) responsible for the analysis, design, and development of a system. **Systems analysis** involves studying the ways an organization currently retrieves and processes data to produce information with the goal of determining what must be done to provide better operations. Here the systems analyst contrives alternative systems to do the job and evaluates each in terms of cost/benefit and feasibility. **Systems design** requires the analyst to decide on the formats of the reports the system will produce, define data-storage methods, plan a method to collect data, and establish the programs the system requires.

Systems development includes actual implementation, programming, testing, training, and use of the new method. Upon completion of the system, the analyst, users, and management evaluate the system to see that it is meeting the goals and objectives originally established.

Systems analyst Person performing the systems study.

Systems analysis The first phase in the systems process, identifying problems.

Systems design The second phase in the systems process, specifying output, input, and data base.

Systems development The last phase in the systems process, including writing, testing, and installation.

ORGANIZATIONAL STRUCTURES

Before initiating a computerized system, the analyst must understand the firm's objectives and organization. We can classify most organizations into three basic types: manufacturing, service, and nonprofit. A manufacturing organization converts raw materials into finished or semi-finished goods for consumers or other businesses, who may, in turn, integrate them into other products. A century ago the majority of all workers in the United States lived on farms, but today only 1 in 10 workers is associated with agriculture. In this century, our society has evolved to rely heavily on manufacturing. A manufacturer needs systems to perform a variety of tasks, from tracking the manufacturing costs per unit to monitoring the operation of an assembly line. General Motors and International Business Machines (IBM) are among the world's largest manufacturing organizations.

During the second half of this century, it appears that society will evolve still further to rely more heavily on service organizations, which sell the goods produced by a manufacturer or perform a specialized task for consumers or other businesses. As do manufacturing organizations, service organizations need such systems as payroll and accounts receivable. Fast-food companies; real estate, dental, and law offices; and video-tape rental firms typify service organizations.

Not-for-profit organizations represent the third type. Although nonprofit or governmental agencies provide services, their not-for-profit orientation separates them from profit-oriented entities. Like other organizations, however, local, state, and national governments use systems—to collect taxes, to pay their employees, and to buy various goods and services from businesses. In fact, the U.S. government buys more electronic office systems, employs more people, and purchases more goods and services than any other organization in the world.

Line and Staff Organizations

Regardless of its type or size, an organization must have a structure that identifies its purpose, lines of authority, responsibility, and accountability. The basic structure used by smaller companies with relatively few employees is a **line structure**. Companies using line structure assign overall responsibility and authority to managers at the top, limiting authority at each lower level. Managers on the same level enjoy equal authority and maintain their departments as independent units.

An **organization chart** depicts lines of authority between individuals and departments. Figure 1.2 illustrates a typical line structure. Stockholders elect a board of directors, which, in turn, hires a company president. The president maintains overall management of the entire company. Beneath the president are a number

Line structure Structure wherein overall responsibility and authority are assigned to top-level management.

Organization chart Flowchart identifying responsibilities and lines of authority.

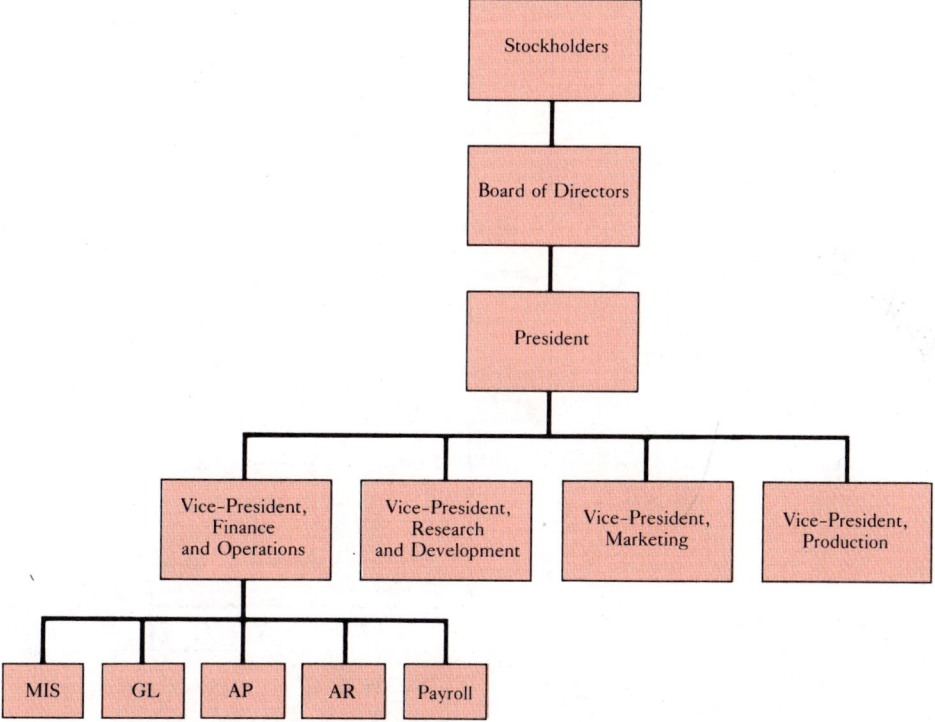

FIGURE 1.2. Organization chart for Dynamic Systems, Inc., which uses the line structure. The vice-president of finance and the vice-president of sales enjoy equal authority.

of vice-presidents, who oversee departments with specific tasks. At the lowest level of this chart are the managers of the various subdepartments. The marketing department, for example, contains sales, advertising, selling, and administrative divisions.

As companies grow, they usually add a **staff structure** to line responsibilities. With a staff structure, certain departments, such as personnel and data processing, are established to serve the other departments. The organization chart in Figure 1.3 shows such a line and staff structure. Notice how each vice-president enjoys direct access to the service department, and retains direct communication with the president. Departments that operate in a staff capacity usually hire highly trained individuals with particular specialties, such as an employee benefits administrator and a data communications expert.

Staff structure Departments in an organization serving other departments.

How Data Processing Fits Into an Organization

Although most organizations have a staff structure, this is not always the case. Sometimes a service department will operate as a subdepartment of one of the other departments. If the data processing department, for example, services only the finance department, it may fall under the responsibility of the financial vice-president (Figure 1.4). This structure enjoyed popularity in the 1960s and early

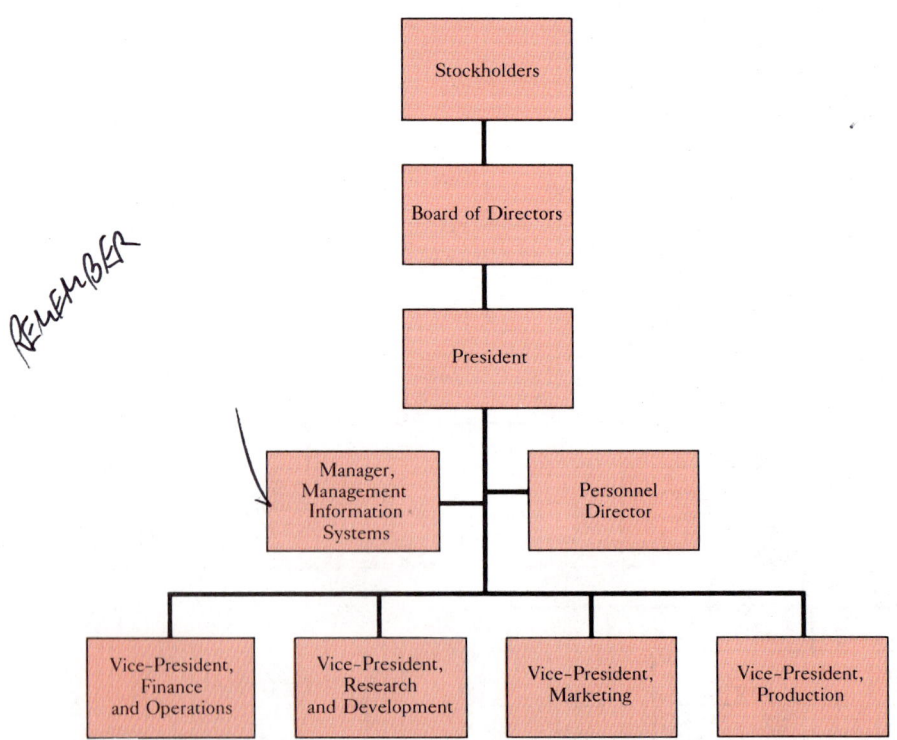

FIGURE 1.3. Line and staff organization chart for Dynamic Systems, Inc. The two staff departments are personnel and management information systems.

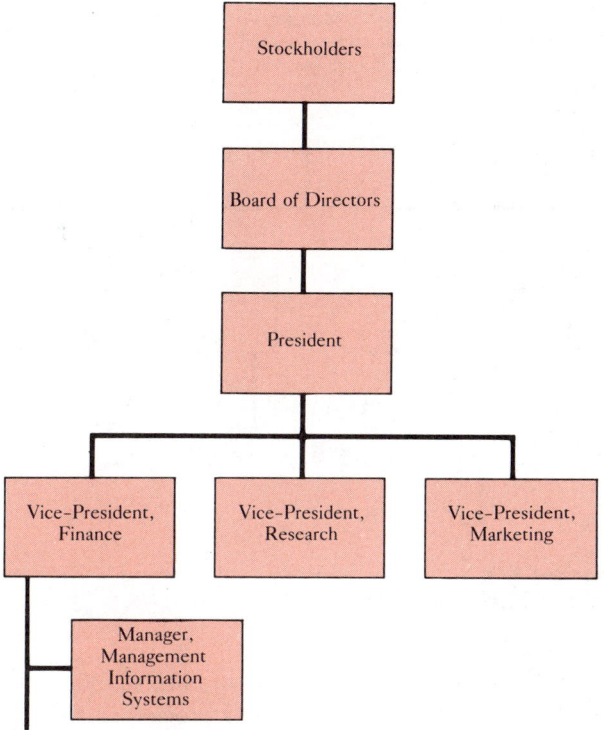

FIGURE 1.4. *Organization chart showing the management information systems department as a subdepartment of the finance department. Other major departments do not have direct access to MIS, but must go through the vice-president of finance.*

1970s when data processing entered most organizations as an accounting tool (payroll, accounts payable and receivable, inventory).

By the late 1970s, the role of the data processing department expanded to provide a wider variety of services to the production, marketing, and sales departments. This growing role of data processing forced the department to become either a staff function or a line function on an equal footing with other departments.

Another possibility is for each department to contain its own data processing department. As the cost of hardware and software drops, this pattern is becoming more prevalent, with data processing more departmental in scope and less centralized. The need for a central computer probably will not disappear, but its function is changing from that of being viewed as the source of an organization's data to that of holding all the organization's data. Perhaps the name "data processing" will change to "information services" to reflect its broader purpose.

THE ROLES OF THE ANALYST AND THE USER IN THE ORGANIZATION

Regardless of a business's particular structure, its data processing department exhibits its own internal structure. In Figure 1.5 we see a typical organization chart for a data processing department. Notice how much it looks like Figures

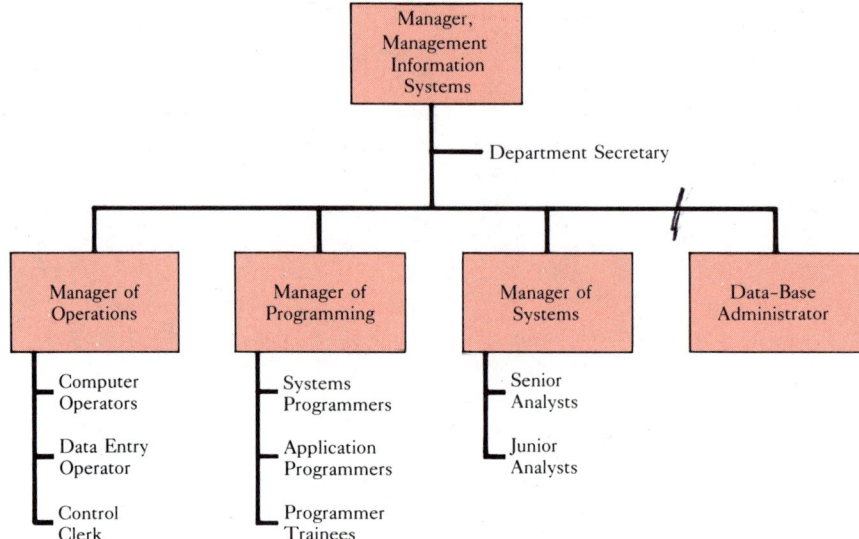

FIGURE 1.5. Organization chart for a typical data processing department. A more modern term is management information systems department, often abbreviated MIS.

1.2, 1.3, and 1.4. Although not the only structure in use today, it does reflect the common practice of separating the various data processing functions (operations, programming, and analysis) into separate departments because the duties of each department are quite diverse.

In some firms, an analyst may officially be a part of a data processing department but have a staff responsibility to a particular department. In a small West Coast county office of education, for example, each analyst reports in a line function to the data processing manager but is assigned to a specific area: elementary schools, middle schools, or high schools.

The systems analyst is responsible for dealing with each department and sub-department within the company and must be aware of the company's organization and how it relates to data processing. Lines of authority should be clear so the analyst can deal effectively with individuals at all levels. In the morning, an analyst may be working with clerical employees, examining their requirements for increased productivity by speeding the data-entry process for payroll; a half hour later the same analyst may be talking with the vice-president of sales, analyzing the need for more detailed reports showing sales by employee, by territory, and by region for the sales force. A week or so later, the analyst may present both needs to the company president. From the analyst's perspective, the clerical workers, the vice-president, and the president are **users** of new or existing systems.

When designing any new system, the analyst must remain acutely sensitive to user needs. From interviews with users, the analyst acquires the first-hand knowledge necessary to achieve the goal of a successful system. Employees may feel threatened by the analyst's appearance in their office. They may fear losing their jobs to automation or that their company is heading toward an impersonal machine age. To be successful, the analyst not only must be aware of user needs, but must

User Person or group initiating the systems study.

be able to cope with a wide range of psychological problems. Nothing can replace rapport with users and the analyst must strive to establish this rapport from the very beginning.

This need for consideration of user needs is well illustrated by Oakes Crafts, which suffered a serious morale problem when it automated its organization. Makers of innovative modern furniture, Oakes had grown from a basement operation to a multi-state enterprise employing over 100 people. In the early days, everyone at Oakes took great pride in old-fashioned hand-made quality, and Dorothy, the bookkeeper, loved to hand-letter each bill that accompanied a shipment. When that became impossible and the company decided to automate its billing, Dorothy threatened to quit, not just because a computer might replace her, but, as she complained, "It will ruin our image as a company that cares." The outside analyst Oakes hired sympathized with her point of view and showed her how a special printer could maintain the custom look but generate bills 10 times faster, which would make her job much more valuable to the company. As the months passed without any evidence of customer dissatisfaction, Dorothy's skepticism lessened, and she found herself actually enjoying being part of the computer age.

If an analyst fails to establish rapport and trust, the user may actually cause the system to fail through lack of cooperation. Dorothy was so loved at Oakes, the owner might have jeopardized his profitability in order to keep her happy. The analyst patiently showed Dorothy the benefits of the printer and she joined in the enthusiasm for the new system.

TECHNICAL DEVELOPMENTS AND INFORMATION SYSTEMS

During the past 30 years, we have witnessed an explosion in the sophistication of computer hardware, software, and procedures. To illustrate how far computer technology has advanced, consider the first computer, ENIAC, produced in 1946, which could perform little more than the addition and subtraction of numbers. Today, as shown in Figure 1.6, microcomputers can do everything from running arcade games to processing payrolls to teaching mathematics.

In the following sections, we will try to give you a sense of the dramatic developments in computer technology, beginning with the earliest, mammoth machines and ending with modern computers and their powerful new applications.

Hardware

ENIAC, the first computer, was a scientific marvel in 1946. It weighed 30 tons and could calculate a few hundred additions per second. Today's IBM Personal Computer, the size of a small suitcase, can add or subtract 20 times faster. A large computer such as the IBM 3080 can perform tens of millions of additions per second. But its capability for storing and retrieving data and its internal transfer rates are the best measure of computer power, not its ability to do arithmetic.

The computer industry has blossomed from a few companies selling huge mainframe computers (Figure 1.7) to today's wide range of firms offering mini- and

WILL SOMEONE PLEASE TELL

ACCOUNTING

Account Keeper
Accounting Plus II
Accounting Plus II Biz Package
Accounts Receivable
Accounts Receivable Balance Forward
Accounts Receivable/Sales Analysis
ACS Basic Accounting System
AMI Client Write-Up
Asset Record System
B/F Accounts Receivable System
Billings Management
Bookkeeper II General Ledger
Bookkeeper II-Depreciation
BPI General Ledger
Business Accounting
Business Check Register and Budget
Business Control System
C PA
Client Accounting System
Construction Accounting
CPA Client Write-Up
Datawrite Client Write-Up System
Delivery Service Automation
Depreciation Calculations and Reports
Executive Accounting System
Financial and Management Accounting
Financial Partner
Fixed Asset Accounting
Fixed Asset Depreciation
Fixed Assets Depreciation Schedules
Fund Accounting System
General Accounting
General Accounting Package
General Ledger
Glector
Insoft Accountant System
Integrated Accounting System
IRAP
Ledger System Business Module
Management – Financial Reporting
MAXILEDGER
Microaccountant Accounting System
MICROLEDGER
MJA Multi-Journal Accounting
Nominal Ledger
One-Type Accounting System I
One-Type Payroll and Accounting
Paysystem Accountant
Peachtree General Ledger
SBCS General Ledger
SNIP – Integrated Accounting
TCS Accounting
TCS Client Ledger
TCS General Ledger
TCS Total Ledger
The Accountant Finance Data Base
The Bookkeeper System
The Boss Financial Accounting
The Business Bookkeeping System
The Controller
The Depreciation Planner
The Software Fitness Program

AGRICULTURE

Adjusted Weaning Weights
BEEFUP-Herd Management Performance
Cattle Feeding Economics
Corn Harvest Losses
Corn vs. an Alternate Crop
Cow-Calf Profitability
Crop Yields
Economics of Corn Production
Farm Management
Farrow-To-Finish Swine Production
Feeder Pig Production
Fertilizer Formulation
Field Population
Field Size
Finishing Feeder Pigs
Job Cost (Crop Cost)
Least Cost Fertilizer Application
Liming Soil
Liquid Manure and Fertilizer
Net Energy for Feedlot Cattle
PEDIGREE-5 Generation Annotated Pedigree
Protein Balancing for Feedlot Cattle
SBCS Agri-Ledger
Selling Wet Corn vs. Dry
Sheep Production Economics
Soil Erosion
Soybean Harvest Losses
Swine Ration Analysis
Swine Ration Formulation

APPLICATION PROGRAM DEVELOPMENT AIDS

A-FORTH
ABT Pascal Tools
APEX-6502 Assembly Language
Apple-80 Disassembler
Assembly Language Development System
AUDEX-Audio Programming Aid
CBASIC Program Maintenance Utilities
CINDEX
Cosapple 1802 Disassembler
CRTFORM Programmer Productivity
Diagnostics II
DISTEL-Disk Based Disassembler
Executive Planning System
Floating Point Dictionary
Forms 2
Key Perfect-Checksum Table Generator
Linkdisk-Disk Utility for Apple Pascal
Linkvideo-Screen Utility
Lower Case Character Generator
MULISP/MUSTAR-80
OGI-Forth-Implementation of FIG-Forth
Pascal Programmer
Pascal Level 1
Pearl III-Rapid Logic Generator
Personal Programmer
Prism/Ads Data Base Generator
Program Development System I
Program Writer for Non-Programmers
Programming Aids 3.3
Quic-N-Easy Application Development
RAID-Real Time Assembly Debugger
Scientific Data Base
SID-Symbolic Instruction Debugger
Stok Pilot-Menu Generator
STRING-80
STRING-BIT
Systems Analyst
Teacher Plus Teaching & Reference Pkg
The BASIC Teacher
The Last One-Program Generator Pkg
The Toolbox Programming Utilities
Tiny-C-Interactive Programming
UCSD Pascal
Unlock Development Tool
V-COM Disassembler Package
Z8000 Cross Assembler

BUSINESS MANAGEMENT TOOLS

Analyst-Business Productivity
Apple Sack General Business Program
Bookkeeper II-Sales Analysis
Business Pac 100
Business Planner
Creative Financial Package
Desktop/Plan
Execuplan Planning & Forecast
Financial Modeling System
Financial Planning Series
Financial Planning/Analysis
Finplan/Financial Planning
FP2020 Financial Planner
FPL-Financial Planning Language
Magic Worksheet
Magicalc-Forecasting Package
Micro-DSS/Finance
Microfinesse-Financial Modeling
Milestone-Critical Path Network Analysis
Optimiser
PFS-Personal Filing System
Personal Report System
Plan 80-Financial Planning & Analysis
Project Boss-Mgr's Cost Control System
Project Planning and Budgeting
Retail Purchasing & Pricing
Salary Planner
Senior Analyst
Supercalc-Electronic Spread Sheet
Support Pkg for Real Estate Mgmt
T/Maker II-Visual Calculating Tool
The Analyzer
The Budget Planner
Universal Business Machine Planning and Forecasting
VisiCalc III
VisiCalc Real Estate Template

CAPITAL PROJECTS PLANNING & CONTROL

Angle Project Scheduling
APM-Project Management System
Jobtrak-Project Tracking
Milestone Project Management
Project Management System
Project Planning

COMMUNICATIONS

Apple Access III
BISYNC-80
BSTAM
Class Data Recorder
CM-900 Burroughs Network Services
Communications Program
Crosstalk Smart Terminal/File Transfer
Data Capture 4.0
Data Transporter Package
Datalink
DTS-3-Serial Data Transfer
Electronic Mail
IBM-CP/M Allows Transfer of Data
IE/Modem
Intercom Communications
METTY-Intelligent Terminal Package
Micro-Courier
Micro-Telegram
Microlink-80-File Transfer Program
Reformatter-CP/M IBM Data Transfer
Remote Console Program
Smarterm-CP/M Terminal Program
Term II-Computer Intercommunications
Term Intercommunications Package
TTY-Communications With Other Computers
U-Net-Shared Resources Network
Ultimate Transfer
Visiterm-Communications Program
VT-100 Emulator
Western Union Interface

DATA MANAGEMENT

ANALYST
CBS-Configurable Business System
CCA Data Mgt System
CM 2020 Configurable Manager
Condor Series 20
Data Management Program
Data Manager
Data Master
Data-View Electronic Filing Cabinet
Database II
Database Management
Datafax
Dataflow-Info Processing
Datastar
Datastore
Datatree
Disk-Edit-Screen Oriented Disk Editor
DMS-Data Mgmt System
FABS II-Rapid Keyed Access
Fast Entry for Tabs Business Modules
FINDAFYL-Reference Retrieval System
FMS 80-Data Base Management System
GBS Database
General Database
HDBS-Hierarchial Data Base
IFO Database Manager
Information File Organizer
Information Master-Data Mgmt System
KTDS-Key to Disk, Data Entry
Linkindex-Pascal Utility
MAG/Base-Data Base Management
Manager-Relational Data Base
MDBS DRS-Micro Database Mgt System
MUMPS-Language for CP/M Database
Optimum Data Mgmt Program
PRISM/IMS-Information Mgt System
RADAR-Random Access Data Acquisition
Reprogrammable Data Base
Scientist-Data Base & Statistical Pkg
Selector III-Data Base Processor
Selector IV-Data Base Mgt
Selector IV-Key Access Info
Selector V-Data Base Mgmt
STATPRD-Integrated Database System
Stoneware Utility Package
Super Kram II – Multi-Keyed Random Access
The Reprogrammable Data Base Program
VisiDex-Data Base Mgt System
VisiFile-Data Base Mgt Package
Whatsit?-Conversational Query/Retrieval

DATA SECURITY SYSTEMS

Absolute Security
Encode/Decode Security System

DISTRIBUTION

ABT Retail Manager
Beer Distributor Management
Inventory, Order Entry, Invoicing
Oil Jobber Management System
Order Entry and Inventory Control
The Store Manager
Wholesale/Retail Distribution System

EDUCATION – BUSINESS

Accounting Tutor
Comparative Buying
Income Meets Expenses
Interactive Typing Tutor
Job Readiness-Assessment & Development
Mastertype-Typing Instruction
Money Mgmt Assessment
Typing
Typing Tutor
You Can Bank On It-Bank Concepts

EDUCATION – CHEM/PHYSICS

Acid-Based Chemistry
Atomic Structure
Chem Lab Simulation
Chemical Equilibrium
Chemistry With A Computer
Fundamental Skills for General Chemistry
High School Chemistry
High School/Jr. College C.A.I. Biology
High School/Jr. College C.A.I. Physics
Organic Nomenclature
Physics

EDUCATION – ENGLISH

A Batch of Endings
Agreement of Pronoun/Antecedent
Alphabetize
Capitalization
Catalog Cards
Commas
Compu-Read
Compu-Spell
Coordination
End Marks
Excess Words
Faulty Coordination
Hearing the Homonyms
Irregular Verbs
Is It "ie or ei?"
Language Drill
Locate Books on the Shelf
Magic Spells
Misplaced Modifiers
Parallel Structure
Possessing the Possessives
Prefixes & Suffixes
Quotations
Reading Level
Readings In Literature
Run On Sentences
Scramble
Sentence Diagramming
Sentence Fragments
Speedreader
Spell-N-Time
Spelling Bee with Reading Primer
Spelling Those Plurals
Still More Nasty Demons
Subject/Verb Agreement
Subordination
The End of the Endings
Those Nasty Demons
Understand the Card Catalog
Understand the Title Page
Use an Index
Use the Table of Contents
Using Adjectives/Adverbs Correctly
Word Scrambler & Super Speller

EDUCATION – MATH

Addition & Subtraction
Algebra I
Basic Math Skills
Compu Math: Arithmetic Skills
Compu-Math Decimals
Compu-Math Fractions
Counting Bee
Decimal Estimation
Division Drill
Drill II
Elementary Math
Fractions
Geometry
Geometry and Measurement Drill
Lessons in Algebra
Matching and Using Numbers
Matching Geometric Figures
Math-Addition & Subtraction
Matrix Mathematics Package
Measurements
Multiplication & Division
Mumath-PO Symbolic Math
New Subtraction
Numerical Analysis Mathematics
Problem Solving
Problem Solving in Everyday Math
Sets
Sign Drill/Typing
Statistical Analysis I Mathematics
Statistics 3.0
Typing Fractions

EDUCATION – MISC.

2ES Courseware
American History Through Biographies
American Indians
Antonyms
Apple Sack 2 Home Education
Approximate Measure
Astronomy I & II
Concentration-Taxing
Counting Calories
Early Civilization
Educational Package
Educator's Disk
Family Fun
Farm and Farm Products
HI-Res Life
History
Home Safe Home
Insects
Light Pen Quiz
Literature
Living Things
Math, Sports, Etc.
Middle Ages
Money
Moptown
Mother Goose Rhymes
Music/Art
Our Bodies
Poison Proof Your Home
Questions & Answers in Biology
Questions & Answers in History
Quizstat
Reverse/Sampling
School Days
Sentence Beginning
Shore Features
Sound
Supermap
Synonyms
Systems of the Body
Teacher Create Series
Teacher Plus
Telling Time
The Basic Teacher Pac
The Earth and It's Composition
The Professional-Teaching Program
The Solar System
Transportation History
Typing
United States
Visual Perception Tests
Weather Fronts
Work Relationships
World Desert Region
World Polar Regions

FINANCE-INVESTMENT & PORTFOLIO ANALYSIS

Analysis 1-Stock Trend Data Analysis
Commoapx System
Computicker
Computrac File Reader
Dow Jones News & Quotes Reporter
Dow Jones Portfolio Evaluator
Dowlog-MC
Electronic Stock Package
Engineer's System For Trading
Forecast I
Forecast II
Fotofolio-Visual Display w/Statistics
Gann's Square of Nine Analysis
Intelligent Investor
Investment Analysis
Market Charter-Technical Analysis
Moneybee Investment Analyst
Options 80-Stock Options Analyzer
Portfolio Master
Quotecharter
Quoteprocessor
Ratortn-Investment Analysis
Stock and Options Analysis
Stock Forecasting
Stock Market Management
Stock Market Utility
Stock Option Analysis
Stock Tracker
Stock Valuation Program
Stocksheets
Strategy M-Monitor Price Change Dynamics
The Clover Method Trading System
The Stock Portfolio Program
Tickertec-Tickertape Program
Wilers 6 Systems Analysis

FOREIGN LANGUAGES

Chinese Lessons
Foreign Words and Phrases
Greek Roots and Prefixes
Japanese Lessons
Latin Roots and Prefixes
The French Hangman
The Russian Disk
The Spanish Hangman

GAMES

Adventures
Alien Rain
Alien Typhoon
Almanac – The Time Machine
Amaze
Analiza
Animal
Anti-Ballistic Missile
Apple Adventure
Apple Bowl
Apple Fun
Apple Panic
Apple Sack 3 – Adventure Pak
Apple Sack 7 – Space Sack
Apple Sack 8 – Game Sack
Apple Sack 9 – Base Star
Apple Stellar Invaders
Apple-oids
Asteron
Astro-Scope
Astrology
Autobahn
Backgammon 20
Battle of Midway
Beer Run
Best of Muse
Biorythms
Blackjack
Both Barrels
Brands
Bridge 2.0
Bridge Tutor
Bubbles, Planetoids and Burnout
Cartels and Cutthroats
Castle Wolfenstein
Chambers of Xenobia
Chebychev 1
Chebychev 2
Chronicles of Osgroth
Civil War
Compu-Math Arithmetic
Compu-Math Decimals
Compu-Math Fractions
Computer Air Combat
Computer Baseball
Computer Bismark
Computer Conflict
Computer Napoleonics
Computer Quarterback
Cops and Robbers
Cosmo Mission
County Carnival
Cyber Strike
Disk Talker
Dr. Chips
Dragon Fire
Dungeon
Executive Fitness
Falcons
Fantasyland 2041
Fastgammon
Flight Simulator
Galactic Attack
Galactic Wars
Galaxy Wars
Games People Play
Gamma Goblins
Gobbler
Golf/Cross-Out
Gorgon
Hammurabi
Head On Game
Hellfire Warrior
Hi-Res Football
Hi-Res Soccer
In The Army Now
Into Ships
Jet Fighter Pilot
Klondike 2000
Lost By Ship
Mastermind
Meteoroids in Space
Micro Othello
Mimic
Mind Games Package
Mission Asteroids
Mystery House
Need an Analyst
Nominoes Jigsaw
Oil Tycoon
Olympic Decathlon
Operation Apocalypse
Orbitron
Outpost
Paddle Fun
Pegasus II
Perception 3.0
Phantoms Five
Planetoids
Plot 3D
Pokeno
Poker Slot Machine
Pool 1.5
Pot 'O Gold I
Pot 'O Gold II
President Elect
Pro Football
Pro Picks
Project Omega
Pulsar II
Race For Midnight
Raster Blaster
Red Baron
Rendezvous
Robot Wars
Sahara Warriors
Sargon II (Chess)
Satellite Trak
Shell Games
Shuffleboard
Skybombers
Skybombers II
Sneakers
Snoggle
Soft Porn
Softside Publications
Space Eggs
Space Warrior
Spellguard
Spelling Bee
Star Cruiser
Star Dance
Star Thief
Startraders
Startrek
Stock
Sub Attack
Tawala's Last Redoubt
Teacher's Pet
Temple of Apshal
Terrorist
Tetrad
The Strip
The Asteroid Field
The Great Escape
The Horse Selector II
The Prisoner
The Scorekeeper
The Shattered Alliance
The Warp Factor
Three Mile Island
Torpedo Fire
Ultima
Voyage of the Valkyrie
War and Games
War Games
Warp Factor
Watch Your Moves
Win at the Races
World's Greatest Blackjack
Wumpus
Xplode

FIGURE 1.6. *This advertisement appeared in national magazines and prominent newspapers. Without any fanfare or flash, Apple simply lists all the applications of its microcomputer. (© Apple Computer, Inc. All rights reserved for reproduction.)*

ME WHAT AN APPLE CAN DO?

GRAPHICS/
COMPUTER-AIDED DESIGN

3-D Surface Plotter Package
A2-3D1 Graphics Family
ABT Barwand Software
Action Sounds & Hi-Res Scrolling
Apple Plot
AppleGraphics II
Artist Designer
Bar Chart (Histogram) Graphics
Business Graphics III
Circuit Designer Graphics
Circuit Simulator
Creativity Tool Box
CURVFIT
Data Plot
E-Z DRAW
FLGDZINE
Graforth – Development Tool
Graph-Fit
Graph-Pak
GRAPHPOWER
Hi-Res Secrets
Line Graphics
MC Painting
ORIFICE
Pascal Animation Tools
Pascal Graphics Editor
Perspective Plot – 3-D Graphics
PGE – Graphics Editing Package
PILOT Animation Toolkit
Polar Coordinate Plot
RGL Real Time Graphic System
Screen Director
Shape Table Generator
Stats-graph
Super Shape Draw & Animate
Tablet Graphics
The Coloring Board Program
The Designer
Topographic Mapping
Ultra Plot
Utopia Graphics Tablet System
VACVESL – Vacuum Vessel Design
VESDZINE – Design of Vessels
VISITREND/VISIPLOT
X-Y Vector Plot Package

HOME MANAGEMENT

Address File
Auto Records'
Checkbook Balancing
Checking Account Management
Chequemate
Diet Analysis
Financial Analyzer
Five Minute Financial Check-Up
Grocery List
Home Finance
Home Inventory File
Home Money Minder
Home Purchase Analysis
Magazine File
Mortgage Analysis
Personal Accounting System I
Personal Expense Record
Personal Finance Manager
Personal Financial Planning
Programmed Exercise
The Personal Check Manager

INCOME TAX

Dow Jones Portfolio Evaluator
Individual Tax Planner
Micro-Tax Individual Tax Package
Micro-Tax Integrated State Income Tax
Micro-Tax Partnership Package
SHORTAX – Tax Planning Package
Tax Planner
Tax Preparer
TRPS – Tax Return Preparation System

INVENTORY CONTROL

ARM-1000 – Rental Business
Basic Business Inventory
Bill of Materials
BPI Inventory Control
Intotory Inventory System
Inventory Accounting
Inventory Control
Inventory Management
Inventory Management for Stock Control
Inventory Pac
Inventory System Business Module
Manufacturing Inventory Control
MATSTAT-Materials Tracking
Order Entry/Inventory Control
Peachtree Inventory System
Point-Of-Sale Retail System
Property Manager for Moveable Equipment
Retail Inventory
Rogis Stock Control for Components
Stock Control
Stock Recording
Stockfile Inventory System
Stockroom Inventory and Purchasing
Structured Systems Inventory Control
TCS Inventory Management
The Order Scheduler

JOB & CONTRACT
COST ACCOUNTING

Billflow
Bookkeeper II-Job Costing
BPI Job Costing
Contract Billing
Contractor Job Cost
Cost Accountant
Job Accounting System
Job Control System
Job Cost Accounting
Project Cost Accounting for Architects
Project Cost Accounting for Engineers
The Software Fitness Job Cost Analyst
Time Recording-Job Cost Analyst
Timerec-Transaction Carry Forward

MAILING LIST &
LABEL PROCESSING

Address Book Mailing List
Apple III Mail List Manager
Apple Mail Sack
Apple Post
Benchmark Mail List
Commercial Mailer
Mail List
Mail80 Mailing List Software
MAILER-Name & Address Management System
Mailing Address
Mailing List Package
Mailing System
MAILMERGE
MAILPRO
Mailroom-Mailing List Management
Master Mailing List
NAD-Name & Address Selection System
Name And Address
Postmaster-Mail Management
Professional Mailout
School Mailer
Small Business Mailing & Filing
Super-M-List Mailing List Program
Ultra Plot/Mailing & Filing System I

MARKETING/SALES ANALYSIS

EASYTRAK-Salesmen Monitoring Package
Marketing Systems – Proposal Developer
Office and Agent Productivity Package
Sales Analysis
Sales Pro Prospect Mgt Package
Sales Tracker
SALESLOG – Sales Mgt Program
SNAP – Questionnaire Design and Printing
TCD Life Insurance Computer System

MISCELLANEOUS

BILL – Building Energy Use
Circuit Analysis
Hand Holding BASIC
Insulate
Mini-Warehouse System
Stepwise Multiple Regression

MUSIC

Alpha Syntauri Music Synthesizer
Apple Music Theory
Apple Sack Music & Graphics
Appleodion Music Synthesis System
Music System
Musicomp
The Electric Duet

ORDER ENTRY/
ACCOUNTS RECEIVABLE

BPI Accounts Receivable Program
Cash Receipts System
Company Sales
Invoice Compiler
Invoicing
Membership Billing
MICROREC
Multi-Property Accounts Receivable
Open Item Accounts Receivable
Order Entry
Order Entry and Billing
Order Entry and Invoicing
Order Tracking System
Peachtree Accounts Receivable
Peachtree Sales Invoicing
Progressive Billing
Purchase Order System
Receivables System Business Module
Receiver
Sales Invoicing
Sales Ledger
Sales Order Processing
Software Fitness Program – A/R System
Structured Systems Accounts Receivable
T-SOP Sales Order Processing
TCS Accounts Receivable Package
TCS Total Receivables
The Biller

PAYROLL PROCESSING

Advanced Payroll Package
After-The-Fact-Payroll – updates records
Apple Payroll System
Bookkeeper II-Payroll
BPI Payroll
Business Basic Payroll System
Contractor Payroll
Jobcost Payroll
Micropayroll
Passive Payroll
Paymaster
Paymaster-Payroll System
Payrecord I
Payroll
Payroll Accounting Package
Payroll Assistant
Payroll I
PeachPay
Piece Rate Payroll System
Post Facto Payroll
Print/Paycheck Accounting System
Run Time Payroll Program
Sheltered Workshop Reporting
Structured Systems Group Payroll
TCS Payroll Package
TCS Total Payroll
Variable Worker's Compensation
WH-347-Accessory program for Jobcost

PERSONNEL MANAGEMENT

AMI Post-Facto Payroll
MICROPERS – Payroll & Personnel Mgmt
Personnel Data Recorder
Personnel Office – Federal Compliance
Personnel Record
Personnel Record/Employee Records System

PROFESSIONAL OFFICE SYSTEMS

AMI Omegabyte Time & Billing
BETA – Stand Alone Time & Billing System
Billkeeper – Professional Billing
Client Billing System
Client Record/Bill Preparation
Datalaw System 3-Law Office Mgmt
DataTime
Dental 80A-Dental Accting & Billing
Dental Billing Package
Dental Office Management
DentalEase
Dentistaid – Dentist Office Management
Insyst (Insurance System)
Legal Billing & Timekeeping System
Legal Clerk – Office Management System
Legal Time Accounting System
Medicaid Day Treatment
Medical Accounting and Billing
Medical Clinic
Medical II – Office Mgmt System
Medical Office Management
Medical Secretary
Medical/Dental Management System
Medical/Manager
MedicalEase
MedPak
Medtrips – Billing & Insurance Forms
PAS – 3-Patient Billing & Accts Receivable
Patient Accounting System
PIP-Payroll/Invoicing Program
Professional Office Management
Professional Time & Billing
PTA – Professional Time Accounting Pkg
Series 8000 Dental Mgmt
Series 8000 Medical Mgmt
Series 9000 Family Dental Management
The Patient Scheduler
Timeclok
Timemaster – Time Accounting
Timesaver Client Billing System

PROGRAMMING LANGUAGES

Ada Compiler
APL/V80 Language
Apple III Business Basic
Apple III Pascal
Apple FORTRAN
Apple Logo
Apple PILOT
ASM 65-Assembler
BASIC A+ – Extended Business Basic
BASIC Compiler
BASIC-80
BASIC/Z – Native Code Compiler
BD Software "C" Compiler
C Compiler
CBASIC 2 Compiler
CIS COBOL
COBOL 80
Cos Assembler
Cos COBOL
Focal 65-High Level Programming
Forth 86
Forth-Language Compiler
FORTRAN 80
FORTRAN IV
Hand Holding BASIC
KBASIC – Microsoft Disk Extended BASIC
Language System with Apple Pascal
LISP-80 Compiler
MAC 8080 Macro Assembler
MULISP Compiler
MULISP/MUSTAR 80
muMath/muSimp 80-High Level Programming
Nevada COBOL Compiler
Pascal Compiler
Pascal/M86
Pascal/MY+With SPP-ISO Standard
PL/1-80-Programming Language
RATFOR – FORTRAN Language
S-BASIC
SSS FORTRAN Compiler
Softronics
Stiff Upper Lisp
TCL Disk BASIC Interpreter
TCL-Pascal
TEC 65-Editing Language
Tiny BASIC High-Level Language
Tiny C
Tiny Pascal
Tiny-C-Two Compiler
Transform II
UCSD Pascal
Whitesmith's Compiler
XPLO-Structured Language
XY BASIC Interactive Process Control

PROGRAMMING UTILITIES

Apple Sak 4 – Utility Package
Basic Utility Disk
Disk Utilities 3
Disk Utility Package
Disk-o-Tape-Pascal
DOS Tool Kit
File Maintenance Package
MAG/Sam Keyed File Mgmt System
MAG/Sort-Record Sort
Masterdisk-disk Sector Editor
MSORT – for COBOL 80
Pascal Utility Library
Pascal – Sort Program
PSORT – Pascal File Sorting
QSORT – Sort/Merge Program
SORT/B – Hybrid Sort
Supersort
Ultrasort

PURCHASING/ACCOUNTS PAYABLE

Accounting Payable
Accounts Payable Business Module
Accounts Payable/Purchase Order
Bookkeeper II – Accounts Payable
Cash Disbursements Posting System
Check Writer
Company Purchases
Contractor Accounts Payable
Disk-O-Check
Micropay-Accounts Payable
Print Check Accounting System
Purchase Ledger
Structured Systems Group Accts Payable
T-POP – Purchase Order Processing

REAL ESTATE

American Software Property Management
Apartment Building Investment Analysis
Apartment Manager
Commercial Property System
Construction Cost/Profit Analysis
Cornwall Apartment Management
Income Property Analysis
Listings
Multi-Property Accounting System
Office/Apartment Real Estate Management
Property Analysis System
Property Management
Property Management System
Property Mgmt – G/L Tenant and Expenses
Real Estate Analysis Program
Real Estate Analyzer
Realty Package
Rent vs. Buy
Rental Manager
Residential Property Management
Tax Deferred Exchange Model
Tenant Processing Package
The Landlord-Property Mgmt System
VisiCalc Real Estate Templates

TIME MANAGEMENT
& SCHEDULING

Agenda Files
APM – Project Scheduling
Appointment Calendar
Color Calendar Package
Datebook Appointment Calendar
Datebook Time Management System
GUARDIAN – Computerized Scheduling
Office Manager – Staff Appointments
Personal Datebook
Professional Secretary
PROSCHED – Project Schedule
Time Manager

WORD PROCESSING

Apple World Oriented Text Editor
Apple Writer II
Apple Writer III
Benchmark – Word Processing System
Docuwriter Text Processor
Easywriter Word Processing
EDITRIX 1.0 – Word Processing
Form Letter Module
Formulex – Business Form Design
Goodspell
Letter Master – Basic Word Processor
Letteright Correspondence Processing
Letterite Word Processing System
Magic Spell – 20,000 Word Dictionary
Magic Wand – Phrase Insertion
Magic Wand – Word Processor
Magic Wand Word Processing System
Magic Window Word Processor
MAIL-MERGE-Wordstar Enhancement
Manuscripter – Word Processor
Master Text Processor
Memorite III Word Processing
Microspell Spelling Corrector
PALANTIR – Word Processing and Accounting
Personal Text Processing
Report Writer – Word Processing
Script III
Secretary – Word Processing
Spellbinder Word Processing
Spellguard
Super-Text Word Processing
Supertext II
TEXTWRITER III – Text Formatting Program
The Word Spelling Checker
VTS-80 CP/M Word Processing
WordIndex
WordMaster – Comprehensive Editor
Word Master Text Editor
WordStar – Word Processing

With these and thousands of other ready-to-use programs to choose from, including the vast array of CP/M® software, you can do more things with an Apple® than any other computer you can buy. So over 1200 authorized Apple dealers have a question for you:
What do you want it to do? apple® The personal computer.

Call **800-538-9696** for the location of the authorized Apple dealer nearest you, or for information regarding corporate purchases through our National Accounts Program. In California, (800) 662-9238. Or write: Apple Computer Inc., Advertising and Promotion Dept., 20525 Mariani Ave., Cupertino, CA 95014. Software products listings courtesy of SOFSEARCH™ Call 800-531-5955, (in Texas, call 512-340-8735)
CP/M is a registered trademark of Digital Research, Inc. Apple is a registered trademark of Apple Computer Inc.

FIGURE 1.6 (Continued)

FIGURE 1.7. An IBM 3081 main-frame computer. This computer can handle hundreds of terminals at the same time and still do batch jobs such as print statistical reports, produce lists of customers, or compile new programs. (Courtesy of International Business Machines Corporation.)

microcomputers (Figure 1.8). In 1980 some two dozen companies sold almost 750,000 machines worth $1.8 billion. The next year there were 35 companies in the field (including IBM). Sales jumped to 1.4 million machines worth almost $3 billion. In 1982 there were more than 100 companies selling 2.4 million machines for $4.9 billion. And this was only the beginning. In fact, total sales of computers and accessories within the 14-month period of October 1982 through Christmas 1983 equaled those of all previous years together. The power of an early computer costing millions of dollars has become available for approximately $1000, and as a result, computer power no longer is limited to big businesses.

One reason for the computer's increasing speed, decreasing cost, and the resulting growth in sales has been the development of more efficient methods of collecting data. Until the mid-1970s, the "IBM" or data processing card developed by Herman Hollerith, the first president of what is now IBM Corporation, was the most widely used medium for entering data into a computer. Now most data are collected at the point of creation, thus bypassing the need for cards, and are sent by means of a terminal directly into the computer for processing (see Figure 1.9).

Terminals have become a familiar part of the U.S. scene. We see them at checkout stands in grocery stores, in department stores, at airports, and even in real estate offices. The power of the computer no longer is confined to the sealed and antiseptic showcase of corporate headquarters, but has reached the worker's job station. By 1985 four out of five workers will use a terminal at least once a day, to enter data about what they are doing or to ask the computer questions, the answers to which will enable them to work more productively. This trend toward cheaper, faster, more powerful hardware should continue for many years.

FIGURE 1.8. An Apple Macintosh microcomputer. This system with a printer costs less than $3000 and can make over 100,000 calculations per second. (Courtesy of Apple Computer, Inc.)

FIGURE 1.9. A terminal keyboard and CRT (cathode-ray tube) screen show data input or output. Some terminals offer color or special sound effects. Terminals vary in price from a few hundred dollars to many thousands of dollars. (Courtesy of International Business Machines Corporation.)

Software

Hardware and software are intertwined; neither can operate independently of the other. Hardware is to software what a television set is to the programs that appear on it.

Unfortunately, advances in software have not progressed as rapidly as those in hardware. Early programmers laboriously wrote instructions to the computer in complex machine language (Figure 1.10). Today's English-like programming languages, such as COBOL, Pascal, and Ada, enable programmers to work more efficiently and to check for errors more easily (Figure 1.11).

Data-base management systems (programs to ease the acquisition, storage, and retrieval of data stored in a computer) represent one of the major software advances. Formerly, to access and process data, the programmer had to know and remember the locations of the data on the disk or tape. Now businesses can buy software that automatically locates the desired data, thereby reducing computer time for analysts and programmers and freeing them to concentrate on users' needs.

Structured Methodology

Ten years ago, most systems analysts operated rather like hired guns brought in to clean up Dodge City. They shot from the hip, using whatever techniques or tools they had learned in school, developed on their own, or adopted from their predecessors. This lack of standardization more often than not led to poorly conceived, designed, programmed, and developed systems, which failed to operate as intended.

Structured methodology Set of rules, guidelines, and tools facilitating the systems process.

Modern **structured methodology**, a set of rules, guidelines, and tools that facilitates effective design, programming, and development, has improved the situation. To understand structured methodology, consider an architect designing a home or commercial office building. To find out what size beam will span a

```
0001:4      MPCW                    BF00000000E002
0003:0      NAMC  (01,0000)         6000
0003:2      STFF                    AF
0003:3      BSET                    960D
0003:5      LT48                    BE27000074000B
0005:0      LT8                     8205
0005:2      STAG                    95B4
0005:4      LT48                    BE04000054000 
0007:2      STAG                    95B4
0007:4      LT48                    BE27000074000C
0009:0      LT8                     B205
```

FIGURE 1.10. Machine-language program for a Burroughs computer. Each variety of computer has its own machine language. The three columns show the memory address of the instruction (0001:4), the actual instruction written by the programmer (MPCW), and the machine version (BF00000000E002).

```
005000  PROCEDURE DIVISION.
005050
005100  000-MAIN-DRIVER.
005200      PERFORM 100-OPENING-MODULE.
005300      PERFORM 200-PROCESSING-MODULE UNTIL DONE.
005400      PERFORM 300-CLOSING-MODULE.
005500      STOP RUN.
005550
005600  100-OPENING-MODULE.
005700      OPEN OUTPUT PRINTER-FILE.
005800      OPEN INPUT VIDEO-DATA-BASE.
005900      MOVE CURRENT-DATE TO LINE-3-DATE.
006000      PERFORM 400-HEADINGS-MODULE.
006100      READ VIDEO-RECORD AT END MOVE "DONE" TO FINISHED.
006200
006300  200-PROCESSING-MODULE.
006200      WRITE PRINT-FILE FROM VIDEO-RECORD BEFORE ADVANCING 1
006250          LINE AT END-OF-PAGE PERFORM 400-HEADINGS-MODULE.
006500      READ VIDEO-RECORD AT END MOVE "DONE" TO FINISHED.
006700
007000  300-CLOSING-MODULE.
007100      CLOSE PRINTER-FILE.
007200      CLOSE VIDEO-DATA-BASE.
008800
008900  400-HEADINGS-MODULE.
009000      ADD 1 TO PAGE-COUNTER.
009100      MOVE PAGE-COUNTER TO PAGE-NUMBER.
009200      WRITE PRINTER FROM HEADING-LINE AFTER ADVANCING PAGE.
```

FIGURE 1.11. A portion of a COBOL program. Lieutenant Grace Hopper (a U.S. Naval Officer and a Phi Beta Kappa from Vassar, with master's and doctor's degrees from Yale University) spearheaded the development of this language in the late 1950s.

known distance, the architect can consult a book called the *Universal Building Code*, which relates distances to beam sizes. Systems analysts, on the other hand, do not enjoy such a handy design reference tool. Instead they employ structured methodology as a sort of "walkthrough" of the system, allowing other analysts to review the design and locate errors before the system is completed.

The rules and guidelines that shape structured methodology are not as formalized as those for programming languages. For the time being, consider structured methodology simply as a collection of:

1. Techniques such as top-down design, where the analyst starts with the most general function and carefully works down to the most specific or detailed function of the system.
2. Tools such as the data dictionary and the data-flow diagram, which graphically and symbolically depict the events that take place in a system.
3. Software such as data-base managers and text editors that ease the task of interacting with the computer and improve the productivity of analysts and programmers.
4. Guidelines for interpersonal relationships such as a walkthrough during which peers review the analyst's work to help detect errors or omissions.

The use of structured methodology offers advantages and disadvantages, but as the following list suggests, the former outweigh the latter.

The advantages of structured methodology are:

1. Lower long-term costs.
2. Improved system reliability, with longer gaps between system failures.
3. More easily enhanced or improved systems.
4. More flexible systems, allowing a wider variety of tasks to be performed.
5. Greater user satisfaction through earlier and constant user involvement.
6. Lower maintenance costs.
7. Reduced likelihood of errors in analysis, design, and development.
8. Early delivery of selected subsystems.
9. More understandable and accessible programs and systems.
10. Wider involvement of more people leading to easier acceptance of the new system.

Dijkstra, Bohm, and Jacopini: The Developers of Structured Programming

The year is 1965, and Edsgar Dijkstra has just set the programming world on its ear with the publication of his landmark paper, "Programming Considered as a Human Activity." Dijkstra said, "I am of the opinion that it is worthwhile to investigate to what extent the needs of Man and Machine go hand in hand and to see what techniques we can devise for the benefit of all of us." Just what was this obscure Dutchman proposing?

Clarity and simplicity, that's what. Dijkstra was the first to question how programs were written, and to propose a more orderly approach. Up until this time programs were mathematically elegant but often looked like Einstein's blackboard by the time they were completed. Sometimes there were mistakes, or the programmer wanted part of the program to repeat; programmers could slap on a handy band-aid called the "GO TO" statement. Dijkstra wrote a letter to the editor of *Communications of the ACM* insisting, "The GO TO statement as it stands is just too primitive; it is too much an invitation to make a mess of one's program," and the critics howled with rage. People are still publishing articles today arguing the pros and cons of the GO TO.

Two Italians, C. Bohm and G. Jacopini, theoretically proved the structured method could work in 1966; and IBM tested it in "The Super-Programmer Project." Dr. Harlan Mills, as chief programmer, undertook to use it on a project which until then could have taken 30 man-years to complete; he finished it singlehandedly in 6 man-years.

This experience led to IBM's developing the *New York Times* information bank (83,000 lines of COBOL averaging 4 errors per 10,000 lines of COBOL), utilizing the chief programmer team concept and structured techniques. The project was not only successful, but programmer productivity using structured methodology improved four to six times over previous approaches.

The disadvantages are:

1. More time spent designing the system, resulting in additional design costs.
2. Need for retraining of analysts and programmers.

Advocates of structured methodology argue that longer design time usually results in shorter programming time and decreased maintenance costs, with a net result of lower overall programming costs.

Structured methodology, therefore, creates correct, reliable, and maintainable systems on a cost-effective basis. Often the word structured is linked with programming. In this capacity, it not only means programming, but the entire systems process: analysis, design, and development.

As we will continually see in this book, change brings with it both opportunities and problems. And as systems analysts are often instruments of change, they must be aware of the importance of interpersonnel relationships.

Oddly, everyone was amazed. Apparently, programmers assumed they had to write mathematically elegant, complex programs to make the computer work efficiently. Now they were learning it was far more efficient to organize, or structure, the programming to create greater reliability. The whole process was not unlike assembling a bicycle. You could take all the parts out of the box and try to assemble it by haphazardly fitting parts together, or you could begin by assembling the frame and wheels, working your way to the gears and accessories. Needless to say, you stood a better chance of riding a bicycle you built the latter way.

Everyone was excited; now you could write programs that were very easy to understand and which could be read from top to bottom, without any GO TO's branching back to some earlier point. Of course, everyone wanted to convert old unstructured programs to structured, but quickly encountered problems. A great deal of code had to be rewritten, flexibility and maintenance suffered, and people realized the proverbial cart was somewhat before the horse. So, structured design was introduced by Stevens, Myers, and Constantine in *The IBM Systems Journal* in 1974. Tom DeMarco and others defined structured analysis in 1974. Since maintenance constitutes about 80 percent of a programmer's time, structured maintenance techniques soon followed.

Schools of thought and methodologies emerged, led by consultants such as Edward Yourdon, Kenneth T. Orr, Milt Bryce, Jean-Dominique Warnier, Michael Jackson, and Daniel D. McCracken. In an article entitled "The Revolution in Programming," published in *Datamation* in 1973, McCracken said, "Structured programming is a major intellectual invention, one that will come to be ranked with the subroutine concept or even the stored program concept."

Source: Essay by Jack Rochester.

THE SYSTEMS LIFE CYCLE

Systems life cycle Expected time a system will operate, from time of conception to removal.

All systems studies involve the same sequence of events, called the **systems life cycle**. It consists of three steps: analysis, design, and development. Taking a problem-solving approach, we first define the problem, then study it to determine alternative courses of action, and, finally, select the most appropriate alternative for which a plan can be prepared, tested, implemented, and evaluated.

The three-step systems process is graphically shown in the data-flow diagram of Figure 1.12. Circles or ovals depict a process that transforms or changes data input to output. A process might be a single program or, for complex processes, a series of programs, or even a manual task such as data entry or visual verification of data. An arrow shows the flow of data. Notice that data-flow diagrams do not concern themselves with what or how data are transformed, nor do arrows indicate anything more than data flow. Thus we see systems analysis as the process of converting a request from a user into the feasibility study. Design then takes the feasibility study and yields system and program specifications. Development converts specifications into a completed system and an evaluation of the completed system.

Since the following sections will constantly refer to it, carefully study the data-flow diagram of the systems cycle in Figure 1.12.

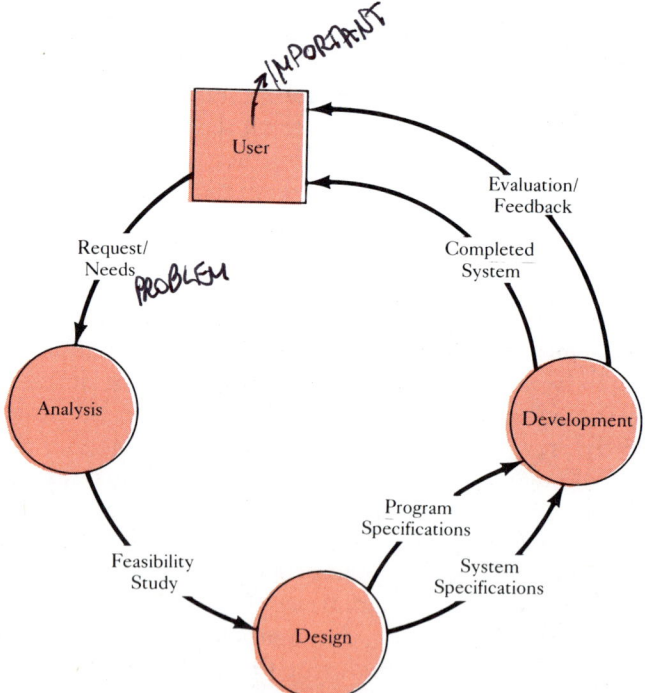

FIGURE 1.12. The systems life cycle in data-flow diagram form. The uncircled items are outputs from a process, such as the specifications resulting from the design stage. Circles represent the rules and procedures that allow one to convert input to output data.

Step 1. Systems Analysis

Systems analysis is a two-step process—preliminary and detailed. The **preliminary analysis** begins when someone perceives a problem, wants an existing system repaired or modified, or wants an entirely new system. A sales manager, for example, might feel that it takes too long to get a sales report or that the sales reports lack vital information, such as sales by region or salesperson. The manager then submits a request for study to whoever is responsible for information processing. If the study seems worthwhile—if it seems that it will eventually lead to a system that will benefit the company—the manager approves the request and assigns an analyst to conduct a preliminary investigation, to determine exactly what the need involves and whether overall benefits will exceed the cost of fulfilling it.

During this problem-definition phase, the analyst first checks out the current system, notes any problem areas, and considers how the user's need might best be satisfied. Problem definition may seem intuitive, but the goal of the activities performed by the analyst centers on answering the question: "What is our problem?"

Eventually the analyst summarizes the information gleaned, including personnel requirements, basic costs, and potential benefits of the new system, in a formal report called a **preliminary report**. In this report, the analyst defines the problem, as perceived from interviews and meetings, and by gathering reviews of written documents from users and management.

If the analyst concludes that further study is warranted, and management approves, **detailed analysis** commences. This study expands the preliminary one to include an analysis of all possible alternative solutions to the problem and an elaborate explanation of the most practical solution to the problem.

The report that results from the detailed study is called a **feasibility study**. This report further expands on the systems study, outlining in far greater detail the following:

1. Problem definition.
2. Scope and objectives of a "new" system.
3. Alternative solutions.
4. Cost and benefit estimates to be derived from alternatives.
5. Potential organizational or policy changes.
6. Description of the new system's major outputs.
7. Recommended alternative and course of action.

Feasibility studies are formally presented to users and managers. If the president of the organization thinks the need extremely important, he or she may play a role in final approval. However, in large organizations, vice-presidents or their subordinates often handle routine systems decisions that primarily affect their departments. The sales vice-president, for example, may approve a new expense account reporting system for a company's sales staff.

Some feasibility studies may lead to a decision to stop systems analysis and continue with the existing system. Decisions to terminate the systems process can

Preliminary analysis Consists of evaluation of request, analysis of request, and management action.

Preliminary report Contains the problem review, findings, recommendations, cost, and schedule.

Detailed analysis Expands beyond preliminary analysis to an elaborate explanation of the solution.

Feasibility study A report including costs, alternatives, schedules, and background information.

result when cost factors outweigh expected benefits, or when management perceives that it will take too long to implement the new system.

Step 2. Systems Design

After decision makers approved the analysis report, the analyst begins the second step in the systems cycle, design, during which he or she determines the various data inputs to the system, how data will flow through it, necessary outputs, and program specifications. The scope of the analyst's activities centers on solving the user's problem.

Analysts enjoy the use of a variety of tools in this second step, such as the data-flow diagram, data dictionaries (lists of terms and their definitions), and the Gantt chart (a graph depicting design events, personnel assignments, and time schedules).

Design is the conversion of a theoretical solution postulated in the feasibility study into a physical reality. The activities performed by the analyst include:

1. Drawing a model of the new system with a data-flow diagram.
2. Devising formats of all the reports the system will generate.
3. Defining the data requirements with a data dictionary and a file or data-base layout.
4. Developing a method to collect the system's data.
5. Writing program specifications.
6. Specifying control techniques for the system's outputs, data base, and inputs.

System specifications The description of the input, output, control, and file designs for a system.

By the end of the design stage, the analyst has written complete **system specifications** in the form of a detailed report with step-by-step instructions that describe the proposed system. These specifications are like the plans an architect gives a contractor before the contractor begins construction.

The system specifications, like the feasibility report, are reviewed by users and management. If the specifications are satisfactory (costs are acceptable, benefits still attainable, and the schedule reasonable), direction will probably be given to proceed to development.

Step 3. Systems Development

The third step in the systems cycle is development, during which the system is built. As in the other two stages, the analyst is involved in:

1. Writing, testing, debugging, and documenting programs.
2. Converting data from the old to the new system.
3. Training the system's users.
4. Ordering and installing any new hardware required by the system.
5. Developing operating procedures for the computer center staff.
6. Establishing a maintenance procedure to repair and enhance the system.
7. Completing system documentation.

The analyst's involvement in each of these activities varies from organization to organization. For a small organization, the analyst may perform all the tasks. In large organizations, specialists may be available to train, order equipment, or convert data.

The development stage ends with an evaluation of the system after it has operated for a period of time. By then most program errors will have shown up and costs should be fairly clear. To be absolutely sure that the system is performing as expected, however, a final **system audit** (review) takes place.

Systems work is never done and systems are never finished; users always want changes or find errors. Evaluation is the feedback part of the cycle that keeps development going as long as the system is in use.

> **System audit** The last check or review of a system to assure it meets objectives and goals.

The Systems Life Cycle Illustrated

Let us return to Mountain Motors, the auto-care group that uses a computer to prepare reminder notices for customers. The controller decides she needs more timely information about the company's cash position. Perhaps a weekly report on all unpaid invoices or bills would enable her to decide which invoices require immediate payment and which do not. If so, she could determine the amount of cash needed at any given time. Accountants call this type of planning cash flow analysis or cash flow projection.

The controller describes her need for a cash flow projection to a systems analyst, thus initiating the systems cycle. The analyst studies the company's accounts payable system carefully to discover how it currently handles invoices and bills. Using experience as a guide, the analyst estimates the new costs Mountain Motors will incur in generating and producing the desired cash flow report. With costs estimated, the user, in this case the controller, will decide whether to continue the systems cycle.

If she asks the analyst to proceed with design, the analyst focuses on the contents of the report in order to propose a format for it (Figure 1.13). The controller may suggest changes in the format, such as adding a line for totals by individual vendor under a caption "Vendor Total. . . ." The analyst modifies the report format to include the new specification and resubmits the report for approval.

Once the user and the analyst agree to the report format, the analyst can determine where and how to acquire the data for it. Since Mountain Motors already uses a system to write checks to vendors, some of the data are readily available. The only unavailable data are the amount of the vendor's discount, say 2 percent, and the number of days for which the discount is effective, perhaps 10 (usually abbreviated as "2/10 net 30"). To calculate the exact amount owed creditors at any given time, Mountain Motors must collect these two important data items and enter them into the computer. Mountain Motors can determine everything else it needs to know by calculation. For instance, Mountain Motors can calculate the number of days that have elapsed since the date of the invoice, and can figure out when a discount lapses by subtracting elapsed time from the allowed discount period.

MOUNTAIN MOTORS, INC.

Accounts Payable Trial Balance
as of April 30, 1984

Vendor Number	Vendor Name	Invoice Number	Date of Purchase	Due Date	Orig. Amount	Discount	Net Amount Due
11-001	ABC Office Supply	A1234	3/21	3/31	50.00	5.00	45.00
11-001	ABC Office Supply	A3433	4/22	5/01	9.00	.90	8.10
			Vendor Total		59.00	5.90	53.10
13-398	K & L Gaskets	35-45	1/19	1/29	90.91	.00	90.91
			Vendor Total		90.91	.00	90.91

Grand total as of April 30, 1984   149.91   5.90   144.01

FIGURE 1.13. *Accounts payable (AP) listing of unpaid invoices. This report shows which invoices need to be paid when. For example, the K & L Gaskets bill is overdue.*

Having established the necessary inputs, the analyst plans the flow of data through the proposed system. At this point, the analyst may discover a need for a master file of vendors so the company can match vendor number with vendor name. Fortunately, Mountain Motors already maintains such a file, so the analyst can plan to store the two new pieces of data in it.

Finally, the analyst determines whether to store and maintain data in conventional files or in a data-base management system, such as IBM's SQL/DS (Structured Query Language/Data System) or Burroughs' DMS-II (Data Management System). Since Mountain Motors uses the SQL/DS base manager for its IBM-computer-based payroll and accounts receivable systems, the analyst would probably choose SQL/DS as the data-storage method. Using SQL/DS will retain systems compatibility and require no new training for the programming staff.

Some systems will require the analyst to specify additional hardware. Here, the analyst must determine exactly what to order, arrange for payment, and schedule delivery, installation, and operator training. Because Mountain Motors can use its existing printer for the new reports, it does not need additional hardware. A small system such as this would consume little additional printer time since the printer can print 10 pages per minute or 600 pages an hour.

At last, the analyst can prepare to develop the new system by drafting a "program specification," a narrative describing the main functions of the needed programs and a recommended programming language. The specifications will include a schedule for completing and testing the programs and directions for locating test data. Once the programs are running smoothly, the users begin training. Several months later, a final audit will help verify that the system works effectively.

After the system has been operating for a while, the user may discover another application for the data. For example, Mountain Motors may request that the report appear in two ways, by vendor number and alphabetically by vendor name. This version of the report allows the user to identify an account name when only the number is known, and vice versa. This possible by-product of the system may not be initially evident but an analyst may be able to anticipate such a future need and plan for it. Flexible systems are always desirable.

CASE STUDY

Fleet Feet Athletic Shoes: An Ongoing Example of the Systems Life Cycle at Work

Fleet Feet, with main offices in Sacramento, California, sells sportswear, including shoes and other athletic attire. It began operation in 1975, with one store, when Sally Edwards and Elizabeth Jansen formed a partnership, and bought a dilapidated Victorian house to remodel into a shoe store. Before long they acquired the vacant lot next door and turned it into a parking lot. At first, they rented out the bottom floor of their building to other small businesses (a used-book firm and a retail plant store), but as their own business expanded with the jogging and fitness craze, they reclaimed the rental space, converting it into company offices. By the summer of 1984, they had opened their 19th store, in Newport Beach, California; their 20th store, in Atlanta, Georgia; and their 21st store, in Oklahoma City.

All Fleet Feet stores operate as franchises whereby people purchase the right to use the Fleet Feet name and exploit the owners' expertise. Sally and Elizabeth help establish the store, train the owners, consult on which goods to buy, and monitor each store's activities twice a month.

With over 20 stores to monitor, Sally and Elizabeth had to hire additional personnel. Rod Farley joined them to assume responsibility for dealing with existing franchises and determining the best-selling products. Jeff Jones manages the original store, and Mark Stensaas keeps the books. Mark places orders for new stock, pays for goods received, calculates payroll and the commissions for the sales staff, and handles all financial transactions. Fleet Feet's organization chart appears in Figure 1.14.

As the firm expanded, manual record keeping became arduous, so the company decided it needed a computer. Initially Sally and Elizabeth bought computer services from their accountant, but within a few years, their business had grown sufficiently to justify acquiring their own system, a Hewlett-Packard 3000 (Figure 1.15). They also purchased applications software for the company's largest and most significant systems: the general ledger, payroll, and inventory. Its other systems (accounts receivable and payable and fixed assets) remained manual.

Although Fleet Feet uses systems common to most retailers, the fact that it has over 20 franchised operations makes the accounting more complex. The parent company must be able to keep track of all these stores through its chart of accounts. Figure 1.16 is a subaccount for the revenues and expenses of each franchisee.

CASE STUDY

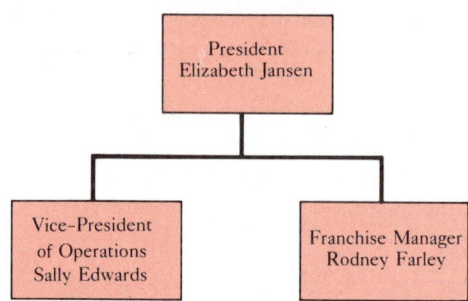

FIGURE 1.14. *Organization chart for Fleet Feet. Since Fleet Feet manages a franchise operation, this chart differs from most organization charts.*

Each store is supervised by the parent company but operates semiautonomously, keeping its own in-store records and reporting summarized data to the central office in Sacramento for posting to the HP 3000. The summary is a list of cash and credit sales, including descriptions of the items sold and copies of all sales invoices. Each store must verify a credit customer's acceptability, then process a credit or charge card payment as if it were a cash sale. The average store grosses $1125 per day over a 5-day workweek. Fifty-two percent of sales are by cash or check, 28 percent are by charge card, and the remaining 20 percent

FIGURE 1.15. *Fleet Feet's Hewlett-Packard 3000 computer. This system has 10 terminals, 1 MB of memory, a 450-line-per-minute printer, a tape drive, and 240 MB of external disk storage. When purchased in 1982, it cost about $80,000. (Courtesy of Hewlett-Packard Company.)*

CASE STUDY

FLEET FEET SPORTS, INC.
PAGE 1

GENERAL LEDGER ACCOUNT LISTING
12/31/83

Account Description

Current Assets (100–120)
101 Cash
102 Marketable Securities
103 Notes Receivable
105 Accounts Receivable
106 Allowances for Bad Debts
107 Inventory
108 Supplies
109 Prepaid Rent
110 Prepaid Taxes
111 Prepaid Insurance

Long Term Assets (141–160)
141 Land
142 Buildings
143 Parking Lot
144 Store Equipment
145 Office Furniture
146 Accumulated Depreciation

Intangible Assets (161–180)
161 Patents
164 Goodwill

Other Assets (181–199)

Liabilities (200–299)

Current Liabilities (200–219)
201 Notes Payable
202 Accounts Payable
203 Salaries Payable
204 Interest Payable
205 Payroll Taxes Payable
206 Income Taxes Payable
219 Other Current Liabilities

Account Description

Sales (400–499)
400 Sales Revenue
401 Sales Returns
404 Sales Discounts

Cost of Goods Sold (500–599)

Operating Expenses (501–599)
501 Sales Salaries
510 Freight-Out
520 Payroll Taxes
523 Utility Expenses
524 Postage Expense
525 Travel Expense
528 Advertising Expense

General/Admn. Expenses
 (550–599)
551 Officer Salaries
552 Office Salaries
553 Administrative Salaries
570 Payroll Taxes
571 Office Supplies
573 Utility Expenses
574 Postage Expenses
575 Travel Expenses
576 Depreciation Exp.
578 Office Equip. Rental
579 Accounting/Legal Fees
581 Building Repair
583 Doubtful Accounts
584 Amortization-Goodwill
585 Amortization-Patents

Other Expenses (600–699)
601 Interest on Notes
602 Interest on Bonds
603 Interest on Mortgage

FIGURE 1.16. The chart of accounts for Fleet Feet demonstrates the way the company tracks its financial information.

CASE STUDY

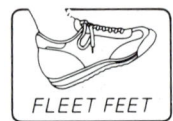

FLEET FEET SPORTS, INC.
PAGE   1

GENERAL LEDGER ACCOUNT LISTING
12/31/83

Account Description

Long Term Liabilities (220–239)
- 222   Bonds Payable
- 223   Mortgage Payable

Stockholders Equity (300–399)
- 301   Capital Stock
- 302   Retained Earnings

Account Description

Other Income (700–799)
- 701   Interest Revenue
- 702   Revenue from Events
- 703   Revenue from Books
- 704   Revenue from Talks
- 705   Revenue from Sponsors
- 706   Revenue from Camps
- 707   Revenue from Bay to Breakers
- 751   Revenue from Franchise #1
- 752   Revenue from Franchise #2
- 753   Revenue from Franchise #3
- 754   Revenue from Franchise #4
- 755   Revenue from Franchise #5
- 756   Revenue from Franchise #6
- 757   Revenue from Franchise #7
- 758   Revenue from Franchise #8
- 759   Revenue from Franchise #9
- 760   Revenue from Franchise #10
- 761   Revenue from Franchise #11
- 762   Revenue from Franchise #12
- 763   Revenue from Franchise #13
- 764   Revenue from Franchise #14
- 765   Revenue from Franchise #15
- 766   Revenue from Franchise #16
- 767   Revenue from Franchise #17
- 768   Revenue from Franchise #18
- 769   Revenue from Franchise #19
- 770   Revenue from Franchise #20

FIGURE 1.16 (Continued)

Data for the Fleet Feet Case Study used with the permission of Sally Edwards and Elizabeth Jansen.

CASE STUDY

FIGURE 1.17. Fleet Feet and its owners: Sally Edwards and Elizabeth Jansen. (Rebecca Gregg, photographer.)

are on store credit. Most sales are in the $30–$42 range, and each store makes between 65 and 70 sales per day. Credit sales are sent to Sacramento for manual processing since the HP 3000 is not programmed to handle the accounts receivable transactions automatically. At the end of the month, the Sacramento office types each customer's bill (there are 250 credit customers) and sends it to the customer for payment. Since the accounts receivable system is a manual one, Fleet Feet uses the balance-forward method to track a customer's purchases, payments, and aged receivables.

The Sacramento office processes all payroll transactions using the HP 3000 payroll system, collecting the employees' time cards at the end of the week and issuing paychecks to the stores before the close of the next pay period. Since employees get weekly checks, everyone is always a week behind in receiving pay. If a store decides to terminate an employee or an employee quits, the store's manager calls Sacramento so the bookkeeper can have the computer calculate wages owed, and then write a paycheck by hand. Most of the stores have two full-time owner–employees who oversee four to seven part-time employees.

The Sacramento office purchases new stock, entering data about it into the HP 3000. Individual stores keep their own inventory records by hand and the home office keeps a complete duplicate set. This dual inventory system allows the home office to locate a particular pair of shoes, for example, for another store, and to compare regional sales patterns. In turn, the home office pays for goods ordered but asks vendors to make deliveries to the specific stores. Fleet Feet buys from approximately 145 vendors whom they try to pay within 30 days. Each store carries about 400 items.

CASE STUDY

Fleet Feet owns the Sacramento Victorian home and the parking lot next door plus two automobiles and a van. The company also owns furnishings, office equipment, the Hewlett-Packard and an Apple Macintosh, inventory, and a small telephone system. The total value of all assets amounts to $2.5 million. Fleet Feet maintains all depreciation and maintenance records by hand.

Fleet Feet's sales for the past 4 years were as follows:

Year of Operation	Sales	Number of Franchises
1	$1,455,231	12
2	$1,657,233	14
3	$1,807,901	15
4	$2,345,331	19

In subsequent installments of the Fleet Feet case study, we will watch the company's systems evolve in step with its growth.

WORKING WITH PEOPLE   Communication Skills

Ben Morjig hovered around the recruiting tables in the Collingsworth College student union, nervously waiting for his interview for a position as a systems analyst with General Products, a manufacturing conglomerate. He was 15 credits short of his bachelor of science degree, but had passed all his programming, mathematics, physics, chemistry, and history courses, and so felt confident about his background. He also felt the fact that he had put himself through college by selling encyclopedias door to door would convince potential employers that he was a motivated, hard-working individual. When he heard his name called, Ben sauntered toward the room where the General Products representative, Judy Celoni, sat waiting.

The interview seemed to go well, and Ben left pleased with his performance. He had taken the initiative throughout, anticipating questions before Ms. Celoni asked them, and emphasizing what he could do for her company. The rejection letter shocked him. What had gone wrong? Were his clothes inappropriate? Had he said something to aggravate the interviewer? Hoping to learn from any mistakes, he called General Products to find out why he had failed and to request a second interview.

Unable to recall his interviewer's name, Ben asked for General Products' personnel department, which located his file and put him in touch with Ms. Celoni. Patiently she described the attributes she was looking for in a systems analyst and why she felt Ben lacked them.

First, an analyst must be "user friendly," someone who enjoys working with people. Ben's description of his sales job indicated he liked to work by himself, with little supervision. Also, an analyst must communicate with people who little understand, and perhaps even fear, computers. Analysts are agents of change, and change threatens most people, so an analyst must remain sensitive to emotions and demonstrate that any new system will be easy to use and improve the user's job performance. Moreover, the analyst must win the confidence of users by speaking their language rather than mystifying technical jargon. As Ms. Celoni pointed out to Ben, he had tried to impress her with his mastery of computer terms and had not asked one penetrating question about General Products.

Effective communication skills include listening, and Ms. Celoni rated Ben as a talker, not a listener. Analysts must be able to pick out subtle nuances. Users may downplay a vital concern or overdramatize a trivial one. Being a good listener means letting the user answer questions rather than leading the user to the answer the analyst expects. Listening implies the willingness to take the time to hear the users' complaints and empathize with them. When Ben pointed out his strong academic science background, Ms. Celoni suggested that an analyst could get as much mileage out of a degree in psychology as one in computer science or information systems. Why hadn't Ben taken courses in English, speech, and communications? Because analysts write many documents and make frequent presentations to users, as well as to management and other personnel, they must write and speak clearly. Poor writing and speaking skills can weaken the effectiveness of an otherwise knowledgeable analyst.

Analysts usually have a bachelor's degree, and sometimes a master's degree, in business, computer science, information systems, or a related subject. They do not usually come from engineering, science, or the humanities, unless they work for such an organization or have other unusual qualifications. Although a degree cannot guarantee a good analyst, the college experience provides a means of acquiring both technical and interpersonal skills.

Last, an analyst must gain experience, usually by working as a programmer. Ben had yet to program outside a college classroom. Not only does programming familiarize the analyst with data processing principles and procedures within organizations, but it allows the building of interpersonal and problem-solving skills as well. An effective analyst should be a technical expert, skillful communicator, businessperson, psychologist, and problem solver, all wrapped up in one.

SUMMARY

Systems surround us: homes have heating, air-conditioning, and ventilation systems; our human bodies are complex muscle, bone, digestive, and circulatory systems; businesses have accounting systems.

Businesses organize themselves around specific departments, each overseeing a system: sales, personnel, research and development, production. With faster and cheaper hardware and more sophisticated software, computerized systems developed in the next decade for these departments will differ dramatically from those of the past. Modern structured methodology yields the higher quality, lower cost systems that users increasingly demand.

The three steps in the systems cycle are analysis, design, and development. Systems analysis examines an existing system with the express goal of making it better. Systems design results in the preparation of system specifications, including the methods for data inputs, formats, data flow, and outputs. Systems development involves programming, testing, training, documenting, and auditing the system.

The systems analyst functions like an architect, beginning the cycle by interviewing the user in a preliminary fashion. Having defined user needs, the analyst draws tentative plans for user approval and conducts a detailed analysis, resulting in system specifications. The analyst and programmer construct the system; the user learns it, runs it for a while, and then evaluates it.

The person who analyzes and designs the new generation of systems enjoys sophisticated tools, but also must skillfully handle interpersonal relationships, effectively communicate verbally and in writing, and maintain a strong grasp of ever-changing data processing concepts.

NEW TERMS

Detailed analysis
Feasibility study
Feedback
Line structure
Organization chart
Preliminary analysis
Preliminary report
Staff structure
Structured methodology
System audit
System specifications
Systems
Systems analyst
Systems analysis
Systems design
Systems development
Systems life cycle
User

QUESTIONS FOR REVIEW AND DISCUSSION

Questions appear in three categories. You can find the answers to the A group in this chapter. The B group requires you to apply the material presented here, whereas the C group necessitates investigation or research.

- A.1. Some new systems will require new equipment. Does the analyst select new equipment during the analysis, design, or implementation phase?
- A.2. Can an analyst design a system without performing a systems study? Explain.
- A.3. Identify the characteristics of a good analyst.
- A.4. Why would the study of psychology help a systems analyst?
- A.5. Some data processing departments are attached to individual departments. Sketch an organization chart for such a structure. What do we call it?
- A.6. What are the three steps in the systems process?
- A.7. List the inputs to each step in the systems process.
- A.8. List the outputs from each step in the systems process.
- A.9. List four benefits of structured design.
- A.10. List three disadvantages of structured methodology.
- A.11. Identify each of the following as either hardware or software.
    - a. COBOL program
    - b. Terminal
    - c. Data-base management system
    - d. Line printer

- B.1. List six systems that currently affect your life.
- B.2. Does an analyst need to be able to program a computer? Explain.
- B.3. In a systems design, why does one sketch report formats before determining data-collection requirements?

B.4. Why are users sometimes reluctant to talk with a systems analyst?
B.5. Does Fleet Feet use line or staff reporting?
B.6. For each of the following functions, identify the person responsible—the data processing manager or the analyst.
   a. Budgeting
   b. Training users
   c. Hiring new staff
   d. Designing files
   e. Supervising operations
   f. Specifying equipment
   g. Dealing with company politics
   h. Managing a systems project
   i. Overseeing a group of systems projects
   j. Developing long-range goals, objectives, and plans

C.1. Systems designed in the 1960s and early 1970s were mostly batch. Today's systems are online. How do these two types of systems differ?
C.2. What other names can you find for "data processing department"?
C.3. Would you say the educational requirements for an analyst are greater than those for a programmer, and if they are, why?
C.4. Find the organization chart for the college you are attending or the firm for which you are working. Is the data processing department line or staff?
C.5. From a recent issue of *Computerworld*, find recruiting advertisements for analysts. What skills and/or experience are required?
C.6. The number of keypunch machines being sold in the United States is declining while the number of terminals is increasing. How does this affect an analyst's job?
C.7. A newer job title is "programmer–analyst." What do you suppose a person with this title does? Would you be more likely to find this person in a large or a small organization?

REFERENCES

Tom Demarco, *Structured Analysis and System Specification*, Prentice-Hall, Englewood Cliffs, N.J., 1979.
B. Dickinson, *Developing Structured Systems*, Yourdon Press, New York, 1981.
Kenneth T. Orr, *Structured Systems Development*, Yourdon Press, New York, 1977.
G. Polya, *How to Solve It*, Princeton University Press, Princeton, N.J., 1971.
Wayne P. Stevens, *Using Structured Design*, Wiley, New York, 1981.
Gerald M. Weinberg, *The Psychology of Computer Programming*, Van Nostrand Reinhold, New York, 1971.

CHAPTER 2

Structured Systems Analysis

- Goals and Preview
- Data-Flow Diagrams
- An Overview of Analysis
- Preliminary Analysis
- Evaluation of a User's Request
- Analysis of the Request
- The Interview
- The Preliminary Report and Management Action
- The Politics of Systems
- Case Study: Camp Beaverbrook's Camper System
- Working with People: Conducting a Meeting
- Summary
- New Terms
- Questions for Review and Discussion
- References

GOALS AND PREVIEW

After reading this chapter you should be able to:

1. Construct a data-flow diagram of a system and level or decompose it into more specific details.
2. Describe preliminary analysis and the analyst's main function during this step of the systems process.
3. List the topics covered in a preliminary report to management at the end of preliminary analysis.

Suppose you develop an idea for a new digital watch. Before you spend a lot of time actually making your watch, you might sketch it and write up a short "sales pitch" to stimulate an investor into providing venture capital for drawing detailed diagrams. Such detailed plans would enable you to determine accurate costs for manufacturing your watches. Your initial sketch provides preliminary analysis, and your detailed plans offer detailed analysis. Detailed analysis of course, could lead you to design a manufacturing facility and actually produce watches.

In Chapter 1, we previewed the systems cycle, following the steps the analyst takes to develop the most effective and efficient system. This chapter and the next will present the analysis phase; later chapters will deal with systems design and development.

DATA-FLOW DIAGRAMS

The systems cycle and its three phases (analysis, design, and development) are symbolized graphically in Chapter 1 by a data-flow diagram, Figure 1.12. This simple example of a data-flow diagram used circles and arrows to show the inputs and outputs for each of the three phases. In addition to such circles and arrows, data-flow diagrams also use squares and three-sided rectangles.

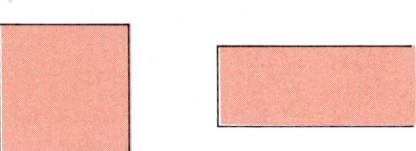

Boxes, or squares, depict sources or destinations of data. In our Mountain Motors auto-care example of Chapter 1, customers receiving reminder notices would be depicted by boxes, Figure 2.1. Three-sided or open-ended rectangles represent files. For example, Mountain Motors has a master file of customers and this file is depicted by this symbol. Thus the customer master file provides data for the reminder system, which prints notices for customers.

We will be using data-flow diagrams to show the systems process throughout this book. Some of these diagrams will have multiple circles, boxes, arrows, or rectangles as we encounter more complex diagrams.

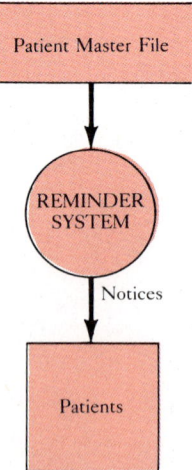

FIGURE 2.1. In a data-flow diagram arrows indicate data inputs or outputs and circles refer to data conversion. Boxes show sources of data and open-ended rectangles stand for files.

AN OVERVIEW OF ANALYSIS

Analysis begins when a user (some member of an organization, such as an accountant or a sales manager) recognizes the existence of a problem and requests either a new report or an improvement to the overall operation of an existing system. We can show the steps in analysis graphically with a data-flow diagram. It will help you better to understand systems analysis itself. As you can see in Figure 2.2, systems analysis contains three sources of inputs:

1. Users describe their needs and objectives in a formal request.
2. The computer services staff or analysts provide computer applications expertise.
3. Management funds and staffs the study of any new system.

It has three outputs:

1. The feasibility study offers a complete description of the system's objectives.
2. The budget and schedule requirements include expected costs and the amount of time it will take to design and implement the system.
3. Analysis documentation details all background information pertinent to the design of the system.

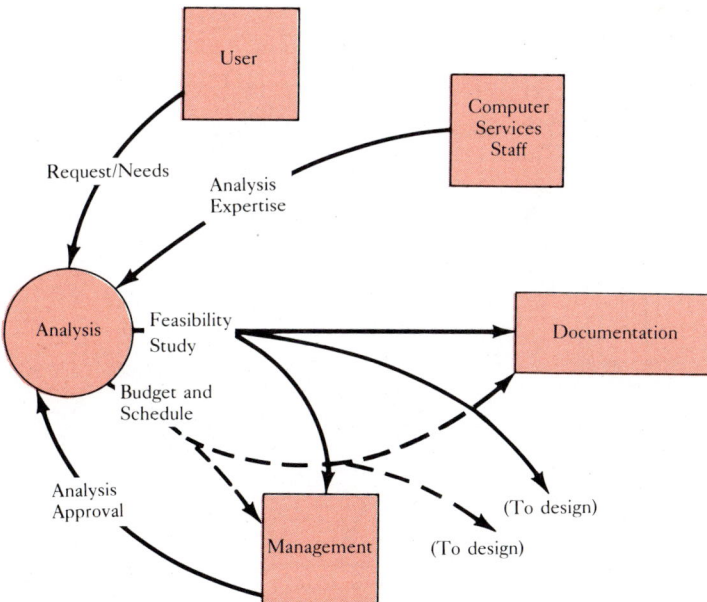

FIGURE 2.2. A data-flow diagram of systems analysis. Users, members of the computer services department, and management provide inputs. Outputs from analysis include a feasibility study, a budget and schedule, and analysis documentation. Often boxes are not drawn but instead are "understood" to be there.

Decomposition The continual breaking down of a task into its elementary components.

Leveling Expanding a data-flow diagram to more specific details.

Since a more detailed analysis follows a preliminary one, we need a data-flow diagram that shows all the details. So we expand the diagram in Figure 2.2 to produce Figure 2.3. We call such an expansion to more specific details **decomposition**, or **leveling**.

Before formally responding to the request that started the analysis, the analyst should develop an overall perspective of an organization's information needs, its structure, and all relevant personnel. The analyst can best accomplish this by reviewing the company's organization chart and relating it to currently operating systems. For instance, the analyst involved with Mountain Motors' auto reminder system will want to discover to whom the appointments secretary reports since it is this person who is most directly involved with the proposed system.

The analyst then can construct a data-flow diagram to illustrate the system graphically. This enables both analyst and user to visualize how the system functions and what modifications may enhance it. Preliminary analysis is a fairly general investigation of the request. Assuming management approval of preliminary rec-

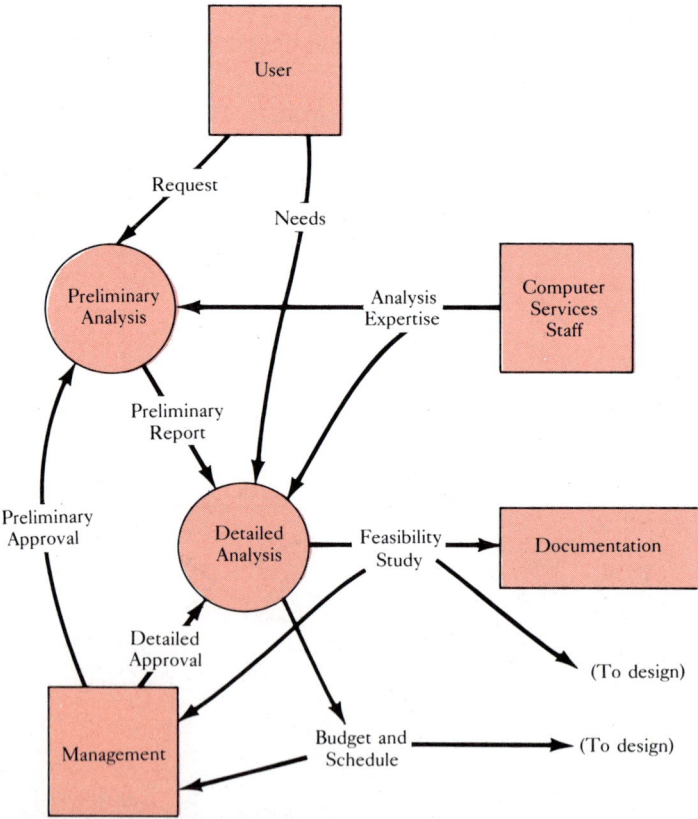

FIGURE 2.3. This data-flow diagram levels, or decomposes, the general diagram in Figure 2.2 to show more detail. Some outputs of preliminary analysis become inputs to detailed analysis. Management, users, and the computer services staff offer inputs to both phases. The detailed phase results in a feasibility study and analysis documentation.

ommendations, the second phase would proceed with the analyst undertaking a detailed study of the current system and how it might be improved. In Chapter 3, you will find a thorough discussion of detailed analysis.

PRELIMINARY ANALYSIS

The data-flow diagram in Figure 2.4 decomposes preliminary analysis into three activities: evaluation of the user's request, analysis of the request, and management's action based on a preliminary report. These activities are, in effect, a logical sequence of problem-solving steps. They apply to all situations, from preparing a new report to reorganizing an entire operation.

EVALUATION OF A USER'S REQUEST

The systems cycle formally begins when a user requests, often by memorandum (Figure 2.5), that an outside consultant or the computer services department conduct an analysis. If the organization hires an outside consultant, that person

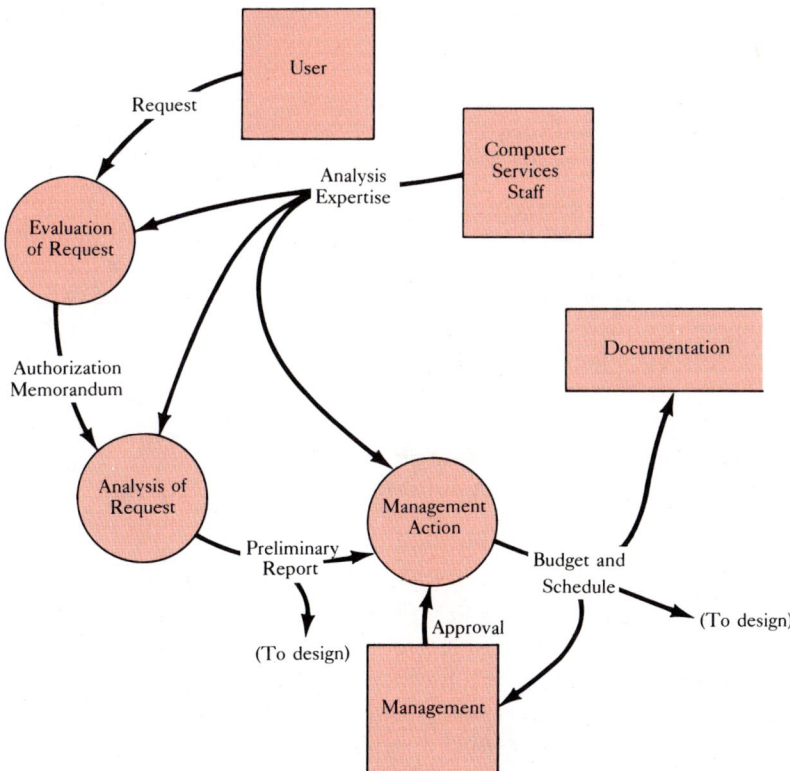

FIGURE 2.4. *A data-flow diagram decomposing the preliminary analysis phase into its components. Users, management, and the computer services staff provide input at various stages.*

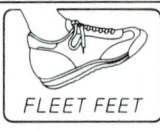

MEMORANDUM July 20, 1984

TO: Kathleen Williams, Manager of Computer Services

FROM: Sally Edwards, Vice-President of Operations, Fleet Feet

SUBJECT: Accounts Payable

  For the past few months I've been watching our accounts payable system and have noticed we're not taking full advantage of discounts offered by our suppliers. I've talked with our bookkeeper, Mark Stensaas, and he assures me we have sufficient cash to make these payments, leading me to feel this isn't a cash flow problem. I think we could save a substantial amount of money by taking advantage of more, if not all, of these discounts. My calculations show we could have saved over $1250 in just March and April.
  Will you perform a study of our accounts payable system to discover how we can take advantage of these "lost" discounts?

SE: jks

FIGURE 2.5. Memorandum from a user requesting a systems study.

receives a formal letter outlining the problem and asking for fee and schedule estimates. In some companies with computer services departments, memos may briefly describe an existing problem or a desired improvement. Some organizations require users to complete a special form for any request (Figure 2.6). This form should include spaces for all relevant facts, such as:

1. The date.
2. The user's name, title, and telephone number.
3. A description of the problem or situation.
4. Comments by the analyst.

  As shown in Figure 2.7 (page 42), the user fills out the top portion of the form and sends it to the computer services department. Thus Sally Edwards, Fleet Feet's vice-president of operations, has requested an analysis of the accounts payable system to determine whether a procedure can be set up to allow the company to take advantage of vendors' discounts for prompt payments. Copies of the memo are sent to the user, to the personnel assigned to the project, and to the analyst's files.
  In evaluating Sally's request, Kathleen Williams, the computer services manager, considers several factors:

REQUEST for SYSTEM SERVICES

Request Date: Requestor: Telephone Number:	Title:

Subject of Request:

Description (Use additional pages if necessary):

Do not fill in this portion of the request

Date Received:
Assigned To:
Action Taken:

File Number:
Approved By:

Signature: _____

Form No. IS-36-1981

FIGURE 2.6. *A typical user's request form. The form allows both user and analyst to add pertinent facts.*

1. Will the organization's goals be better achieved?
2. Is the user requesting a system analysis of a critical or pressing problem?
3. Does the computer services staff have the expertise to work on the problem?
4. What will be the impact of the system on existing hardware?
5. Will the system achieve any savings or cost avoidances?
6. Can the system be developed in an acceptable amount of time?
7. What is the backlog of work, and how would this request affect it?

Using rough estimates, intuition, and perhaps some quick research, the computer services manager must answer the seven questions. If the answers seem mostly positive, the manager will authorize the second stage of preliminary analysis. Mostly negative answers could cause the study to be halted.

Most of the seven questions attempt to answer what may seem to be an obvious question: "What is the problem?" The answer to these questions is not: "Here is the solution." Rather it is a formal recognition and definition of a problem with no preconceived or implied solution. There may not be a solution to the problem or the solution may not require the application of a computer.

Assuming an overall positive reaction to the request and problem definition,

```
                REQUEST for SYSTEM SERVICES
  Request Date:      July 20, 1984         Title:
  Requestor:         Sally Edwards
  Telephone Number:  371-8865                VP Operations, Fleet Feet

  Subject of Request:   Accounts Payable System

  Description (Use additional pages if necessary):
     For the past few months I've been watching our accounts payable
  system and have noticed we're not taking full advantage of discounts
  offered by our suppliers.  I've talked with our bookkeeper, Mark
  Stensaas, and he assures me we have sufficient cash to make these
  payments, leading me to feel this isn't a cash flow problem.  I think
  we could save a substantial amount of money by taking advantage of
  more, if not all, of these discounts.  My calculations show our
  savings could have amounted to over $1250 for just March and
  April.
     Will you perform a study of our accounts payable system to
  discover how we can take advantage of these "lost" discounts?

  Do not fill in this portion of the request

  Date Received:   July 24, 1984
  Assigned To:     Frank Pisciotta, Analyst II
  Action Taken:    Request approved, to begin
                   by August 1 and be completed by August 15.

  File Number:     84-023
  Approved by:     Kathleen Williams, Manager, Computer Services

  Signature:  Kathleen Williams                          Form No. IS-36-1981
```

FIGURE 2.7. A completed user's request form. The user completes the request portion and sends it to the computer services department. The manager evaluates the request, and then makes the assignments as shown here.

the computer department manager assigns an analyst to the task (Figure 2.8). To confirm this decision and alert staff to it, management distributes a second memo to all involved personnel. In addition to informing employees about the impending study, the memo solicits their cooperation, and serves to eliminate rumors and misconceptions that might impede the project. The request has progressed from the idea stage to the point where preliminary analysis is actually underway.

ANALYSIS OF THE REQUEST

Preliminary analysis continues with thorough research. Two steps are involved. First, the analyst, Frank Pisciotta, must determine whether the user has identified the real problem or need. For example, Sally Edwards is requesting an analysis to make it easier for her to identify those due invoices that offer discounts for prompt payment. Frank discovers that since its origin, Fleet Feet has paid all bills on the 24th of each month. After examining Fleet Feet's monthly checking account records, Frank determines that there is sufficient cash to permit bill payment at

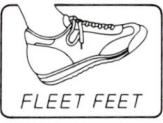

MEMORANDUM July 28, 1984

TO: Operations Personnel and all Vice-Presidents

FROM: Sally Edwards, VP Operations

SUBJECT: Authorization for Accounts Payable Study

Our systems analyst, Frank Pisciotta, has been assigned to study our accounts payable system to determine how we might take full advantage of discounts offered by our suppliers for prompt payment of bills.

I hope you will assist Frank in any way you can. Since he will want to talk with some of you in the next 2 weeks, please make room in your already tight schedules for him.

Thank you in advance for your cooperation.

SE: jks

FIGURE 2.8. *Letter of authorization sent to the personnel involved with the system under study. This memo confirms management's approval to proceed with preliminary analysis.*

any time during the month. He concludes that Sally's request indeed has merit: The real problem was simply one of tradition, not accounts payable.

Second, the analyst must estimate the costs of a detailed study. For Fleet Feet's accounts payable request, this requires a review of past invoices, profit and loss statements, balance sheets, bank statements, stockholders' reports, and other historical documents. This quick review should roughly determine whether a new, or revamped, system is warranted to solve the problem. Not all problems are best solved by computerization. For example, when Sally and Elizabeth requested a computerized fixed-asset system for Fleet Feet, they soon learned it would not be cost effective for a firm such as theirs with only 10 pieces of equipment. Manual calculation of depreciation records in this case is still the best method.

THE INTERVIEW

One of the best sources of information when researching any request is the people the system will affect. An analyst usually consults the personnel department and examines organization charts (Figure 2.9) and job descriptions (Figure 2.10) in order to select the right personnel to interview. For example, Fleet Feet's organization chart reveals that James Taylor, the finance manager, and his staff are responsible for the accounts payable system. Frank Pisciotta, who has been assigned to conduct the preliminary analysis, therefore will interview the key people

44 PART I SYSTEMS ANALYSIS

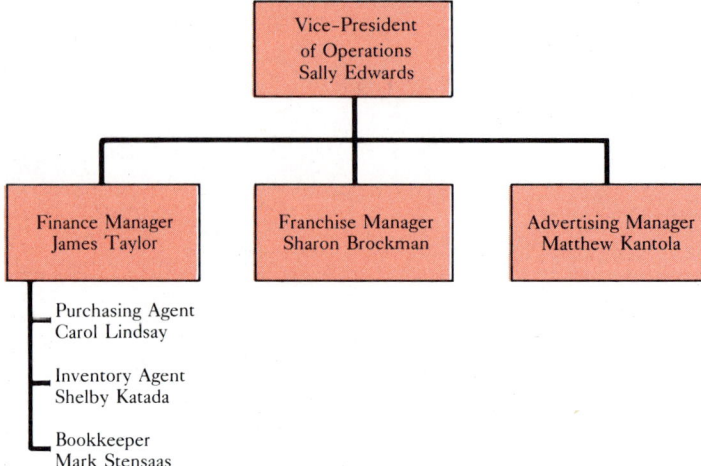

FIGURE 2.9. Organization chart for Fleet Feet's operations department.

TITLE: Bookkeeper

REPORTS TO: Finance Manager

DUTIES AND RESPONSIBILITIES:

1. Post accounting transactions to receivables, payables, and general ledger.
2. Prepare a list of overdue accounts needing follow-up contact.
3. Prepare a list of vendors to be paid to take advantage of discounts.
4. Train new employees in the department.
5. Lead an office staff of two clerks.
6. Perform related work as assigned.

PRIOR EXPERIENCE: At least 3 to 4 years of progressively responsible experience keeping books for a private business. Ability to post a set of books with transactions of all types, analyze data, draw conclusions, speak and write effectively, work well with others, and conduct oneself in a professional manner. Some background in computing is mandatory. Must be able to operate a 10-key adding machine and a computer terminal. Must be bondable.

EDUCATION: Two-year community college degree in accounting or similar area. Prior experience can be substituted for education. High school diploma mandatory.

SALARY RANGE: $15,350 to $18,460 per year, depending on experience.

FIGURE 2.10. A sample job description the analyst might find in Fleet Feet's personnel files.

in Taylor's department: Carol Lindsay, Shelby Katada, and Mark Stensaas. Analysts should follow protocol, first contacting managers or supervisors to ask their cooperation. Therefore, Frank would talk to Taylor before approaching his staff.

Preparation

In some cases, selecting the correct people to interview may not be so easy because an organization chart reveals how the firm is supposed to run, not how it actually may work. If Frank had any doubts about the accuracy of Fleet Feet's chart, he could follow an invoice from its receipt until it was paid, noting along the way all those who handled it. A list of these people would confirm the actual operation of the system and indicate the necessary people to interview.

It helps to plan **interviews** carefully, making an appointment for a specific time and place and providing in advance a written summary of questions the analyst wishes to ask (Figure 2.11). This memo allows the interviewee to gather relevant information and to consider his or her responses. As with most aspects of systems analysis, careful planning helps eliminate inaccurate guesses and saves valuable time. One of the most common complaints within organizations is that too much time is wasted in meetings, so analysts should try to make their meetings as efficient as possible.

Interview A meeting between analyst and relevant personnel for fact finding.

With the list of questions well formulated, the actual interview focuses on collecting answers. Effective interviews have three components: warm-up, questioning, and closure.

Warm-up

During the **warm-up**, the analyst seeks to establish rapport with the interviewee, striving to create a relaxing atmosphere and making the interviewee a partner in the process. Topics for opening discussions can range from the weather to the outcome of the Monday night football game. If the analyst has had past dealings with the interviewee, reminiscences can help to establish the needed rapport.

Warm-up Establishing rapport during an interview.

Analysts conduct interviews with a wide range of people with varying educational and cultural backgrounds: corporate officers, middle managers, clerks, assembly-line workers. In all instances, it is imperative that the warm-up establish the analyst as the leader in the interview.

Questioning

Once the interviewee seems relaxed, the analyst sets the scene for **questioning**, beginning with a review of the reason for the interview, citing management's authorization, and stating the problem. Questions should follow the outline prepared beforehand, but should adapt to the answers given by the interviewee.

Questioning The fact-collecting component of an interview.

During questioning the analyst also should be a good listener, eliciting complete answers and paying close attention to everything said. One must allow sufficient time for full responses, and try not to interrupt them unnecessarily. A good interviewer avoids trying to provoke an expected response or to answer for the person being questioned. If an interviewer attacks an answer or places a value

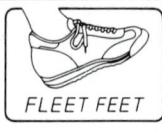

MEMORANDUM August 3, 1984

TO: James Taylor, Finance Manager

FROM: Frank Pisciotta, Analyst II

SUBJECT: Accounts Payable System Study

 This confirms our meeting for August 8 at 3:00 p.m. in your conference room. As you are aware, I am analyzing our AP system to determine how we can take advantage of supplier discounts. I want to cover the following topics.

1. On the average, how many invoices do you pay each month?
2. What is the average balance in Fleet Feet's checking account?
3. Are there any time periods during which we need extra cash to meet specific needs?
4. How long does it take an invoice to get from the warehouse to your office?
5. Once an invoice is in your possession, how long does processing take?
6. Are invoices complete when you receive them or do you have to contact the warehouse for missing data? Give me some typical examples of missing data.
7. Since all checks must be signed by the vice-president, how long does it take for them to be returned to you for mailing?
8. After the signed check comes back to you, how long before you mail it?
9. Do you have all the necessary equipment and staff to perform the duties assigned to your office?
10. Are you aware of any other problems in the AP system?

FP: jks

FIGURE 2.11. To prepare people for interviews, an analyst should send a memorandum sufficiently in advance to allow them to gather relevant information and thoroughly consider their responses.

on it, the interviewee may feel threatened, and the initial rapport may be lost. Furthermore, the interviewer should minimize computer jargon and buzzwords, and rely instead on common language people use in their daily work. Remember that the interviewee, not the analyst, is the expert on the subject under discussion. Facial expressions and body language should not reveal the interviewer's reactions, attitudes, or feelings about answers. If clarification of a response seems necessry, the interviewer may ask, "Just to make sure I understand your answer, do you mean. . . ?"—not simply assume he or she knows what the other person is trying to say.

 If the list of questions is long, they are technically complex, or they require detailed answers, the interviewer should take brief notes summarizing the responses, but not a verbatim transcript. Some analysts make video- or audio-tape recordings of interviews, to make sure nothing escapes their attention. Because many people feel uncomfortable with a microphone or camera present, one should obtain their permission before using such equipment.

Closure

When all the planned questions have been answered, the interviewer terminates the discussion (**closure**), perhaps by asking whether the interviewee has any questions. The interviewer then thanks the other person and indicates whether or not a second interview might be necessary.

Closure The conclusion of an interview.

Follow-up

Shortly after the interview, the analyst should summarize (**follow-up**) the findings in writing, forwarding a copy to the person interviewed (Figure 2.12). A summary:

Follow-up Summary of findings sent to the interviewee.

1. Produces a record of information gained from the interview.
2. Allows those interviewed an opportunity to check for accuracy.
3. Permits the analyst to express appreciation formally, which helps maintain rapport between analyst and user.

Interviews afford the analyst an opportunity to personalize the fact-finding process, helping both parties see each other as people rather than just voices over a telephone or faceless authors of impersonal memoranda. Two disadvantages of the interview are the time it takes to schedule, conduct and follow up on one, and the danger of personal bias on the part of either participant.

THE PRELIMINARY REPORT AND MANAGEMENT ACTION

After examining the request and estimating the cost of a detailed analysis, the analyst presents findings to management. Incorporating information gleaned from job description, organization charts, files, and interviews, the analyst submits a written preliminary report, which usually contains four sections.

1. Problem review.
2. Findings.
3. Recommendations.
4. Cost and schedule estimates.

Whenever possible, analysts should pose alternatives so management can weigh the costs and benefits of more than one solution to a given problem. Since most business decisions involve trade-offs between costs and benefits, management is already accustomed to making decisions this way. The decision may range from halting the process to requesting a detailed analysis. Before presenting the report orally (Figure 2.13) to the requesting user and appropriate department managers,

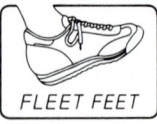

MEMORANDUM August 9, 1984

TO: James Taylor, Finance Manager

FROM: Frank Pisciotta, Analyst II

SUBJECT: Summary of August 9 interview

I greatly appreciate your time yesterday, since I know how busy you are. Your advice was invaluable and confirmed some of my feelings about our AP system. My notes show we covered the following.

1. We are paying the following number of invoices:

Month	Number of Invoices Paid	Average Value
April	78	$455.66
May	92	603.90
June	84	505.23

2. Fleet Feet's average checking account balances for April, May, and June were: $20,331.56; $22,099.51; and $23,347.96.
3. Fleet Feet needs extra cash ($3656.55) the first week of each month to pay utilities and the mortgage on the building and parking lot.
4. It takes 6 days for an invoice to get to your office from the warehouse.
5. Once an invoice is in your possession, it takes an average of 7 days to process it.
6. All necessary information appears on the invoice when you receive it from the warehouse. Occasionally invoices do arrive stained as a result of careless handling.
7. The process of securing signatures on checks take 2 days.
8. Once the check comes back signed, you mail it within 24 hours.
9. You feel you need an additional staff person and a typewriter and calculator to keep up with the work load.
10. You do not see other problems in the AP system.

FP: jks

FIGURE 2.12. An example of a memorandum summarizing an interview.

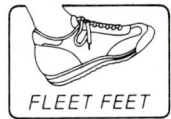

FLEET FEET

MEMORANDUM August 15, 1984

TO: Sally Edwards, Vice-President of Operations

FROM: Frank Pisciotta, Analyst II

SUBJECT: Preliminary Report on Accounts Payable System

Problem Review:

We have now completed our preliminary investigation of the accounts payable system, which we undertook to ascertain the feasibility of taking advantage of vendors' discounts.

Frank Pisciotta interviewed the four key individuals involved with the AP system; summaries of these interviews have been circulated to everyone involved. We also interviewed two of our suppliers, Nike and Converse, to verify potential discounts.

Findings:

During the interviews and our search we learned the following.
1. It takes 16 days for the average invoice to be paid.
2. The average value of each invoice during April, May, and June of this year was about $520.
3. For this same 3-month period, an average of 85 invoices were processed each month.
4. The AP system is functioning smoothly. Our credit rating with the two suppliers interviewed is good.
5. The work load of Mark Stensaas' staff is quite heavy, and we have acquired no new equipment in his area for the past 4½ years.
6. Mark's staff has increased by two people in the last 4 years while our business volume has expanded tenfold.
7. While the majority of our suppliers offer a 2 percent discount for payment within 10 days, at least two offer a 3 percent discount for payments made within 5 days.

Recommendations:

On the basis of these preliminary findings, we recommend the following:

1. A detailed analysis should begin immediately to determine the feasibility of computerizing our manual accounts payable system. Based on 85 transactions per month with an average value of $520 and a 2 percent discount for payment within 10 days, there is a potential savings of $884 per month ($520 × 85 × 0.02) or $10,608 per year.

FIGURE 2.13. *A preliminary report to which management can react.*

> 2. Mark Stensaas and his staff should receive special thanks for their work.
> 3. Mark Stensaas should be authorized to sign checks. This would save 3 days of processing time. Mark would have to be bonded, at a cost of a few hundred dollars.
>
> Cost and Time Schedule:
>
> The computer services staff is ready to perform the detailed analysis and can begin 4 days after management's approval. Frank Pisciotta would continue the analysis and expects it will take him 5 weeks to complete. The cost of a detailed analysis would be
>
> 1. Five weeks for Frank Pisciotta $3192.00
> 2. Secretarial aid (30 hours)   210.00
> 3. Two weeks for other Fleet Feet staff
>   for interviews and discussions   1300.00
>   Total $4702.00
>
> FP: jks

FIGURE 2.13 *(Continued)*

the analyst circulates it for review. This allows everyone concerned to mull over the analyst's findings, discuss them among themselves, and investigate questionable statements. In Fleet Feet's case, Sally Edwards would confer with her staff, discussing the analyst's recommendations, cost estimates, and proposed schedule for detailed analysis.

Although the person who requested the study (Sally Edwards in our example) presides over the oral presentation, the systems analyst (Frank Pisciotta) leads the discussion, responding to management's questions. The analyst should pay special attention to communication and never condescend to the audience or use unnecessary jargon. If the preliminary report indicates that a new system would not solve the problem or would cost too much, management may halt the process to avoid wasting time and money. If, on the other hand, the report favors a new system, detailed analysis will probably begin.

In some organizations, the manager responsible for the new system makes such a decision independently; other businesses use consensus decision making, letting all interested parties contribute. Businesses in the United States tend to use the former method; the "Japanese" style of management prefers the latter. Using the consensus approach to decision making, Fleet Feet accepts Frank Pisciotta's recommendations and decides to proceed with a detailed analysis of its accounts payable system.

The first two steps in preliminary analysis (evaluation and analysis of request) provide a cursory look at a problem. The examination is purposefully superficial, with no attempt to solve the problem, but rather to define its scope and magnitude. The third step allows management to discontinue any further examination before large amounts of staff time and funds are spent.

The Politics of Systems

No arena of human interaction can exist without a certain amount of subjectivity and emotion, or what we commonly call the "politics" of a situation. In the field of data processing we seem to have reached a critical stage in this regard. Just as the National Organization of Women has made feminist issues a factor every politician must face, so have Management Information Services (MIS) departments become a new political force within organizations. While MIS departments may have enjoyed tremendous clout when they first arrived on the scene, they may find themselves now saddled with outdated systems that grew up in the '60s and '70s. There, system "usurpers," both within and outside the organization, have sprung up to meet the demands of a changing environment. During the '70s, the threat came primarily from timesharing firms which offered to fill the gap; in the '80s, microcomputer retailers have begun marketing their new systems directly to users. In addition, software vendors, consulting firms, and self-proclaimed systems "experts" are diluting many MIS department's once sacred authority, and users armed with increased computer awareness and the power of desktop computing, are demanding, and in many cases assuming, more control of their data processing requirements.

To complicate matters, systems developed during the '50s and early '60s, were often statistically and scientifically oriented and staffed with operations research professionals whose job it was to implement and support the accounting and management needs of an organization. In many firms, such systems remain the mainstay of MIS, but they find themselves increasingly challenged to respond to business needs that are less statistical or scientific. No longer can "the eggheads who run the computer" claim to hold the mystical keys to the information kingdom.

As many organizations move from centralized mainframe to distributed micro-based computing, an MIS department finds its exclusive control eroding. The old guard which once sat smugly behind a barrier of expertise and jargon may laughingly say that DDP stands for "disorganized, not distributed data processing," but unless they learn to work closely with top management to accommodate changes in computing, they may find themselves stuck with no control at all. Today's computer professionals must:

1. Analyze business needs and determine the firm's information requirements independent of technology.
2. Adapt evolving technology to new systems.
3. Keep in mind that, above all, systems exist to support people and organizations that often resist neat, clean technical formulas.

People, not systems, make organizations successful. To be sure, the computer age requires even more sophisticated technical know-how, but no degree of technical expertise can singlehandedly solve the complex problems of people working in groups to achieve goals. Since programming will eventually move to the user areas (except for "core," firm-wide systems), the distinction between MIS professionals and end users will blur. Users will be more systems savvy, and MIS professionals will become more people-oriented.

The stakes are high. As the saying goes, "a little knowledge is a dangerous thing," and users armed with micros can undermine MIS to the considerable detriment of the firm. "On my Apple I can do this in five minutes" cry users, at times mistakenly pushing MIS to complete a three-month project in two days. On the other hand, defensive MIS departments can ignore the real reason to streamline systems and stick to a level of technical detail more appropriate for a nuclear reactor than a shoe store's payroll. Top management cannot avoid responsibility by trying to stay above the fray. It must face and wrestle with the issues and remove the roadblocks that keep the organization from effectively achieving its goals. Politics invade everything we do. We can try to ignore this fact, but doing so will only make a messy situation messier. As long as politics do exist, one is far wiser to master rather than fight them.

Source: Samuel H. Solomon, "The Politics of Systems," *Datamation*, pp. 212–218, Dec. 1983. Reprinted with permission of DATAMATION® magazine, © Copyright by Technical Publishing Company, A Dun & Bradstreet Company, 1983—all rights reserved.

CASE STUDY

Camp Beaverbrook's Camper System

Ron and Lynn Garrison, who own and operate Camp Beaverbrook, a summer youth camp, want to schedule activities according to the interests of their 700 campers. From camper applications (Figure 2.14), the Garrisons compile areas of interest (archery, canoeing, horseback riding, marksmanship, computing) into a campers' interest roster. Because it takes so much time to do this job by hand, Lynn asks a data processing consultant for help (Figure 2.15). The consultant, Art Greene, agrees to work for the Garrisons (Figure 2.16) if they accept his fees and proposed schedule. Ron and Lynn agree to hire him and tell him so by letter (Figure 2.17).

Art Greene initiates an analysis by sketching a data flow diagram depicting the flow of data into and out of the system (Figure 2.18). It contains a single input, 700 camper applications, and two outputs, the campers' interest roster and rejected camper applications.

A preliminary data-flow diagram such as Figure 2.18 provides a broad overview that Art can expand by leveling or decomposing into specific components. For Camp Beaverbrook, Art expands the preliminary data-flow diagram to include two specific tasks: (1) checking application completeness and accuracy, and (2) categorizing applications according to both campers' chosen week(s) of attendance and interests (Figure 2.19).

Next, Art Greene considers two alternatives for beginning his analysis of Beaverbrook's interest roster system: interviewing Ron and Lynn, or studying their camper file. Anticipating a quiet day at the office, he chooses the latter, and while studying the prior year's applications and interest rosters, he notices asterisks and pluses after some campers' names, but not others (Figure 2.20).

CASE STUDY

CAMPER'S APPLICATION

Camper's name:_____ Camp age:___ Grade next fall:___ B/date:___/___/___
                                                                                                                               mo. day yr.

Address:_____City:_____Zip:_____Home Phone:(___)_____

Parents' Names:_____,_____Occupations:_____,_____Emerg. Phone:(___)_____

Who is authorized to pick my child up from camp? Mother:___, Father:___, Other:___

Session(s) Desired: #_____ Dates From:_____ To:_____ Alt. Session: #_____

Sex of child:___; Sisters (ages):_____ Brothers (ages):_____; New___ or Old___ to CBB.

Approximate Height:_____, Weight:_____. Hobbies or interests:_____

Any severe allergies?_____ Any restrictions?_____

My camper will take the chartered bus to camp: Yes:___ No:___, From: Orinda___ Napa___

In signing this application I certify that my child agrees to abide by the camp's rules and regulations, is cooperative and is free from habits that would make an undesirable camper. My child and I recognize that in all programs offered at CBB there is an element of risk and possible injury. I therefore hold harmless the camp owners, management, and employees of responsibility for accidents.

In the event of an emergency and I cannot be reached, I give permission to transport my child to the nearest physician or hospital for necessary treatment. I give permission for Camp Beaverbrook to use photographs of my child for camp promotion.

I wish to enroll_____ in the session(s) indicated and enclose $100 deposit per session.

Signed:_____ (Parent/Guardian) Date:_____

Please fill out separate application for each child.
To NEW Beaverbrook familes: Where did you first hear about CBB?_____

All other pertinent details will be sent on receipt of this application and deposit.
THANK YOU!

FIGURE 2.14. Camp Beaverbrook's application form. Beaverbrook advertises in popular magazines for new campers and sends the application form to respondents in February of each year.

    A few days later, Art schedules an interview with Ron and Lynn and discovers that the camper interest roster is far more than a simple list of names and interests. Asterisks identify those campers with special health conditions (asthma and epilepsy, for example). Pluses indicate which campers need transportation to and from camp.
    Had the consultant relied solely on his examination of the campers' files for his preliminary

CASE STUDY

October 2, 1984

Mr. Arthur Greene
Camper Systems
1233 North J Street
Pittsburg, CA

Dear Mr. Greene:

    Bob and Helen Stein gave us your name as a consultant who specializes in summer camps and data processing. They were very pleased with your work for them 2 years ago and felt you might be able to assist us with our problems.

    Our high mountain wilderness camp serves about 700 campers each summer. We are not a specialty camp (solely dedicated to computers) but instead try to serve a wide variety of campers with differing interests. Our main problem centers on scheduling each camper into the week in which the activities are best for that camper. I have enclosed a camper application form so that you can see what paperwork prospective campers complete before we accept and schedule them into Camp Beaverbrook.

    In view of the campers' various interests, we feel a computerized solution to our scheduling problems might be in order and would like to hire you to perform a study. Please send us your fee schedule as soon as possible, as we are anxious to begin.

Sincerely yours,

Lynn Garrison

Lynn Garrison

FIGURE 2.15. *Lynn Garrison's letter to data processing consultant Art Greene. The consultant was recommended to the Garrisons by their friends, Bob and Helen Stein, who run a camp similar to theirs and had successfully used the consultant 2 years earlier.*

Correspondence and data courtesy Ron and Lynn Garrison, Owner/Directors, Camp Beaverbrook.

CASE STUDY

October 11, 1984

Mrs. Lynn Garrison
Camp Beaverbrook
Beaverbrook, CA

Dear Mrs. Garrison:

I always am happy to hear that my work for others was satisfactory and appreciate the comments by Bob and Helen Stein.

Recently I completed a study similar to the one you requested for another summer camp in Los Angeles. My charge to them for a preliminary investigation was $475 plus expenses, which amounted to $187.50. The study took two visits to their camp and lasted 2 weeks. I estimate my fees to you would be about the same. My normal charge-out rate is $47.50 per hour plus expenses (travel, lodging, and food).

For my fee I shall perform the following functions:

1. Review your application form to ensure the correct data are being collected.
2. Interview selected members of your staff about scheduling.
3. Examine your files for backup information and statistics about past summer campers and their interest selections.
4. Calculate the cost to perform a detailed investigation of your scheduling system.

I can begin work within 5 days of your acceptance of these terms and await your reply.

Sincerely,

Art Greene

Art Greene

FIGURE 2.16. Lynn's letter to the consultant brings the following response.

CASE STUDY

October 22, 1984

Mr. Arthur Greene
Camper Systems
1233 North J Street
Pittsburg, CA

Dear Mr. Greene:

 Ron and I both agree that your list of tasks and charges is acceptable and wish you to start work immediately. Unfortunately there is a meeting of the Western Summer Camp Owners Association the first week in November in Disneyland, which we will be attending. Perhaps we can start the second week. Please call us to schedule a meeting time and place.

Sincerely yours,

Lynn Garrison

Lynn Garrison

FIGURE 2.17. *The Garrisons agree to hire the consultant and authorize him to proceed with the study.*

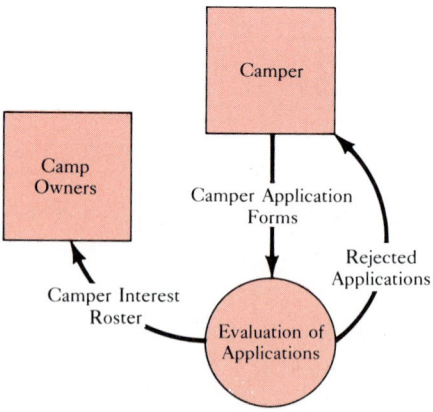

FIGURE 2.18. *The preliminary data-flow diagram for the Garrisons' camper roster has a single input and two outputs.*

CASE STUDY

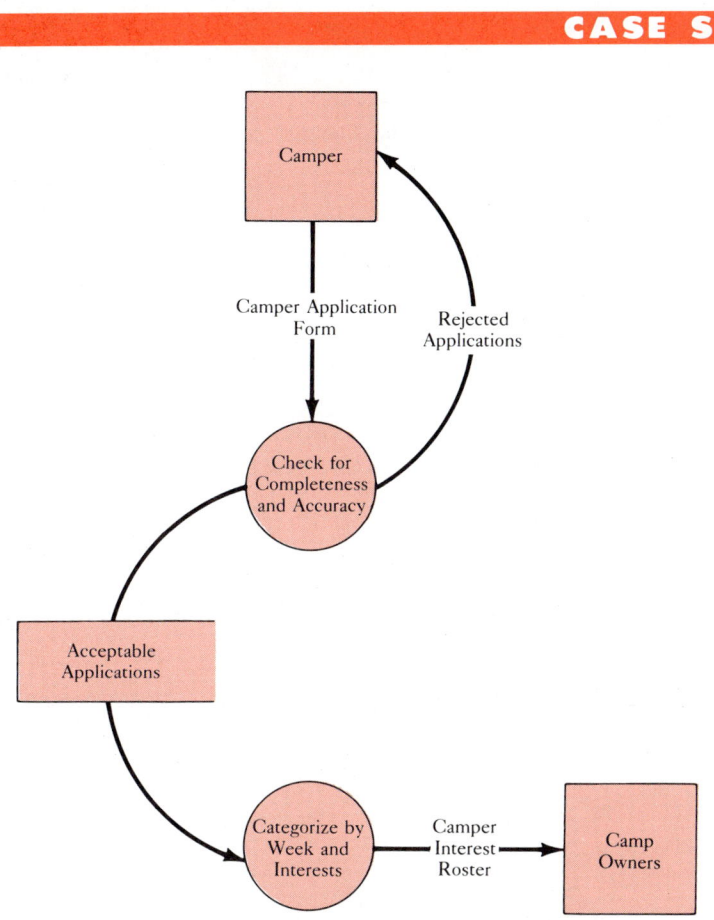

FIGURE 2.19. *Art Greene's data-flow diagram decomposed to a second level.*

analysis, he would have overlooked some vital information. Fortunately, his interview with Ron and Lynn leads him to discover that Ron's original description of his need is deceptively simple, and that he really wants more than just an interest roster.

Furthermore, as often happens during systems analysis, Lynn Garrison casually wonders aloud whether Art might not recommend computerizing the currently manual Beaverbrook payroll system. With a staff of only 30 summer employees, Lynn worries that computerizing the payroll might cost more than it is worth. Art agrees to look into it, but asks to delay that project until he has finished the camper roster system.

Within 2 weeks, Art is able to give the Garrisons a preliminary report outlining his findings and suggestions for their camper interest roster (Figure 2.21). Ron and Lynn feel good about the consultant's preliminary report and cost estimate and authorize him to continue to detailed analysis (Figure 2.22). Art agrees to start the detailed analysis in a week's time at the agreed-to price (Figure 2.23).

CASE STUDY

Camp Beaverbrook
Summer of 1984 **Major Interest:** Computing

Camper Interest Roster
Week: 1 **Page:** 1

Camper Name	Secondary Interests	
Anderson, Marion	A, V, H	+
Asplund, Mary	A, R	
Bello, Diane	H	*
Braz, Janet	A	
Castle, Irene	R	
Duke, Ron	A, V	
Ewert, Judy	V	+
Fritz, Ron	H	
George, Tony	H	+
Hernandez, Janet	H	
Hopkins, Dale	A	+
Jimenez, Jose	V	+
Kosik, James	R	
MacDonnell, Bill	R	*
Mc Kay, Peter	V	
Nelthorpe, Judi	H	+
Oreno, Marty	V	
Roy, Adelene	V	
Russ, Jerry	H, V, R	
Shackelton, Craig	H, R	*
Smith, Stan	V	
Southard, Gail	H	
Tsuda, Elizabeth	A, R	
Tuft, Lynn	H	
Ward, Nancy	A, R	+
Winton, Donald	H	
Yoshikawa, Henry	H, V, R, C	*
Yunk, Robyn	H	
Zimmerman, Christine	H	*

FIGURE 2.20. Camper interest roster built each summer by Lynn and Ron Garrison. The asterisks and pluses indicate special needs (transportation or health problems).

CASE STUDY

November 26, 1984

Mrs. Lynn Garrison
Mr. Ron Garrison
Camp Beaverbrook
Beaverbrook, CA

Dear Mrs. and Mr. Garrison:

This letter will serve as a report of my findings on your camper interest roster and scheduling problems. I hope you will find the recommendations satisfactory and will authorize me to proceed with the study. I have calculated an estimated fee and time frame for such a study.

During my interviews, examination of your application form, study of your filing system, and tour of your facilities, I learned the following:

1. You receive 90 percent of your applications 2 months after your advertisements appear in national magazines.
2. Most campers attend for a 3-week period and want to participate in the following activities.

Volleyball	50%	Swimming	80%
Hiking	70%	Archery	35%
Rowing	20%	Marksmanship	25%
Canoeing	40%	Computing	15%

3. You do have a scheduling problem, which is in part attributable to the fact that your application form does not collect the correct camper interests.
4. Your manual scheduling process is costing you $5000 per year and takes 250 hours to complete. In addition, you spend over $670 per year on long distance telephone calls to potential campers with scheduling problems.
5. You have a very high (75 percent) camper return rate and this can be attributed to your excellent manual scheduling system.
6. You will need to hire an additional office worker next year to keep the schedules and camper enrollments up to date if no action is taken.

Based on these preliminary findings, I recommend the following course of action.

1. Perform a detailed study, to begin immediately, with the goal of investigating computerizing your entire camper enrollment, interest, and payment record-keeping system.
2. Begin your advertising campaign a month earlier than last year. This early start will give you a longer time in which to schedule the campers for 1985.
3. Redesign your camper application form to gather the missing data items we discussed at our last meeting.

FIGURE 2.21. Art Greene's preliminary report to Lynn and Ron Garrison.

CASE STUDY

> I am willing and wish to perform the detailed analysis. I expect it will take me until January 15, 1985, to complete this project and report my findings to you. My fee for this study is based on my new rate for 1985 of $50 per hour plus expenses. The detailed analysis will probably take 40 hours at a cost of $2000.
>
> If you have any questions or comments, please call me during the daytime hours.
>
> Sincerely yours,
>
> *Art Greene*
>
> Art Greene

FIGURE 2.21 *(Continued)*

> December 5, 1984
>
> Mr. Arthur Greene
> Camper Systems
> 1233 North J Street
> Pittsburg, CA
>
> Dear Mr. Greene:
>
> Lynn and I were very pleased with your preliminary report on our scheduling problems. While the fee to conduct the detailed analysis is somewhat higher than we expected, we would like you to continue. This letter is to direct you to begin the detailed analysis of our camper preference system. We hope that you will complete the work before the 15th of January so we can use it in the summer of 1985.
>
> Sincerely yours,
>
> *Ron Garrison*
>
> Ron Garrison

FIGURE 2.22. *Lynn and Ron's response to the preliminary report authorizing the consultant to continue to detailed analysis.*

CASE STUDY

December 11, 1984

Mrs. Lynn Garrison
Mr. Ron Garrison
Camp Beaverbrook
Beaverbrook, CA

Dear Mrs. and Mr. Garrison:

   I am pleased you have found my preliminary report satisfactory and wish me to proceed with a detailed analysis. My charges are based on my normal rate and on the work I performed for another camp on a similar project. I shall try to complete your work before January 15, and I hope under budget. Please remember that I am performing a detailed study and you will have to approve and direct me to design a camper scheduling system upon approval of the detailed study. I think we could have a system operational in time for your 1985 camper year.

Sincerely,

Art Greene

Art Greene

FIGURE 2.23. Consultant agreeing to perform the detailed analysis.

WORKING WITH PEOPLE   Conducting a Meeting

Muttering under her breath, Heidi Kantola walked back to the office she shared with Douglas Manufacturing's other systems analysts. After working on a preliminary report for a new system over the past 3 months, Heidi had finally finished the task and had scheduled a meeting to present her recommendations to Douglas' department managers. Although she began the meeting with great enthusiasm, it turned into a disaster. While people entered the meeting room, Heidi stood at the door, handing out a three-page detailed agenda and a 20-page report she had spent days preparing. Hoping to impress as many people as possible, she had invited assistant managers as well as department heads, and she soon ran out of reports. The small room did not contain enough chairs for everyone, so before Heidi could call the meeting to order, she had to find more chairs. Scuffling feet, rattling chairs, and the shuffling of papers delayed the start of the meeting. When it finally began, Heidi asked if anyone had questions about her agenda and report.

   Of course, no one had finished reading them yet. When the head of the advertising department asked a question about word processors, an item not on the agenda, Heidi spent 15 minutes providing a detailed answer while others in the room began to chat among themselves about the upcoming company picnic. When the national sales manager excused himself to tend to important business, five other people sneaked out of the room. Heidi struggled to establish control but finally adjourned the meeting when the conversation in the room drifted to the weekly football office pool and the point spreads among the teams playing that week. No one discussed the third item on the agenda, much less her recommendations.

Fearing she had made a fool of herself, Heidi approached Joe Sandvan, a retired Douglas data processing manager with a reputation for handling people skillfully. Joe had hired Heidi, so she felt he might agree to help her recover from her setback.

As Joe listened to Heidi describe her meeting and all the problems she had encountered, he jotted down notes and soon offered several suggestions.

Joe began by pointing out that Heidi had not planned the meeting well. Should she have arranged for a larger room with sufficient seating, or should she have restricted attendance to department heads? Heidi wished she had done the latter. Why hadn't Heidi distributed the agenda a few days earlier, giving those attending a chance to prepare? Heidi hadn't thought of that. Joe went on to point out that a meeting's leader should strive to avoid losing the audience's attention and should deflect irrelevant questions or those of interest to only one or two people.

Meetings cost time and money. A one-hour meeting for eight people, with an average cost per person of $15 per hour in lost productivity, amounts to $120.

Considering the expenses involved and the reluctance of most people to attend meetings, Joe recommended the following 10 "rules" for more effective meetings:

1. Avoid calling a meeting if a memo and a series of telephone calls would serve your purpose.
2. Refrain from calling a meeting to decide something you could or should decide yourself.
3. Invite only personnel vital to the discussion.
4. Insist on punctuality. If you're 2 minutes late for a meeting with 20 people, you've wasted 40 minutes.
5. Keep the purpose of the meeting in mind.
6. Draft an agenda. Even a lengthy agenda, if well constructed, could mean a short meeting.
7. Circulate the agenda sufficiently ahead of time so those involved can prepare for the meeting.
8. Set time limits for each item on the agenda. Discussion, like work, expands to fill the time available.
9. See that the meeting leader states the issues, sticks to the agenda, lets everyone have a voice in the discussion, restricts the discussion to the subject, and sums up at the end.
10. Check the conference room to make sure it is available, clean, supplied with paper, pencils, and an easel or chalkboard.

While some of these rules may seem just common sense, many meetings fall apart because someone breaks one or more of them.

Studies of group dynamics in meetings show that small groups of 8 to 10 people come to consensus more rapidly than larger groups of 15 or more. Very large groups of 40 or more can present impossible hurdles.

Joe concluded by saying, "My first meeting was even worse than yours. It took me 10 years to learn that the key word for every meeting is 'less': less meetings, less participants per meeting, less time in each meeting."

Source: Reprinted from COMPUTER DECISIONS, July 1982, pp. 167–168, copyright 1982, Hayden Publishing Company.

SUMMARY

The systems cycle goes through three phases: analysis, design, and development. Systems analysis begins when a user or manager requests a study of either an existing system or a projected one.

We can divide analysis into two stages: preliminary and detailed. During preliminary analysis, the analyst evaluates the user's request, estimates costs to perform the detailed analysis, and lists objectives of the system. These findings are presented in a preliminary report.

To arrive at a preliminary report, the analyst interviews key personnel in the organization. Preceding the actual interview, the interviewer develops a list of questions. Interviews have three components: warm-up, questioning, and closure. Some time after the interview the analyst sends the interviewee a report outlining the findings.

After completing research, the analyst schedules a meeting with users and management at which the analyst presents the preliminary report orally, allowing ample time for discussion, and striving toward a decision, which may range from halting the project to proceeding to detailed analysis.

If management approves the preliminary report, the systems cycle advances to the next task: detailed analysis.

NEW TERMS
Closure
Decomposition
Follow-up
Interview
Leveling
Preliminary report
Questioning
Warm-up

QUESTIONS FOR REVIEW AND DISCUSSION

Questions appear in three categories. You can find the answers to the A group in this chapter. The B group requires you to apply the material presented here, whereas the C group necessitates investigation or research.

A.1. List the inputs to preliminary analysis.
A.2. List the outputs from preliminary analysis.
A.3. List the inputs to detailed analysis.
A.4. List the outputs from detailed analysis.
A.5. Which of the four key people in Fleet Feet's organization should the analyst interview?
A.6. What does leveling a data-flow diagram mean?
A.7. What should be reported to management in the preliminary report?
A.8. Who makes the final decision whether to continue the analysis or terminate it?
A.9. When you interview a user or manager, how should you make the contact?
A.10. List the topics covered in a preliminary report.

B.1. List three factors the analyst should consider in determining the cost of a detailed investigation.
B.2. What are the four symbols used in drawing a data-flow diagram and what does each represent?

C.1. What are the usual goals of a new or improved system?
C.2. What outside stimuli do users have to initiate the systems cycle?
C.3. List three reasons for analysis to be terminated.

REFERENCES

Tom DeMarco, *Structured Analysis and System Specifications*, Yourdon Press, New York, 1979.
B. Dickinson, *Developing Structured Systems*, Yourdon Press, New York, 1981.
Steve Eckols, *How to Design and Develop Business Systems*, Mike Murach and Associates, Fresno, Calif., 1983.
Glenford J. Myers, *Software Reliability*, Wiley, New York, 1976.

CHAPTER 3 — Detailed Analysis and the Feasibility Study

- Goals and Preview
- Review and Assignment
- Fact Finding
- **Fifteen Key Questions to Ask Any Software Supplier**
- The Feasibility Study
- Presentation of Findings to Management
- Case Study: Detailed Analysis of Camp Beaverbrook's Camper System
- Working with People: The Need for Accounting in an Analyst's Training
- Summary
- New Terms
- Questions for Review and Discussion
- References

GOALS AND PREVIEW

After reading this chapter you should be able to:

1. Describe detailed analysis.
2. List the analyst's main functions during this stage.
3. Outline methods for gathering data.
4. ~~Define fact finding.~~
5. State the components of the feasibility study.
6. List the participants in a management review.

Most people in the market for a new car carefully consider whether or not it really suits their needs. Even before they climb behind the wheel to test drive it, they check its specifications. Does it offer enough room for a family of four? Will the trunk hold luggage for a 2-week vacation? Will the engine provide enough power on the freeway? Most important, can they afford it? If they feel satisfied after reviewing the specifications for the new car, they will probably begin earnest comparison shopping.

Preliminary analysis is like the initial evaluation of a new car. If for any reason the proposed system does not seem feasible, workable, or affordable, the organization stops

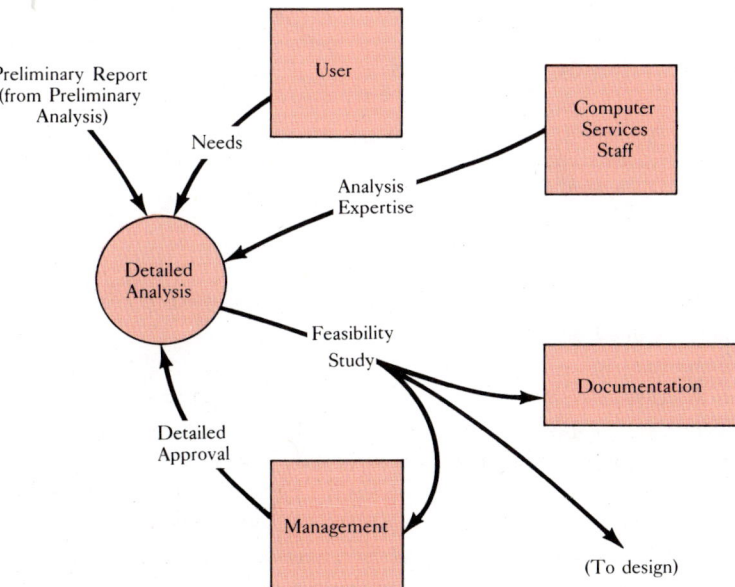

(a) As the name implies, detailed analysis is a more thorough investigation of a system than preliminary analysis. Like preliminary analysis, it requires inputs from users, the computer services staff, and management. In addition, the analyst works with the information contained in the preliminary report.

FIGURE 3.1. Detailed analysis has four inputs and two outputs. We can decompose, or level, this phase of analysis into three subtasks.

68 PART I SYSTEMS ANALYSIS

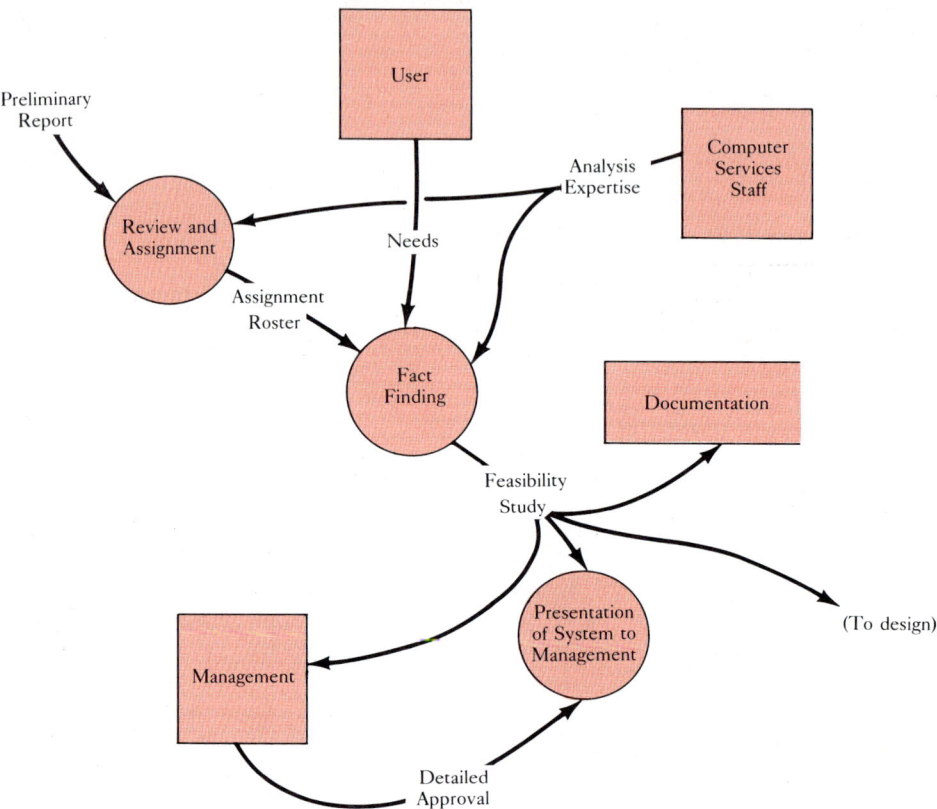

(b) The three major activities of detailed analysis are review and assignment, fact finding, and presentation of the system to management. The two outputs are analysis documentation and a feasibility study, which leads to systems design.

FIGURE 3.1 *(Continued)*

the process; but if the system does survive initial scrutiny, the organization will probably proceed on to detailed analysis.

Detailed analysis (Figure 3.1a) involves thorough investigation of the way a system currently works with the aim of finding and solving any existing problems. It follows a series of procedures for determining what data an organization collects, how it processes the data, and in what ways that processing might be improved. To obtain this information, the analyst adds detail to the preliminary study, interviewing personnel at all levels within the organization, preparing questionnaires for completion by those not interviewed, studying the forms and procedures used, and even contacting hardware or software vendors, calculating costs and benefits, and reporting findings with alternatives and a recommended course of action. Those participating in the preliminary analysis—users, management, and the computer services staff—also play roles in detailed analysis.

The data-flow diagram in Figure 3.1b decomposes detailed analysis into its components: review and assignment, fact finding, and presentation of the system to management. Note the two major ouptuts of detailed analysis, analysis documentation and a feasibility study, which ultimately provide input to systems design (Chapters 5–10).

REVIEW AND ASSIGNMENT

The Preliminary Report

Review begins with careful reconsideration of the preliminary report, even if its author is the one proceeding to detailed analysis. If someone else takes over, or a good deal of time has elapsed, the review acquaints the new analyst or new team members with earlier findings. Since it is assumed at this stage that the decision to continue was the proper one, the review simply reacquaints the analyst with the problem and the findings of the preliminary analysis.

In a large organization, the study may involve a team of analysts headed by a team or project leader who divides up the tasks and assigns responsibilities to individuals. In our continuing case involving Fleet Feet, only one analyst, Frank Pisciotta, will assume responsibility for detailed analysis.

Scheduling the Study with a Gantt Chart

Quite soon the analyst or team leader draws up a detailed schedule outlining anticipated activities. Scheduling may seem a fairly simple task, but it can become time consuming and complicated, and, as with most large projects, good planning does a great deal to ensure success. Taking into account vacations, potential illness, regular meetings, and job responsibilities, analysts much schedule users, management, computer services staff, outside vendors or suppliers, and themselves into specific times.

Most analysts use a **Gantt chart** for scheduling activities. As you can see in Figure 3.2, this chart lists events vertically down the left-hand column with time periods appearing horizontally across the top. For each activity, the analyst draws a hollow bar across the chart, indicating when that task should start and stop, and then later adds a solid bar or fills in the hollow bar to depict actual time consumed. Filling in the hollow bar makes it possible to determine visually whether the work is progressing as it should or is ahead of or behind schedule. Gantt charts always show personnel assigned to specific tasks but do not show the degree of each person's involvement. Thus we cannot tell whether Frank is devoting 100 percent of his time to the study or is mixing it in with his regular duties.

So far, we have described the analysis stage as though analysts perform a series of tasks, one at a time, with no thought of time restrictions. In practice, detailed schedules in the form of Gantt charts guide the analysis stage. These schedules serve two purposes: (1) they indicate to management whether the time allotted for the project is reasonable, and (2) they guide the analysts in performing their tasks efficiently, thus providing a measure of control over the project.

The Gantt chart for systems analysis parallels the eight major steps of detailed analysis and indicates large blocks of time (weeks or months). Each analyst on a large project will prepare more detailed Gantt charts, showing the number of hours needed to complete more specific tasks. Some versions of a Gantt chart permit the analyst to indicate the percentage of each activity completed, time overruns, activities completed early, and the cost of each activity.

Gantt chart Scheduling tool that uses horizontal bars to depict project schedule and progress.

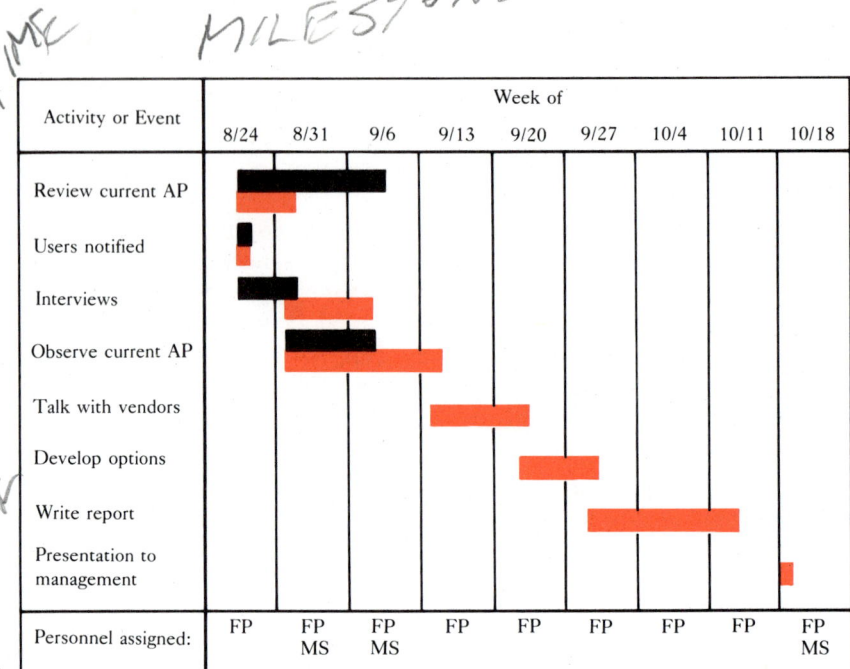

FIGURE 3.2. *A Gantt chart depicts time, measured in weeks, on the horizontal axis, and events on the vertical axis. Hollow bars indicate when an event is supposed to occur and solid bars show actual times. Hollow bars are filled in as the activity or event is completed.*

Gantt charts are valuable tools because they can be understood by both computer and noncomputer people. Management can easily see to whom tasks have been assigned, and management as well as the lead analyst can track each assignment's progress.

We learn from Figure 3.2 that Frank Pisciotta has scheduled the detailed analysis phase of Fleet Feet's accounts payable system to start the week of August 20. At that time he will begin reviewing the current AP system and interviewing users. The chart also shows that he plans to observe the current system during the weeks of September 10 and 17, followed by discussion with vendors during the week of the 17th. Management can expect him to present his final report the week of October 15.

Authorization and Notification

After approving the assignment of tasks and the schedule for their completion, management notifies all appropriate personnel via memo (Figure 3.3). The memo outlines decisions based on the preliminary report and alerts members of the organization to the fact that someone will be conducting a detailed system study.

In our Fleet Feet example, the vice-president of operations and the manager of computer services coauthor the memo. They briefly describe the results of the preliminary analysis, advising personnel of one immediate result of that phase:

CHAPTER 3 DETAILED ANALYSIS AND THE FEASIBILITY STUDY 71

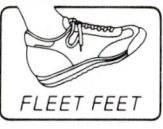

MEMORANDUM   August 24, 1984

TO: All Operations Personnel and Vice-Presidents

FROM: Sally Edwards, VP Operations
           Kathleen Williams, Manager of Computer Services

SUBJECT: Detailed Study of Accounts Payable System

   Yesterday we decided to continue the preliminary study of our accounts payable system with a detailed analysis. Frank Pisciotta will conduct this study and expects to start work early next week.
   Frank will want to talk with many of you again and we hope you will provide him with the usual high level of assistance. During this phase of the investigation, Frank will be collecting sample documents, interviewing people, holding group discussions, and observing the current system.
   The preliminary study revealed that Mark Stensaas and his staff are overworked and we expect to lessen the load placed on his people with the new system. The study also showed that Mark should be authorized to sign checks, so his signature will appear on your payroll checks from now on.

SE: jks

FIGURE 3.3. *As soon as management approves the assignment of tasks and the schedule, it notifies all involved individuals about the impending detailed analysis.*

the bookkeeper will now be authorized to sign checks. Aware that detailed analysis of their accounts payable system has begun, Fleet Feet's staff can prepare to help the analyst by gathering and organizing necessary information. Coauthoring the memorandum lends weight to it, letting all parties know that this project has high-level authority behind it.

FACT FINDING

Fact finding means learning as much as possible about the present system. To do this, the analyst interviews personnel, prepares questionnaires, observes the current system, gathers forms and documents currently in use, determines the flow of data through the system, and clearly defines the system requirements.

Fact finding Learning as much as possible about the present system.

Interviewing

By studying Fleet Feet's organization chart, analyst Frank Pisciotta can confidently schedule interviews with key personnel involved with the system. Earlier he conducted preliminary interviews; now he will engage in detailed interviews with all the people who actually operate the system. Rapport is crucial. Not only will these people use the newly developed system, but they also may be the ones most afraid of change, especially if they feel the computer might replace them.

Like an investigative reporter trying to discover the who, what, when, why, and how of a story, the analyst should conduct the interview in such a way that people provide honest descriptions of their jobs. The following questions can help accomplish this goal.

Who is involved with what you do?
What do you do?
Where do you do it?
When do you do it?
Why do you do it the way you do?
How do you do it?
Do you have suggestions for change?

Interviews help gather vital facts about existing problems, such as lack of quality control or sufficient security, but they also allow the analyst to involve people in change, easing them into it. After all, it is the users' system, not the analyst's.

Questionnaires

Questionnaires economically gather data from both large and small groups of people. Properly constructed, they do not take long to complete and statistical results can be quickly tabulated. Development of a questionnaire requires thought and planning, and usually more than one draft is necessary.

Questionnaire design is critical. Questions should be short, easy to understand, unbiased, nonthreatening, and specific (Figure 3.4). Suppose, for example, that question 5 on the Fleet Feet survey were to read, "In what ways were Fleet Feet reps discourteous to you?" When worded in that way, the question directs the supplier to think only about bad experiences and will probably prompt a negative answer, which will bias the survey. One could reword this question to read: "Have you received courteous attention from Fleet Feet reps? If not, please explain." Similarly, questions that reflect the analyst's preformed opinions will taint the survey. To make sure questions will stimulate needed information, the analyst can test them with one or two outsiders before widespread distribution. Prepaid return envelopes accompanying questionnaires sent to outsiders help assure prompt response.

The analyst should send questionnaires to everyone involved with the system. A questionnaire works particularly well when the analyst must gather data from a large number of people, when the analyst must ask everyone the same questions,

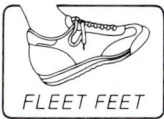

Fleet Feet Vendor Survey

We are in the process of evaluating our accounts payable system to speed the payments to our suppliers. Please complete the following form as soon as possible and return it in the enclosed prepaid envelope.

1. Name of your firm:

2. Name of person completing this form:

3. Title of person completing this form:
   a. Business manager
   b. Accounts receivable clerk
   c. Controller
   d. Salesperson for our account
   e. Other

4. Have you experienced any late payments from Fleet Feet? If so, when did they occur and how late were they?

5. When you contact a Fleet Feet representative, is he or she courteous? Are you satisfied with the way your problems or questions are handled? (Circle your response.)

   Dissatisfied   Satisfied

   1   2   3   4   5

6. How does Fleet Feet compare to your other customers?
   a. About the same
   b. Better than most
   c. Worse than most
   d. No experience

7. Would you offer Fleet Feet a better discount rate if payment were made within 5 days? If so, what might your terms be?

8. Who is the person in our organization you contact most frequently?

9. Comments (please write them here or on the back of this form).

FIGURE 3.4. *Questionnaire used by an analyst to gather data from suppliers.*

or when facts must be collected from people, such as suppliers, who do not work for the organization.

Questions can follow four formats:

1. *Multiple choice:* This gives respondents a specific set of potential answers. This format is ideal for computer tabulating (question 3 in Figure 3.4).
2. *Open ended:* Respondents must answer the question in their own words. Space is provided under each question for the response (questions 1, 2, 8, and 9 in Figure 3.4).
3. *Rating:* This is similar to multiple choice except that respondents must rate their satisfaction (question 5 in Figure 3.4).
4. *Rank:* Rank requires respondents to prioritize their responses from high to low or on a percentage basis (question 6 in Figure 3.4).

Aware that most people do not spend a lot of time responding to questionnaires, Frank decides to mix question formats, including follow-up questions, within the original questionnaire to permit elaboration of certain responses (question 7 in Figure 3.4). By so organizing his questionnaire, Frank gives his respondents the opportunity to express their opinions freely, and yet answer quickly through the use of multiple-choice, rating, and ranking questions.

When all the questionnaires are returned, the data must be tabulated (Figure 3.5). Frank's survey reveals that the company could save a substantial amount of money by paying bills faster because the discount rate would increase from 0 to 2 percent to over 4 percent, outweighing the interest or other income Fleet Feet might have earned by keeping the money longer. With these data, Frank can estimate the possible savings resulting from a automated accounts payable system. Since purchases amount to well over $1.2 million per year, a 3.3 percent savings equals $37,000 per year. As Frank's research indicates that discounts for early payment saved $25,000 during the previous year, the company might save an additional $12,000 by paying earlier. Frank realizes that paying earlier can be accomplished either by hiring additional staff or by computerizing payments.

If the results of a questionnaire survey are incomplete or confusing, the analyst may want to contact selected outsiders by telephone or in person. This requires tact, of course, and an understanding that the analyst's own pressing need may not concern outsiders in the least.

Observing the System

The analyst may want to observe the existing system personally by following a transaction, such as in invoice, through it. Direct observation allows the analyst to verify his or her understanding of the system. Instead of getting second-hand impressions about a specific task, the analyst can experience the actual process. However, he or she must remain outside the flow as an observer, so as not to introduce biases or changes in actual procedures. Observing a system requires caution; when people know they are being observed, they usually behave differently, working more efficiently and at higher speeds to impress the analyst.

RESULTS OF FLEET FEET VENDOR SURVEY

Survey sent to 127 vendors on September 6; 57 responses received by September 21.

1. Name of your firm:

2. Name of person completing this form:

3. Title of person completing this form:
    - Business manager   15
    - Accounts receivable clerk   12
    - Controller   9
    - Salesperson for our account 9
    - Other   12

4. Have you experienced any late payments from Fleet Feet? If so, when did they occur and how late were they?
    - NO: 20
    - YES: 37
   Average: 45 days; longest: 98 days

5. When you contact a Fleet Feet representative, is he or she courteous? Are you satisfied with the way your problems or questions are resolved?

   Dissatisfied   Satisfied

1	2	3	4	5
5	7	8	22	15

6. How does Fleet Feet compare to your other customers?
    a. About the same   20
    b. Better than most   10
    c. Worse than most   15
    d. No experience   12

7. Would you offer Fleet Feet a better discount rate if payment were made within 5 days? If so, what might your terms be?
    - YES: 50
    - NO: 7
    - Average terms: 3.3 percent for 5 days compared with 2.2 percent for 13 days now

8. Who is the person in our organization you contact most frequently?
    - Mark Stensaas   22
    - Sharon Smith   15
    - Sally Edwards   10
    - Elizabeth Jansen   5
    - Other   5

9. Comments (please write them here or on the back of this form).

   Most of the comments centered on the advantage of a faster accounts payable system and were encouraging us to make a switch.

FIGURE 3.5. Frank Pisciotta's tabulation of Fleet Feet's survey.

> **Fifteen Key Questions to Ask Any Software Supplier**
>
> Buying a product from a new business, such as the explosive software industry, poses unusual problems. Customers cannot evaluate decades of performance history by the company, and, not enjoying the benefits of an objective "consumer report" on new products, they often feel at the mercy of fast-talking salespeople. Therefore, it's important for people in the market for software to ask some really tough questions. Relevant ones. Any reputable supplier should be able to answer the following 15 questions without backpedaling.
>
> 1. *Range of Products.* Can you offer us a complete range of software systems designed to work together? Or will we have to piece together a patchwork of systems to fully computerize our organization?
> 2. *Decision Support Systems.* Are your systems just record keepers, or can they really help us make decisions? Can we pull together information from any of our integrated systems? In exactly the form we want it?
> 3. *In-House Development.* Can you provide business software for both mainframe and microcomputers? Do you develop this software yourself or do you simply market it for another company?
> 4. *OnLine.* Are your systems truly online so all of our information is current? How many of your systems are online? How secure are they?
> 5. *Debugging and Testing.* Will my company have to be the one that discovers the bugs in your brand new system? Just how long have your systems actually been used, and how have they been tested?
> 6. *Updates.* Will you update your systems as technology advances and regulations change? What are some of your most recent updates? Will you keep us current on regulatory change?
> 7. *Flexibility/Adaptability.* Are your systems really adaptable to our unique needs? Or will we have to change or add to them ourselves to get the features we want?

In some instances, the analyst may find it useful to visit another organization with a computerized system similar to the one under study. Finding a comparable installation may pose a problem, however. Some competitive organizations may not want to share their experiences, others may be too large or too small for accurate comparisons, and still others may be unwilling to waste employees' time demonstrating their system. Whenever visiting another organization, an analyst should follow the rules of etiquette: make an appointment, research the organization beforehand, know what he or she wants to see, and write a follow-up, thank-you letter.

Hardware and software vendors can also supply valuable information. Computer sales representatives will gladly share their experiences with potential clients, and software firms will send brochures describing their programs. Although very useful, information from such sources should be reviewed carefully because vendors are more interested in promoting their products than in solving your problems. To

8. *History/Performance*. How long have you been in business? What are your revenues? What is your growth record? Where will your company be in five years from now? Can you show me an annual report?
9. *Other Customers*. How many systems has your company installed? How many of these were installed in the past six months? How many of your earlier customers are still using—and liking—your systems?
10. *Security*. Are your systems secure? Do you provide password type protection and to how many levels? What other type of security provisions do your systems have?
11. *Networking*. Can you link our executives' personal computers directly to the mainframe, so they can get their own information? Is that software available right now?
12. *Training/Support*. How will you make sure our own people thoroughly understand your system? Do you have educational centers near us, or will we have to travel all the way across the country to find one? Will you be there to help during installation and after?
13. *In-House Specialists*. How many of your people specialize in software for my industry? How many accountants work for you? Human resource specialists? Manufacturing experts?
14. *Special Features*. Do your systems have built-in features that make them easier to use? What happens if someone needs help figuring out a feature? Do you have online documentation that's easy to understand?
15. *Upgrading*. As my business changes will your system be flexible enough to change with it? Or will we have to pay a lot to revamp it? Or even regenerate it?

Source: Management Science America, Inc. advertisement in *Computerworld*, July 4, 1983, p. 84, "15 pointed questions to ask MSA or any software supplier."

minimize this difficulty, try to visit an installation where you can test a salesperson's claims.

THE FEASIBILITY STUDY

Diagramming the Logical System

Armed with interview results, tabulated questionnaires, and experience through personal observations, the analyst is ready to describe the current system in narrative form, with a data-flow diagram, or with a system flowchart. Returning to our Fleet Feet case study, we see that Frank selects data-flow diagrams (DFD). He begins with the accounts payable system, using a context DFD (Figure 3.6). A **context DFD** defines the system under study in a general form, showing:

Context data-flow diagram Diagram showing the system in its most general form.

78 PART I SYSTEMS ANALYSIS

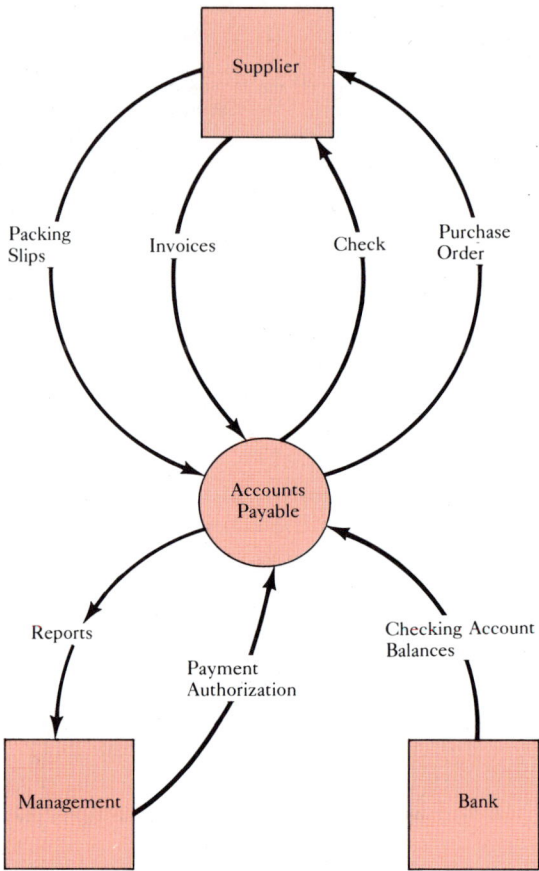

FIGURE 3.6. *A context data-flow diagram depicts Fleet Feet's accounts payable system in its broadest perspective, not showing any of the details or internal processes.*

Inputs to AP: Packing slips, invoices, checking account balances, payment notifications.

Outputs from AP: reports to management, checks to suppliers.

A context DFD does not show any details but is an overview drawing of the system. It is an excellent diagram to share with management whose interest is general in nature. Context DFDs place a boundary around the system under investigation, saying that this is what will be examined—nothing more and nothing less.

After developing a context DFD, Frank turns his attention to the details of accounts payable. In Fleet Feet's case, management reviews inventory reports and determines what to order from suppliers; orders are placed by the accounting department using a purchase order/requisition; on delivery, merchandise and packing slips enter the warehouse; and the packing slips are sent to the accounting department, which receives invoices directly from suppliers, while merchandise stays in the warehouse or goes to a franchise outlet. Accounting clerks compare purchase order requisitions with invoices and packing slips to make sure all invoiced items have actually arrived, and then post the purchase to the supplier's ledger card. Around the 15th of each month, the accounting department receives Fleet Feet's canceled checks and checking account balances from the bank. Then, on the 24th of the month, management tells the accounting department which suppliers to pay, and payment checks are written, signed by an authorized individual, and mailed to the suppliers. At the end of each month, the accounting department prepares a report of balances due suppliers and an inventory report for management evaluation.

These detailed activities by the accounting department, management, warehouse personnel, the bank, and suppliers add up to six major activities (Figure 3.7):

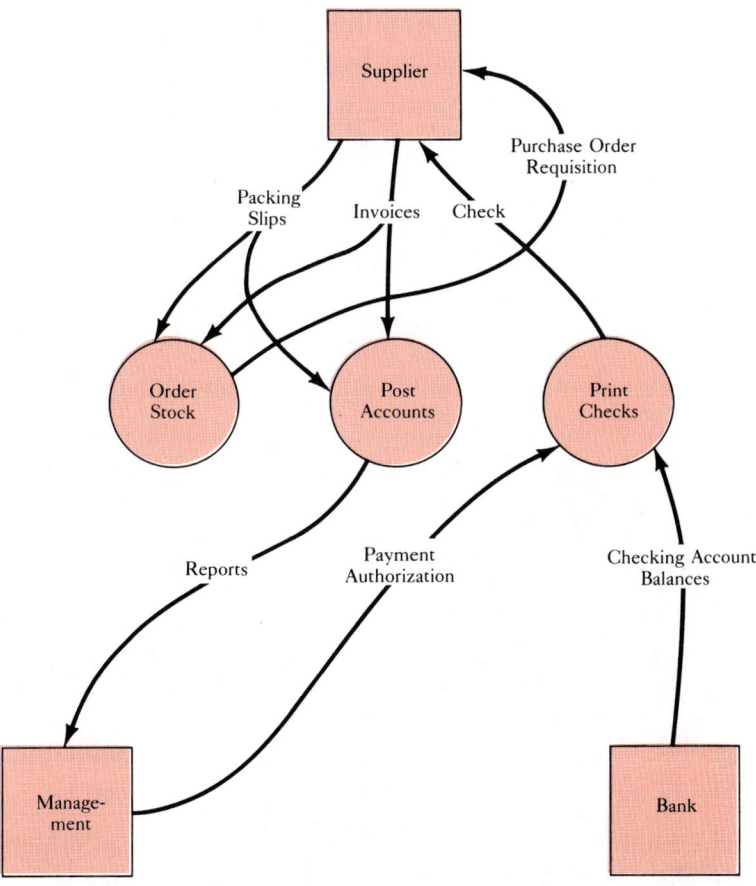

FIGURE 3.7. This data-flow diagram depicts Fleet Feet's accounts payable system. Invoices and packing slips arrive in the warehouse and then go to the accounting office for payment and filing, the bank sends account balances, and management receives reports and authorizes payments.

1. Generation of reports.
2. Ordering of stock.
3. Printing of checks.
4. Posting of accounts.
5. Reconciliation of bank statements.
6. Authorization of payment.

During the design phase of the systems process, Frank will study each of these activities further, leveling the data-flow diagram of Figure 3.7 into far more detail.

To draw the analysis DFD:

1. Look at the system from the inside to the outside.
2. Identify the activities.
3. Locate the data flows.
4. Show the relationships between activities.
5. Find the internal inputs or outputs that exist within the system.
6. Level complex processes in the DFD into simpler ones.
7. Look for duplication of data flows or data stores (files).

VENDOR: 000002						CHECK NO. _____
OUR INV. NO.	YOUR REF. NO.	INVOICE DATE	INVOICE AMOUNT	AMOUNT PAID	DISCOUNT TAKEN	NET CHECK AMOUNT
001013		04/25/84	150.00	150.00	.00	150.00
001014	HJ	05/23/84	263.00	200.00	.00	200.00
					CHECK TOTAL	350.00

CHECK NO.	CHECK DATE	VENDOR NO.
000301	06/25/84	000002

FLEET FEET

CHECK NO. _____

CHECK AMOUNT
$*******350.00

PAY TO THE ORDER OF

RED LINE FREIGHT
350 COMMERCIAL ROAD
BIGTOWN, TEXAS 99996

NON-NEGOTIABLE

(a) Checks are used to pay suppliers, invoices show what Fleet Feet owes on a particular purchase, and packing slips detail what actually arrived.

FIGURE 3.8. Sample check, invoice, and packing slip collected during the fact-finding phase.

		1 2 3 4 5 6 7 8 9 10 11 12 13 14 15 16 17 18 19 20 21 22 23 24 25 26 27 28 29 30 31					
NAME	FAR WEST NIKE SHOES						
STREET	1072 HIGHWAY 72						
CITY	REDDING CA.				BALANCE FORWARD		1 073 28
DATE	VENDOR	DETAIL	DEBIT	DISCOUNT	CREDIT		BALANCE
12/4/84	FWNS		124 22	2 48			1 195 02
12/10/84	PAYMENT				973 —		222 02

ACCOUNTS PAYABLE LEDGER

(b) Fleet Feet's ledger card for Nike shoes. There is a handwritten card for each supplier. When merchandise arrives, the accounting department notes the date and the amount, records payment, and calculates a new balance.

FIGURE 3.8 *(Continued)*

Analysis DFDs, reduce the time it takes to analyze a system. Coupled with context DFDs they show the system in a graphical sense that users, managers, and technical people can understand.

Sample Documents

While determining the flow of data, the analyst collects samples of all relevant documents. In our example, Frank Pisciotta collects sample checks, invoices, packing slips, and other accounts payable forms (Figure 3.8). To create a record of all purchases from and payments to suppliers, Fleet Feet's manual system requires that someone prepare a ledger card for each supplier.

The assembled documents help Frank understand what data the new system must collect and process. For example, the company can easily obtain the following data from the invoice itself:

1. Supplier name, address, and telephone number.
2. Invoice number.
3. Invoice date.
4. Items shipped.
5. Terms of invoice.
6. Amount of invoice.

From the packing slip, it can obtain:

1. Supplier name.
2. Shipping date.
3. Date goods are received.
4. Freight charges.
5. Invoice number.

Packing slips are carbon copies of invoices omitting certain data, such as the dollar value of the shipment. The warehouse clerk checks the merchandise received against the packing slip to be sure everything is in the carton and notes any discrepancies. Then the packing slip goes to accounting for comparison with invoices to be sure that the company received what it is paying for.

The ledger card offers two categories of facts—supplier data and purchase/payment history:

1. Supplier name.
2. Supplier address.
3. Supplier telephone number.
4. Date of transaction.
5. Description of transaction.
6. Amount of invoice or payment.
7. Discount.
8. Balance due the supplier.

Each check sent to a supplier contains the following data:

1. Invoice number.
2. Check number.
3. Amount of payment.
4. Payment date.

In addition to these documents, Frank obtains copies of reports prepared by the accounting department. The monthly supplier balance report lists each of Fleet Feet's suppliers, and the amounts owed them. The weekly inventory report shows the variety of items Fleet Feet sells and the quantities on hand.

The data dictionary Frank builds from the forms, together with the data-flow diagrams and other documents of the system, gives Frank background to start his planning for the new AP system.

Alternatives, Costs, and Benefits

During fact finding, Frank acts as a researcher, gathering facts, figures, and documents, and coming to grips with the entire scope of the problem. Now he must decide what can be done, what it will cost, and the benefits expected to be derived from the new system.

The first step is to generate a list of alternative solutions to Fleet Feet's accounts payable problem. Frank sees solutions ranging from doing nothing to installing a fully computerized AP system. Finally he settles on four alternatives:

1. Do nothing, leaving the existing system alone.
2. Hire more staff, partially automate the system, but continue with essentially a manual system.
3. Purchase AP software from an outside software supplier.
4. Design, program, and install a customized AP system.

Frank examines each in detail, listing the costs, savings, effects, and benefits.

The first alternative, do nothing, means a loss to Fleet Feet of $10,000–$12,000 per year in lost discounts. This alternative provides no potential savings or benefits. As time goes on, the existing AP system will get worse, payments to suppliers will be later, bookkeeping will struggle to keep up, and morale will drop. On the bright side, it involves no out-of-pocket payments for new equipment or software, and no new personnel. Alternative 2, to hire staff and partially automate, would employ one new person immediately, and another in a year. Frank estimates costs, including salaries, equipment, desks, and space, at $12,600 the first year, and $33,600 the second. He forecasts that hiring new people would stabilize the AP situation for the present, and perhaps permit a few discounts to be taken, but that Fleet Feet will face the same problem again in 4 years if business continues to grow as it has. The bulk of the discounts would still be lost.

Purchasing software, the third alternative, does permit Fleet Feet to recover the lost discounts. While researching software, Frank finds AP systems ranging in price from $500 to $15,000. The lower priced systems permit only one user at a time (unacceptable in Fleet Feet's situation) and the high priced ones, besides being too expensive, tie AP to the general ledger. Frank locates an AP system for the HP 3000 computer that is successfully operating in 40 businesses, and watches a demonstration. He notes expected costs:

1. AP software from MCBA—$7500.
2. Modifications to software for franchises—$3700.
3. Equipment, two terminals—$3500.
4. Yearly software update—$2000 per year.
5. Training of staff—$1250.

Frank estimates the software can be installed and operational within 3 months, allowing Fleet Feet to begin taking advantage of lost discounts in a short time period. The software will mean the overtime now paid to bookkeeping employees will disappear. Frank next calculates the savings and benefits:

1. Overtime pay for employees—$4200 per year.
2. Discounts from suppliers—$12,000 per year.

Frank's last alternative, writing customized software, permits the most flexible solution, allowing Fleet Feet to have the exact software it needs to accommodate a franchise operation. Frank estimates costs as:

1. System design—$5000.
2. Programming—$8000.
3. Training, installation, conversion—$2400.
4. Maintenance of software—$900 per year.
5. Equipment, two terminals—$3500.

This alternative permits suppliers to be paid within the 5- to 10-day window discount period. Some changes will have to be made in staffing in the warehouse and the bookkeeping area. Frank estimates the new system can be operational in 6 months. He estimates savings and benefits to be exactly the same as for alternative 3.

With costs, savings, and benefits enumerated, Frank now calculates totals for each alternative. He recognizes some costs and benefits occur only once while others will recur annually. There are, of course, other costs, such as for supplies and terminal maintenance, but they are minimal compared with the costs Frank has found.

From his past experience, Frank knows systems have an expected life of 4 to 6 years. He decides to reject alternative 1 completely. It does not solve Fleet Feet's problems and costs $12,000 per year in lost discounts. Using 5 years as an average, Frank examines the cost and benefits of the other alternatives by adding the one-time and yearly costs, as shown in the following table.

Cost Category	Alternative		
	2	3	4
Total one-time costs	0	$15,950	$18,900
Total yearly cost	$46,200	$ 2,000	$ 900
Total yearly benefits	0	$16,200	$16,200
Annual cost savings	$46,200	$14,200	$15,300

From this chart, Frank rejects alternative 2 because of its high annual cost savings relative to the other alternatives.

Frank is tempted to multiply the annual cost savings by five to calculate the cost savings for the 5-year life of the new AP system. However, recalling a college business mathematics course, he knows the fallacy of simply doing this. Since these benefits will occur in the future, and costs will occur now, future benefits must be discounted into present-day dollar amounts. Frank finds the formula for discounting from his college textbook:

$$P = \frac{F}{(1 + I)^n}$$

where P is the present value of the benefits, F is the future value of the benefits, I is the interest rate, and n is the number of years. Using 10 percent as the interest rate (Fleet Feet earns that rate on its interest-bearing checking account), Frank builds the following table for the two remaining alternatives under consideration.

Year N	$(1 + 0.10)^n$	Alternative 3 F	Alternative 3 P	Alternative 4 F	Alternative 4 P
1	1.10	$14,200	$12,909	$15,300	$13,909
2	1.21	14,200	11,736	15,300	12,645
3	1.33	14,200	10,669	15,300	11,495
4	1.46	14,200	9,699	15,300	10,450
5	1.61	14,200	8,817	15,300	9,500
Totals			$53,829		$57,999

With the present values of the future benefits computed, Frank subtracts the one-time costs, giving the net present value of each alternative:

Net present value of alternative 3: 53,829 − 15,950 = 37,879
Net present value of alternative 4: 57,999 − 18,900 = 39,099

Thus alternative 4 costs $1220 less than alternative 3 and represents the "best" financial solution, assuming all estimates are correct. The cost of purchased software is not appreciably greater than for custom written software so there is no clear-cut decision based on costs and benefits. Both alternatives solve the problems of "lost" discounts. Both permit Fleet Feet to recover costs in a little over a year since the first-year costs are between $16,000 and $19,000 whereas first-year benefits return over $16,000.

Besides financial considerations, Frank must consider other factors in deciding which alternative to recommend to his superiors. Both alternatives permit better service to suppliers, management information for better decision making, greater utilization of Fleet Feet's computer, and increased productivity of the accounting department, and both keep Fleet Feet ahead of its competition. Looking further into alternative 3, Frank finds two potential drawbacks: the purchased AP software does not interact with Fleet Feet's general ledger, and costs to convert the software for franchise operation, according to the references Frank called, may be underestimated. After careful consideration, Frank recommends alternative 4, writing customized software. Although many computer professionals today advocate buying accounting software, Frank views Fleet Feet's franchise operation and its Georgia-to-California base of operation as not fitting into a standard AP purchased software solution.

Reporting Findings

When all necessary facts, figures, documents, data-flow diagrams, questionnaires, and observations are complete, the analyst can write the final report. The format of the final report, called the feasibility study, parallels that of the preliminary report. As you can see in Figure 3.9, it starts with a restatement of the problem and its importance, followed by a list of the study's objectives, a review of the analyst's findings, tallies of expected costs and savings, and the analyst's recommendations. In our example, Frank suggests that the company design, program, and install its own AP system, because, after weighing all the options, it will cost less in the long run, be more efficient, and provide more flexibility.

In a large organization, the analyst may use a standardized form for the final report; in smaller organizations, the analyst simply chooses the most logical format. In any case, the analyst distributes the typed, photocopied report to the managers who will decide whether to adopt, modify, or reject the recommended solution.

PRESENTATION OF FINDINGS TO MANAGEMENT

After management has thoroughly considered the feasibility study, it calls a meeting to discuss the study and to choose a course of action. This meeting should take place a few days after the study's distribution and should be conducted by the manager of the computing services department or whoever requested the analysis. The analyst plays a major role and should be well prepared to answer questions and supply needed information. In fact, the analyst should rehearse the presentation in order to identify and improve upon weak areas.

If the analyst leads the meeting, he or she must exercise control. The following rules are helpful.

1. Never read the feasibility study aloud; instead, summarize it, while trying to lead the audience to support the study's recommendations.
2. Use visual aids, such as chalkboards, flipcharts, slides, photographs, and overhead transparencies.
3. If appropriate, demonstrate equipment or software to show how it will work.

Often one key individual must be convinced, and this person will influence the others to follow. If all goes well, the meeting will end with a decision to implement the analyst's recommendations.

After the meeting, management notifies all appropriate staff members of its decision. If management has decided to proceed to the design stage, the notifi-

CHAPTER 3 DETAILED ANALYSIS AND THE FEASIBILITY STUDY 87

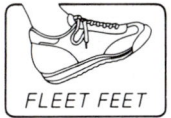

MEMORANDUM September 12, 1984

TO: Sally Edwards, Vice-President of Operations

FROM: Kathleen Williams, Manager of Computer Services

SUBJECT: Final Report of the Detailed Analysis of the Accounts Payable System

We have completed the detailed analysis of our accounts payable system. This report outlines our objectives, findings, and recommendations.

System Objectives:

How can Fleet Feet speed payments to vendors? Can we accomplish this goal and still lessen the heavy work load in the bookkeeping department?

Findings:

1. Processing time of invoices has dropped by 2 days since Mark Stensaas was authorized to sign vendor and payroll checks. Unfortunately this does not save enough time to permit us to take advantage of discounts.
2. The monetary loss of not taking discounts is significant. For the calendar year 1983, Fleet Feet could have saved $8234, or around $686 per month.
3. The usual discount offered by vendors is 2 percent for payments made within 10 days. More than half of our vendors reported in a survey that they would give a 3.3 percent, 5-day discount. Our calculations show this could boost our savings to over $12,000 for 1985.
4. The present manual system includes excessive paperwork. Invoices are hand posted to both the ledger card and the stub portion of the check to the vendors.
5. We currently generate no reports that automatically could trigger payments. Mark Stensaas depends on his memory to know which to pay when.
6. We face rising costs. Any new franchises will require us to hire additional personnel. Our current pay rate for a clerical position in the accounting department exceeds $1200 per month. We will need one new person within the next 3 months, and a second within a year.
7. We do not adequately identify our vendors. We currently list them by name only. Frequently we post invoices to the wrong ledger card and suffer a finance charge for late payment.

Alternatives

Four possible solutions could eliminate the problems in our accounts payable system and expedite payments. Each offers different costs, savings, and effects on our operation. None of the alternatives take into account effects of investment-tax credits for equipment or depreciation as a savings. All costs are pretax costs.

FIGURE 3.9. *Final report or feasibility study.*

Option 1. Do nothing. Leave the system alone.

    Costs: Loss of potential discounts   ($10,000–$12,000/yr)

  Savings: None

   Effects: The manual AP system can only get worse. Payments to vendors will get slower as our business volume grows. Bookkeeping will struggle to keep up and morale will suffer.

Option 2. Hire staff: One additional person in now, another in a year. Both people would work in bookkeeping to maintain the work load at its current level. Note that second year cost is high due to first person working remainder of this year.

Costs:	Loss of potential discounts	$10,000–$12,000/yr
	Two employees: First year	12,600
	Second year	33,600

  Savings: None

   Effects: The AP system would stabilize but not improve. In 4 years we would be back to where we are now.

Option 3. Purchase AP software. An accounts payable software system written by MCBA (Micro Computer Business Applications) is available and would fulfill most of our needs. This system does not have the ability to deal with franchises but is compatible with our HP 3000 computer. This software is installed in 40 businesses in the United States and Canada.

Costs:	MCBA accounts payable software	$7500
	Modifications to MCBA system for franchises	3700
	Two terminals (in warehouse and bookkeeping)	3500
	Update of MCBA on a yearly basis	1000/yr
	Training of staff	1250
Savings:	Overtime pay for bookkeeping employees	$ 4200/yr
	Discounts from vendors	12,000/yr

   Effects: Vendors could be paid within the 5- to 10-day discount period. Some changes would be necessary in the warehouse and office areas. The system can be installed and operating in 3 months. Good training manuals are part of the system.

Option 4. Design, program, and install our own AP system. This alternative gives us the most flexibility: we can tailor our reports to our needs and accommodate franchise operations.

Costs:	System design	$5000
	Programming	8000
	Training and installation	1400
	Maintenance of system	900/yr
	Two terminals (in warehouse and bookkeeping)	3500

FIGURE 3.9 *(Continued)*

CHAPTER 3 DETAILED ANALYSIS AND THE FEASIBILITY STUDY **89**

> Savings: Overtime pay for bookkeeping employees $ 4200/yr
> Discounts from vendors 12,000/yr
>
> Effects: Vendors could be paid within the 5- to 10-day discount period. Some changes in the warehouse and office would have to be made. The system can be installed and operating in 6 months. The problems associated with rising costs, excessive paperwork, work load, and vendor identification can be alleviated.
>
> Options 3 and 4 do not show the cost of the computer, since we own our HP 3000. Neither of these two choices requires us to modify or upgrade the HP 3000, nor will they damage system performance on existing applications. We have sufficient processor time and disk-storage space to add the accounts payable system to the HP 3000.
>
> *Recommendations and Rationale*
>
> The Computer Services Department recommends that management approve Alternative 4 for the following reasons.
> 1. The cost of a customized system is not appreciably greater than the cost of the MCBA software.
> 2. It will solve the problems in the current manual AP system.
> 3. MCBA's estimate may be low because the company wants our business. Other users have found actual costs to run higher than MCBA predicted.
> 4. The new system would interact with our current general ledger system. MCBA's would not.
> 5. The payback for the new system is a little more than a year.
>
> *Intangible Benefits*
>
> Alternative 4 also provides Fleet Feet with the following nonmonetary benefits.
> 1. Our vendors will receive better service.
> 2. Information not now available to management can be provided in a monthly report, which will help us make better decisions.
> 3. Our firm will be able to stay ahead of the competition.
> 4. The bookkeeping department will feel higher job satisfaction.
> 5. We can more fully use our HP 3000 computer.
>
> FP:jks

FIGURE 3.9 *(Continued)*

cation memo explains the plan briefly and establishes an overall schedule (Figure 3.10). Even if management decides to maintain or modify the current system, it should still issue a memo, or people will wonder why the company wasted time with a study that produced no results.

After a decision to proceed, analysis ends and design begins. The analyst will organize all the memoranda, questionnaires, interview documents and forms, data-flow diagrams, and reports from both the preliminary and detailed analysis into one file, which becomes the **analysis documentation**.

Analysis documentation All the written reports produced during the first phase of the systems process.

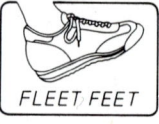

MEMORANDUM September 14, 1984

TO: Operations Personnel and All Vice-Presidents

FROM: Sally Edwards, VP Operations
Kathleen Williams, Manager of Computer Services

SUBJECT: Decision on a New Accounts Payable System

As you are all aware, our accounts payable system has been under extensive study for the past 2 months. Yesterday the study revealed that the existing manual system needs to be replaced. We decided that the best long-term results would be for us to develop our own accounts payable system for our HP 3000 computer. In making this decision, we took into account many factors, including speed of processing, the overworked accounting staff, and the desire to modernize. The development of our new computerized AP system will begin immediately and it should be in operation by next spring.

FP:jks

FIGURE 3.10. Management communicates its decision to the staff through a formal memorandum. This notice should go to all those affected by the action. In our example, Sally Edwards and Kathleen Williams send a memo to everyone involved with the accounts payable system.

CASE STUDY

Detailed Analysis of Camp Beaverbrook's Camper System

Lynn and Ron Garrison's authorization letter to Art Greene flashes the green light for Art to proceed with a complete study of Camp Beaverbrook's needs. Although Art thinks the study will be relatively simple, he does not want to omit anything, so he constructs a schedule of events for himself and his clients (Figure 3.11).

Art then writes to the Garrisons, requesting a meeting and asking them to fill in specific data on a questionnaire (Figure 3.12). To inform the clients about impending events, Art encloses his Gantt chart. Although he still has copies of some documents from the preliminary analysis, Art has requested all forms the camp uses.

Art must meet with the Garrisons' office worker, Diane Bell, who can best show him how paperwork actually flows through the current system, how she handles applications with

Correspondence and data courtesy Ron and Lynn Garrison, Owner/Directors, Camp Beaverbrook.

CASE STUDY

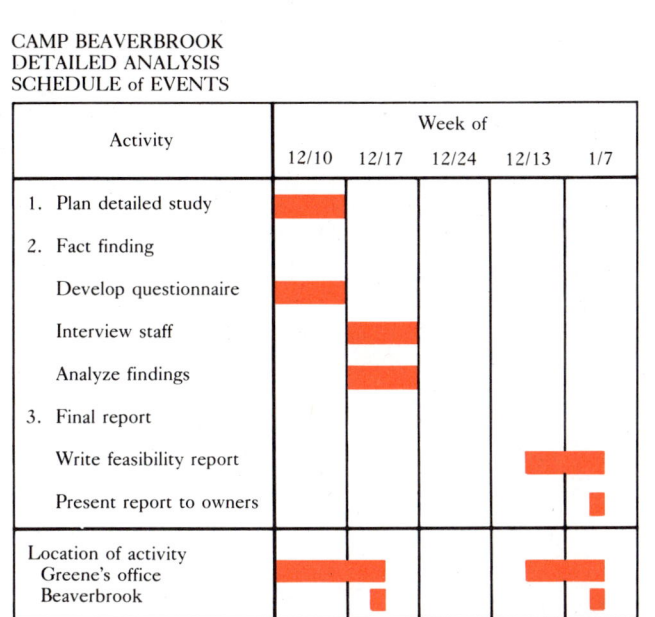

FIGURE 3.11. Art Greene's Gantt chart to schedule activities for himself and his client.

data missing, and how she deals with future campers' deposits. His meeting with Diane will also enable him to find bottlenecks in her operation—times when she is inundated with work, when she has nothing to do, or when she finds one task preventing her from performing some other vital task.

After receiving the questionnaire, Lynn and Ron begin their research. Like many small businesspeople, they spend most of their time running their business rather than filling out forms and keeping neat, accurate records. They base most decisions on guesses or feelings about their business rather than on facts. Together, Ron, Lynn, and Diane collect all the data needed to complete the questionnaire. As shown in Figure 3.13, they uncover some surprising facts while gathering the data. For example, they discover a great increase in camper requests for computing, hiking, and backpacking and decreased interest in swimming instruction and marksmanship. Diane remembers reminding the Garrisons of the loss of interest in marksmanship and the popularity of computing (with the purchase of two Atari 800s 2 years earlier), but the owners did not act on that information by adjusting their counselor hiring requirements. They now decide to hire extra counselors in computing and to highlight this interest area in their next national advertisement.

Art's discussions with the Garrisons proceed as planned. As the owners report their observations on the changes in camper interests, Art reassures them that such observations and changes often occur when owners and managers methodically analyze their operations. In addition to their original camper roster system, the Garrisons request a mailing label system for advertising future camp sessions to former campers. Art takes copies of the documents with him for reference and analysis.

Crucial to Art's detailed study is his observation of Diane Bell's paperwork flow. After spending several hours watching and talking with her, Art formulates a data-flow diagram

CASE STUDY

of her application handling. While developing the data-flow diagram, he becomes aware that the counselor application system parallels the one for campers. During previous discussions with the Garrisons, neither Ron nor Lynn had mentioned this, nor had they discussed the problems it causes. Consequently Art and Diane develop two data-flow diagrams, showing camper and counselor systems side by side (see Figure 3.14).

Diane points out two major bottlenecks she encounters each spring: differentiating campers' from counselors' applications, and the time involved in personally responding to each applicant. Beaverbrook keeps a file for each camper and counselor applicant, including his or her application form, health release, emergency notification form, and a ledger card showing payments and charges. Art notices much duplication of data: each form includes the applicant's name and address and parents' names and addresses, all of which Diane must type on cards.

As completed camper application forms arrive in Diane's office, she begins compiling rosters by recording the camper's name, preferred camp session, and activity interests. Eventually she types her handwritten rosters and duplicates them at the last possible moment, early in June.

December 7, 1984

Mr. and Mrs. Ron Garrison
Camp Beaverbrook
Beaverbrook, CA

Dear Lynn and Ron:

I started my detailed analysis this morning by laying out a schedule, which I have enclosed along with a questionnaire for you to complete. As you can see from the schedule, I shall be visiting you during the second and fifth week of the study. My plan is to drive to your camp for a late morning meeting, arriving at 8:45 to 9:00 a.m. I would like to interview your office assistant, Diane Bell, for a couple of hours in the afternoon, and then reconvene with you.

To help us work quickly and minimize costs, please complete the questionnaire before I arrive. If you can think of other pertinent information, forms, or applications, please put them together with those I have requested.

See you on Tuesday.

Sincerely yours,

Art Greene

Art Greene

FIGURE 3.12. *Art Greene's letter and questionnaire serve two purposes: to inform his clients of his plans and to request that they provide specific data. To complete the questionnaire, the Garrisons must review data they collected in previous years.*

CASE STUDY

QUESTIONNAIRE FOR CAMP BEAVERBROOK December 7, 1984

1. Draw an organization chart of your camp's staff structure. Include the name and title of each individual. If responsibilities are not evident from their titles, please describe their duties.

2. How many campers participated in the following activities for each of the summers listed?

Activity or Event	Summer of		
	1983	1982	1981
1. Hiking			
2. Swimming			
3. Backpacking			
4. Computing			
5. Archery			
6. Other (please list)			
a.			
b.			
c.			
d.			

If more room is needed, use a separate sheet.

3. How many applicants did you have in the following years?
   1983
   1982
   1981

4. How many campers had health or transportation requirements in the following years?

YEAR	HEALTH	TRANSPORTATION
1983		
1982		
1981		

5. Photocopy the following documents.
   a. Camper registry or history
   b. Winter reminder notices
   c. Accounting ledger sheets for 1983
   d. Camper emergency form
   e. Employee application form
   f. Employee time sheet
   g. Advertisements appearing in national magazines

FIGURE 3.12 *(Continued)*

CASE STUDY

QUESTIONNAIRE FOR CAMP BEAVERROOK December 11, 1984

1. Draw an organization chart of your camp's staff structure. Include the name and title of each individual. If responsibilities are not evident from their titles, please describe their duties.
   ———See attachment.

2. How many campers participated in the following activities for each of the summers listed?

Activity or Event	Summer of 1983	Summer of 1982	Summer of 1981
1. Hiking	500	458	407
2. Swimming	509	527	560
3. Backbacking	245	190	153
4. Computing	123	15	0
5. Archery	107	114	112
6. Other (Please list)			
a. Riding	55	34	51
b. Marksmanship	23	67	89
c. Chess	29	45	25
d.			

3. How many applicants did you have in the following years?

   1983   1056
   1982   1012
   1981   879

4. How many campers had health or transportation requirements in the following years?

YEAR	HEALTH	TRANSPORTATION
1983	45	247
1982	34	238
1981	39	267

5. Photocopy the following documents.———See attached sheets.
   a. Camper registry or history
   b. Winter reminder notices
   c. Accounting ledger sheets for 1983
   d. Camper emergency form
   e. Employee application form
   f. Employee time sheet
   g. Advertisements appearing in national magazines

   ———Additional documents: Employee W2 form, blank paycheck, and roster distributed to camp counselors.

FIGURE 3.13. Lynn and Ron Garrison's research for the consultant's questionnaire reveals some surprises: a 25 percent increase in applicants, no requests for computing in 1981 but 123 in 1983, and dramatic decreases in campers desiring archery and marksmanship instruction.

CASE STUDY

Name_____ Social Sec. #_____-____-_____
Present Address_____
          Mailing Address   City   State   Zip   Area Code & Phone No.
Permanent Address_____
          Mailing Address   City   State   Zip   Area Code & Phone No.
Birth Date: Mo.:_____ Day:_____ Year:_____ (Not required if over 21) Date:_____
Driver's Lic. #_____ Your present occupation_____
If student, are you planning to return to school full time in the Fall?_____

EDUCATION

Name & Location	Dates of Attendance	Date of Grad.	Major
High School			
College			
Other			

CAMP EXPERIENCE (as an employee) List most recent camp first:

Position	Camp	Director	Address	Dates

Date available for employment: From_____ To_____
Do you eat from all 4 food groups daily?_____ Any dietary restrictions?_____
What age group do you prefer to work with at camp?_____ Right/Left-Handed_____
What are your hobbies &/or interests?_____
Have you ever been arrested & convicted?__yes__no. If yes, explain._____
Have you ever been discharged or asked to resign from any job?__yes__no (If yes, give details on separate sheet.)

REFERENCES (Three, including former employers. No relatives or peers.)

Name	Address	Position
1.		
2.		
3.		

(b) Counselor application form used by Camp Beaverbrook. Each attempts to pinpoint the applicant's specialties: hiking, swimming, canoeing, computing.

FIGURE 3.13 (Continued)

CASE STUDY

> I hereby apply for summer employment at Camp Beaverbrook understanding that our standards require that staff do not use drugs, smoke or chew tobacco, or drink alcoholic beverages while in the employ of the camp. I understand that it is not a time for bringing or acquiring a boy or girl-friend, since it limits the attention promised to and required by campers. As a staff member I am willing to dedicate myself to the well-being of children. I further agree to provide Beaverbrook with a health exam report upon my arrival, if selected. In our camp, *Camp is for the Camper*.
>
> Signature:_____ Date:_____

FIGURE 3.13 (Continued)

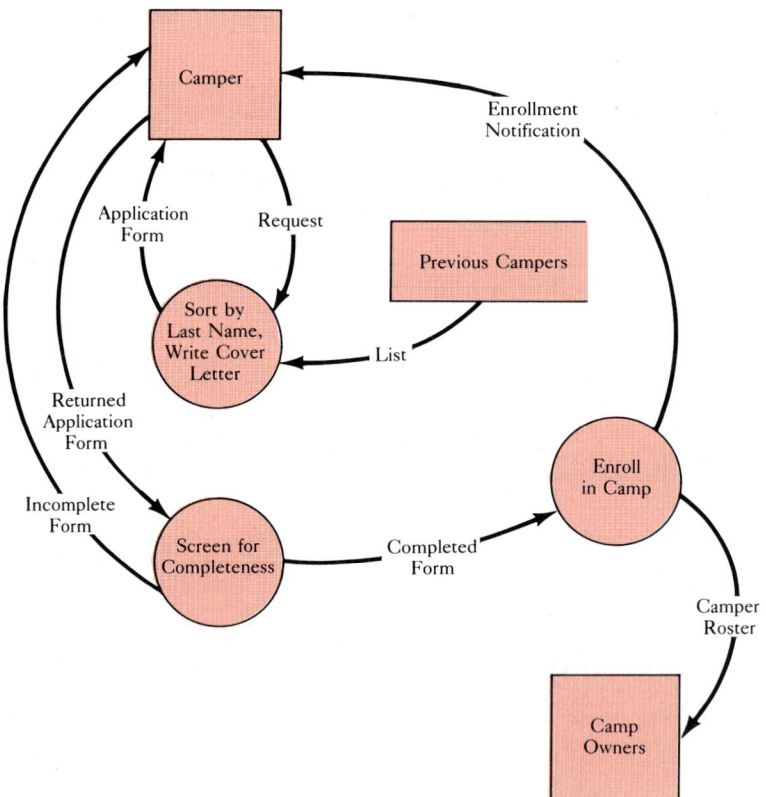

FIGURE 3.14. (a) Data-flow diagram of the camper.

CASE STUDY

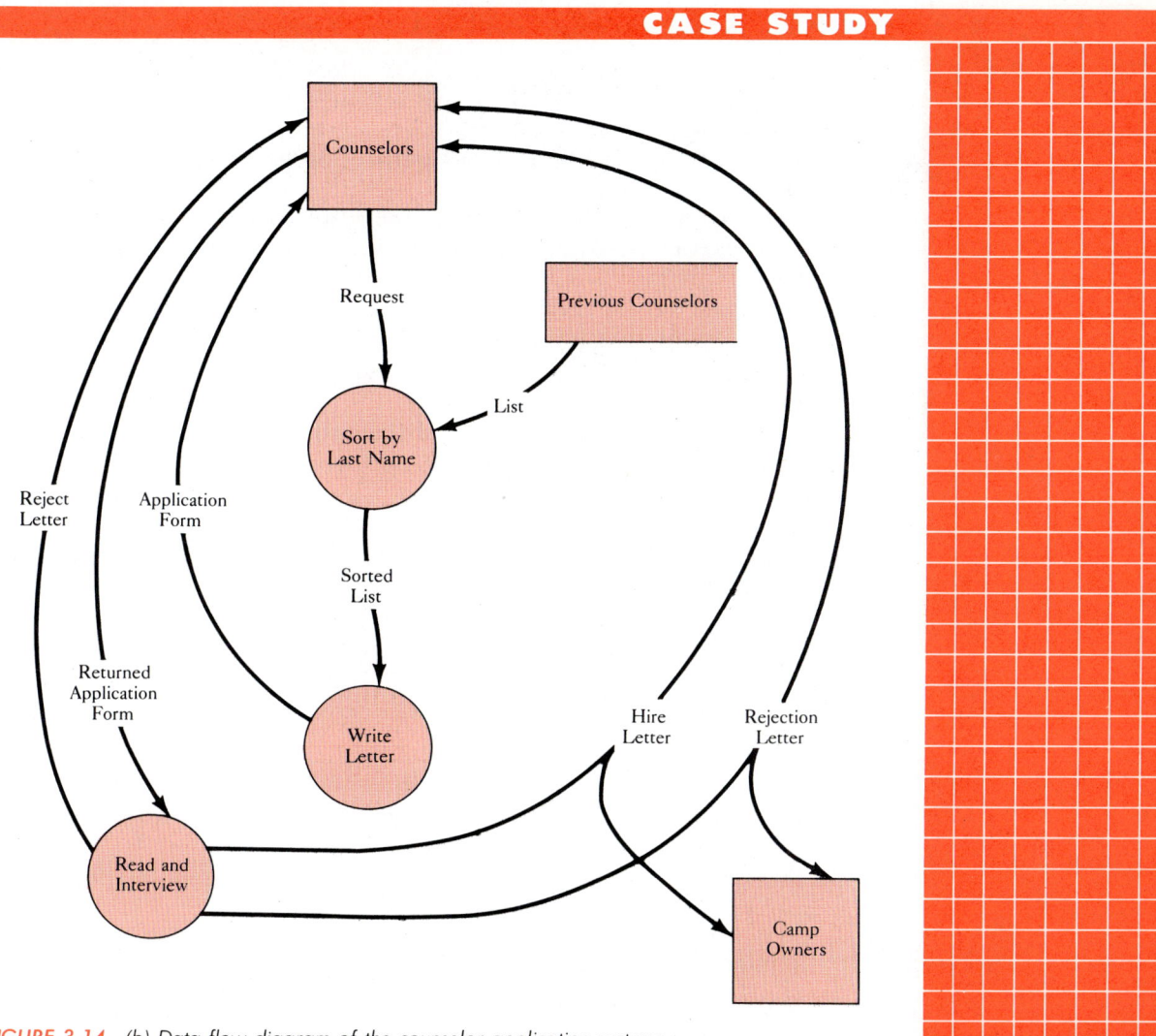

FIGURE 3.14. (b) Data-flow diagram of the counselor application systems.

CASE STUDY

Armed with the gathered information and the two data-flow diagrams, Art again meets with the Garrisons. After discussing the similarities between camper and counselor applications, Ron and Lynn agree with Art's recommendation that they add counselor applications to the system.

Returning to his office, Art reviews the information gained during his visit to Beaverbrook. Rather than rushing into a final report, he decides to document the Garrisons' needs.

Word processing: Mailing labels, letters to campers and counselors, advertising.
Camper files: List of each camper and pertinent information about this person (name, emergency telephone number, health problems, parents' names, transportation needs, charges, payments).
Counselor files: List of each counselor's capabilities and personal data (age, rate of pay, sex, previous experience, address, telephone number, specialty areas).

From previous experience, Art has learned that it always pays to let a little time elapse between the detailed anlaysis and the final report, and so he puts Beaverbrook's system on hold for a week. Then, after balancing and weighing all the alternatives, Art concludes that a microcomputer with word processing and file processing software would best suit Camp Beaverbrook needs. He writes to the Garrisons again, enclosing a copy of his feasibility study (Figure 3.15). The Garrisons agree with Art's recommendations and direct the consultant to acquire the necessary hardware and software stipulated in alternative C of the feasibility study (Figure 3.16).

CASE STUDY

January 4, 1985

Mr. and Mrs. Ron Garrison
Camp Beaverbrook
Beaverbrook, CA

Dear Lynn and Ron:

The enclosed report cites four options that you can pursue. Each has different costs and benefits. Some will solve both your camper and counselor problems while others will not. It is my recommendation that you take the third alternative because of its limited cost and yet positive benefits for Beaverbrook.

I look forward to your remarks and decision.

Sincerely yours,

Art Greene

Art Greene

CAMP BEAVERBROOK Camper and Counselor System
Report of Detailed Findings

by

Art Greene
Camper Systems
Pittsburg, California

Made to
Mr. and Mrs. Ron Garrison
Owners of
Camp Beaverbrook
Beaverbrook, California

January 2, 1985

FIGURE 3.15. Art Greene's letter and feasibility study.

CASE STUDY

Statement of Problems

Camp Beaverbrook currently hand-processes all camper and counselor application forms. This process is time consuming and expensive considering the number of papers that must be read and evaluated. The owners feel a computerized system might be able to reduce the costs as well as speed the process.

Findings

1. Both counselor and camper applications are processed too slowly. Beaverbrook may be losing counselor candidates to other camps because of the slow manual system now in use. Camper assignment to specific weeks operates well but is very costly.
2. The filing system is quite inefficient. Too much data must be hand-transcribed to various forms: emergency release, health information, payment ledger card.
3. Data are duplicated on all the forms and a single change in an address must be hand-posted to at least three other forms.
4. Communication with potential campers and counselors is slow and expensive. Each letter is individually typed. This does personalize Beaverbrook's correspondence, but it is costly.
5. There are no reports showing which campers owe what amounts of money. Ledger cards show these data but the data is never summarized nor are reminders sent to parents of campers asking for early payment to guarantee their child's week at camp.
6. Access to campers' names and addresses is severely restricted. It is difficult and quite expensive to maintain mailing lists for advertisements and special mailings.

Solution Possibilities

Five possible solutions exist.

Option A: Do nothing.
Continue to operate Camp Beaverbrook as you have for the last 35 years.

- Costs: None
- Benefits: None
- Effects: None of the problem areas will be resolved. Camper and counselor application systems will continue to function as they do now.

Option B: Hire staff.
Add a new person to your office staff who will work 7 months a year (February through August). This person's task would be to assist Diane Bell in typing, filing, and building the camper rosters.

- Costs: Salary and benefits of $1200/month for an annual amount of $8400. Office furniture (desk, chair, typewriter, supplies), a one-time cost of $2800.
- Benefits: Speed up in the processing of camper and counselor application forms.
- Effects: Does not solve problems with the filing system, duplication of data, financial reporting, or access to names and addresses.

FIGURE 3.15 *(Continued)*

CASE STUDY

Option C: Buy microcomputer and software.
Far West Camp Owners Association recommends "Camper" software. This software was designed and developed specifically for camp owners who want to prepare camper rosters, address labels, and health, emergency, and payment data. Also keeps data on counselors—names and skills.

Costs: Software—$2300 including 2 days' on-site training.
Hardware—Apple computer, hard-disk, letter-quality printer: $7649.
Benefits: Solves all problems associated with campers and counselors.
Effects: Should buy word processing software ($200) to aid in letter writing. Software and hardware can be bought immediately and you can be operational a week after your purchase.

Option D: Buy software and time and computer time sharing
Buy same software as in alternative C but run it on an online time-sharing computer service bureau.

Costs: Terminal—$985
Computer time—Estimated to be $450 per month for $5400 a year.
Software—$2300 including 2 days' on-site training. Service bureau already has word processing software and makes it available to users at no cost.
Leased telephone line—$230 per month for $2760 per year.
Benefits: Solves all problems associated with campers and counselors.
Effects: Software can be bought immediately and you can be operational a week after your purchase.

Option E: Write new software to run on a microcomputer. Design and develop software especially for Camp Beaverbrook that will solve camper, counselor, and word processing requirements.

Costs: Software—Estimate is $5800 including training.
Hardware—Apple computer, hard-disk, letter-quality printer: $7649.
Benefits: All problems resolved.
Effects: Time to design and develop software estimated to be 6 months. Close involvement of Garrisons and Diane Bell required, taking 100+ hours.

Recommendations

Each alternative has its own group of advantages and disadvantages. It is my recommendation that you choose alternative C. This choice is based on the following eight reasons.

1. Software can be used immediately.
2. Costs are within reason and certainly less than alternatives B, D, and E when more than a year is considered.
3. "Camper" is used by 38 other camp owners and none are reporting difficulties. A call to six of these users reveals that all would buy it again if they were to start over.
4. Equipment and software can be depreciated in a single year and qualifies for a 10 percent investment tax credit. Tax advantages may reduce costs by 50 percent.

FIGURE 3.15 *(Continued)*

CASE STUDY

5. Apple computer with Z-80 card can support word processing. Recommend WordStar.
6. Apple computer can be programmed to track counselor applications via dBase II database manager at additional cost of $475.
7. Computerized payroll available for Apple at extra cost.
8. Hardware can be moved home in winter. Can also be used by campers.

The total cost for alternative C with all applications (except Payroll) is:

Hardware: (computer, CP/M, hard disk, letter-quality printer)	$7649
Software:	
"Camper" and training	2300
WordStar	200
dBase II	475
	$10,624

FIGURE 3.15 (Continued)

January 17, 1985

Mr. Art Greene
Camper Systems
Pittsburg, CA

Dear Art:

We have both read and reread your report and agree that alternative C is the best choice for us and Camp Beaverbrook. It offers us immediate start-up capability and the very real potential of being operational for this year's group of campers and counselors.

To begin this process, we want to contract with you to be our agent and buy the hardware and all the software you listed in alternative C. Furthermore, we want you to design and develop a counselor system using the dBase II software you described. Please give us an estimate of your fees and expected time frame.

It was a surprise to find the detailed analysis was under budget and completed early. Enclosed is our check for $425.50.

We both want to thank you for the excellent job you did for us. Hiring a consultant was a sound decision and we know we saved money in the process. If you ever need a recommendation, please use our names.

Sincerely yours,

Ron Garrison
Lynn Garrison

Ron and Lynn Garrison

FIGURE 3.16. The Garrisons' response to Art's feasibility study.

WORKING WITH PEOPLE   The Need for Accounting in an Analyst's Training

After graduating magna cum laude with a degree in computer science, Matthew Reid accepted a job as a junior systems analyst at the headquarters of Riteway Hardware, an Illinois home-building supply chain. Matthew looked forward to his first assignment and walked confidently into his meeting with Shelby Ray, Riteway's controller.

As Ms. Ray described her needs for a new accounts receivable system, which should include conversion from balance forward to open item, Matthew found himself sinking deeper into his chair. From his only exposure to accounting, a one-semester introductory course during his second year at college, he knew aging was an important concept, but he couldn't recall the difference between balance forward and open item. While his mind danced with bits and bytes and modems, Ms. Ray talked casually about debits and credits and postings. Not wanting to appear ignorant, Matthew tried to look attentive, but he left the meeting frustrated and bewildered.

He immediately consulted with Dale Hardt, Riteway's senior analyst. "I spent 4 years studying COBOL, computer architecture, and data-base management systems, but now I feel like a complete idiot!"

Dale told Matthew about her own early experience with the company. Like Matthew, she had found her accounting background embarrassingly weak, but she had done something about it. Dale asked an accountant friend to give her a weekend crash course in accounting in exchange for some free programming on the accountant's new microcomputer system.

"Believe it or not," she told Matthew, "I found the subject fascinating and ended up taking two night courses in it. Not only did I learn to understand people in Shelby Ray's department, but I gained confidence and expertise in helping solve their problems."

Dale outlined several options for Matthew. He could:

1. Quit the accounts receivable project and let another analyst take his place.
2. Try to bluff his way past Shelby Ray.
3. Do what Dale had done, perhaps reading a book she recommended.
4. Ask that Shelby Ray allow Dale to help him with the project so he could benefit from her experience.
5. Find a software firm specializing in accounting to tutor and help him.

After debating the pros and cons of each option, Matthew decided to combine two of them. First, he would spend evenings and weekends reading about accounts receivable in the book Dale suggested. Second, he would try to convince Ms. Ray that he and Dale together could convert the accounts receivable system more quickly. It worked. Six months later, Riteway had its new system, as well as a much more effective analyst.

SUMMARY

If management approves the results of a preliminary analysis, the analyst conducts a detailed analysis, gathering facts about the old system, outlining objectives for a new one, estimating costs, listing possible alternatives, and making a recommendation. Before making a recommendation, the analyst may interview people in the organization, talk with hardware or software vendors, collect typical forms currently in use, personally observe the existing system, and perhaps send questionnaires to other people, even outsiders, involved with the system. A tool useful in this phase of the systems cycle is the Gantt chart, which specifies personnel and time spans for various activities.

Management evaluates the analyst's final report of the feasibility study and decides whether to halt or proceed. If management decides to proceed, the analyst ends the analysis phase and begins the design phase.

NEW TERMS
Analysis documentation
Context data-flow diagram
Fact finding
Gantt chart

QUESTIONS FOR REVIEW AND DISCUSSION

Questions appear in three categories. You can find the answers to the A group in this chapter. The B group requires you to apply the material presented here, whereas the C group necessitates investigation or research.

A.1. List the inputs to the detailed analysis phase.
A.2. List the outputs from the detailed analysis phase.
A.3. Which people in Fleet Feet's organization should the analyst interview?
A.4. Make a list of three do's and three don'ts you should follow when interviewing a user.
A.5. What questions should an analyst ask personnel during the fact-finding phase of the detailed analysis?

B.1. List four vendors an analyst might contact to gather information about terminology.
B.2. Why is a vendor number better to use than a vendor name in identifying invoices?
B.3. What is the purpose of an invoice number on the packing slip?
B.4. Why would Fleet Feet keep the name of the contact for each vendor?
B.5. Calculate the total cost of all five alternatives that Camp Beaverbrook faces with regard to its camper system.
B.6. Are there any duplicated data on a ledger card?
B.7. What do you suppose would happen if management decided on Frank Pisciotta's first alternative?

B.8. What will eventually happen if management decides on alternative 2?
B.9. What will happen if management decides on alternative 3?

C.1. Make a list of audio or visual aids an analyst can use in presenting the feasibility report to management.

REFERENCES

Tom DeMarco, *Structured Analysis and System Specification*, Yourdon Press, New York, 1979.

Steve Eckols, *How to Design and Develop Business Systems*, Mike Murach and Associates, Fresno, Calif., 1983.

Alan L. Eliason, *Online Business Computer Applications*, Science Research Associates, Inc., Chicago, 1983.

CHAPTER 4 — Tools for the Systems Designer

[Handwritten annotations: "GANTT PROCESS CHART T-F + MULT. CHOICE"]

- Goals and Preview
- The Need for a Structured Approach
- Flowcharts *[handwritten: —SYMBOLS]*
- HIPO
- Pseudocode
- **The Phenomenon of the Spreadsheet**
- Data-Flow Diagrams
- Data Dictionary
- Decision Tables and Trees
- Warnier-Orr Diagrams
- Nassi-Shneidermann Chart
- Structured Walkthroughs
- Case Study: Data-Flow Diagrams, the Data Dictionary, and Pseudocode for Fleet Feet's Accounts Payable System
- Working with People: New Tools Versus Old
- Summary
- New Terms
- Questions for Review and Discussion
- References

GOALS AND PREVIEW

After reading this chapter you should be able to:

1. Define structured methodology.
2. Read a VTOC (visual table of contents), flowchart, data-flow diagram, decision table, Warnier-Orr diagram, and Nassi-Shneidermann chart.
3. Read a data dictionary definition and pseudocode.
4. Describe the tools available for the systems process.
5. Define a module and list the rules for composing one.
6. State the purpose and role of a structured walkthrough.

Imagine buying a bicycle so complicated to assemble that only the original designer could put it together or fix it for you. Such a bicycle might represent a unique work of art, but it would be impractical for everyday use. Yet, until recently, much the same situation existed in systems analysis and design. With few guidelines to show them the way, each analyst brought his or her own set of skills to the development of systems such as accounts payable. As a result, many complex computer-based information systems bear one analyst's personal stamp. Such peculiar computer systems may contain many thousands of lines of code in their programs, making it difficult for anyone else to fix or modify the system. Worse, even the person who originally wrote the program may not clearly recall all the intricacies of the system a few months later.

Chapters 1, 2, and 3 covered the systems process in general and explored preliminary and detailed analysis more thoroughly. In this chapter, we present the tools that help standardize the process. Just as an architect uses a drafting board, ruler, and drawing instruments to create plans, an analyst can employ certain tools to create systems. Some tools are graphic (the data-flow diagram), whereas others are verbal (the walkthrough). Users demand simple, flexible, efficient computerized systems, and such systems require analysts to learn more about the array of tools at their disposal.

THE NEED FOR A STRUCTURED APPROACH

Doug Hickman, programmer analyst for Data Managers, a computer software firm, had just put the finishing touches on an accounts receivable system at Bennett's Department Store. He felt relieved to have finished the huge project. Though he had recently accepted a new job with a big computer firm, he had wanted to complete this assignment before moving on. Proud of his clever design and the obscure programming techniques he had built into the system, Doug briefed his replacement, Virginia Smith, on it, assuring her of its efficiency and beauty.

It became Virginia's job to supervise the first run of the programs through the computer at Bennett's. Unfortunately errors began to appear midway through the run: totals were incorrect, counts of records processed did not agree with the number of records entered, and rounding errors created false finance charges. When Virginia stopped the program to search for the cause of the errors, she grew increasingly frustrated, because the harder she studied Doug's design and his programs, the less she understood them. She simply could not unravel Doug's obscure programming technique. After spending several days unsuccessfully pondering Doug's work, Virginia was forced to redesign and reprogram major sections of Bennett's AP system, costing Data Managers a lot of time and money, not to mention rapport with its client.

A half a dozen years ago, this bleak picture of system design and programming was quite common. Today, however, the use of structured methodology, which places more emphasis on the design phase, helps us more fully to standardize the systems process. Structured methodology allows us to divide any system into small sections or modules, each with its own special function or purpose. Among other benefits, this permits us to closely scrutinize manageable building blocks for errors all along the way.

The Top-Down Approach

Top down (1) Testing approach that checks highest level modules first, and then lower level ones. (2) Dividing a system's major modules into increasingly detailed modules.

Crucial to structured methodology is the **top-down** concept, which establishes the components within a system's structure. When an analyst studies a system, he or she creates a hierarchy or order among the components of that system. Earlier we decomposed the systems process into manageable modules—analysis, design, and development—and we further divided each of these modules into even smaller components (Figure 4.1).

Imagine a landscape artist working on a painting of a forest. First the artist sketches the general idea of the painting—three large boulders in the foreground, a towering peak rising above the forest. Once the artist establishes the size of the larger components and determines their relationship in space, he or she can begin to add veins to the granite, branches to the trees, and snow to the mountaintop. Eventually, of course, the artist will use very fine brushes to paint small details: pine needles, a squirrel perched on a branch, a hawk circling high overhead.

Notice how the artist followed a decreasing set of priorities from general to more detailed concerns. In much the same way, a systems analyst begins with the general idea of a system (what the user wants it to accomplish), then proceeds to ever smaller details, following a descending pattern of priorities.

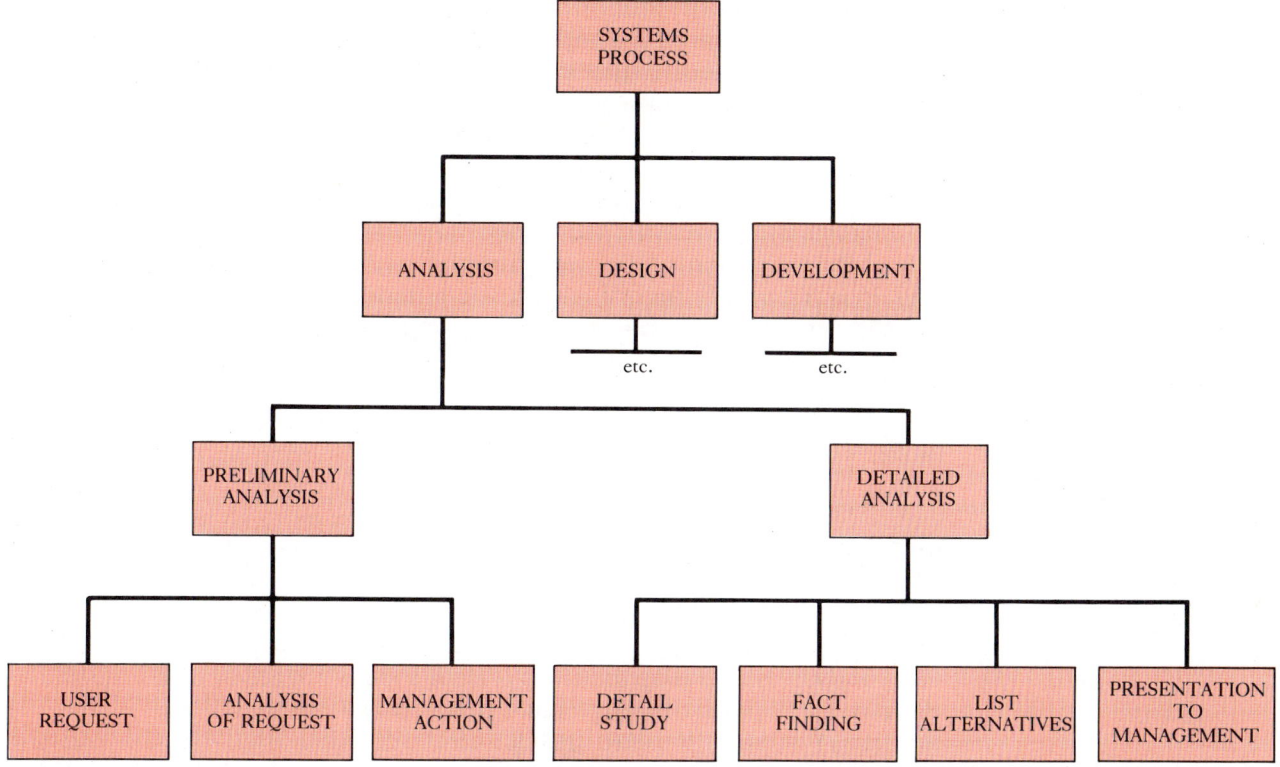

FIGURE 4.1. *A top-down view of the systems process. We will add components under design and development in later chapters.*

The systems analyst starts with a big picture: the user wants a payroll system. Initial study reveals the payroll system must be able to handle hourly, salaried, and commissioned employees. More detailed study discloses that hourly employees fall into four subgroups: day, evening, graveyard, and split shifts, with each shift having a unique pay program the payroll system must be able to calculate.

Such subdividing of tasks into subtasks graphically demonstrates the top-down approach. The task listed at the top encompasses all of those listed below, and each subtask represents a smaller component of the one above. Tasks at the top "boss" those at the bottom. Thus the task with the highest authority in Figure 4.1 is systems process, followed by three subordinate tasks of equal authority: analysis, design, and development. Within analysis the preliminary and detailed components command more authority than user request, analysis of request, and management action (which enjoy equal authority).

Modules

Once we have decomposed a system into its components (called functional decomposition) using the top-down approach, we can study each **module,** which is a unit or entity with a single specific function. The front wheel of a bicycle, for

Module A discrete or identifiable single-function program unit.

example, is a module by itself, whereas the rear-wheel module includes the power train and therefore can be decomposed into two modules: the wheel and the power train. In Fig. 4.1 we encounter analysis, design, development, preliminary analysis, and detailed analysis modules. Some modules are "children" to "parents," just as the rear wheel and power train of a bicycle are slaves to the pedals. They are useless unless attached to the frame. So it is with the systems process. All three modules (user request, analysis of request, and management action) are children to systems analysis, the parent.

Analysts have adopted the modular concept from programming where each module is responsible for a single routine or task. For example, the task of calculating net pay is a module in a payroll system, and calculating discounts for early payments is a module in an accounts payable system. Program modules must remain single purpose, with single entry and exit points (Figure 4.2). A program cannot jump into the middle of a module, nor can it leave the module except via its sole exit. Modules that perform a single logical function are called **cohesive**.

Cohesion A measure of a module's strength to accomplish a single task.

Programs are thus a collection of cohesive modules linked together in a hierarchical or treelike manner. Some modules are parents to child modules and a parent module has control only over those child modules subservient to it. We measure the control and interdependence among modules and call this measurement **coupling**. There are five types of coupling: data, stamp, control, common, and content. Modules that process data and send the data to other modules exhibit data coupling. Stamp coupling takes place if a module receives more data than it can process. Control coupling occurs between two modules if one conrols the function of another. If many modules access the same data, we have common coupling.

Coupling The measure of control and interdependence among modules.

The last type of coupling is content, a module referencing the inside of another module or leaving from the middle of itself. From the structured perspective, content-coupled modules are the worst—they violate the important single-entry and exit concept. However, most programming languages make it difficult to program content-coupled modules.

Since some modules are parents, just how many child modules should each

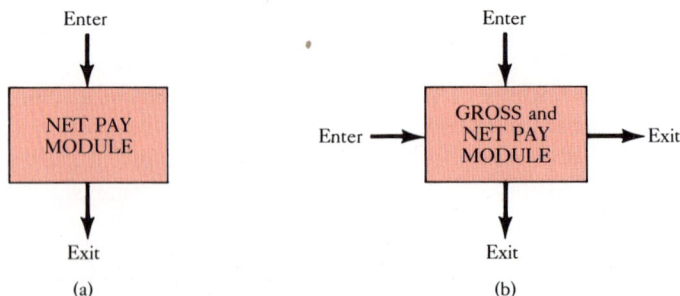

FIGURE 4.2. Structured methodology requires the use of modules for systems analysis, design, and development. The single entry and exit concept requires that each module have only one entrance and exit. (a) A valid module. (b) Module is invalid because it has two entrances and two exits, and is multiple purpose.

control? **Span of control** refers to the number of subservient modules controlled by a parent module. Spans of five to nine are considered ideal. Spans of one or two are too few, and spans of 12 to 15 are too many.

Span of control Refers to the number of modules subservient to a parent module.

FLOWCHARTS

One of the oldest tools available to the analyst is the **flowchart,** which graphically describes a system. As shown in Figure 4.3a, standard flowcharting symbols have specific meanings: an oval indicates the beginning and end (start and stop), parallelograms depict data input or output, circles show an entry or exit to another part of the flowchart, and rectangles reveal processing activities. Arrows or flowlines link the symbols together and show the order, or hierarchy, of events. Figure 4.3b is a simple flowchart of the steps taken to assemble a bicycle.

Flowchart Graphic representation of the method of a problem's solution.

Many types of flowcharts exist, but analysts find two especially useful: the system flowchart and the program flowchart. The former provides an overview of the system and the latter details specific actions within the system.

The following guidelines for drawing flowcharts help ensure effectiveness and correctness.

1. Avoid crossing lines when drawing arrows.
2. Draw the chart from top to bottom.
3. Use standard symbols for their specified purpose only.

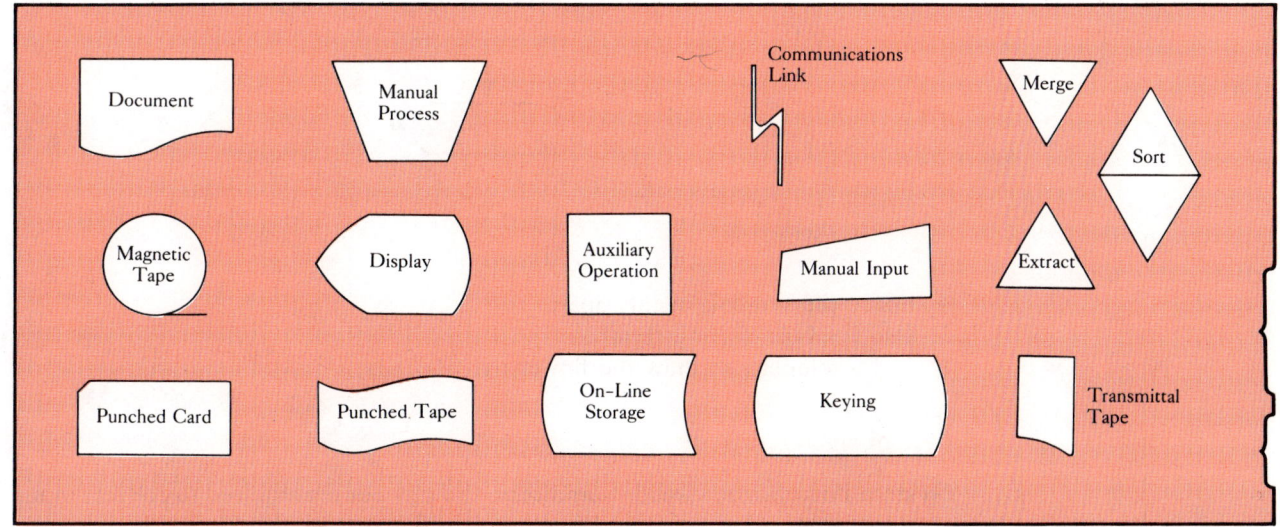

(a) ANSI standard symbols and their meanings.

FIGURE 4.3. Standard flowcharting symbols recommended by the American National Standards Institute (ANSI). The institute sets standards for many other applications: number of threads per inch on bolts, sizes and gauges of metals, diameters of wire, and so on.

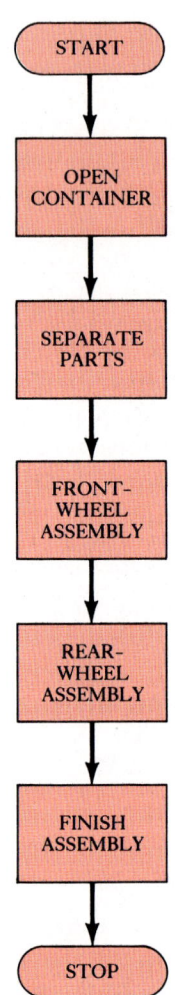

(b) Flowchart of bicycle assembly.

FIGURE 4.3 *(Continued)*

4. Space the symbols evenly apart.
5. Use symbols of consistent size.
6. Use a template to draw the flowchart.

System Flowcharts

System flowchart A flowchart showing the interaction among programs.

System flowcharts, visually summarize the procedures required to convert input data to output information and, include additional symbols. An upside-down triangle (with the vertex pointing down) depicts actions for sorting and merging data files; circles with horizontal lines at the bottom represent tape files; the document symbol represents input or output on paper (printed reports or source documents); and trapezoids show clerical operations, such as storing tapes in a vault.

A system flowchart offers a graphic tool that even noncomputer people can

CHAPTER 4 TOOLS FOR THE SYSTEMS DESIGNER **113**

comprehend. We construct and read one from the top down and from left to right. System flowcharts do not use the oval for start or stop. Arrows point readers in the proper direction. Figure 4.4 flowcharts an accounts payable system. Invoices and data changes are entered and verified; then accepted invoices are placed in a file and rejected ones are printed in an error report. The system proceeds by

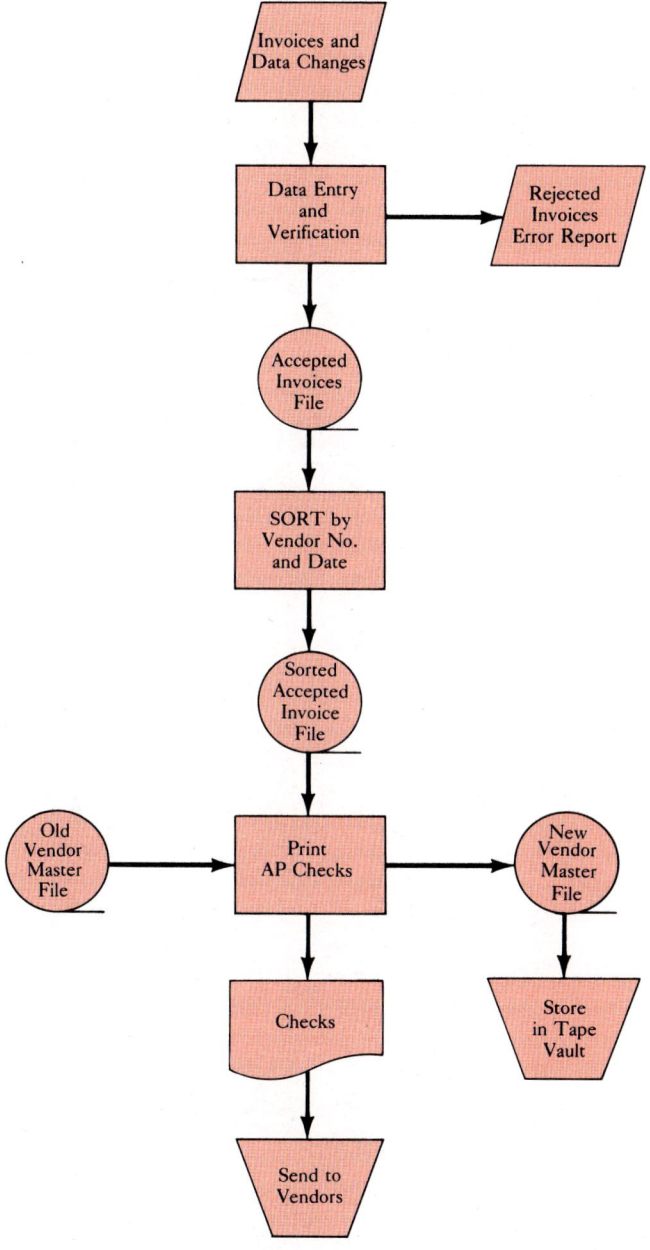

FIGURE 4.4. An accounts payable system flowchart begins with an invoice from a vendor and ends when a check goes to the vendor for payment.

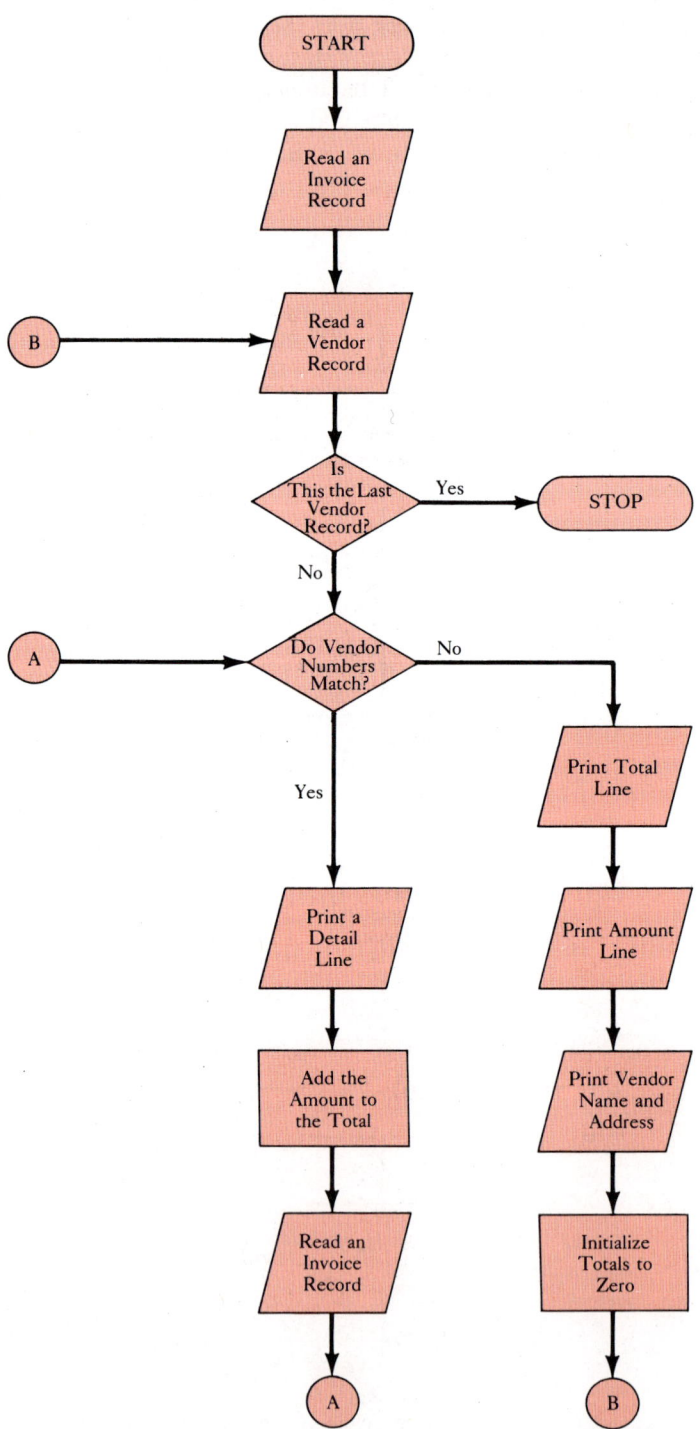

FIGURE 4.5. One reads a detail flowchart from the top down and from left to right. It employs four symbols: ovals for beginning and ending the flowchart; parallelograms for data input or output; diamonds for decision points; and circles to connect one point within the flowchart to another.

sorting the data by vendor number and date. Finally, it reads vendors' accounts and prints out checks for the vendors who are owed money. Of course, the company will store the files for subsequent use, such as processing the next week's invoices and data changes.

Program, or Detail, Flowcharts

Another common tool is the **program,** or **detail, flowchart.** Sometimes called block diagrams, detail flowcharts sketch the logic of a particular computer program within a system, enabling them to serve as an outline for a programmer to follow when finally programming a system. Program flowcharts assist the analyst and the programmer and are not as understandable to noncomputer people as system flowcharts.

Program flowcharts may employ five main symbols. An oval shows the beginning or end of the flowchart, a parallelogram indicates data input or output, a diamond shows a decision point or choice between two alternatives, a rectangle depicts a processing or calculational step, and a circle connects two points in the flowchart. The detail flowchart in Figure 4.5 sketches the logic required to print the check in our accounts payable system.

In recent years, some professionals have criticized both types of flowcharts because they do not sufficiently illustrate the concepts of hierarchy and modularity that are so crucial to structured methodology. Flowcharts are time consuming to draw, difficult to maintain, and do not readily support top-down design or modules. In many situations, flowcharts are constructed after writing programs or designing systems, rather than before. Some devotees to structured concepts even go so far as to say that flowcharts are prehistoric. Nevertheless, they still provide a useful tool for the analyst's toolbox.

Detail flowchart A type of flowchart depicting the steps necessary to write a program.

HIPO—HIERARCHY PLUS INPUT PROCESS OUTPUT

Hierarchy plus input process output, or **HIPO** (pronounced hypo), consists of two types of diagrams: the visual table of contents, or **VTOC** (pronounced vee tock), and the input process output, or **IPO** (pronounced eepo or eye po). Together these diagrams assist in designing programs and their functions.

Following the structured approach that begins with generalities and descends to details, VTOC diagrams break a system or program down into increasingly detailed levels. Therefore, the name of the system appears at the top of the VTOC, the names of the major functions within the system lie on the second level, and even smaller subfunctions lie on the third and succeeding levels.

When used to diagram a program, the VTOC arranges the program modules in order of priority, and it reads from the top down and from left to right. Each module of the program appears as a rectangle which contains a brief description of the module's purpose (two to four words, beginning with a verb followed by an object, i.e., "compute net pay").

The VTOC for the correct assembly of a bicycle might include five major tasks: open the carton, and remove the parts, group similar parts, assemble the wheels, finish assembling the bicycle (Figure 4.6a). Compare this diagram with the one in Figure 4.3b. Both indicate hierarchy but the VTOC offers a more complete

Hierarchy plus Input Process Output (HIPO) A set of charts emphasizing the functions of a system or computer program.

Visual Table of Contents (VTOC) Method for diagramming program modules in order of priority.

Input Process Output (IPO) A detail chart listing inputs, processing steps, and outputs of a module.

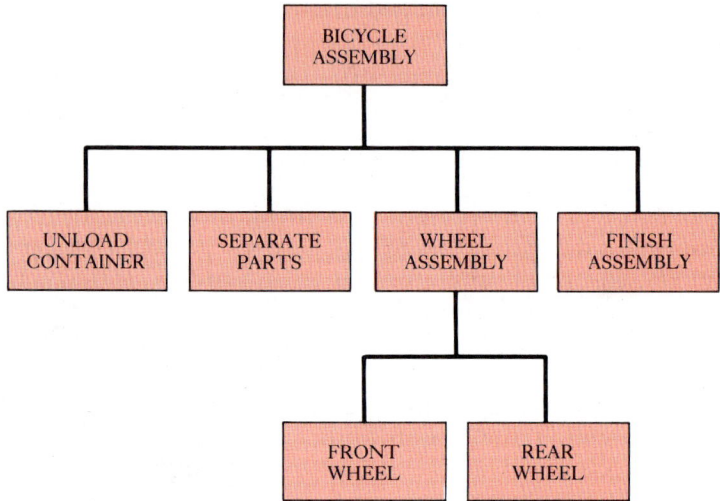

(a) VTOC for the modules to assemble a bicycle.

SYSTEM: Bicycle Assembly		**Author:** M. Bell
MODULE: Finish Assembly		**Date:** 12/02/84

INPUT	PROCESS	OUTPUT
1. Frame	1. Bolt front- and rear-wheel assembly to frame	1. Completed bicycle
2. Front-wheel assembly	2. Attach handlebars	
3. Rear-wheel assembly	3. Attach chain	
4. Seat	4. Attach seat	
5. Chain		
6. Handlebars		

(b) IPO for the finish assembly module in Figure 4.6a.

FIGURE 4.6. The VTOC and IPO for the assembly of a bicycle.

picture of the overall process. When assembling something, no matter how clear the instructions are, it helps to refer to a picture of the finished product. Within the VTOC, each task must be performed in the order specified, and each task may involve several subtasks. For example, wheel assembly involves the separate subtasks of front- and rear-wheel assembly.

An IPO chart defines the inputs, processing, and outputs for each module in the program. Figure 4.6b is an IPO for the "finish assembly" module, inputs for which include the frame, front-wheel assembly, rear-wheel assembly, seat, handlebars, and chain. The processing requirements are bolting wheel assemblies to the frame, and attaching handlebars, chain, and seat. The output is the completed bicycle.

Constructing a VTOC

Now let us learn how to prepare a VTOC for the computerization of a manual accounts payable system. We first assign all the major outputs (reports) to modules as shown in Figure 4.7. We name modules according to their function so the reader can tell exactly the purpose of the module (again using the verb-object format).

After naming, we choose a number for the system (1), we give each module a sub or level number, beginning with 1 for the most general or highest level function. Therefore, we would identify the overall system as 1.0, the accounts payable check module as 1.1, the numeric vendor list as 1.2, and so on (Figure 4.8). Such a number system clarifies the relationships between modules, and allows anyone reading it easily to locate detailed IPO charts with corresponding numbers.

After assigning level numbers to each module, we can consider whether further decomposition is necessary. Take the accounts payable check module, for instance. It must be decomposed to a lower level because it represents two tasks (remember, a module must be single purpose), one for each part or stub of the check (Figure 4.9). The upper stub is the remittance advice, which contains the date, number, discount, balance, and total of each invoice covered by the check. The lower stub

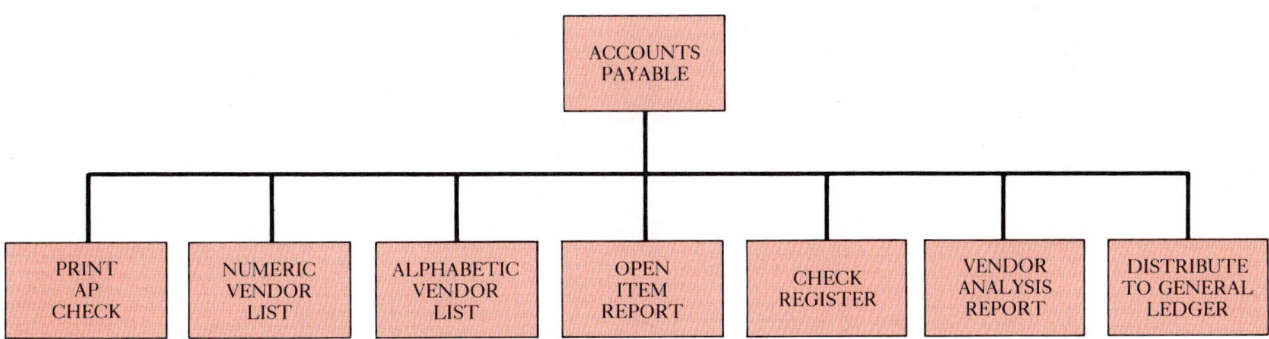

FIGURE 4.7. The VTOC for an accounts payable system begins with assignment of modules to each major output report.

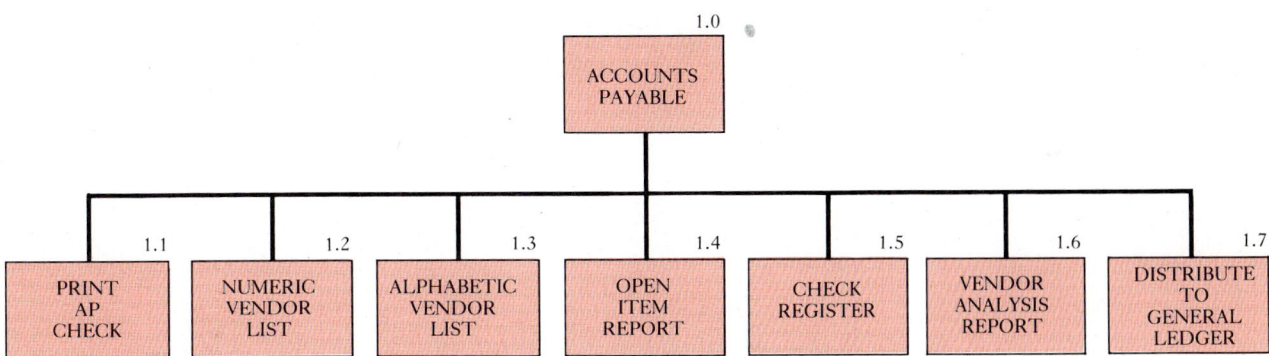

FIGURE 4.8. The VTOC for the AP check-printing system with level numbers.

118   PART I   SYSTEMS ANALYSIS

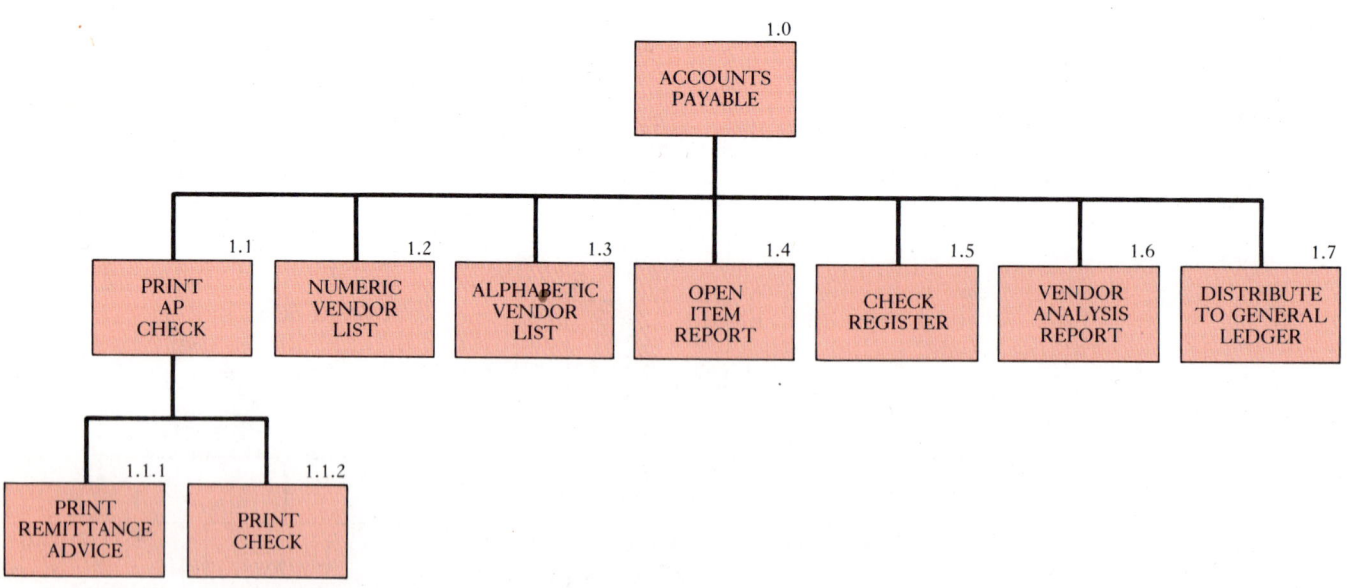

(a) Two-part check sent to vendors.

(b) This VTOC reveals the two component modules within the PRINT AP CHECK module.

FIGURE 4.9. Check sent to a vendor for payment of invoices. The check is divided into two parts, one containing remittance advice, and the other the actual check.

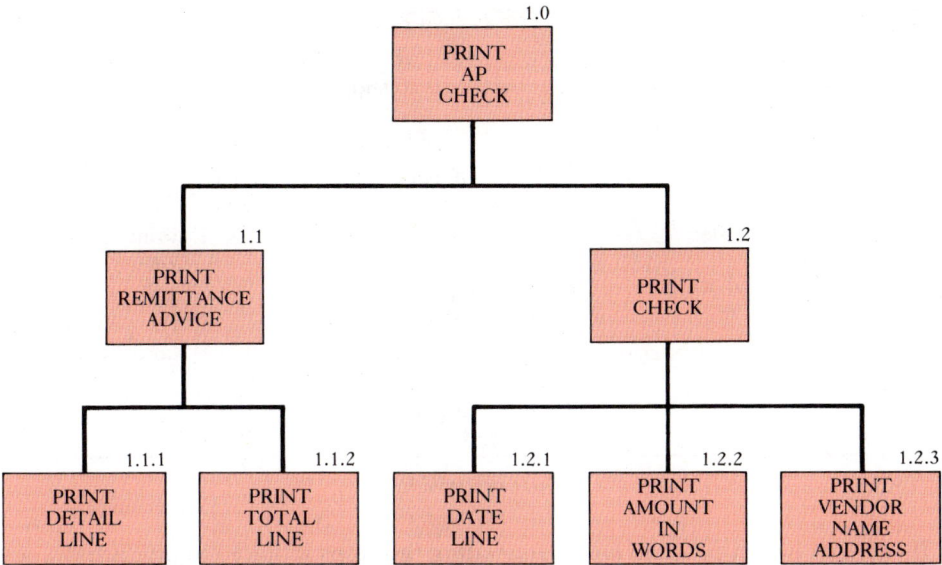

FIGURE 4.10. Module development of the AP check system from the programmer's perspective.

is the check itself, complete with check number, dollar amount, and vendor name and number. Applying the top-down concept to the check module, we can add another level of modules: one for printing the remittance advice and one for printing the check itself (Figure 4.10). We assign this third level of modules a third set of numbers (1.1.1 and 1.1.2). Decomposition ends when all modules are single purpose, single entry, and single exit.

As far as the analyst is concerned, decomposition of the accounts payable check-writing system can end at the third level of modules. However, the programmer who eventually receives the HIPO chart probably would decompose the modules to even lower levels, thus establishing a new series of numbers (Figure 4.10). We examine this new leveling further when we discuss programming a system in Chapter 12.

Programmers sometimes call their VTOCs structure charts, hierarchy charts, or tree charts. When finally programming the accounts payable check system, the programmer would begin at the top left, programming each level before moving down and to the right. Thus the programmer first would code "remittance advice" (1.1), followed by "print detail" (1.1.1), and then "print totals" (1.1.2). Having coded all of module 1.1, the programmer would tackle module 1.2, beginning with "print date line" (1.2.1), moving on to "print amount in words" (1.2.2), and finishing with "print vendor name address" (1.2.3).

Figure 4.10 shows three levels of modules, but complex systems may require many more. Regardless of their number, modules should receive unique and brief names that contain just enough detail for readers to understand their purposes.

IPO—Input Process Output

Let us add the second part of the HIPO diagramming system, the IPO chart. Whereas the VTOC diagram graphically shows an overview of the system, the

SYSTEM: Accounts Payable	Date: 9/12/84	Author: Gary Hatch
MODULE: 1.0	Name: AP Check	
DESCRIPTION: Prints the stub-over-check sent to suppliers.		

INPUT	PROCESS	OUTPUT
1. Vendor master file	1. Read Invoice record	1. Print remittance advice on top stub
2. Invoice file	2. Match with vendor	2. Print check on bottom stub
	3. Total amount	
	4. Print detail line	
	5. Print total line	
	6. Print date line	
	7. Print vendor name and address	

FIGURE 4.11. The IPO chart shows detail or program logic.

IPO charts depict program logic, illustrating the steps required to produce desired outputs.

As Figure 4.11 shows, the top of the IPO chart identifies the module with its number, title, a brief description, date, and the analyst's name. The chart itself is divided into three units: data input (the names of the files used), processing activities that will require programming, and output, which, in the case of our accounts payable system, would be the printed remittance advice (1.1) and the check (1.2). In the body of the chart, we use a narrative form to describe the input, process, and output as a list of activities. It simply lists activities, not necessarily ordering them in the sequence they should occur.

The HIPO system forms valuable system documentation and helps the analyst prepare reports as the system is being designed and developed. These charts offer several advantages. First, they can be drawn or modified rapidly. Second, they allow the analyst graphically to convey the system to noncomputer people. Third, standardized symbols enable some future analyst to grasp the system quickly. Finally, HIPO charts facilitate efficient schedules because they make it easy to estimate the time it will take to program a module, thus simplifying programming assignments. The VTOC offers the analyst an alternative to the system flowchart whereas the IPO replaces the program or detail flowchart.

PSEUDOCODE OR STRUCTURED ENGLISH

Pseudocode A concise English-like description of what users want the computer to do.

Structured English A synonym for pseudocode with limited vocabulary and limited syntax.

Pseudocode is a relatively new tool that provides a concise, step-by-step, English-like description of what users want the computer to do. Also called **structured English,** pseudocode uses simple imperative sentences; it omits most punctuation and all adjectives and adverbs.

The pseudocode for finish assembly of a bicycle might read:

1. Bolt front-wheel assembly to frame.
2. Bolt rear-wheel assembly to frame.
3. Attach handlebars.

4. Attach chain.
5. Attach seat.

Like an IPO or a detail flowchart, pseudocode produces a detailed description of a module, using only three types of sentences: sequence, selection, and iteration (repetition). We call such sentences **control structures** since they control the flow of the module's details.

The first control structure, **sequence,** shows an event or action and subsequent events or actions without interruption between events. Figure 4.12a illustrates sequence control structure using pseudocode.

The second control structure, **selection,** provides a means for testing a certain condition to determine which of two possible events or actions should take place next (Figure 4.12b). We also call this the IF-THEN-ELSE structure, because IF stands for the test, THEN indicates event(s) that should occur if the test proves

Control structure A pattern for building the logic of a computer program.

Sequence Control structure where each action follows the next in a linear fashion.

Selection A control structure that allows tests to be made for events or conditions.

1. Open container and remove parts.
2. Separate parts.
3. Assemble wheels.
4. Finish assembly.

(a) Sequence control structure shows one action or event followed by another without interruption.

IF (Brake-Type = Hand) THEN

    Bolt brake handles to handlebars.
    Run cables to front and rear brakes.
    Attach cables to front and rear brake calipers.

ELSE

    Bolt coaster brake to frame at rear wheels.

ENDIF.

(b) Selection control structure uses the IF-THEN-ELSE along with indenting actions.

WHILE (there are cartons of bicycles) DO
    Open container and remove parts.
    Separate parts.
    Assemble wheels.
    Finish assembly.
ENDDO.

(c) Iteration control structure provides for the repeating of events. This might apply to a bicycle assembly line.

FIGURE 4.12. *Pseudocode sentences fall into one of three control structures: sequence, selection, or iteration. Each has the single entry and exit points required by structured methodology.*

Iteration A control structure that allows repetition as long as a condition remains true.

true, and ELSE indicates event(s) that should occur if the test proves false. A selection ends with the words "ENDIF."

The third control structure, **iteration,** repetition, or loop (Figure 4.12c), sometimes goes by the name DO/WHILE. Loops are groups of event(s) we want the computer to perform a given number of times or until a certain condition prevails. This structure has all subservient instructions repeated as long as a condition holds true.

Let us apply pseudocode to the accounts payable check logic we diagrammed with a VTOC, an IPO, and a detailed flowchart. As Figure 4.13 shows, pseudocode indents sentences that are subordinate to an iteration sentence. Thus the four sentences in the middle of Figure 4.13 (If, Add, Print, and Read) depend on the "While this customer do" iteration sentence. Similarly, the entire center of the pseudocode depends on the "While not end of file do" sentence. Unfortunately there are no hard and fast rules governing the construction of proper pseudocode for a problem because each unique problem requires its own unique psuedocode solution.

However, analysts may follow these 10 general guidelines when writing pseudocode:

1. Each sequence control structure must contain at least one sentence.
2. Selection control structures may be IF-THEN, without ELSE. This occurs

```
AP Check Module

READ invoice record
READ vendor master record
WHILE (not end of file) DO
        Initialize total of amounts to zero
        WHILE (this customer) DO
                IF (discount available) THEN
                        Calculate discount amount
                        Subtract discount amount from invoice amount
                ELSE
                        Set discount amount to zero
                ENDIF
                Add amount to the total
                Print data about this invoice
                READ invoice record
        ENDDO.
        Print the date line
        Print the amount line
        Print the vendor name and address lines
        Update vendor master record
        READ vendor master record
ENDDO.
```

FIGURE 4.13. Pseudocode for the AP check module containing all three control structures. Records from the invoice and master file are read before any processing activities begin. Looping or iteration is achieved with the "While (not end of file) do" command. Most file-processing routines will involve a similar command that causes all the records in the file to be read.

when a condition leads to only one possible activity. If the condition is false, then nothing should happen. In such cases, you may omit the ELSE in the pseudocode or write a phrase indicating nothing should happen.
3. An iteration can specify a critical condition (repeat five times) or the instruction to repeat an activity until a file is complete ("While not end of file do").
4. Subordinate sentences in an iteration do not necessarily have to be performed (if we have no invoices for a customer, we do not print a check to that customer).
5. Indent subordinate sentences.
6. Show the end of a selection or iteration structure with an ENDIF or an ENDDO.
7. If the pseudocode for a module is longer than one page or CRT screen, divide it into two or more subservient modules.
8. Place parentheses around conditions in an IF or WHILE statement.
9. Avoid weak verbs. Instead pick strong action verbs: combine, read, write, or calculate.
10. Consider using an editor or computer word processing program for writing and maintaining the pseudocode.

Both computer and noncomputer people can easily learn to use pseudocode. Analysts who currently employ detailed flowcharts to depict program logic can quickly adapt them to pseudocode, which also follows the rules of structured methodology—single entry and exit points, and single-function modules. As a system evolves, so can the corresponding psuedocode.

The terseness of pseudocode poses its major disadvantage. Its lack of adverbs and adjectives and its limited number of verbs can make it quite cryptic. Some COBOL programmers feel pseudocode wastes time because it so closely resembles the COBOL language itself, and they prefer to write their COBOL programs without first developing the pseudocode.

The Phenomenon of the Spreadsheet

Until recently even rather minor business decisions took some technical expertise. If the owner of a hardware store wanted to know the impact on profit of a two-day storewide 20 percent discount sale, he'd have had to ask his bookkeeper or accountant to "run some numbers," in other words, calculate three or four different quantities of items sold and the relative profitability of each item. Today, however, even the least number-oriented manager can answer such a question virtually at the push of a button with an electronic spreadsheet. These powerful tools have become the most popular software product in history. Of the nearly 3.75 million microcomputers installed nationwide, nine hundred thousand incorporate a spreadsheet. That number could reach 3.0 million by the end of 1984, and the United States Department of Commerce estimates that by 1986 one out of every four white collar workers will be using one.

Why has the electronic spreadsheet ignited such interest? It all started back in 1976 when Dan Bricklin, a programmer involved in the development of word processing and text manipulation for Digital Equipment Corporation, returned to Harvard Business School and undertook a complex financial analysis assignment requiring extensive manual calculations and recalculations. Bricklin wondered whether he could apply what he knew about the computer's ability to handle words and text to this difficult and time-consuming project.

Bricklin and another programmer, Bob Frankston, formed a company called Software Arts, and together developed VisiCalc for the APPLE II. Afterwards they hired a software publishing house, Personal Software (now VisiCorp), to market their product, and VisiCorp and VisiCalc catapulted into prominence.

More than 50 different relatively inexpensive spreadsheet programs have appeared since then, and although they differ widely in terms of capability and sophistication, they all enable managers to easily compile, modify, and/or recalculate the sort of complex budgets, action and profit plans, job cost analyses, sales forecasts, expense reports, and cash flow analyses that would have formerly taken a trained accountant many hours to perform by hand.

VisiCalc implemented a basically simple idea: simulating the manual spreadsheet (or ledger sheet) format on a CRT screen. Bricklin and Frankston divided the screen into rows and columns (cells) into which they could insert values. Users could change a value in any row or column, then ask the computer to calculate the effects that value would have on all other rows and columns. What would take hours to do by hand could now be done in seconds, making it possible for users to "play" with data and test the outcomes of hypothetical decisions.

Newer so-called "second generation" spreadsheets have dramatically surpassed the memory limitations of the first generation, and include such user-friendly features as extensive prompts, built-in instructions, and displays of the status of each spreadsheet location. New spreadsheets also offer expanded formatting capabilities, which provide automatic assignment and adjustment of cells, as well as thorough entry validation. Users can even consolidate multiple spreadsheet models and employ state-of-the art graphics and database management.

Several of the newer products have eliminated the need to understand operating systems and unfamiliar jargon, while still others employ "soft function" or "action" keys, instead of confusing combinations of conrol keys to effect a process. Some have even been adapted for minicomputers and mainframes, an ironic turn of events, because most state-of-the-art computer products began in large-scale environments and cascaded downward to the micros.

Spreadsheets have become so integral for personal computers for home and office that some hardware manufacturers, such as Osborne and Coleco, have built them into their machines to gain a competitive edge. Regardless of the enhancements spreadsheets will offer in the coming years, they have become powerful soldiers in the computer revolution.

Source: dataBASE, vol. 7, no. 4, Aug. 1983. Reprinted by permission.

CHAPTER 4 TOOLS FOR THE SYSTEMS DESIGNER

DATA-FLOW DIAGRAMS

We discussed data-flow diagrams in Chapters 1 and 2 but they are such useful tools that they deserve a more detailed discussion. **Data-flow diagrams** offer alternatives to the VTOC and the system's flowchart. Unlike IPO's and detail flowcharts, however, data-flow diagrams do not supply detailed descriptions of modules but graphically describe a system's data and how the data interact with the system. Sometimes we call them data-flow graphs, bubble charts, or Petri networks.

To construct data-flow diagrams, we use arrows, circles, open-ended boxes, and squares (Figure 4.14*a*). Arrows show the flow of data between the other symbols. Like the rectangle in flowcharts, circles stand for a process that converts data into information. An open-ended box represents a file of data, and a square shows a source of data (such as an invoice or a customer's receipt), or a sink (a recipient) of data. Squares, circles, and files receive names, as do arrows, unless they point to or from a file line. Compare the data-flow diagram in Figure 4.14*b* with the system flowchart in Figure 4.10, both of which depict the program for printing AP checks.

The following seven rules govern construction of data-flow diagrams:

1. Arrows should not cross each other.
2. Squares, circles, and files must bear names.
3. Decomposed data flows must be balanced (all data flows on the decomposed diagram must reflect flows in the original diagram).

Data-flow diagram Graphical description of a system's data and how the processes transform the data.

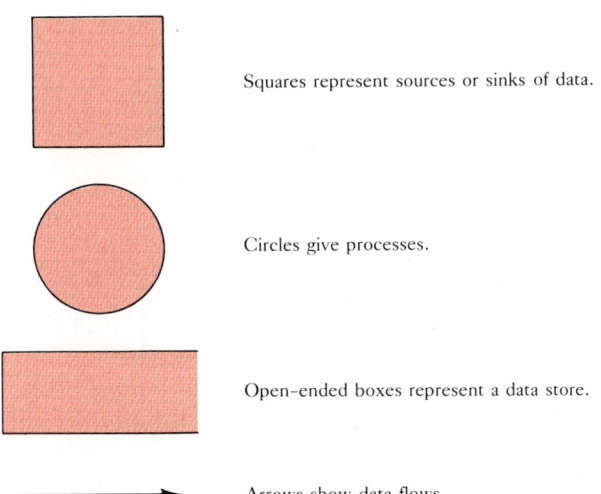

Squares represent sources or sinks of data.

Circles give processes.

Open-ended boxes represent a data store.

Arrows show data flows.

(a) Data-flow diagrams can be drawn with four symbols.

FIGURE 4.14. *A DFD shows flow of data rather than processes, procedures, or controls. Notice how the components bear numbers, just as modules do. However, they do not match the VTOC numbers because the two tools do not parallel one another perfectly.*

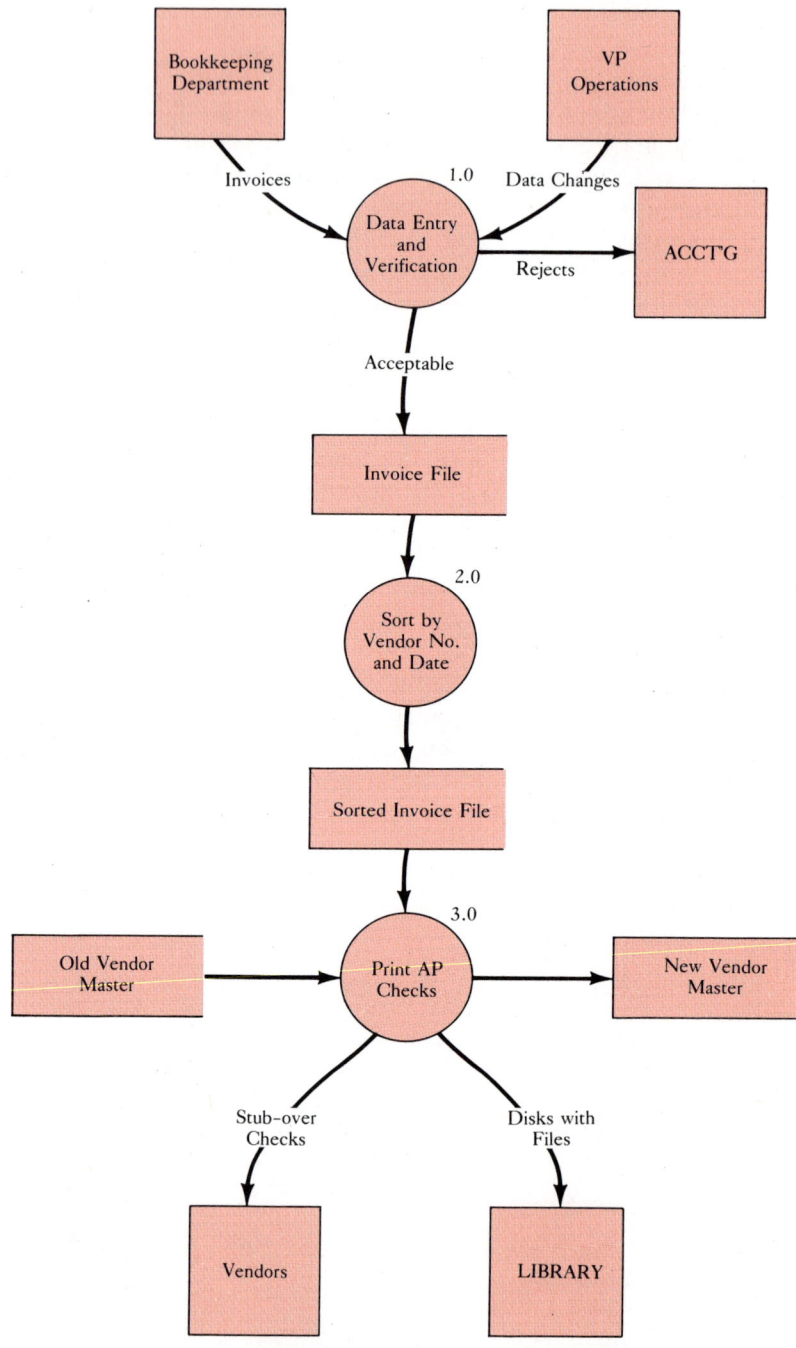

(b) Accounts payable data-flow diagram showing inputs of invoices and data changes. Outputs are stub-over checks.

FIGURE 4.14 *(Continued).*

CHAPTER 4 TOOLS FOR THE SYSTEMS DESIGNER **127**

4. No two data flows, squares, or circles can have the same name.
5. Draw all data flows around the outside of the diagram.
6. Choose meaningful names for data flows, processes, and data stores. Use strong verbs followed by nouns. Avoid words such as collect, open, and close.
7. Control information such as record counts, passwords, and validation requirements are not pertinent to a data-flow diagram.

If too many events seem to be occurring at a given point, an analyst can decompose a data conversion (circle). Figure 4.15 shows how we might decompose the AP check circle in Figure 4.14*b* into two circles. The new data conversions form a parent-child relationship with the original data conversion: the child circle in Figure 4.15 belongs to the parent in Figure 4.14. Devotees of data-flow diagrams insist that no other analyst's tool expresses so fully the flow of data. After all, don't computer people begin with data flow rather than the processing of the data? Another strong advantage is the balancing feature that builds in an error-detection system other tools lack. For example, if a parent data-flow diagram shows three inputs and two outputs, the leveled child diagrams taken together must have three

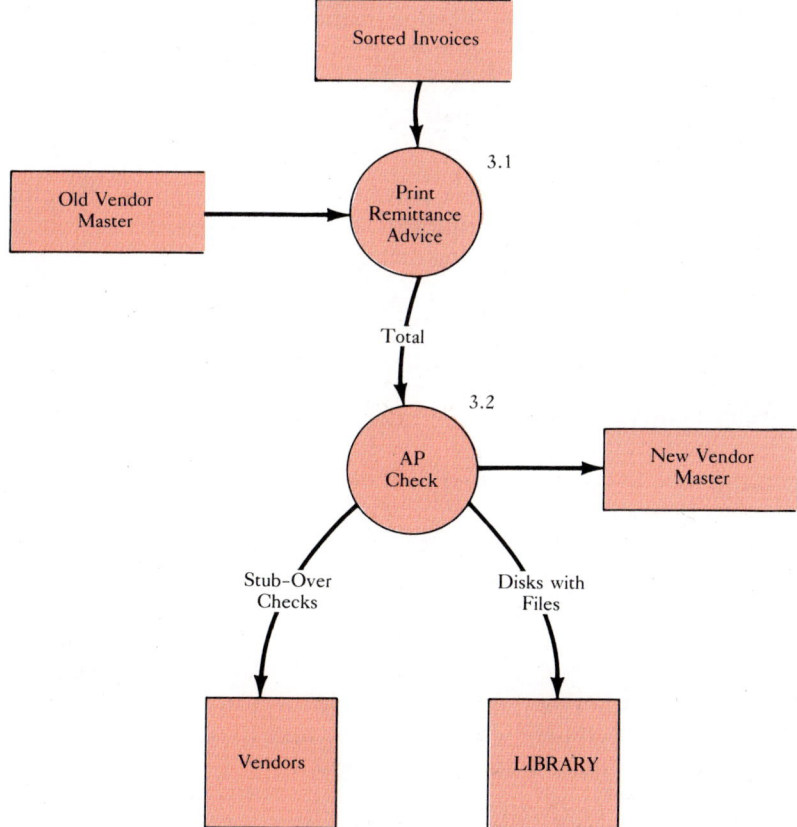

FIGURE 4.15. We can decompose DFDS into more detailed DFDs. We number the decomposed DFD to show how it relates to its parent circle.

inputs and two outputs. If there is an imbalance between parent and child data-flow diagrams, an error exists in either the parent or child diagram.

On the other hand, some professionals feel that the data-flow diagram's inattention to processing is a disadvantage. Flowchart arrows lead the reader from one processing activity to another, whereas data-flow diagrams show only the data inputs and outputs, omitting processing descriptions. Detractors also suggest that data-flow diagrams are difficult for both users and analysts to learn. However, though new and different, data-flow diagrams do offer yet another valuable tool to the analyst.

DATA DICTIONARY

Data dictionary Definition of each term, data item, or file used during analysis or design.

A **data dictionary** defines each term (called a data element) encountered during the analysis and design of a new system. Data elements can describe files, data flows, or processes. For example, suppose you want to print the vendor's name and address at the bottom of a check. The data dictionary might define "vendor's name and address" as follows

Vendor name and address = Vendor name +
Street +
City +
State +
Zip

This explicit definition becomes a part of the data dictionary that ultimately will list all key terms used to describe various data flows and files.

A data dictionary uses the following major symbols:

```
=   Equivalent to
+   And
[]  Either/or
()  Optional entry
```

Four rules govern the construction of data dictionary entries:

1. Words should be defined to stand for what they mean; use VENDOR NUMBER not XYZPQR or DATA13. Capitalization of words helps them to stand out and may be of assistance.
2. Each word must be unique; we cannot have two definitions of vendor name.
3. Aliases, or synonyms, are allowed when two or more entries show the same meaning; a vendor number may also be called a customer number. However, aliases should be used only when absolutely necessary.
4. Self-defining words should not be decomposed.

We can even decompose a dictionary definition. For instance, we might write:

Vendor name = Company name,
Individual's name

which we might further decompose to:

Company name = (Contact) +
Business name

Individual's name = Last name +
First name +
(Middle initial)

After defining a term, say VENDOR NUMBER, we list any aliases or synonyms, describe the term verbally, specify its length and data type, and list the data stores where the term is found (Figure 4.16). Some terms may have no aliases, may be found in many files, or may be limited to specific values.

Some self-defining or obvious words and terms may not require inclusion in the data dictionary. For example, we all know what a Zip code and a middle initial are. Data dictionaries seldom include information such as the number of records in a file, the frequency a process will run, or security factors such as passwords users must enter to gain access to sensitive data. Rather, data dictionaries offer definitions of words and terms relevant to a system, not statistical facts about the system.

Data dictionaries allow analysts to define precisely what they mean by a particular file, data flow, or process. Some commercial software packages, usually called Data Dictionary Systems (or DDS), help analysts maintain their dictionaries with the help of the computer. These systems keep track of each term, its definition, which systems or programs use the term, aliases, the number of times a particular term

DATA ELEMENT NAME:	VENDOR-NUMBER
ALIASES:	None
DESCRIPTION:	Unique identifier for vendors in the accounts payable system.
FORMAT:	Alphanumeric, six characters.
DATA FLOWS:	Vendor master Accounts payable open item Accounts payable open adjustments Check reconciliation
REPORTS:	Alphabetic vendor list Numeric vendor list A/P transaction register Open item Vendor account inquiry Cash requirements Pre-check-writing Check register Vendor analysis

FIGURE 4.16. Data dictionaries list the completely decomposed term defined as well as other pertinent facts.

is used, and the size (number of bytes of memory allocated) of the term, and can be tied to commercial data-base managers.

DECISION TABLES AND DECISION TREES

Decision tables and trees were developed long before the widespread use of computers. They not only isolate many conditions and possible actions, but they help ensure that nothing has been overlooked. Decision tables and trees do not use "computerese" and may be familiar to you from some other discipline (management, advertising, or accounting, for example).

Decision table A chart with four sections listing all the logical conditions and actions.

A **decision table** depicts both simple and complex module logic. It is a verbal chart listing all the theoretically possible conditions and actions that the analyst must consider. As shown in Figure 4.17a, a decision table contains <u>five</u> main sections. The top section allows space for the title, date, author, system, and pertinent comments.

Condition stub A list of all the necessary tests in a decision table.

The **condition stub** displays all the necessary tests or conditions. Like the diamond in a flowchart or the IF in pseudocode, these tests require yes or no answers. The condition stub always appears in the upper left-hand corner of the decision table, with each condition numbered to allow easy identification.

Action stub A list of all the processes involved in a decision table.

In the lower left-hand corner of the decision table we find the **action stub,** where one may note all the processes desired in a given module. Actions, like conditions, receive numbers for identification purposes.

Condition entry A list of all the yes/no permutations in a decision table.

The upper right corner provides space for the **condition entry**—all possible permutations of yes and no responses related to the condition stub. The yes or no possibilities are arranged as a vertical column called rules. Rules are numbered 1, 2, 3, and so on. We can determine the number of rules in a decision table by the formula:

$$\text{Number of rules} = 2 \wedge N = 2^N$$

where N represents the number of conditions and $\wedge$ means exponentiate. Thus a decision table with four conditions has 16 ($2\wedge 4 = 2 \times 2 \times 2 \times 2 = 16$) rules, one with six conditions has 64 rules, and eight conditions yield 256 rules.

Action entry Indicates via dot or X whether something should happen in a decision table.

The lower right corner holds the **action entry.** X's or dots indicate whether an action should occur as a consequence of the yes/no entries under condition entry. X's indicate action; dots indicate no action.

Returning to the assembly of a bicycle, let us assume we must assemble a variety of containers full of parts. Since a bike can have either hand caliper or foot coaster brakes, the decision table must show the two conditions and five actions (Figure 4.18a). The two conditions necessitate four condition entries, and the five actions produce 20 possible action entries.

When we build the yes or no rules for the condition entry, we must construct all possible patterns of y's and n's. An arrangement that guarantees thoroughness is to place two y's in succession followed by two n's. In the second row, we place alternating pairs of y's and n's.

A decision table with four conditions ($2\wedge 4 = 16$) would have 16 different sets of y's and n's and would result in the following pattern of yes and no responses.

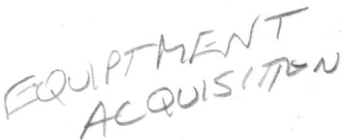

CHAPTER 4 TOOLS FOR THE SYSTEMS DESIGNER   131

TITLE:	DATE:
Author:	System:
Comments:	
Condition Stub	Condition Entry
Action Stub	Action Entry

(a) Five sections of a decision table.

SCHEDULE X—Single Taxpayers Not Qualifying for Rates in Schedule Y or Z				SCHEDULE Y—Married Taxpayers and Qualifying Widows and Widowers If you are a married person living apart from your spouse, see page 5, paragraph 1(d), of the instructions to see if you can be considered as "unmarried" for purposes of using Schedule X or Z.								SCHEDULE Z—Unmarried (or legally separated) Taxpayers Who Qualify as Heads of Household (See page 6)			
Use this schedule if you checked the box on Short Form 1040A, line 1.				Use this schedule if you checked the box on Short Form 1040A, line 2 or 5.				Use this schedule if you checked the box on Short Form 1040A, line 3.				Use this schedule if you checked the box on Short Form 1040A, line 4.			
				Married Taxpayers Filing Joint Returns and Qualifying Widows and Widowers (See pages 5 and 6)				Married Taxpayers Filing Separate Returns							
If the amount on line 15 is:		Enter on line 16:		If the amount on line 15 is:		Enter on line 16:		If the amount on line 15 is:		Enter on line 16:		If the amount on line 15 is:		Enter on line 16:	
Over—	But not over—		of the amount over—	Over—	But not over—		of the amount over—	Over—	But not over—		of the amount over—	Over—	But not over—		of the amount over—
$20,000	$22,000	$5,230+38%	$20,000	$20,000	$24,000	$4,380+32%	$20,000	$20,000	$22,000	$6,070+48%	$20,000	$20,000	$22,000	$4,800+35%	$20,000
$22,000	$26,000	$5,990+40%	$22,000	$24,000	$28,000	$5,660+36%	$24,000	$22,000	$26,000	$7,030+50%	$22,000	$22,000	$24,000	$5,500+36%	$22,000
$26,000	$32,000	$7,590+45%	$26,000	$28,000	$32,000	$7,100+39%	$28,000	$26,000	$32,000	$9,030+53%	$26,000	$24,000	$26,000	$6,220+38%	$24,000
$32,000	$38,000	$10,290+50%	$32,000	$32,000	$36,000	$8,660+42%	$32,000	$32,000	$38,000	$12,210+55%	$32,000	$26,000	$28,000	$6,980+41%	$26,000
$38,000	$44,000	$13,290+55%	$38,000	$36,000	$40,000	$10,340+45%	$36,000	$38,000	$44,000	$15,510+58%	$38,000	$28,000	$32,000	$7,800+42%	$28,000
$44,000	$50,000	$16,590+60%	$44,000	$40,000	$44,000	$12,140+48%	$40,000	$44,000	$50,000	$18,990+60%	$44,000	$32,000	$36,000	$9,480+45%	$32,000
$50,000	$60,000	$20,190+62%	$50,000	$44,000	$52,000	$14,060+50%	$44,000	$50,000	$60,000	$22,590+62%	$50,000	$36,000	$38,000	$11,280+48%	$36,000
$60,000	$70,000	$26,390+64%	$60,000	$52,000	$64,000	$18,060+53%	$52,000	$60,000	$70,000	$28,790+64%	$60,000	$38,000	$40,000	$12,240+51%	$38,000
$70,000	$80,000	$32,790+66%	$70,000	$64,000	$76,000	$24,420+55%	$64,000	$70,000	$80,000	$35,190+66%	$70,000	$40,000	$44,000	$13,260+52%	$40,000
$80,000	$90,000	$39,390+68%	$80,000	$76,000	$88,000	$31,020+58%	$76,000	$80,000	$90,000	$41,790+68%	$80,000	$44,000	$50,000	$15,340+55%	$44,000
				$88,000	$100,000	$37,980+60%	$88,000					$50,000	$52,000	$18,640+56%	$50,000
				$100,000	$120,000	$45,180+62%	$100,000					$52,000	$64,000	$19,760+58%	$52,000
				$120,000	$140,000	$57,580+64%	$120,000					$64,000	$70,000	$26,720+59%	$64,000
				$140,000	$160,000	$70,380+66%	$140,000					$70,000	$76,000	$30,260+61%	$70,000
						$83,580+68%	$160,000					$76,000	$80,000	$33,920+62%	$76,000
												$80,000	$88,000	$36,400+63%	$80,000
												$88,000	$100,000	$41,440+64%	$88,000
												$100,000	$120,000	$49,120+66%	$100,000
													$140,000	$62,320	

(b) A tax table is a type of decision table people use on a periodic basis.

FIGURE 4.17. A decision table contains five sections. We are all familiar with airline, bus, train, or tax tables, which are all examples of decision tables.

TITLE: Bicycle Assembly	DATE: Sept. 25, 1984
Author: Gary Hatch	System: Bicycle
Comments: More than one carton of parts needs to be assembled.	

	1	2	3	4
1. Last carton?	y	y	n	n
2. Hand brakes?	y	n	y	n
1. Open container	•	•	x	x
2. Stack parts	•	•	x	x
3. Assemble wheels	•	•	x	x
4. Finish assembly	•	•	x	x
5. End of assembly operations	x	x	•	•

(a) Decision table for bicycle assembly.

TITLE: AP Check	DATE: Sept. 25, 1984
Author: Gary Hatch	System: Accounts Payable System
Comments: Two files are to be read until the end of file.	

	1	2	3	4	5	6	7	8
1. End of vendor master file?	y	y	y	y	n	n	n	n
2. End of sorted invoice file?	y	y	n	n	y	y	n	n
3. Do vendor numbers match?	y	n	y	n	y	n	y	n
1. Read a vendor master record	•	•	•	•	•	•	•	x
2. Read an invoice record	•	•	•	•	•	•	x	•
3. Add amount to total	•	•	•	•	•	•	x	•
4. Print invoice detail line	•	•	•	•	•	•	x	•
5. Print date line	•	•	•	•	•	•	•	x
6. Print amount in words	•	•	•	•	•	•	•	x
7. Print vendor name/address	•	•	•	•	•	•	•	x
8. End of module	x	x	•	•	•	•	•	•

(b) AP check decision table. We call this type of table limited entry because the condition entry contains yes or no responses for each rule.

FIGURE 4.18. Decision table for bicycle assembly and the AP check module. Note the bifurcated yes/no pattern in the decision entry.

CHAPTER 4   TOOLS FOR THE SYSTEMS DESIGNER   **133**

TITLE: AP Check				DATE: Sept. 25, 1984			
Author: Gary Hatch				System: Accounts Payable System			
Comments: Two files are to be read until the end of file.							

	1	2	3	4	5	6	7	8
1. Vendor master file?	End	End	End	End	More	More	More	More
2. Sorted invoice file?	End	End	More	More	End	End	More	More
3. Vendor numbers?	End	More	End	More	End	More	End	More
1. Read a vendor master record	•	•	•	•	•	•	•	x
2. Read an invoice record	•	•	•	•	•	•	x	•
3. Add amount to total	•	•	•	•	•	•	x	•
4. Print invoice detail line	•	•	•	•	•	•	x	•
5. Print date line	•	•	•	•	•	•	•	x
6. Print amount in words	•	•	•	•	•	•	•	x
7. Print vendor name/ address	•	•	•	•	•	•	•	x
8. End of module	x	x	•	•	•	•	•	•

(c) AP check written as an extended-entry decision table.

TITLE: AP Check				DATE: Sept. 25, 1984			
Author: Gary Hatch				System: Accounts Payable System			
Comments: Two files are to be read until the end of file.							

	1	2	3	4	5	6	7	8
1. Is vendor master file at end?	y	y	y	y	n	n	n	n
2. Sorted invoice file?	End	End	More	More	End	End	More	More
3. Vendor numbers?	End	More	End	More	End	More	End	More
1. Read a vendor master record	•	•	•	•	•	•	•	x
2. Read an invoice record	•	•	•	•	•	•	x	•
3. Add amount to total	•	•	•	•	•	•	x	•
4. Print invoice detail line	•	•	•	•	•	•	x	•
5. Print date line	•	•	•	•	•	•	•	x
6. Print amount in words	•	•	•	•	•	•	•	x
7. Print vendor name/ address	•	•	•	•	•	•	•	x
8. End of module	x	x	•	•	•	•	•	•

(d) AP check written as a mixed-entry decision table.

FIGURE 4.18 (Continued)

	1 2 3 4 5 6 7 8
1. End of vendor master file? 2. End of sorted invoice file? 3. Do vendor numbers match?	y y y y n n n n y y n n y y n n y n y n y n y n
1. Read a vendor master record 2. Read an invoice record 3. Add amount to total 4. Print invoice detail line 5. Print date line 6. Print amount in words 7. Print vendor name/address 8. End of module 9. Go to next module	• • • • • • • x • • • • • • x • • • • • • • x • • • • • • • x • • • • • • • • x • • • • • • • x • • • • • • • x x x • • • • • • x x • • • • • •

TITLE: AP Check DATE: Sept. 25, 1984
Author: Gary Hatch System: Accounts Payable System
Comments: Two files are to be read until the end of file.

(e) AP check written as an open-ended decision table.

FIGURE 4.18 *(Continued)*

Limited entry A type of decision table listing a y or n response for each condition.

Extended entry Type of decision table displaying values to be tested in the condition entry.

Mixed entry A type of decision table mixing values in the condition and action entries.

Open ended (1) A type of decision table that permits access to another decision table. (2) Questionnaire items that respondents must answer in their own words.

The first row therefore will have eight y's followed by eight n's. The second row (corresponding to the second entry in the condition stub) has four y's, four n's, four y's, and four n's. The complete four-condition entry would read:

```
y y y y y y y y n n n n n n n n
y y y y n n n n y y y y n n n n
y y n n y y n n y y n n y y n n
y n y n y n y n y n y n y n y n
```

This form ensures that the analyst includes all combinations without duplication.

If very many conditions exist (four conditions result in 16 condition entries, six conditions in 64), decision tables can become unwieldly. To avoid lengthy decision tables, analysts must remove redundancies and yet still take precautions not to overlook anything. On occasion, two or more rules may be combined to reduce or eliminate redundancy. In Figures 4.18*a* and 4.18*b*, rules 1 and 2 cause the last action in the action stub to occur. Therefore, these two rules could be combined to eliminate redundancy. To indicate redundancy, we put a dash (—) in the condition entry to show that this condition stub is irrelevant and can be ignored.

The decision table in Figure 4.18*b* depicts the AP check module. Compare it with Figure 4.11 (IPO), Figure 4.5 (detail flowchart), and Figure 4.13 (pseudocode). Although this format is fairly typical, in practice you will encounter several different kinds of decision tables. Figure 4.18*b* is called **limited entry,** because the condition entry contains yes or no responses for each rule.

An **extended-entry** decision table displays the values of the tested conditions in the condition entry (Figure 4.18*c*). A **mixed-entry** decision table combines the value and yes or no (Figure 14.18*d*), while an **open-ended** one allows an action

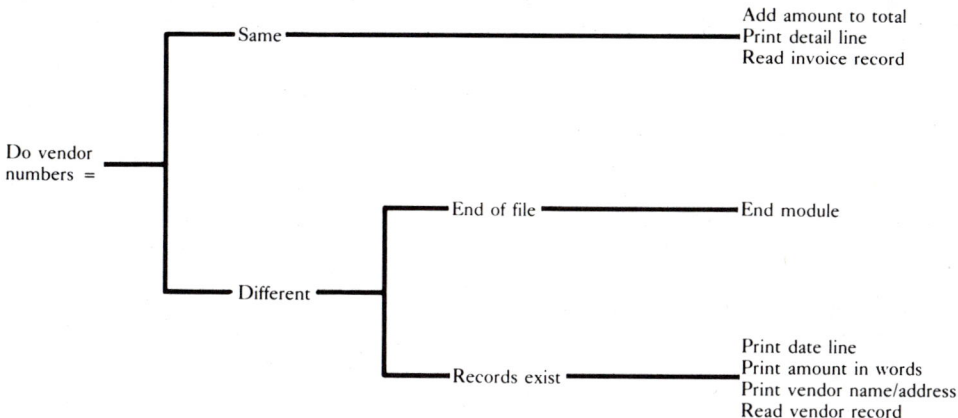

FIGURE 4.19. *Decision trees are graphic equivalents of decision tables.*

entry specifying an additional decision table (Figure 4.18e). An analyst may want to use one of these other types of decision tables to make the table more readable for a user or manager or to decompose a large (seven conditions leading to 128 rules) table into a series of smaller ones.

Decision trees turn a decision table into a diagram (Figure 4.19). This tool is read from left to right, decisions result in a fork, and all branches end with an outcome. Figure 4.19 shows the decision tree for printing the accounts payable check. Trees can be easily read by nontechnical users who find decision tables too complex. Users readily grasp branches, forks, and outcomes.

Decision tree A graph representation of a decision table.

WARNIER-ORR DIAGRAMS

Warnier-Orr diagrams (pronounced warn yay or) are yet another tool aimed at producing working and correct programs. The Warnier-Orr diagram takes its name from its codevelopers, Jean-Dominique Warnier and Kenneth Orr. Unlike VTOCs, pseudocode, or flowcharts, which read from the top down and then from left to right, the Warnier-Orr diagram reads from left to right, then from the top down. Whereas a flowchart requires many symbols, Warnier-Orr diagrams employ brackets, circles, parentheses, dots, and bars. Diagrams can depict data-dictionary-type definitions (Figure 4-20a) or detailed program logic (Figure 4.20b).

The Warnier-Orr uses brackets to group related elements following the sequence control structure. Technically the symbol for a bracket is "[" and a brace is "{" but Warnier-Orr calls "{" a bracket. Thus we see that three elements in Figure 4.20a make up the single element "Systems Process." Two elements, preliminary and detailed, make up Analysis.

Iteration structure (called repetition in the Warnier-Orr notation) is depicted by a parenthesis to the left of the series of elements to be repeated (Figure 4.20b). A number inside the parentheses indicates the amount of times the iteration should be performed. Thus we will repeat the four bracketed elements until there are no more bicycles to assemble.

Warnier-Orr diagram A design tool that uses brackets to group related items or operations.

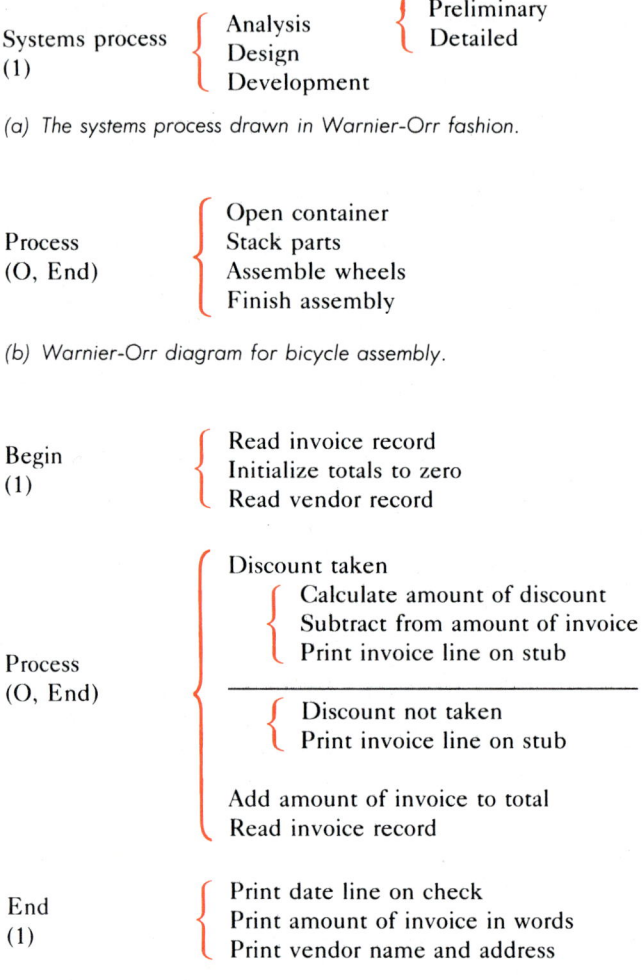

(a) The systems process drawn in Warnier-Orr fashion.

(b) Warnier-Orr diagram for bicycle assembly.

(c) Warnier-Orr diagram for the accounts payable stub-over-check module. This diagram shows the three control structures of sequence, selection, and repetition.

FIGURE 4.20. A Warnier-Orr diagram reads from left to right, and then from the top to the bottom.

A plus sign enclosed in a circle indicates an alternation or selection structure. A bar separates the decision into true (above the bar) and false (below the bar); see Figure 4.20c.

For our AP check-printing logic, the Warnier-Orr diagram (Figure 4.20c) has brackets surrounding the repetitive operations and a decision point inside the process section to determine whether a discount will be taken. This diagram also has beginning and ending logic that our bicycle example does not require.

Warnier-Orr diagrams show the beginning, processing, and ending parts of the detailed logic quite explicitly. In keeping with the structured methodology, they have a single entry and a single exit, they support the three control structures, and, compared with other tools, they employ few symbols. Disadvantages of the Warnier-Orr system include its left-to-right, top-down construction (the opposite of all other tools, which are top down, left to right) and its focus on processing versus data flow.

NASSI-SHNEIDERMANN CHART

Another detail or logic technique is the **Nassi-Shneidermann chart,** sometimes called the Chapin chart. This system was developed by and named for I. Nassi and B. Shneidermann in the early 1970s and differs markedly from those we

Nassi-Shneidermann chart A tool that divides rectangles into halves to depict logical structures.

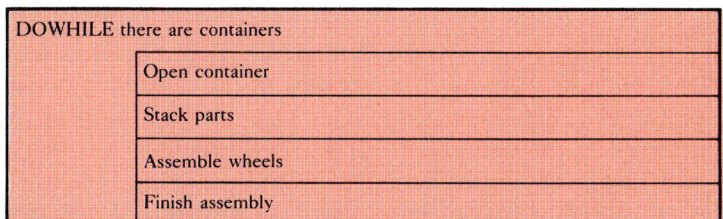

(a) Nassi-Shneidermann chart for bicycle assembly.

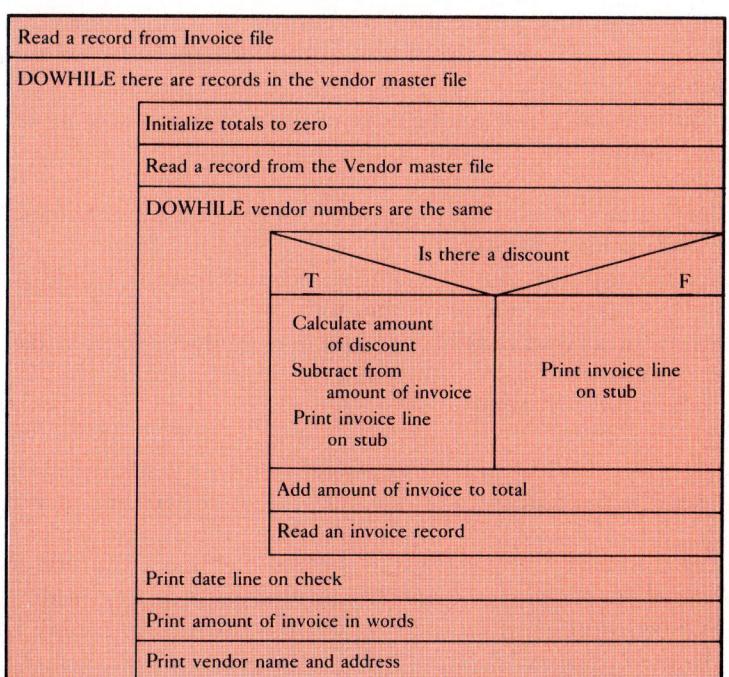

(b) Nassi-Shneidermann chart for AP check-printing logic.

FIGURE 4.21. Boxes, rectangles, and DOWHILE comprise the Nassi-Shneidermann chart. The DO-WHILE can appear at the top of the box or in the middle.

have examined thus far. It uses rectangles divided into halves with an angular line for selection, a horizontal rectangle for sequence, and the word DOWHILE for iteration. The bicycle-assembly chart appears in Figure 4.21a, and the AP check-printing system with discounting in Figure 4.21b.

Nassi-Shneidermann charts are winning wider acceptance in the computer industry because they so simply illustrate complex logic. Perhaps as analysts evaluate the various tools at their disposal, these charts will gain even more followers. As with other structured tools, they are single entry and single exit, and efficiently accommodate modules. However, they are not as useful for conveying system flow as they are for detailing logic development and they are difficult to maintain.

STRUCTURED WALKTHROUGHS

Structured walkthrough A peer review of a system conducted to uncover errors.

The last tool we will examine offers a dramatically different approach. A **structured walkthrough** is a controlled peer review of the developing system during which the analyst or programmer can obtain an honest appraisal of the system. Management does not attend this review process and should never use it to evaluate the staff. Otherwise people may feel reluctant to be perfectly candid.

Reviewers receive copies of programs, data-flow diagrams, flowcharts, data dictionaries, and other pertinent material prior to the walkthrough and are asked to provide their immediate reactions. The emphasis is on detection of errors, not on correcting them. Walkthroughs often include personnel at all levels: junior programmers and programmer-analysts, as well as analysts. Nontechnical staff members, with the exception perhaps of data-entry personnel, do not participate in the walkthrough.

Conducted during all phases of the systems cycle, these reviews strive to achieve the following goals:

1. To uncover and remove errors, omissions, and misunderstandings in specifications, terminology, or coding.
2. To familiarize the staff with the system and to seek expert advice.
3. To serve as a teaching device for the staff.
4. To motivate the analyst and the programmer to produce higher quality systems.
5. To promote uniformity across the department and systems.
6. To serve as a project management tool.

Analysis walkthrough System review at the end of analysis conducted to uncover analysis errors.

Design walkthrough System review at the end of design conducted to uncover design errors.

The focus of the walkthrough differs depending on the phase being reviewed. **Analysis walkthroughs** center on the analyst's understanding of the problem, the correctness of the data-flow diagram depicting the system under study, and the proposed solution. Attendees at this walkthrough include users, the analyst, and other members of the computer services staff participating in the analysis.

Design walkthroughs concentrate on completeness, ascertaining whether all components mesh as planned. Reports must give users wanted data, the data base must store all required data elements, data collection must capture data (checking for errors as the data are collected), programs must be identified and modules defined, equipment must be specified and ordered, and costs must be under control. Attendees at this walkthrough include people from the analysis and programming staffs.

Development walkthroughs ensure that the completed system will comply with the system as originally planned. This walkthrough is held just before the system is turned over to the user and should review manuals, training materials, and system documentation. Again, users, the analyst, and computer services staff (manager, programmer, and operations) attend this meeting.

During the walkthrough session, the reviewee gives a brief overview or tutorial of the material, and the main programmer controls and monitors the discussions, recording errors, omissions, or suggested improvements. Prerequisites to a successful session include limiting the number of participants (no more than six or fewer than three), distributing materials at least a day in advance, limiting the session to an hour, emphasizing the error-detection function of the session, and assuring the absence of management personnel.

The leader of the session must watch out for trivial discussion, overcriticism, lack of preparation by the participants (reviewers and reviewee), and the aggressive superprogrammer who will try to take over the walkthrough, and also must remember that the session is not a meeting or a walk-on. If it develops that the materials are sufficiently deficient to warrant it, the leader should terminate the session and reschedule it when the deficiencies have been repaired.

Analysts unaccustomed to having their work examined by their peers may feel threatened by a walkthrough, but the results always prove valuable. Some fears often expressed are that the walkthrough is a witch hunt conducted by the reviewers or an indictment against the reviewee, that the atmosphere is one of conflict and hostility, or the reviewers are inquisitors. However, if conducted in an objective and nonjudgmental fashion, walkthroughs can even be enjoyable because a true professional always wants to learn and will not suffer ego damage upon hearing *constructive* criticism. The walkthrough also provides an opportunity, of course, to show off impressive work.

Development walkthrough Review of the completed system to determine if it complies with original plan.

CASE STUDY

Data-Flow Diagrams, the Data Dictionary, and Pseudocode for Fleet Feet's Accounts Payable System

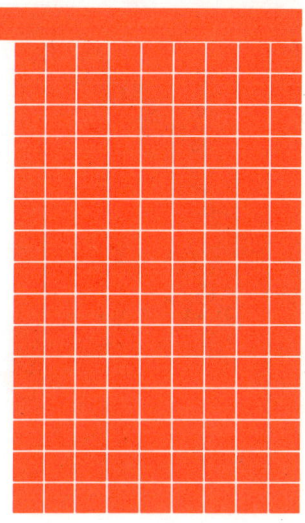

When initially assigned to the analysis and design of Fleet Feet's new accounts payable system, Frank Pisciotta first described the flow of data through it. After weighing the various tools at his disposal, Frank chose the data-flow diagram because it not only would graphically depict data flow, but it was the tool with which he and the Fleet Feet management were most familiar. During his 2 years with the company, he had presented data-flow diagrams so frequently that most employees and managers felt comfortable with them.

The Fleet Feet AP procedure would start when an invoice arrives at the warehouse from a vendor, and it would end when Mark Stensaas' bookkeeping staff sends a check to the vendor. First, Frank drew an overview of the system, labeling it 1.0 (Figure 4.22). Though

CASE STUDY

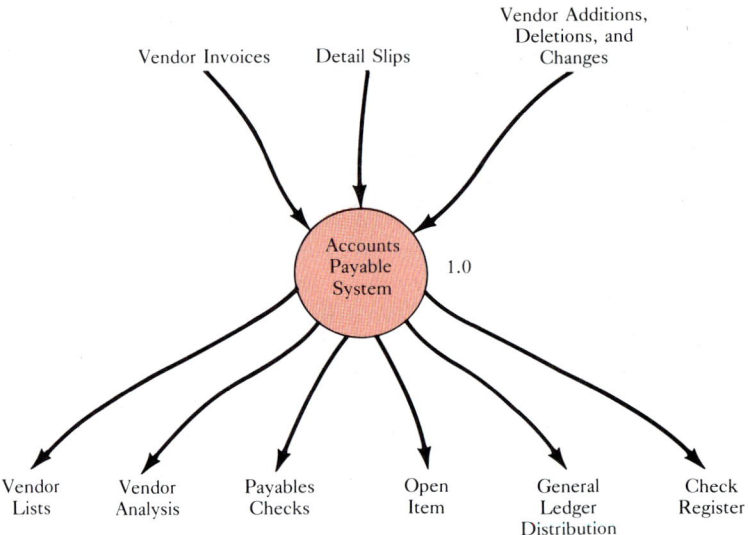

FIGURE 4.22. *Data-flow diagram for Fleet Feet's accounts payable system. This DFD would be especially useful to management because it shows data flow from a broad perspective.*

this diagram was too broad to be useful for actual system design, it would be clear and meaningful to Fleet Feet's workers and managers. Next, Frank decomposed the diagram into the components shown in Figure 4.23. Each new circle received its own number (1.1, 1.2, 1.3, etc.), lending visual emphasis to the dependent relationships between various processes.

Since two data files would provide input into the system, Frank wrote a data dictionary for each (Figure 4.24). Frank decided to create data dictionaries for the same reasons he chose the data-flow diagram: their effectiveness, his personal bias, and his colleagues' familiarity with them.

Finally, Frank converted the diagrams and dictionaries into pseudocode (Figure 4.25). For consistency he linked the pseudocode descriptions to the data-flow diagram with process numbers. Frank selected pseudocode to show the logic of the system because its verbal nature makes it accessible to noncomputer people. Before coming to Fleet Feet, he had relied on flowcharts, but he switched to pseudocode because he could write the logic with his Hewlett Packard 3000 text editor, which allows him to update the code easily. Frank also used the HP 3000 text editor to record concise data dictionary definitions.

The three resulting documents became part of the systems documentation, which Frank put in a folder for storage. He retained early drafts of his data-flow diagrams and dictionaries in the same folder to create a permanent historical record.

CHAPTER 4 TOOLS FOR THE SYSTEMS DESIGNER **141**

CASE STUDY

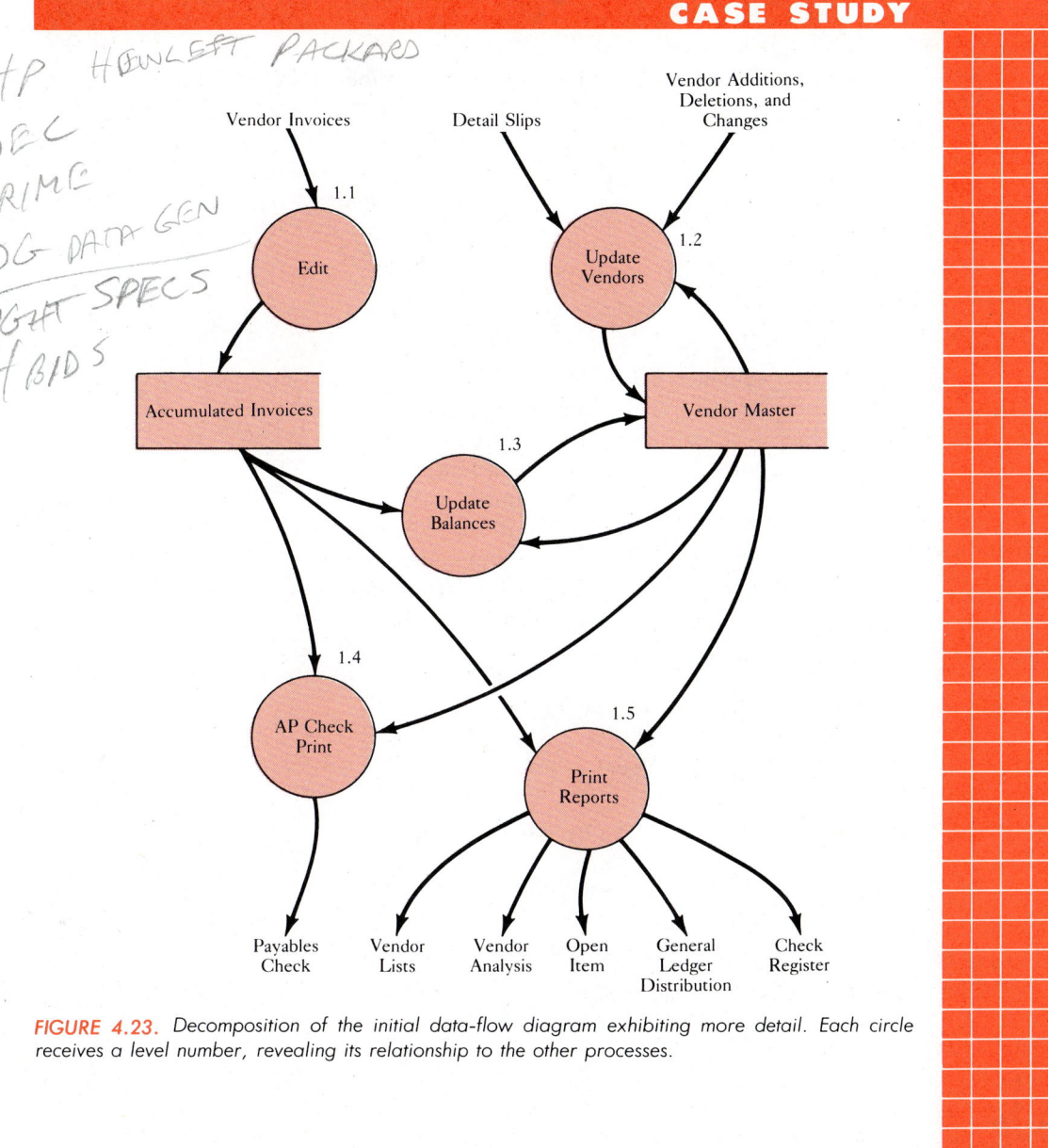

FIGURE 4.23. Decomposition of the initial data-flow diagram exhibiting more detail. Each circle receives a level number, revealing its relationship to the other processes.

CASE STUDY

Vendor master file = Vendor number +
Vendor name +
Vendor address +
Discount percentage +
Vendor balance +
Telephone number +
Date opened +
Terms +
History

Invoice file = Vendor number +
Invoice date +
Invoice number +
Invoice due date +
PO number +
Amount due +
Discount amount +
General ledger account number +
Batch number +
Date entered

Date = Month +
Day +
Year

Month = (1 through 12)

Day = (1 through 31)

Year = (1982 through 1989)

FIGURE 4.24. *Data dictionary for the two inputs to Fleet Feet's AP system. As with the DFD, Frank has decomposed some items to display more detail.*

CHAPTER 4 TOOLS FOR THE SYSTEMS DESIGNER   **143**

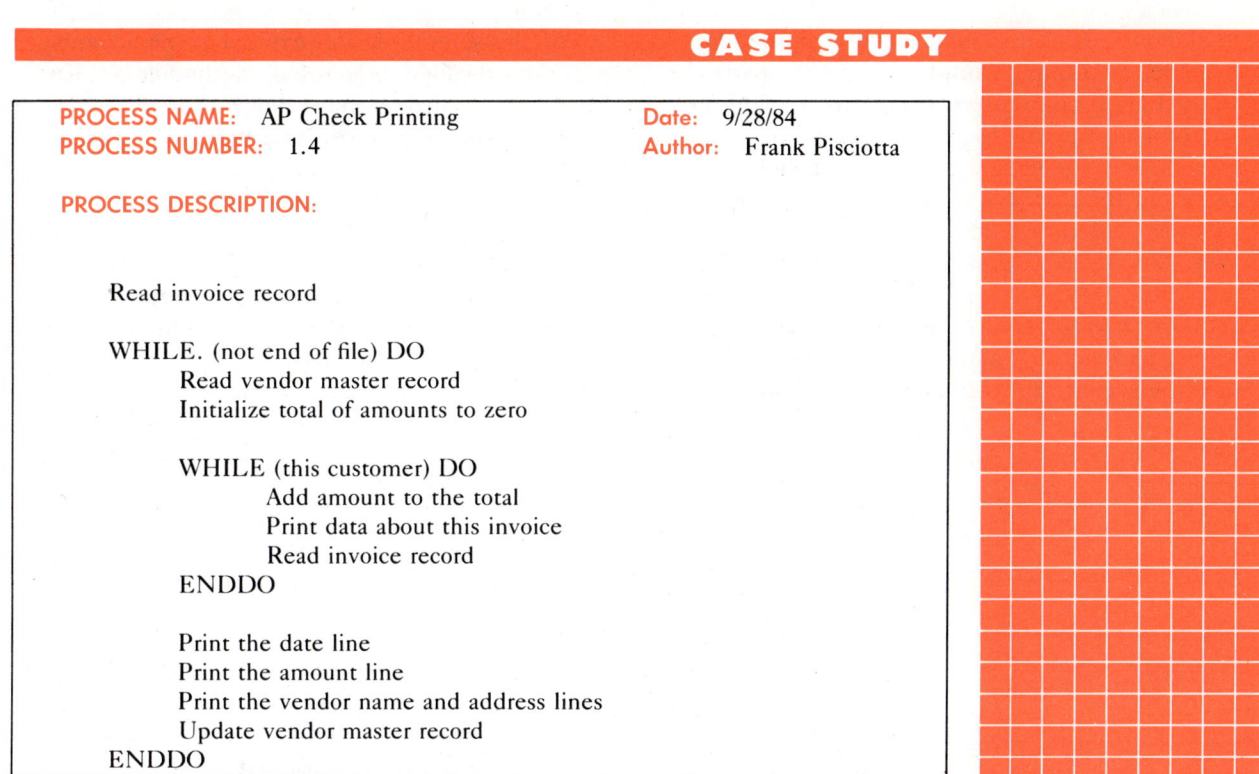

FIGURE 4.25. Pseudocode for one process within the AP system.

WORKING WITH PEOPLE   New Tools Versus Old

Helen Langley, data processing manager for the Bay Area Telephone Company, had decided to send her key analyst, Kurt Wurst, to a data-base management systems training seminar in Baltimore. Helen hoped the knowledge Kurt would pick up would help her staff switch from a record file system to a data-base environment.

Kurt looked forward to the seminar and carefully read the manual that the seminar leader, a software vendor, sent after Kurt enrolled in the program. As part of the seminar, each participant would be required to actually analyze, design, and program a small system. Since he had designed many such systems in the past, Kurt felt confident about doing so again and hoped the new software would add an exciting new element to the task.

By the end of the first day of the seminar, Kurt felt he had learned a lot, and he happily took the assignment back to his hotel room, where he spent 3 hours drawing and refining his flowcharts. When he returned the next morning, he noticed that the person sitting next to him had not used flowcharts at all, but an amazing array of tools with which Kurt was only superficially acquainted: a HIPO, a decision table, and a Warnier-Orr diagram. But as the old method had always worked for him, Kurt decided to stick with it rather than resort to some "new-fangled fad."

As the second day's sessions progressed, the seminar leader distributed sample data to the participants, asking them to use it testing their solutions of the problem. Kurt's 15-page solution failed, but his neighbor's 4-page one passed. Kurt took little solace from the fact that most of the other participants also failed. He felt confused and embarrassed. When questioned by the seminar leader, most of the "failures" revealed that they had stayed with their flowcharts, whereas Kurt's successful neighbor had employed modular problem solving, the three control structures, and top-down development.

As the seminar leader explained the advantages of the so-called structured methodology, Kurt's feeling of embarrassment slowly changed to one of appreciation. He vowed that he would master the new techniques, so he would really have something to show to his colleagues at Bay Area Telephone.

Noticing how the seminar leader carefully avoided poking fun at flowcharts by simply offering the new tools as useful alternatives, Kurt decided to use the same nonthreatening approach with his colleagues. He knew they, like himself, would learn much more rapidly without the anxiety or guilt that accompanies the fear of being left behind.

SUMMARY

Increased productivity is a key to success in most organizations, but that only happens when an organization improves its efficiency. To boost the efficiency, and hence the productivity, of analysts and programmers, a number of tools are available. The following chart compares some of the tools that should be included in any analyst's toolbox.

Category	HIPO		Flowcharts		Pseudocode	Data-flow Diagrams	Dictionaries	Nassi-Shneidermann	Warnier-Orr	Walkthrough
	VTOC	IPO	System	Detail						
Analysis	Yes	No	Yes	No	Yes	Yes	Yes	No	No	No
Design	Yes	Yes	Yes	No	No	Yes	Yes	No	No	Yes
Development	Yes	Yes	Yes	Yes	Yes	Yes	Yes	Yes	Yes	Yes
Graphic	Yes	Yes	Yes	Yes	No	Yes	No	Yes	Yes	No
Verbal	No	No	No	No	Yes	No	Yes	Yes	Yes	Yes
Single entry and exit	Yes	Yes	No	No	Yes	No	NA	Yes	Yes	NA
Modularity	Yes	Yes	No	No	Yes	Yes	NA	Yes	Yes	NA
Sequence	Yes	Yes	No	No	Yes	NA	NA	Yes	Yes	NA
Selection	Yes	Yes	No	Yes	Yes	NA	NA	Yes	Yes	NA
Iteration	No	Yes	No	Yes	Yes	NA	NA	Yes	Yes	NA
Top down	Yes	No	No	No	Yes	NA	NA	Yes	Yes	NA
Year of origin	1960s	1960s	1930s	1930s	1970s	1970s	1970s	1960s	1970s	1970s

These tools strive to:

1. Increase the productivity of analysts and programmers.
2. Reduce system and program errors.
3. Reduce system and program maintenance and enhancement costs.
4. Make systems and programs easier to understand and use.

They attempt to make the resultant software more reliable, less expensive, easier to use, and less difficult to change, and to put it in the user's hands on time.

NEW TERMS

Action entry	Design walkthrough	Open ended
Action stub	Detail flowchart	Pseudocode
Analysis walkthrough	Development walkthrough	Selection
Cohesion	Extended entry	Sequence
Condition entry	Flowchart	Span of control
Condition stub	HIPO	Structure chart
Control structure	IPO	Structured English
Coupling	Iteration	Structured walkthrough
Data dictionary	Limited entry	System flowchart
Data-flow diagram	Mixed entry	Top down
Decision table	Module	VTOC
Decision tree	Nassi-Shneidermann chart	

QUESTIONS FOR REVIEW AND DISCUSSION

Questions appear in three categories. You can find the answers to the A group in this chapter. The B group requires you to apply the material presented here, whereas the C group necessitates investigation or research.

A.1. Draw six system flowcharting symbols.
A.2. List the best tools for programmers.
A.3. Describe the tools an analyst would use for design.
A.4. What verbal tools are available to the analyst?
A.5. What do the symbols +, [ ], and ( ) stand for when used to draw a data dictionary definition?
A.6. What graphic tools are available to the analyst?

B.1. Modify the detail flowchart in Figure 4.11 to include the possibility of a discount.
B.2. Modify the IPO chart in Figure 4.8 to include the possibility of a discount.
B.3. If module 1.2.1.2 in Figure 4.7 were decomposed into three new modules, what would their module numbers be?

B.4. If the process circle 3.2 in Figure 4.15 were decomposed into two new processes, what would their process numbers be?
B.5. Write the data dictionary definition for a Zip code.
B.6. Write the data dictionary definition for a vendor name.
B.7. Write the data dictionary definition for a vendor number.

C.1. Write a decision table to compare the date of a transaction with today's date and determine whether payment is over 30, 60, or 90 days past due.
C.2. There are many other tools available to a systems analyst. Learn about two of them and include them in the summary comparison chart.
C.3. Locate a bus, airplane, or train schedule and identify the five main areas on it that correspond to a decision table.
C.4. Draw a system flowchart to present the systems analysis phase of the systems process.
C.5. Draw a detail flowchart that compares the date of a transaction with today's date and determines whether payment is over 30, 60, or 90 days past due.
C.6. Write the pseudocode that compares the date of a transaction with today's date and determines whether payment is over 30, 60, or 90 days past due.

REFERENCES

C. Bohm and G. Jacopini, "Flow Diagrams, Turing Machines and Languages with Only Two Formation Rules," *Communications of the ACM,* pp. 366–371, May 1966.

Tom Demarco, *Structured Analysis and System Specification,* Prentice-Hall, Englewood Cliffs, N.J., 1979.

B. Dickinson, *Developing Structured Systems,* Yourdon Press, New York, 1981.

E. W. Dijkstra, "GO TO Statement Considered Harmful," *Communications of the ACM,* March 1968.

Robert T. Grauer, *Structured Methods Through COBOL,* Prentice-Hall, Englewood Cliffs, N.J., 1983.

David Higgins, *Designing Structured Programs,* Prentice-Hall, Englewood Cliffs, N.J., 1983.

HIPO—A Design Aid and Documentation Technique, IBM Publication GC20-1851-1.

G. A. Miller, "The Magical Number Seven, Plus or Minus Two: Some Limits on Our Capacity for Processing Information," *Psychological Review,* vol. 63, pp. 81–97, 1956.

G. J. Myers, *Composite Structured Design,* Van Nostrand Reinhold, New York, 1978.

Lawrence J. Peters, *Software Design: Methods and Techniques,* Yourdon Press, New York, 1981.

Edward Yourdon and Larry L. Constantine, *Structured Design,* Prentice-Hall, Englewood Cliffs, N.J., 1979.

E. Yourdon, *Structured Walkthroughs,* 2nd ed., Prentice-Hall, Englewood Cliffs, N.J., 1979.

PART II

Systems Design

CHAPTER 5

Systems Design – A Broad Perspective

- Goals and Preview
- Design Overview
- **PERT Charts**
- Review and Assignment of Tasks
- Case Study: Review and Assignment of Tasks for the Design of Fleet Feet's Accounts Payable System
- Working with People: How Important Is Keeping Current with the State of the Art?
- Summary
- New Terms
- Questions for Review and Discussion
- References

GOALS AND PREVIEW

After reading this chapter you should be able to:

1. Define and list the tasks performed during systems design in order of occurrence.
2. Explain two scheduling tools used by systems analysts.
3. List the components of system specifications in a data dictionary format.
4. Describe the components of program specifications in a data dictionary format.
5. Explain the roles of both users and management in systems design.

In Chapters 2 and 3, we examined analysis, the first phase of the systems cycle, and we decomposed it into its two subphases—preliminary and detailed analysis. Then Chapter 4 introduced us to a number of useful analytical tools, including data-flow diagrams, the HIPO system with its VTOC and IPO, structured English, decision tables and trees, structured walkthroughs, and Warnier-Orr and Nassi-Shneidermann charts.

Now we are ready to explore the second phase, design, which is the detailed planning of a new or improved system (Figure 5.1). Just as a construction company needs detailed plans to erect a building, so do systems analysts need good blueprints to construct effective computer systems. Design creates those blueprints.

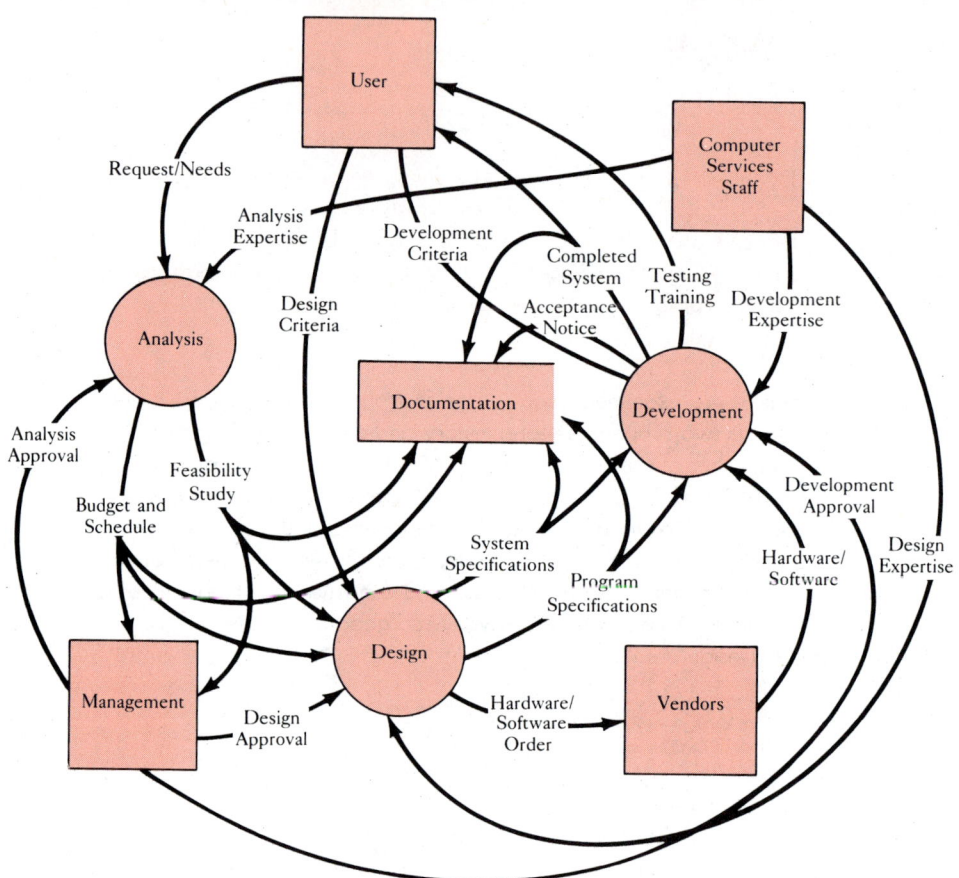

FIGURE 5.1. Analysis results in a feasibility study, which becomes a major input to design.

DESIGN OVERVIEW

During system design, analysts often develop two or more options from which, after weighing all relevant factors, they must select the most effective. For example, suppose an analyst has tackled the design of a system to print adhesive mailing labels. How many printing options can you imagine? You can print the labels singly, two across ("up," as computer people call it), three up, or even four up (Figure 5.2). The analyst should consider several factors before choosing one approach over the others. What about costs? Four-up labels are more expensive. How about complexity of programming? Four-up poses more difficult programming problems. Time to print the labels? Four up prints fastest. Such choices, called trade-offs, confront analysts at every step along the way to a sound design. The best choice may not always be the cheapest, easiest, or fastest. It depends a great deal on the user's needs. One user may have a strictly limited budget but a lot of time, whereas another may be willing to spend more money to get the labels quickly.

Design begins when management approves the feasibility study produced during detailed analysis and authorizes the necessary funds and personnel to continue. It concludes when management approves the design and authorizes development of the actual system.

To gain a broad perspective of systems design, consult the data-flow diagram in Figure 5.1. Notice the inputs to design:

1. A feasibility study, developed during detailed analysis, which includes costs, alternatives, schedules, and background information.

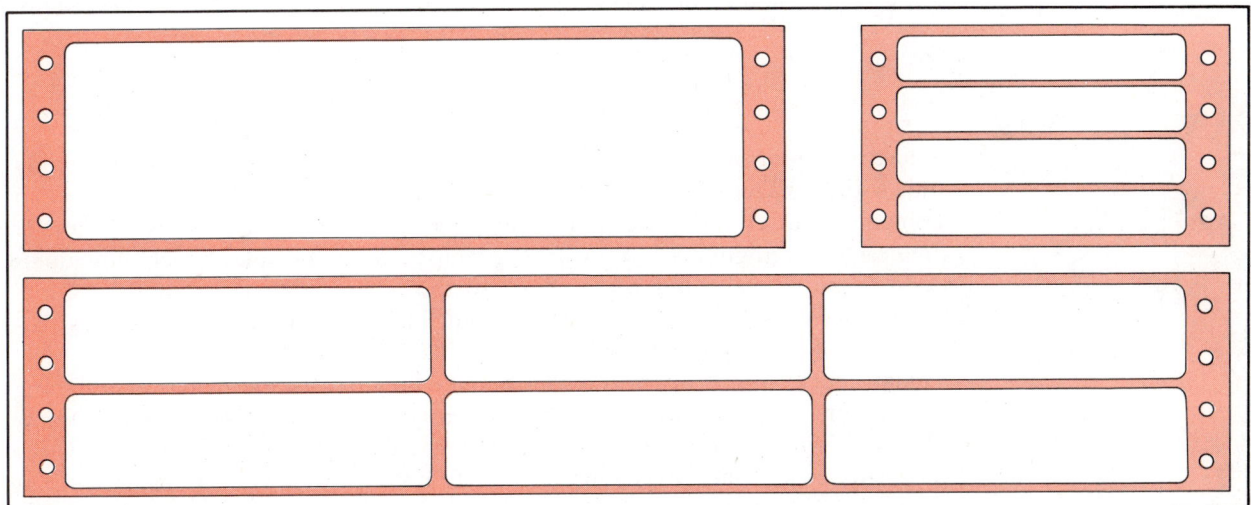

FIGURE 5.2. Labels can be purchased in a variety of widths, heights, and number of labels across, or "up." Each type of label has different costs, programming requirements, and printing speeds. The analyst must choose the right type of label for the application in question.

2. The computer services staff, which understands the necessary hardware and software.
3. Management, which makes decisions and authorizes development.
4. The users, who will actually benefit from the system.

Now consider the outputs from design:

> **System specifications** The description of the input, output, control, and file designs for a system.

1. **System specifications.** A general description of the input, output, control, and file designs required by the system.

> **Program specifications** A functional description of each program the system needs.

2. **Program specifications.** A description of every program the system needs in order to be properly developed.

> **Hardware order** A contract with a vendor to supply equipment.

3. **Hardware** or **software order.** Contracts with vendors for required programs and equipment.

> **Software order** A contract with a vendor to supply programs.

We can decompose design into nine component stages (Figure 5.3). Note how outputs from one stage become inputs for the next. Refer to Figure 5.3 as we briefly examine each of these elements. During the first component, the analyst determines who is going to work on the design and how long each component will take. Output design examines every report the system produces, determining its format and contents. Data-base, or file, design has the analyst resolve what data the computer will store and how it will store the data as well as how the computer will retrieve and transmit the data.

As its name implies, input design requires the analyst to ascertain how the data will be collected. Process design fixes how the computer will convert raw data into information. Program definition lists each program required by the system and module design fleshes out each program, giving greater detail. Package design brings together all the specifics of the system. The last component, design review, gives management another chance to decide whether to proceed with the system.

Review and Assignment of Tasks

Before launching any other activities, the analyst first reviews the feasibility study and, with management's support, assigns tasks to individuals. This results in an assignment roster that matches individuals to tasks, and projects the amount of time each task might take. A Gantt chart graphically can display which individuals perform which tasks and when. Because the organization must continue functioning during systems analysis and design, management always authorizes participation of its people. The owner of a retail floral shop who sends 550 statements a month, for example, might allow the bookkeeper to assist the analyst in designing an accounts receivable system because time spent away from routine chores now ultimately will save the firm time and money. However, since the owner does not post daily charges and payments to the AR system, the owner probably will not participate personally.

For our mailing label system, which is quite simple, the analyst may decide that only the users need to be involved and that the design will take 2 days.

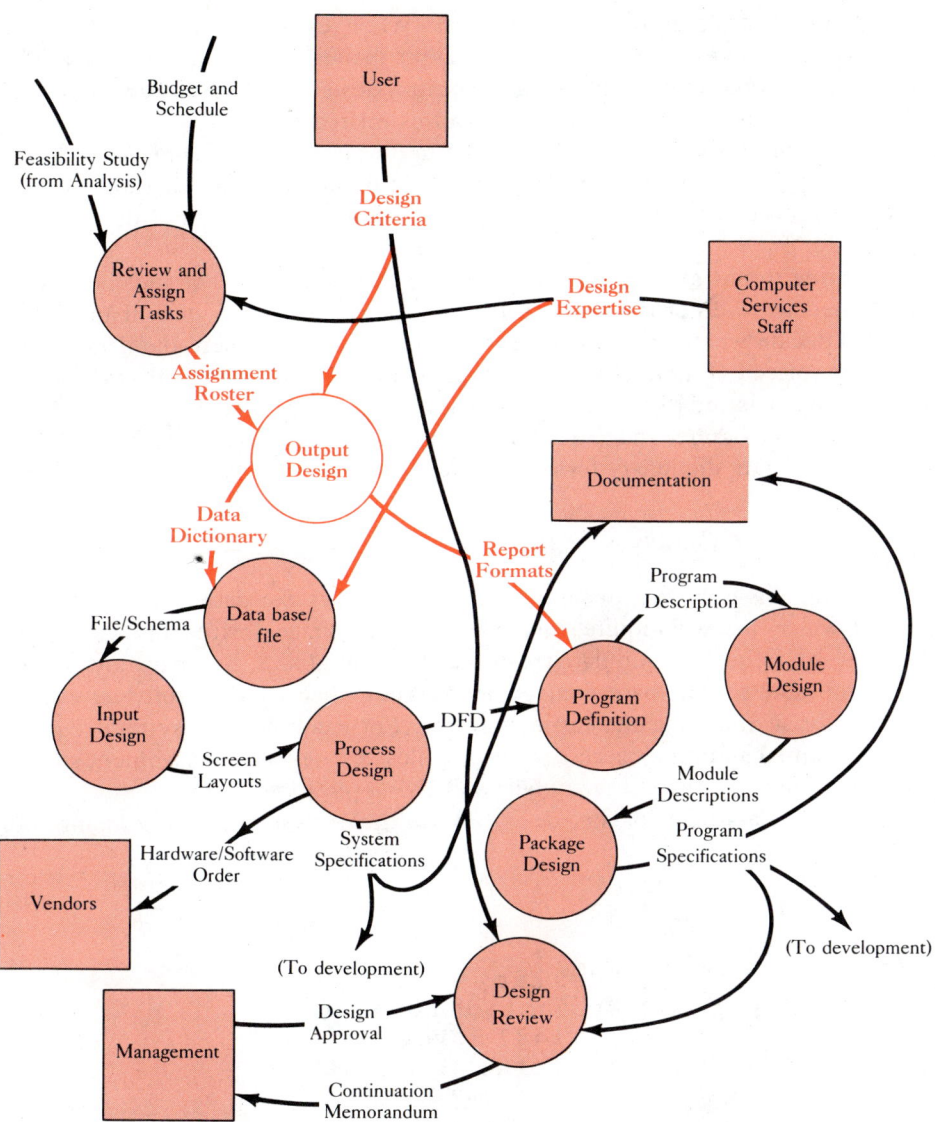

FIGURE 5.3. *Design can be decomposed into nine stages. Ultimate outputs include hardware and software orders and program specifications.*

Output Design

The second stage, **output design,** determines the content and format of the various reports that users want printed on paper or displayed on terminal screens. Because their knowledge of the jobs should influence layout, users provide valuable ideas at this point. To ensure the usefulness of reports, analysts usually sketch samples

Output design Design stage where the analyst determines the content and format of reports.

of preliminary designs for users' approval. When everyone has agreed on the content and format of the reports, the analyst makes a list of required data elements. In addition to report formats, the output design generates a data dictionary containing concise definitions of all key terms related to the new system.

Last, output design requires the analyst to devise a series of controls to do the following: prevent reports from being used improperly, verify that all the data have been processed, and ensure the accuracy and completeness of the system's results.

Returning to our mailing label example, at this point the user and the analyst should decide whether the labels should be one, two, three, or four up. They will also decide what data must appear on each label, and where the data will be physically placed (Figure 5.4a). In this case, the user and the analyst decide to use one-up labels and to have each label show the addressee's name, title, address, city, state abbreviation, and Zip code. The list of data items required by the labels is stated in a data dictionary format, Figure 5.4b.

Data-Base, or File, Design

Data-base design
Design stage where the analyst decides on the system's file storage needs.

During **data base,** or **file, design,** the analyst studies the data dictionary to determine which data will require storage on external storage devices (disks or tapes). Analysts still frequently stipulate traditional sequential files, but many are turning more and more to data-base management systems because this software can efficiently assist in storing the retrieving data. For example, if a system requires both sequential and direct access to certain files, a data-base management system (DBMS) with this capability may best suit the application.

Sometimes data-base, or file, design forces the analyst to call in a member of

```
Mrs. Mary Williams
Production Manager
211 Hill Street
New York, NY  10016
```

(a) *This label is 2 inches high and 4 inches across, thus fitting most common sizes of envelopes.*

$$\begin{aligned}
\text{Address Label} = \; &\text{Addressee name } + \\
&\text{Addressee title } + \\
&\text{Street address } + \\
&\text{City } + \\
&\text{State } + \\
&\text{Zip code}
\end{aligned}$$

(b) *Data dictionary showing what appears on a label.*

FIGURE 5.4. *A one-up label is chosen for the mailing label system.*

the computer services staff, such as the data-base administrator, who specializes in designing such files. Output from this stage is the **schema** (the data-base design) or file layout. A schema lists each field to be stored on a disk, a description of the field's purpose, and the type of data (numeric or alphanumeric) the field will hold.

As with output design, the analyst again is responsible for designing a control system. Here the controls aim to prevent tampering with the data stored by the computer or to enable data to be reconstructed if files are damaged or destroyed.

For our mailing label system, the analyst determines that the computer will store an identifier (customer number), name, title, street, city, state abbreviation, and Zip code (Figure 5.5) on a disk. Besides identifying the actual data fields, the analyst specifies how the labels will be printed—perhaps in ascending order by Zip code or alphabetically by name.

Schema The definition or description of a data base, including fields and records.

Input Design

Having determined the schema, the analyst proceeds to **input design.** Input can come from terminals, online cash registers, point-of-sale terminals, or wands that read the universal product codes printed on most supermarket items. Once again the analyst works closely with the user to determine the most efficient way to collect required data. Results of the input design are record layouts or CRT screen designs (if the analyst has selected terminals).

Again the analyst must establish controls. Here the controls try to catch faulty data before they enter the system. Input controls commonly involve verification or validation of data, counting records, or accumulating totals to see if they match hand-calculated ones.

If the computer we want to use for printing mailing labels does not have a key-to-disk for data input, we design data collection around a CRT device (Figure 5.6). Screen layouts for the CRT show the format of the screen as the user will eventually see it, although it contains no actual data, only places where an operator will enter data once the system is fully operational.

Input design Design stage where the analyst decides how the system will collect data.

Field Name	Description	Length	Type
Name	Addressee name	25	Alpha
Title	Title of addressee	20	Alpha
Street	Street address of addressee	25	Alpha
City	Addressee's city	15	Alpha
State	Abbreviation of state	2	Alpha
Zip	Zip code of addressee	9	Numeric

Retrieval: Ascending by addressee name
Ascending by Zip code

FIGURE 5.5. A system design follows nine steps. Each step provides valuable information for the next. In this case, the data dictionary definition of the adhesive label permits the analyst next to define the data the computer will store for each addressee. File layouts show the fields, their names, lengths, and types, and possible retrieval requirements.

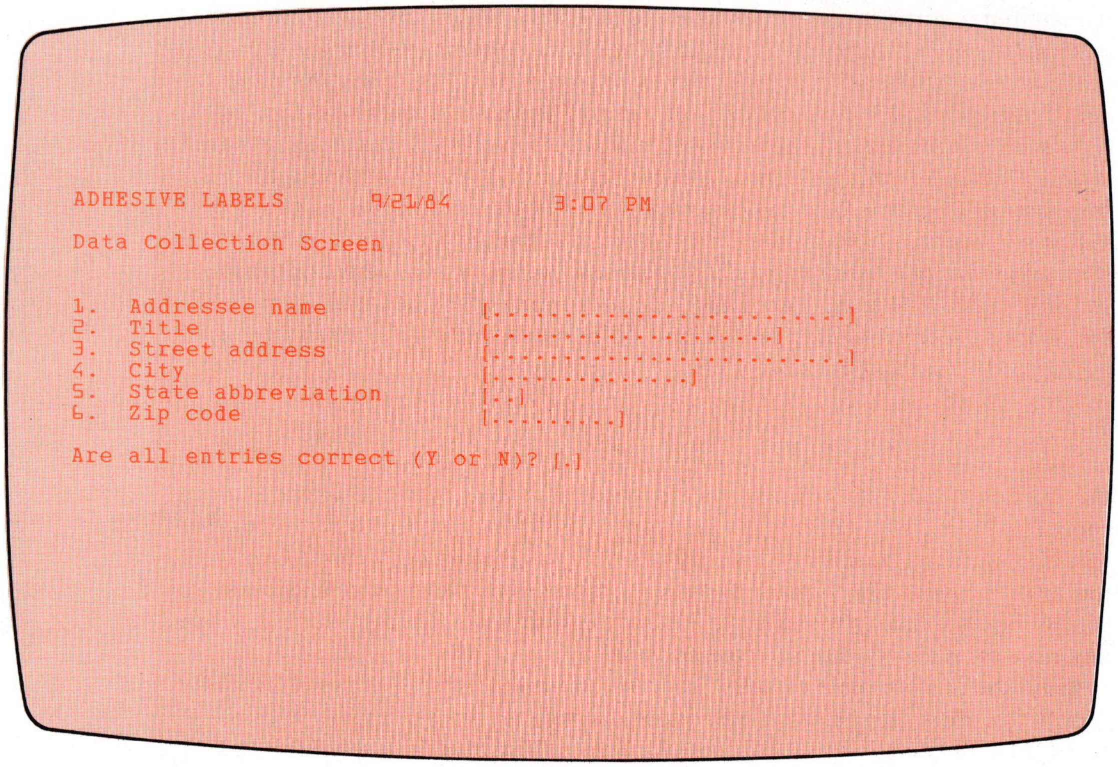

```
      ADHESIVE LABELS        9/21/84        3:07 PM
      Data Collection Screen

      1.    Addressee name          [.........................]
      2.    Title                   [......................]
      3.    Street address          [..........................]
      4.    City                    [...............]
      5.    State abbreviation      [..]
      6.    Zip code                [.........]

      Are all entries correct (Y or N)? [.]
```

FIGURE 5.6. *Screen layout allows places for data to be entered. Square brackets surrounding dots depict maximum field size and location of data on the screen.*

Process Design

Process design Design stage during which the analyst decides how the system will function.

Now the analyst tackles **process design.** Process design determines how the system will function. In other words, will it run in batch mode where transactions or other data occur in groups or at periodic times, will it be an online system processing data interactively or as the data are collected, or will it mix batch and online, perhaps updating in batch but permitting users to examine data at any time?

Outputs from process design are hardware and software orders and system specifications. The new system may require special hardware, such as an optical reader for input, and new software, which may be available in prepackaged form from a vendor. Again, a specialist such as a data-base administrator may be able to help at this point.

If our label system requires that we print 5000 adhesive labels per week with the files growing at the rate of 740 new records per week, the analyst may advocate printing labels in batch mode at night when the demand for the computer is minimal. Online data entry could still be used to keep the master files updated.

Once the analyst has defined and described all of the system components, he or she collects them in the form of a report known as the system specifications. Following data dictionary rules for defining terms, the analyst might write:

SYSTEM SPECIFICATIONS = REPORT FORMATS +
DATA DICTIONARY OF DATA ELEMENTS +
FILE DESIGN OR DATA BASE SCHEMA +
SCREEN FORMATS +
HARDWARE ORDER +
SOFTWARE ORDER

Program Definition

As you might expect, **program definition** describes the purpose of each computer program in the proposed system. The analyst builds these definitions by consulting the system specifications and the feasibility study. During this phase of design, the analyst uses such tools as the VTOC, system flowchart, data-flow diagram, and data dictionary. Although these tools may have been used earlier during analysis when writing the original feasibility study, the evolving design usually requires their continuing refinement to provide an accurate picture of the system.

Program definition output is a description of each program in the system, including required inputs, files, and outputs; Figures 5.7 and 5.8. For example, our adhesive label system requires four programs. The first collects data from the CRT terminals and uses the DBMS (which provides access to the data base on the disk) to store the data in the data base. That is, it allows the person sitting at a terminal to enter data directly into the data base. The second and third programs sort the data by Zip code or name, and the fourth and final one prints the labels themselves.

> **Program definition** A detailed description of each individual program in the system.

Module Design

Program definition is followed by **module design,** which breaks down each program into specific activities or modules. For example, we could subdivide the program for printing adhesive labels into three modules: opening, processing, and closing, each of which is single purpose and has single entry and exit points (Figure 5.9). Valuable tools for completing this step include the flowchart, structured English, decision tables, and the Warnier-Orr and Nassi-Shneidermann diagrams. Although an analyst always hopes to be free to select the ideal tools, the choice is often restricted by departmental policy. For example, some organizations may specify a VTOC to represent the system and structured English to detail program logic.

Outputs from module design are module descriptions. Simple programs may contain as few as three modules, while complex programs may contain hundreds. A general ledger system for one bank (using an online terminal) required almost 600 modules, resulting in over 12,000 lines of COBOL.

> **Module design** Determining the nature and purpose of each module a program requires.

Package Design

At the **package design** stage, the analyst prepares a complete report on all materials developed for each program. We call such a report program specifications, and it

> **Package design** Bringing together all design specifics.

Field Name	Validation Requirements
Addressee Name	Alphanumeric and present
Title	Optional entry, if present must be alphanumeric
Street	Must be present
City	Must be present
State	Two alphabetic characters. Must be present
Zip	Must be numeric and present

Data-base password: ONEUP

FIGURE 5.7. *Controls for our mailing label system might include validation, passwords, presence (must not be missing), and data type (alphanumeric or numeric).*

is the primary output of design. Using the data dictionary format, the analyst might outline program specifications as:

PROGRAM SPECIFICATION = SYSTEM SPECIFICATION +
DATA FLOW DIAGRAM +
PROGRAM DEFINITIONS +
MODULE DESCRIPTIONS +
TEST PLAN +
SYSTEM COSTS

Test plan The method of assuring the system's correctness.

All the elements, except the test plan and system costs, should be familiar to you. A **test plan** is a series of procedures aimed at guaranteeing program accuracy. It may involve running sample data through the program to see whether anticipated results are obtained. The analyst tests modules from each program (called field testing) or can run real data through the new system (called parallel testing), comparing results of the new one with those of the old.

Estimating how much it will cost an organization finally to develop a system is not an easy task. Methods for estimating costs range from guesses based on prior experience to sophisticated formulas.

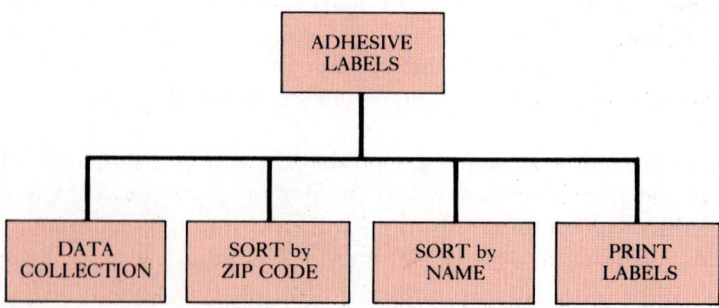

FIGURE 5.8. *During program definition, the analyst decides the number and purpose of each program the system requires. Here a VTOC shows that the system needs four programs.*

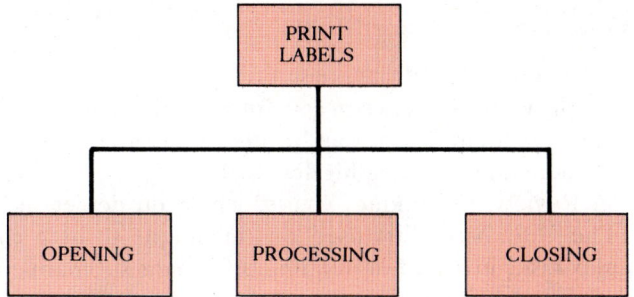

PRINT-LABELS.
    Perform OPENING-MODULE.
    Perform PROCESSING-MODULE until
      no more records in the data base.
    Perform CLOSING-MODULE.
    Stop Program.

OPENING-MODULE.
    Open printer, data base.
    Read a record from the data base.
    Advance paper to top of next label.

PROCESSING-MODULE.
    Print name line.
    If title is present, then Print title line.
    Print Street line.
    Print City, State, and Zip line.
    Advance paper to top of next label.
    Read a record from the data base.

CLOSING-MODULE.
    Close printer and data base.

FIGURE 5.9. *Program definition and module design deal with the specifics of each program in the system. A VTOC shows the "PRINT LABELS" program and pseudocode the contents of each module in "PRINT LABELS."*

Design Review

Design culminates with management action. After the system's users study the package design to make sure it fulfills their needs, management studies it to determine whether it (1) achieves the organization's goals, (2) involves acceptable costs, and (3) can be completed within a reasonable time. If everyone approves, the system progresses toward development.

PERT Charts

PERT chart A graphic tool to assist task scheduling and estimate completion time. An acronym for Program Evaluation and Review Technique.

A Gantt chart provides a useful overview of a project's schedule. It does not, however, indicate the relations between activities; that is, it does not show whether a certain task depends upon another and cannot occur until the first concludes. To illustrate such dependencies graphically, analysts frequently use **PERT** (Program Evaluation Review Technique) charts, which do depict the relationships among tasks. The U.S. Navy first used this technique to help control the exceedingly complex tasks involved in building the Polaris submarine in the middle 1950s. Since then, other military organizations, construction companies, steel fabricators, and a variety of other organizations have turned to PERT as a valuable tool for keeping track of dependent tasks within large-scale projects.

To prepare a PERT chart, an analyst must identify each important task or activity, and then consider how they all relate. Having established the sequence of tasks, we can estimate the time it might take to complete them, allowing for three different possibilities:

Optimistic: The shortest time (achieved by assigning as many people as possible to the activities).
Most likely: The normal time (achieved by assigning the usual number of people to the activities).
Pessimistic: The longest time (achieved by a single person working part-time on the activity).

In applying the PERT concept to systems design, we can identify and order nine tasks. If we assign optimistic, likely, and pessimistic times to each task, we build the following chart:

Activity	Description	Optimistic Time (days)	Likely Time (days)	Pessimistic Time (days)
1	Review and assignment of tasks	1	4	6
2	Output design	5	10	15
3	Data-base, or file, design	5	9	12
4	Input design	6	8	12
5	Process design	3	4	7
6	Program definition	2	3	10
7	Module design	3	4	5
8	Package design	2	4	6
9	Design review (MANAGEMENT INPUT)	1	3	5

We can construct another table to show how activities depend on one another:

CHAPTER 5 SYSTEMS DESIGN—A BROAD PERSPECTIVE 161

Activity	Description	Preceding Activity	Following Activity
1	Review and assignment of tasks	None	2, 3
2	Output design	1	3
3	Data-base, or file, design	1, 2	4, 5
4	Input design	3	5
5	Process design	3, 4	6
6	Program definition	5	7, 8
7	Module design	6	8
8	Package design	6, 7	9
9	Design review	8	None

Notice how this chart illustrates the precedence of one event over another. For instance, the chart shows that events 1 and 2 precede event 3, while events 4 and 5 follow event 3. One could also draw this chart as a linear progression:

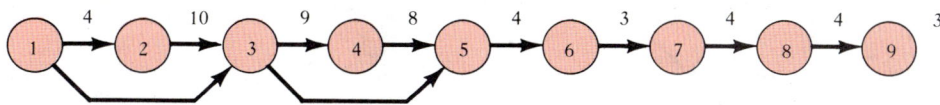

Since most projects take surprising turns, statisticians have developed a formula to calculate reasonable expectations based on the three time categories:

Expected Time = (Optimistic time + Pessimistic time + 4 * Likely time)/6
                = (28 + 78 + 4 * 49)/6
                = 302/6
                = 50.33 days

Another technique for calculating expected time is the **critical path method (CPM)**, which shows the progress of the project from start to finish. It necessarily depicts the longest path time. In our example, the critical path would be 1–2–3–4–5–6–7–8–9.

Drawing facts from the CPM, the precedence of tasks, and the most likely times to perform them, we can calculate the earliest and latest start times for each event as shown in the chart on the next page.

Thus if we hope to complete the design phase of the systems process in 49 days, event 9 can start as early as day 33 but must be begun no later than day 46.

Critical path method (CPM) Planning and scheduling tool showing relationships and schedules for a project.

Activity	Description	Earliest Start Time	Latest Start Time
1	Review and assignment of tasks	1	1
2	Output design	5	5
3	Data-base, or file, design	5	15
4	Input design	14	24
5	Process design	22	32
6	Program definition	26	36
7	Module design	39	39
8	Package design	29	42
9	Design review	33	46
	Completion	36	49

In complex situations, the critical path need not be sequential. In the following PERT chart, we see a critical path (longest way from start to finish) that is 1–2–4–5.

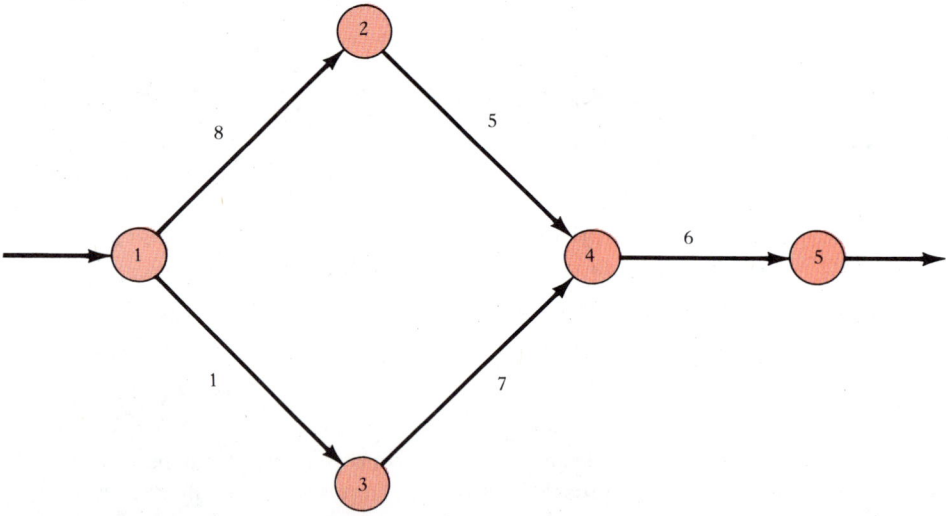

The elapsed time is 19 days. Event 1 starts on day 1, event 2 on day 9, event 4 on day 14, and event 5 on day 20. Event 3 is not on the critical path. Any event lying outside the critical path can start at any time as long as it ends before the event that follows it starts. Thus since event 4 occurs on day 14 and it takes 7 days to complete event 3, this noncritical path event can begin on days 2, 3, 4, 5, 6, or 7.

Because the possible combinations of components can be so numerous, and considering we could include costs for the optimistic, likely, and pessimistic times, computer programs have been developed to calculate the critical path.

Since the PERT/CPM technique applies to situations that have nothing to do with computers, it has become a valuable management tool.

REVIEW AND ASSIGNMENT OF TASKS

Design begins with a review of the feasibility study and an assignment of tasks (Figure 5.10). Careful review is necessary because someone other than the original analyst may have written the feasibility study or a great deal of time may have elapsed since the study was completed. In any case, analysts should reread fea-

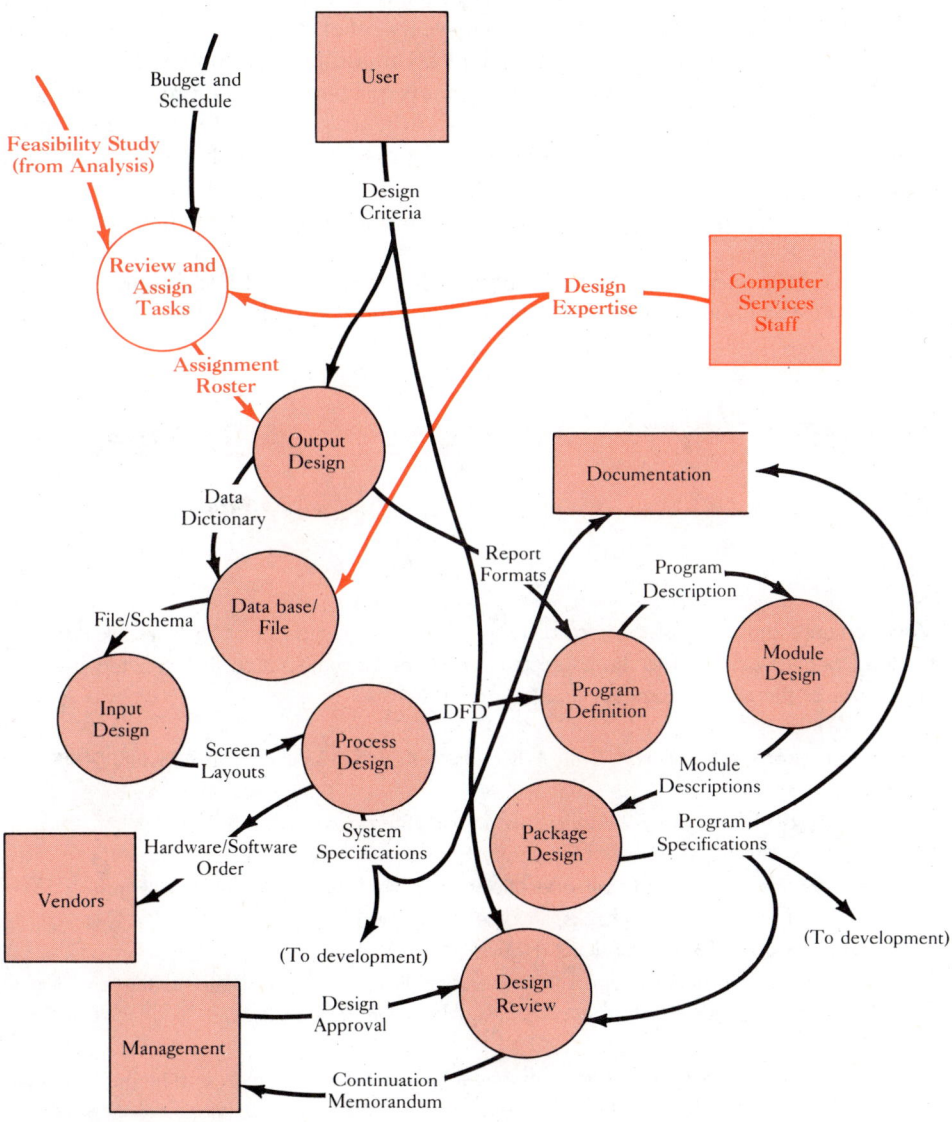

FIGURE 5.10. *The first step in design begins with a review of the feasibility study and yields an assignment roster and schedule.*

sibility studies at this time to familiarize themselves again with the goals, requirements, costs, schedules, and other factors relating to the proposed study.

Assignment of design tasks depends primarily on the availability of personnel. In a large organization, the lead analyst may employ an entire staff of people with expertise in various areas: designing forms, administering data bases or data communications, providing security. In smaller organizations, the analyst may take full responsibility for designing the project. In either situation, the project may be so complex that the analyst will need help from a number of sources, both inside and outside the firm.

If the project requires a team of analysts, the lead analyst determines who will perform which duties and when. Teams of analysts may be used on very large or complex systems that may take many "man-years" of effort. One community college whose registration program definition had 50 modules and 9000 COBOL statements used the team approach. The lead analyst scheduled one programmer, two analyst-programmers, a data-base administrator, and a clerk for the registration system. The completed system had over 100 programs, more than 75,000 lines of COBOL, and took 6 "man-years."

No matter how large or how small the system, the analyst estimates the time required to perform the nine components of preliminary as well as detailed design and chooses the individuals to be involved. Both Gantt and PERT charts are tools the analyst can choose to assist in the scheduling of people and events.

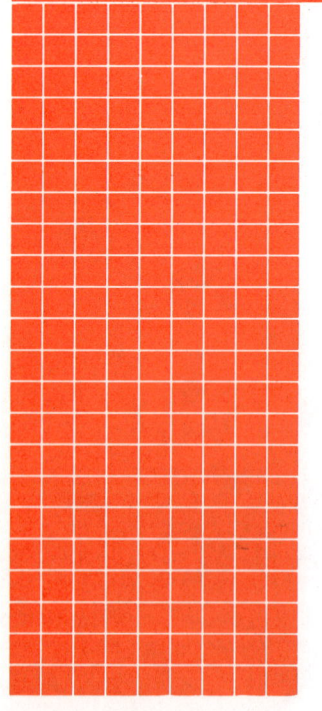

CASE STUDY

Review and Assignment of Tasks for the Design of Fleet Feet's Accounts Payable System

Management's approval of the feasibility study initiates the design phase. Frank Pisciotta, the lead analyst assigned to the project, reviews the documentation that resulted from the first phase of the systems process.

1. Original user's request from Sally Edwards to Kathleen Williams, dated July 20 (Figure 2.5).
2. Summary of the August 8 interview with Mark Stensaas (Figure 2.12).
3. Preliminary report of August 15 to Sally Edwards (Figure 2.13).
4. August 24 memorandum to all operations personnel and vice-president (Figure 3.3).
5. Samples of ledger cards, checks, invoices, and packing slips.
6. Feasibility study of September 12 (Figure 3.9).
7. Memorandum alerting all operations personnel and vice-presidents to the decision to proceed toward the installation of a new AP system (Figure 3.10).

Frank must now assign tasks to the personnel he wants to assist him, and he must schedule events so as to complete the design on time and within budget (6 months and $5000 as stipulated by the feasibility study and approved by management).

CASE STUDY

Frank begins by listing the major activities (see following table). Then, on the basis of experience and discussion with other team members, he estimates the amount of time needed to complete each activity. Frank's Gantt chart appears in Figure 5.11.

Activity	Estimated Time (Working Days)
1. Review and assignment of tasks	4
2. Output design	10
3. Data-base or file design	9
4. Input design	8
5. Process design	5
6. Program definition	6
7. Module design	7
8. Package design	4
9. Design review	4
Total	57

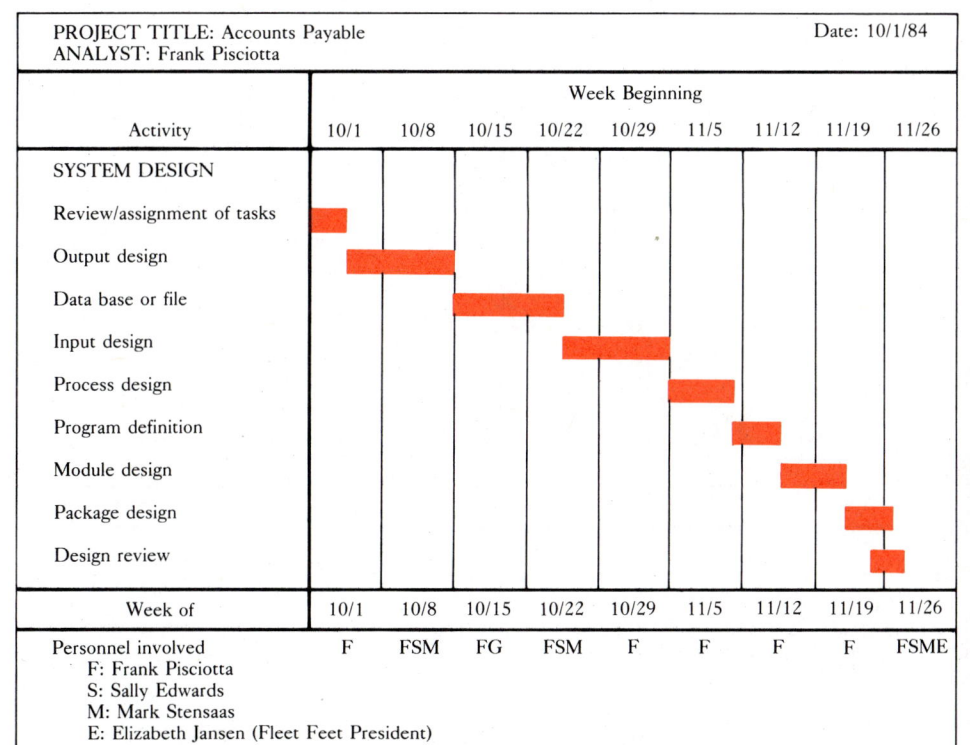

FIGURE 5.11. Original schedule of events for the design phase of Fleet Feet's accounts payable system.

CASE STUDY

By breaking the project down into nine steps and estimating how long each step will take, as shown in the table, Frank now totals the time for each step, arriving at 57 working days (almost 12 calendar weeks or 3 months) as the duration of the design phase of the systems process. Since the time established in the feasibility study and approved by management had been projected at 52 days, Frank will have to make up time by bringing in other analysts, thereby increasing costs. Could he avoid this by undertaking some design activities simultaneously? Process design conceivably could take place at the same time as program definition, but program definition could not take place simultaneously with module design since we must know the purpose of each program before breaking it down into modules. In Fleet Feet's situation, perhaps the packaging of the design could occur during a portion of module design, thereby shaving 2 days off the schedule. Frank could also ask people involved to hone their estimates and achieve the originally scheduled 52 days.

After cutting the schedule to the approved length, Frank revises his Gantt chart (Figure 5.12), this time including noncomputer people who will be assisting him and his staff. For example, Frank has decided to ask a salesperson from Moore Business Forms to review

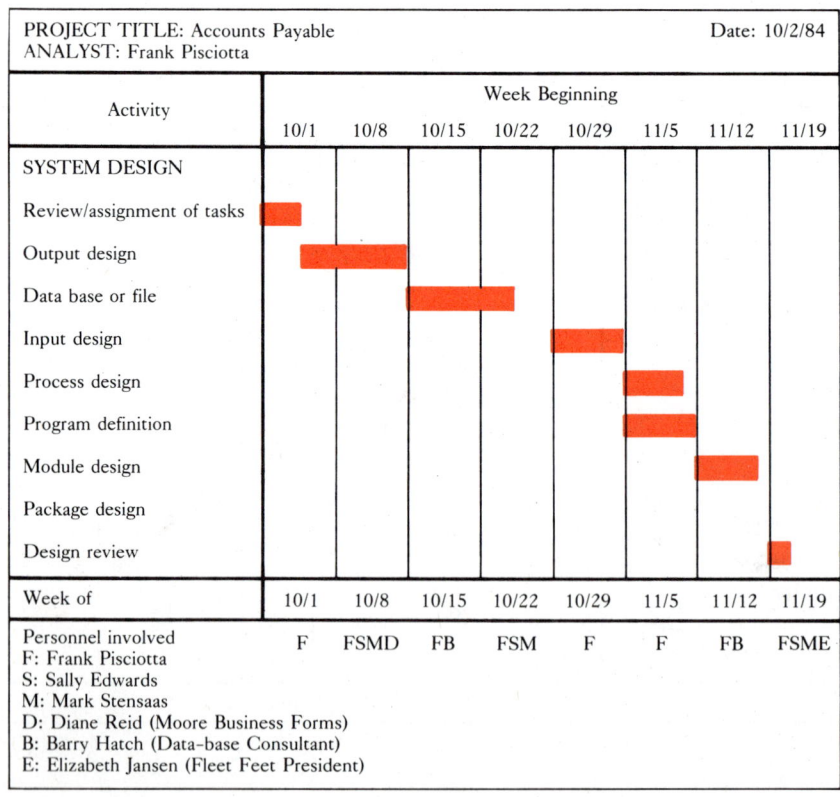

FIGURE 5.12. Modified schedule of events for the design phase of Fleet Feet's accounts payable system. Personnel include members of the computer services staff and Fleet Feet employees as well as suppliers and a consultant.

CASE STUDY

check designs. Since Moore designs many forms for other organizations and is usually willing to share experiences and costs with a potential client, the salesperson can make design suggestions to reduce the cost of producing a form, suggest colors to make the form more attractive, or propose an existing standardized form that will eliminate many costs. Frank has also decided to hire Barry Hatch, a consultant in data-base design, who can help Fleet Feet achieve the optimum performance, minimal disk-drive storage requirements, strong program design, and sound security for its data-base management system.

Management's eventual approval of the design will involve many individuals: the user, Mark Stensaas; the person who requested the study, Sally Edwards; and the president of the company, Elizabeth Jansen. Since approval will lead to the most expensive stage of the systems process, development, it is important that decisions be considered carefully by all involved.

Frank allows 2 months to design the system and 4 months for its development. The $5000 budgeted for design should cover Frank's salary, secretarial assistance, and the consultant's $60 per hour for 30 hours.

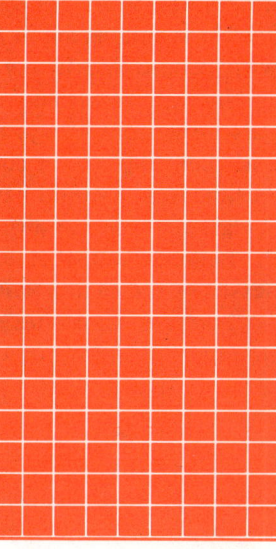

WORKING WITH PEOPLE   How Important Is Keeping Current with the State of the Art?

"We had a big shootout at the office last week," confided Bob Hitchcock, a 52-year-old analyst, to his car-pooling friend Sandra. "The new kids wrote Don Price a memo telling him how out of date I am. They said I use obsolete techniques. I've been in this field since its beginning, but sometimes I feel as though I'll never catch up. It changes too fast! I have a feeling my job's in jeopardy. What should I do?"

Sandra pondered the question. She was an ambitious young accountant who had passed her certified public accountant exams but had not yet accumulated the 4000 hours of work experience required for the coveted CPA. She responded to Bob Hitchcock's worry by agreeing that computers and electronics are two of the most rapidly changing fields in U.S. enterprise.

"How could anyone keep up with it?" she asked.

"Well," replied Bob, "the young kids seem to have less trouble than I do."

Sandra laughed. "That's just because they're fresh out of school. But, no matter how current they are, textbooks can't replace your 15 years of experience. Maybe you just need a little more book learning."

A few days later, Don Price stopped Bob on the hall and asked him to stop by his office at 4 o'clock. For the rest of the day, Bob could do nothing right. He accidentally purged a 1000-line COBOL program from the disk and had to have it recovered from yesterday's backup tape. He was irritable with telephone callers, and his lunch upset his stomach. Finally, the dreaded hour arrived. He felt like a prisoner on death row taking those final few steps.

Don greeted Bob with a smile. "I guess you know why I asked you to stop by."

"I suppose it's about the memo," Bob replied.

"You're really upset about it, aren't you?" Don queried.

"Well, wouldn't you be if you were in my shoes?"

"I suppose so," responded Don. "Let's talk it

out. Where do you feel your strengths and weaknesses lie, Bob?"

"My strengths are in my ability to communicate with people and in my writing skills. I have excellent rapport with my users and my training manuals are written with just the right amount of detail and they are free of computer jargon. As for my weaknesses, I would have to say I don't have up-to-date technical knowledge," answered Bob.

"I agree. And what are the strengths and weaknesses of the memo writers?"

"Just the opposite of mine," Bob said. He frowned. "They mean well, I think. They're really bright and eager. But some of them are in for a rude awakening when they try to use their new theories on a real system."

"Why not offer to exchange your strong practical experience for a little training in new techniques?"

Bob seemed to mull this over. "You're right. I've been lazy. Now's the time to learn some new tricks."

Bob left Don's office feeling much better. Don was right. Technical skills are important, but there are other skills that can make or break a practicing analyst, and some of them don't come from books or school.

SUMMARY

Systems design is the second stage in the systems cycle. During this stage, the analyst formally defines the system through two documents: system specifications (output from the process design) and program specifications (output from the package design).

The system specifications provide output design (the format of all required reports, including control factors), the data-base, or file, design (how the data will be stored on disks or tapes and security provisions), input design (the methods of collecting, verifying, and validating the data), and process design (the mode of processing). System specifications may also generate an order for necessary hardware or software.

From system specifications, the analyst derives program specifications, the guidelines a programmer eventually will follow when programming instructions for the computer. These specifications define the scope and purpose of each program to be written as well as the major activities that will occur in each program.

With both system and program specifications available, management can act. Working together, managers, users, and analysts decide whether the design achieves the goals of the system.

If management approves the system, the analyst makes a schedule of each activity that needs to take place, the personnel to be involved, and the expected cost of this portion of the systems cycle. Gantt charts provide an excellent device for showing how long an activity might take and which personnel are assigned to the activity.

NEW TERMS

Critical path method	Output design	Program specifications
Data-base design	Package design	Schema
Hardware order	Process design	Software order
Input design	Program definition	Test plan
Module design		

QUESTIONS FOR REVIEW AND DISCUSSION

Questions appear in three categories. You can find the answers to the A group in this chapter. The B group requires you to apply the material presented here, whereas the C group necessitates investigation or research.

A.1. Give two examples of systems design tools.
A.2. What is a hardware order?
A.3. What is a software order?
A.4. Do program specifications include COBOL-type statements or are they written in pseudocode?
A.5. List all the inputs to the design phase.
A.6. Where do users have input to the design phase?
A.7. At what point does management have input to the design phase?
A.8. Where does the computer services staff have input to the design phase?
A.9. List all the outputs from the design phase.

B.1. Write a data dictionary definition of a feasibility study report (written during analysis) that would serve as input to the review and assignment of tasks.
B.2. Draw the design phase of the systems analysis cycle as a VTOC chart.
B.3. Write a data dictionary definition of program definition.

C.1. Draw a Gantt chart for the following. Assume activities occur in the order listed.

Activity Name	Estimated Time to Complete
Write programs	22 days
Test and debug programs	34 days
Document programs	12 days
Train users	6 days

C.2. In the summary of Chapter 4, there is a chart comparing an analyst's tools. Redraw this chart to include Gantt charts.

REFERENCES

Tom DeMarco, *Structured Analysis and System Specification*, Yourdon Press, New York, 1979.
Steve Eckols, *How to Design and Develop Business Systems*, Mike Murach and Associates, Fresno, Calif., 1983.
Glenford J. Myers, *Software Reliability*, Wiley, New York, 1976.
Meilir Page-Jones, *The Practical Guide to Structured Systems Design*, Yourdon Press, New York, 1980.
Wayne P. Stevens, *Using Structured Design*, Wiley, New York, 1981.
Edward Yourdon and Larry L. Constantine, *Structured Design*, Prentice-Hall, Englewood Cliffs, N.J., 1979.

CHAPTER 6 *Output Design*

- Goals and Preview
- Output Design: General Guidelines
- Selecting the Best Output Medium
- Formatting Reports
- Types of Reports
- Designing Reports for Printers
- Designing Reports for CRTs
- Guidelines for CRT Screen Designs
- Control of System Outputs
- Building a Data Dictionary
- Case Study: Designing the Reports for Fleet Feet's Accounts Payable System
- Working with People: Involving the User from the Start
- Summary
- New Terms
- Questions for Review and Discussion
- References

GOALS AND PREVIEW

After reading this chapter you should be able to:

1. ~~List the characteristics of at least three kinds of printers.~~
2. List the four types of reports.
3. Describe the four basic report formats.
4. Write a data dictionary.
5. Explain three design criteria for CRT reports.
6. ~~Establish the criteria for specifying a printer.~~
7. List three control techniques for monitoring system outputs.

The design phase, you will recall, involves an orderly sequence of steps, beginning with review and assignment of tasks and ending with package design (Figure 6.1). During the first stage, output design, an analyst determines what data the application needs and how the data are organized and presented. Just as a model airplane builder would not purchase materials and start assembling them until after designing the plane, so a systems analyst would not collect and try to process any data before designing the output of the system.

When designing a system's output, the analyst must make several interrelated and interdependent decisions. The first involves selecting the output medium. A variety of output media are available, including hard copy printers, cathode-ray tubes (CRTs), punched cards, tapes or disks, audio devices, and microfilm. Decisions concerning report formats and the necessary information for each report depend on which medium best suits a given application. An analyst can later determine such details as how often users need reports generated and which people within the organization should receive them.

In this chapter we describe the output media and their uses and learn how to format reports for each. We conclude with an explanation of how to prepare a data dictionary, a major product of output design.

Output design precedes input design, and for some people this process seems backward. We do the steps in this order to establish just what data we need to collect. We do not want to collect data and then not use the data, and the only way we will know what data to collect is first to establish what we want. In many ways this apparent reversal is like programming; we do not start writing BASIC, COBOL, or Pascal without first drawing a flowchart or writing pseudocode.

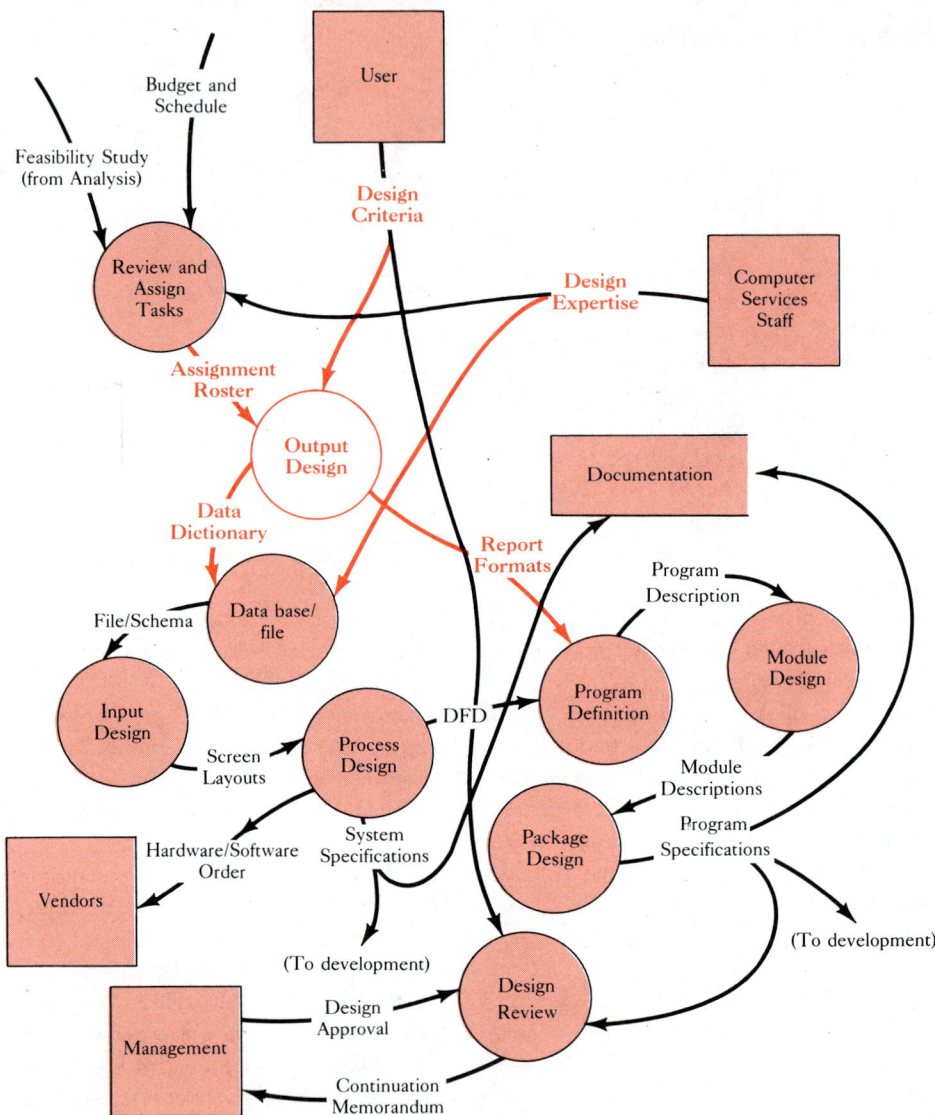

FIGURE 6.1. Output design is the second of the components of preliminary design.

OUTPUT DESIGN: GENERAL GUIDELINES

We encounter computer output all the time: the morning newspaper, utility and other bills, application forms, and course grades, to name a few. These items are useful and understandable because an analyst has carefully designed them while paying special attention to our needs.

Every business and government computer system produces some kind of report, and many systems produce quite a number of them. No matter what the content or use of a report, though, it must be designed with certain basic goals in mind. The following guidelines apply to any report.

1. The information should be clear and accurate, yet concise, and include only relevant data.
2. The report should be easy to understand, with titled, descriptive headings for columns of data, the date, and numbered pages, and it should appear on standard-size paper.
3. The report should be arranged logically, so users can easily locate needed information.
4. The report should be designed for an output medium that best suits the user's needs. For example, a stockbroker who needs instant information may require a CRT, whereas a sales manager who consults monthly figures for reference may require a printout.

Business reports fall into two broad categories, internal and external, a distinction that relates to the ultimate use of the reports. **Internal reports** are used by managers and others in an organization, often for decision-making purposes. Some such reports are simple alphabetic lists of customers and vendors; others are detailed analyses of items ordered from vendors, the amounts owed, and their due dates (Figure 6.2). Still others contain confidential and sensitive information such as outstanding bank loans and investments or employee pay rates.

Because internal reports remain within an organization, their appearance and style are not always as critical as for external reports. However, the accuracy, clarity, and conciseness of internal reports are just as crucial. An accounts payable report, for example, should clearly indicate due dates so the company can take advantage of purchase discounts, but it does not have to appear on bond paper with margins justified.

External reports, sent to customers, clients, stockholders, vendors, government agencies, and others who are usually unfamiliar with the system, must be clean and attractive, as well as accurate and concise. Since such reports present the company's image to the outside world, they must give the impression that the company is well organized, stable, and professional. Figure 6.3 shows one type of external report.

The computer output for some reports becomes input to an ensuing processing activity by the computer. Thus this type of report supports both internal and external use. We call this special type of external report a **turnaround document.** Like an external report, it lists information for outsiders (items ordered and the amount due for customers of a retail store), but it also contains information for

Internal report
Report used by managers and others within an organization.

External report
Report sent to customers, clients, or others outside of the organization.

Turnaround document
Computer outputs that serve as input for a subsequent activity.

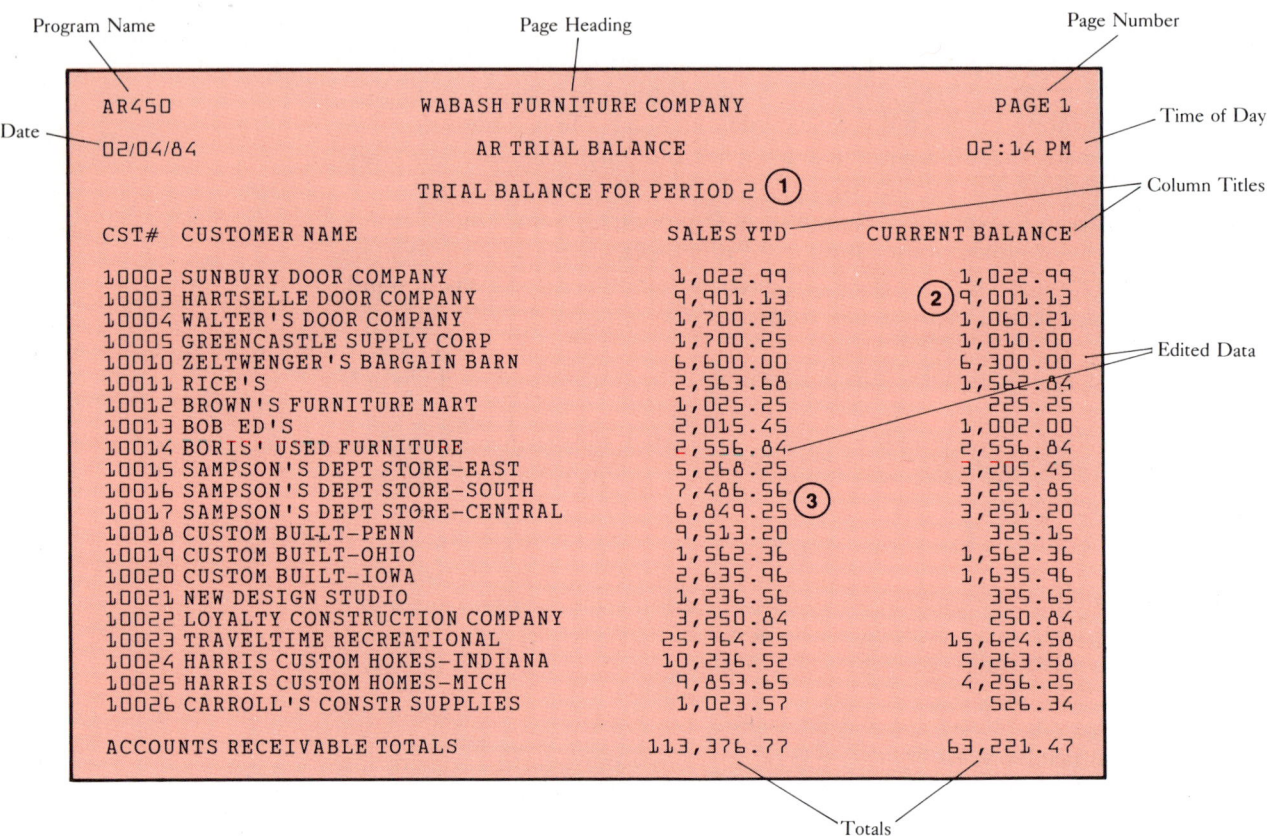

FIGURE 6.2. *Internal reports remain within the organization. Usually produced on plain paper, they provide information needed to perform a particular job. This report contains all the characteristics of good report design: headings, date, time, page number, details, and totals.*

use by the organization (payment received). As you can see in Figure 6.4, this dual use can be accomplished by designing a report with two easily separated sections, one for the customer to keep and another to be returned to the company. The organization usually preprints the customer's account number on the turn-around portion of the document to facilitate cost-effective data collection. A good example of this type of report is the form used by most monthly book clubs, part of which a customer must return each month to indicate whether or not he or she wishes to buy an advertised book.

SELECTING THE BEST OUTPUT MEDIUM

Early in the design phase, the analyst must select the best medium for presenting information to users. In recent years a wide range of media has become available, including many different kinds of printers and terminals, personal computers linked to central data banks, microfilm, tape, disks, audio or sound generators,

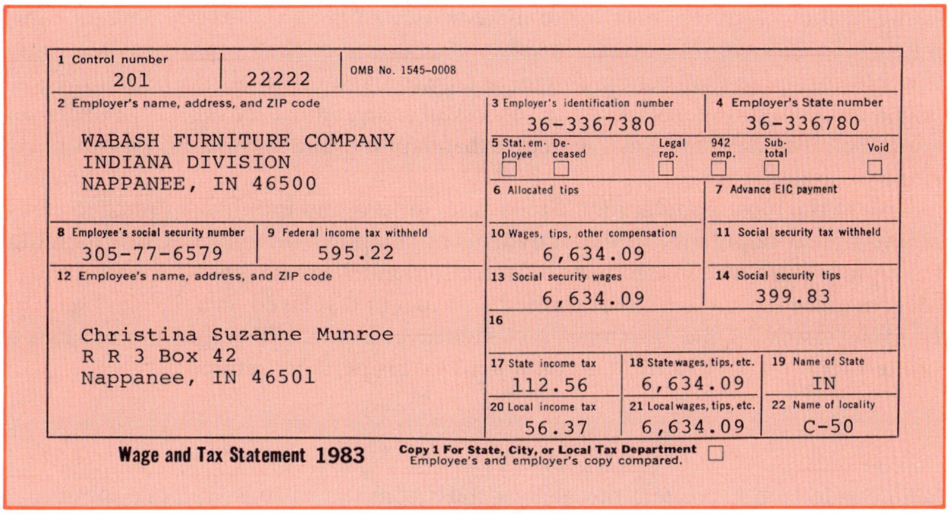

FIGURE 6.3. External reports circulate outside the organization. They are normally printed on standard or preprinted forms with column or category headings that apply to all situations. Spaces under the headings will be filled with appropriate data when the form is printed. The all-too-familiar W-2 wage and tax statement comes to most of us in January of each year.

data communications links, optically scannable forms, and plotters. Each offers special capabilities and speeds, and has limitations, all of which the analyst must consider while carefully weighing the needs of an organization. For a payroll application in one organization, an analyst might select reels of tape to transport data from the employer's computer to the bank that credits net pay to an employee's account. For another organization a telephone data communication link

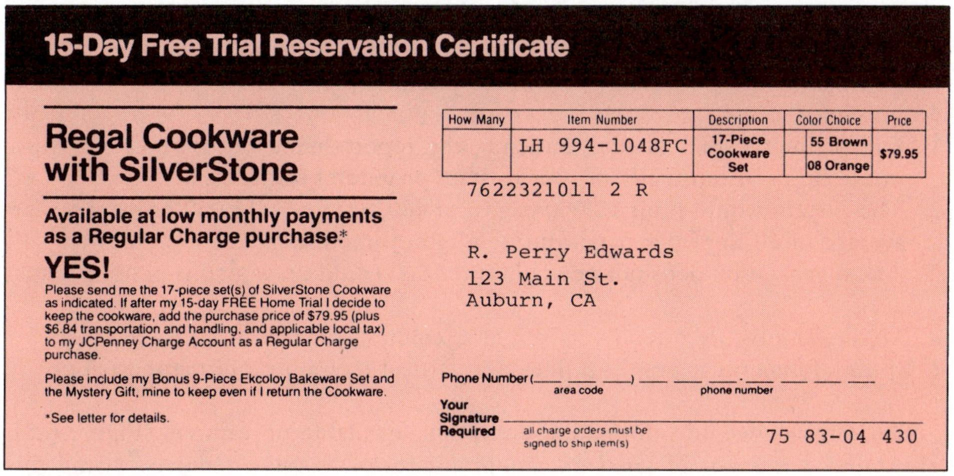

FIGURE 6.4. Turnaround documents create both a record for the customer and a means for the customer to communicate with the organization.

between the computers may make more sense. Although a telephone data communication link would provide for instant crediting of the employees' accounts, it would cost more than sending a reel of tape through the mail. Mailing, however, would delay crediting. As usual the successful analyst must try skillfully to balance a number of trade-offs—cost, speed, expandability, and compatibility—with existing equipment.

Two of the most popular media today are printers and terminals. Punched cards enjoyed great popularity in the early days of computer systems but this medium has rapidly given way to terminals, which afford much greater speed and flexibility. Most modern reports are either printed on paper or displayed on a terminal screen. In some instances the user may mix media: a terminal for rapid inquiry and a printed report with historical account activity for periodic reference.

Printers

Before selecting a printer, the analyst must consider all the system's input and output requirements, as well as the company's budget. Some printers may satisfy most requirements but may be unable to perform other functions.

Printer selection trade-offs focus on the system requirements.

1. Amount of output per week or month, which determines the speed of the device, always allowing for growth.
2. Quality of output based on the format of the report, the number of copies needed, and the type and cost of paper.
3. Existing printers available in the firm's complement of peripheral computer equipment.

We can see this trade-off factor in practice in our Saginaw Medical Group example. Pete Duisenberg, Saginaw's analyst, was asked to select the printer for his firm's new computerized office system, which would be sending out bills, processing MediCare forms, maintaining patients' health records, tracking drug inventories, and processing payroll. Before he even thought about the ideal printer, Pete first listed the requirements of the system.

1. Except for the bills sent to patients, the reports need not be elegant. Most would be used internally or sent to state or federal agencies.
2. The system would print 12,000 pages, or 480,000 lines (12,000 pages times an average of 40 lines per page) during a typical month.
3. The organization does not need a printer that could draw graphs or other visual aids.
4. Medical forms are often oversized (132 columns wide).
5. Billing forms must permit 6 lines per vertical inch on preformatted paper.

Pete then collected information on options available on printers (Figure 6.5). He found a number of good ones appropriate for Saginaw's needs. Using this information, Pete outlined the major differences and trade-offs to consider in Saginaw's printer selection.

	Line-at-a-Time				Character-at-a-Time (Serial)			
	Speed	Price Range	Full character Dot character	Special Paper	Speed	Price Range	Full character Dot character	Special Paper
IMPACT								
Chain/train band	200-2,000 LPM	$3,600 to $71,000	Full	No	30-46 CPS	$2,000 to $4,550	Full	No
Drum	125-3,000 LPM	$9,000 to $49,000	Full	No				
Ball					10-100 CPS	$800 to $9,000	Full	No
Thimble					55 CPS	$2,300 to $3,200	Full	No
Daisy wheel					30-55 CPS	$1,825 to $3,600	Full	No
Matrix	125-500 LPM	$3,700 to $20,000	Dot	No	30-400 CPS	$1,000 to $6,450	Dot and over-lapped dot	No
NONIMPACT								
Electrostatic	275-3,600 LPM	$4,300 to $14,200	Dot and over-lapped dot	Yes	160-225 CPS	$500 to $1,600	Dot	Yes
Ink jet	3,000 LPM	$110,000	Dot over-lapped	No	300 CPS		Dot	No
Laser-Electrophotographic	13,000 LPM	$310,000	Full	Yes				
Magnetic	90-180 LPM	$3,100 to $4,100	Dot	No				
Thermal					10-140 CPS	$1,200 to $3,500	Dot	Yes
Xerographic	2,800-4,000 LPM	$66,300	Full	No				
Laser-xerographic	8,000-18,000 LPM	$295,000	Full	No				

FIGURE 6.5. This chart compares the speed, price, method of printing, and paper requirements of a variety of printers. (Source: Perry Edwards and Bruce Broadwell, *Data Processing: Computers in Action*, 2nd Edition, 1982, Wadsworth Publishing Co., Belmont, Calif., p. 331. Reprinted by permission.)

1. Character formation
   - Letter quality (similar to a typewriter).
   - Formed or dot matrix (each character represented by a pattern of dots).
2. Speed
   - One character at a time (1–400 characters per second).
   - One line at a time (200–3000 lines per minute).
   - One page at a time (up to 25,000 lines per minute).
3. Print method
   - Impact: Uses ribbons to imprint letters. Is very noisy while operating. Can print carbon copies.
   - Nonimpact: Letter characters are made using lasers, ink jets, xerographics, or electrostatic or electrothermal mechanisms. Is very fast and quiet; cannot print carbon copies. Some printers can produce charts and graphs as well as numbers and letters.
4. Characters per inch
   - Paper width can vary from 3 to almost 15 inches, so the number of characters printed per inch directly affects the number of words possible on each line. The most common paper widths are 80 or 132 characters per line.
   - Most printers place 10 or 12 characters per inch.
   - Some dot-matrix printers can compress print, printing 136 characters in the space normally occupied by 80.
5. Friction versus tractor feed—Sprockets pull the paper past the printing mechanism, keeping the paper in correct alignment.
6. Lines per inch—6 or 8 lines per vertical inch of paper.
7. Costs—range from $250 (for an impact, dot-matrix printer that prints one character at a time) to $750,000 (for a nonimpact, letter-quality printer that prints a page at a time).

Like all organizations the Saginaw Medical Group had a limited budget. Although a letter-quality printer would enhance the appearance of reports, a dot-matrix printer could produce acceptable quality at a much lower cost. Internal reports did not need to be beautiful, nor did external reports to patients and state or federal agencies.

Given the need to print approximately 480,000 lines per month, a device printing a maximum of 200 characters per second would be too slow, because it would take 426 hours to print 480,000 lines. Pete settled for a line-at-a-time printer. But how many lines per minute did the application demand? Using 200 as a starting point, he found that it would take 40 hours to print 480,000 lines. That would handle this year's volume, but since Saginaw wanted room for growth, Pete doubled the number of lines printed per minute to halve the time.

(480,000 lines)/(400 lines per minute) = 1200 minutes or 20 hours per month

That allowed plenty of margin for growth or unanticipated extra printing.

The Saginaw clinic did not need a printer that could create graphs or pie charts, but it did sometimes use oversized forms (132 columns wide). Since medical billing forms are designed for 6 lines per vertical inch, Saginaw would not need an 8-line-per-inch printer. Considering the clinic's overall needs, Pete recommended a 400-line-per-minute, dot-matrix, impact, 132-column, 6-line-per-inch printer.

Paper

Having selected a printer, an analyst next chooses the type and quality of paper. Important considerations include cost, the number of copies needed, and the purpose of, or audience for, the reports.

On the basis of these factors, the analyst selects:

1. The weight and bond of the paper.
2. Plain paper or preprinted forms.
3. If preprinted forms, the number to order at a time.
4. A duplication method.

Regardless of its quality or use, most printer paper is continuous fanfolded (Figure 6.6), with sprocket holes on the left and right edges that enable the printer to pull the paper through the printing mechanism. The pages are perforated at the top and bottom so an operator can easily separate or burst them and strip off the sprocket hole edges. If a company cannot afford the time to burst the paper manually, machines can be bought that perform this task automatically (Figure 6.7). Other machines are available that automatically decollate and remove carbon paper.

Paper quality varies dramatically according to both weight and bond. Paper weight is the number of pounds per ream (500 sheets) of 17-inch by 22-inch paper. Onionskin paper weighs 9 pounds per ream; standard paper, about 40–50 pounds; and heavy card stock, 120 pounds. **Bond** refers to the percentage of fiber in the

Bond A measure of the fiber content in paper.

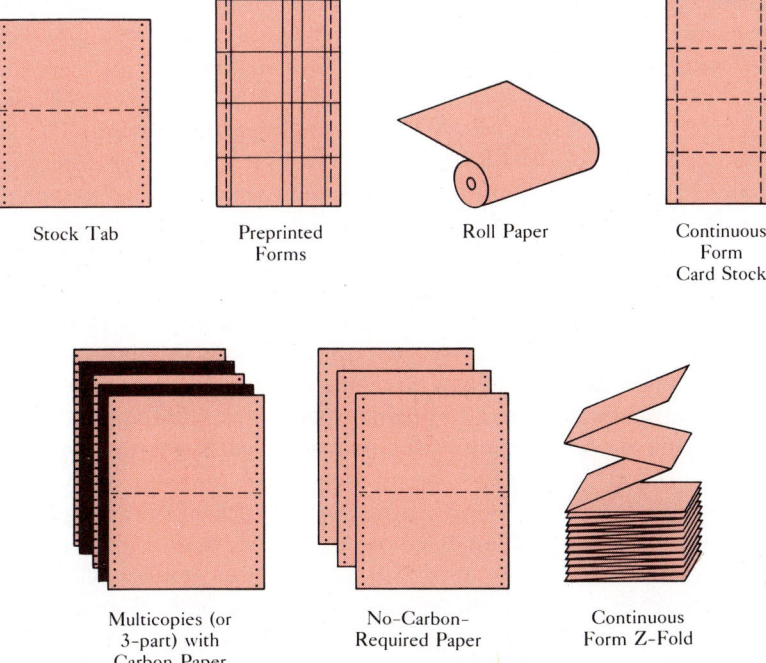

FIGURE 6.6. Continuous fanfolded paper comes in boxes of 3000 sheets.

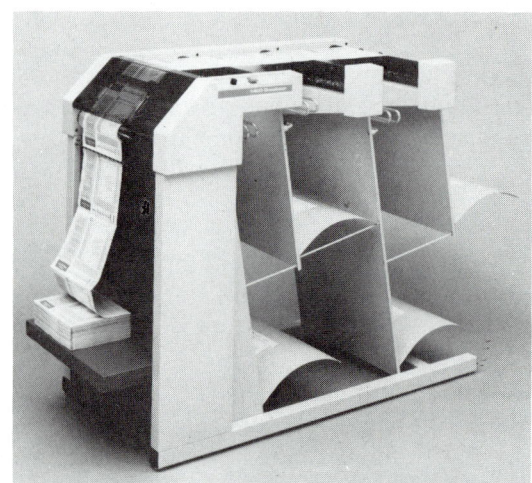

FIGURE 6.7. Bursters and decollators may operate separately or be linked together. (Courtesy of Moore Business Forms.)

paper. The higher the percentage of bond, the higher is the quality of the paper: a 20-pound, 40 percent bond paper is a medium-weight, fairly high-quality paper; a 10-pound, 10 percent bond paper is a lightweight, lower quality paper. If an organization will be using the reports internally, the latter may suffice; if appearance is important, the former may be more appropriate.

Plain paper is less expensive than preprinted forms of the same weight and bond. If an organization requires certain preprinted forms, vendors will quote prices for setup and production. Setup involves making the plate required for printing and usually represents a one-time cost, unless the design changes. Production costs include typesetting, printing, handling, and delivery. The per-page price for production depends on quantity. A thousand forms cost more per page than 50,000, so the analyst should try to gauge appropriate quantities accurately over a reasonable period of time; also, of course, the form might change. Laser printers can provide savings in this area since the printer itself prints the form at the same time it prints the report.

Since most applications demand that reports and other printer outputs be duplicated, the analyst must determine the costs of various duplicating methods by considering the number, frequency, and quality of the output needed. Impact printers can print carbon copies, but carbon paper is expensive and requires decollating, or bursting, of the carbon sets and removal of the carbon paper. One type of paper, called **NCR** (No Carbon Required), produces second, third, or fourth copies via a special chemical coating on the back of each page, except the last. Such paper is about 20 percent more expensive than regular carbon paper but does not require decollating. Given the advances in photocopying technology and falling prices for quality machines, photocopying usually offers an economical alternative to carbon copying; also, the shelf life of NCR paper is limited to a few months.

NCR A type of paper that uses special chemical coatings to produce copies. An acronym for "no carbon required."

Printers have many advantages for users: inexpensively produced reports, easy filing, no need for special viewing equipment, and flexible formats. Disadvantages include relatively high cost and output delay, as well as bulky output and the resulting need for large storage space.

CRTs

Terminals, called CRTs (cathode-ray tubes) or VDTs (video display terminals), are rapidly gathering favor for many applications. Terminals come in all shapes and sizes and offer many different features. Here are the most important options currently available.

1. Appearance of material on the screen:

    - White letters on a black background
    - Black letters on a white background
    - Green letters on a black background
    - Black letters on a green background
    - Gold or amber letters on a black background
    - Letters and other symbols in color

    Eyestrain is a major complaint by users of CRTs.
2. Styles and optional equipment:

    - Built-in memory
    - Built-in floppy disks
    - Attachable to printers
    - Attachable to a cassette tape
    - Tilting screens
    - Nonreflective screens
    - Detachable keyboards
    - Augmented keyboards or function keys
    - Special function keys definable by the user

    CRTs with tilting and nonreflective screens coupled with detached keyboards are preferred by users who spend many hours in front of the CRT.
3. Terminal-to-computer hook-up capabilities:

    - Direct wiring to a computer
    - Linkage through a data communication link
    - Self-operative until there is a need to transmit collected data

    Direct connection is the least expensive method of terminal-to-computer linkage.
4. Speeds:

    - From 110 baud (about 10 characters per second) to 19,200 baud (about 55,000 characters per second)

    Users with high-speed terminals are more productive than those with low-speed ones in spite of the extra cost of the terminal.

As do printers, terminals share several common features (Figure 6.8). The screens on most terminals manufactured since the middle 1970s can display 1920 characters arranged in 24 rows of 80 characters each. When designing reports for terminals, analysts can plan for this 24 by 80 window, no matter what type of terminal eventually is chosen. Newer terminals sometimes display a 25th, or status, line at the bottom of the screen; still others offer a wider screen or more rows. A status line tells users what the terminal is doing at a given moment, whether it is ready for entry, is sending data to the computer, or is awaiting a response from the computer. Wider screens, up to 132 characters per line, permit the terminal to display wider computer forms. Additional rows, useful for word processing, allow the user to see more data at a time.

An organization can use terminals to collect data, to display information, or both. In all cases the analyst must consider format. Possibilities include:

1. *Blinking video*. Flashing data-entry points. Blinking helps the user rapidly locate the cursor on the screen.
2. *Inverse video*. White letters appear on a black background instead of black letters on a light background. Useful for showing data-entry errors to the terminal user.
3. *Secured video*. Data input is not displayed on the screen, thus allowing confidential entries. Most often used for entry of a user's password.
4. *Scrolling or paging*. Information rolls off the top of the screen but can be recalled for reference. This especially helps programmers who must list long programs and want to see them as a complete unit.
5. *Forms or block mode*. Data are sent to the computer in groups of characters at a time. Users enter all the data about a transaction before sending, which allows them to correct any errors before the computer processing takes place.
6. *Dual intensity*. Two different levels of brightness allow one level for data and the other for a form. A background image (such as an invoice) appears while data are being entered onto it. This is especially useful for standardized forms.
7. *Screen splitting or windowing*. The terminal monitors multiple tasks at a time or can be linked to separate computers, with each machine communicating with the terminal. The terminal operator can even split the screen into sections, assigning each computer a separate area.
8. *Hard copy*. A printer attached to the terminal can produce a permanent record of displayed data. This option can be useful for programmers working away from the office or users who need to give receipts to customers.

Terminals grow more popular every day, and all new computers allow terminal hook-ups. Costs of terminals range from as low as $395 for a terminal with little memory to over $10,000 for one with a full range of video, intensity, forms, screen splitting, and hard copy capabilities.

The primary advantage of terminals is their compactness; they can sit conveniently near users, making the system readily available. Disadvantages include the lack of a printed document. In addition to the cost of the terminal itself, an organization must calculate the costs associated with running a terminal environment, such as extra hardware, security, and software. Maintenance costs for terminals can be as low as $15 a month.

FIGURE 6.8. This Hewlett-Packard 150 microcomputer/terminal shows green letters on a black background, can have 256–640 kilobytes of memory, has 12 function keys (across the top row) and a separate keypad for numbers, and features a touch-sensitive screen. It can function as a terminal and as a personal computer running most of the popular word processing, data-base, electronic spreadsheet, and graphics software. As a terminal it can be tied to a variety of brands of computers. (Courtesy of Hewlett-Packard Company.)

FORMATTING REPORTS: HEADINGS, FOOTINGS, DETAILS, AND TOTALS

Most reports, regardless of the medium used to generate them, contain all or some of the following design elements: report heading, page heading, control heading, column heading, detail line, control footing, page footing, and report footing. These elements serve two purposes: (1) they make the report easier to read, and (2) they provide a way to control the content of the report. Consult Figure 6.9 as you read the following paragraphs.

A **report heading** is the title of the report itself. It appears only at the beginning of the report, sometimes on a separate or title page. It may include pertinent information such as the name of the company and the date.

Page headings appear immediately after the report heading. In this category we find page numbers, time of day the report was printed, or the name of the program generating the report.

Control headings are captions and titles that separate one group of data from another. Reports that do not contain distinct groups of data will not need a control heading.

Column headings are captions that appear over vertical units of data and identify what the data underneath represent.

A **detail line** displays the data for a single transaction. If the report will circulate outside the organization, someone will probably edit the detail line to make it look more attractive. Editing a numeric field may involve inserting commas between digits, or printing currency symbols and decimal points. For example, 1234.90 might be edited to read $1,234.90.

Report heading Titles that appear at the beginning of a printed document that shows titles.

Page headings Appear at top of each page and may display dates, page numbers, or program name.

Control headings Captions or titles separating one group of data from another.

Column headings Captions appearing over vertical units of data.

Detail line Displays data for a single transaction.

184 PART II SYSTEMS DESIGN

```
07/14/80   03:42 PM              WABASH FURNITURE COMPANY                                    PAGE 4
LABOR DISTRIB. REPORT
PAY PERIOD ENDING 06/27/84
SOUTHWEST DIVISION DEPT 20—LOADING #2  ①
                  ********DOLLARS**********        *********HOURS*********
CODE DESCRIPTION       TOTAL    REGULAR   PREMIUM    OTHER    TOTAL   REGULAR   PREMIUM   OTHER
1001 VACATION PAY     500.00  ② 0.00       0.00    500.00     0.00  ③ 0.00      0.00     0.00
2003 PARTICLE CORE    214.00    214.00     0.00      0.00    40.00    40.00     0.00     0.00
2006 PANEL JACKETS    210.00    210.00     0.00      0.00    40.00    40.00     0.00     0.00
2010 MACHINING        214.00    214.00     0.00      0.00    40.00    40.00     0.00     0.00
2022 MAINTENANCE  ④   261.60    261.60     0.00      0.00    40.00    40.00     0.00     0.00
2030 GENERAL PF       210.00    210.00     0.00      0.00    40.00    40.00     0.00     0.00
                                                                                          0.00
**** DEPARTMENT TOTALS ****  1,609.60  1,109.60   0.00   500.00   200.00   200.00   0.00   0.00
                  ⑤
```

- Report Heading
- Page Heading
- Control Heading
- Detail Lines
- Page Footing

```
07/14/80   03:42 PM              WABASH FURNITURE COMPANY                                    PAGE 5
LABOR DISTRIB. REPORT
PAY PERIOD ENDING 06/27/84
SOUTHWEST DIVISION
                  ********DOLLARS**********        *********HOURS*********
CODE DESCRIPTION       TOTAL    REGULAR   PREMIUM    OTHER    TOTAL   REGULAR   PREMIUM   OTHER
**** DIVISION TOTALS **** 12,454.60  9,401.60  0.00  3,053.00  960.00  960.00   0.00     0.00
                  ⑥
```

- Control Footing

```
07/14/80   03:42 PM              WABASH FURNITURE COMPANY                                    PAGE 6
LABOR DISTRIB. REPORT
PAY PERIOD ENDING 06/27/84
****FINAL TOTALS****
                  ********DOLLARS**********        *********HOURS*********
CODE DESCRIPTION       TOTAL    REGULAR   PREMIUM    OTHER    TOTAL   REGULAR   PREMIUM   OTHER
**** FINAL TOTALS ****  12,454.60  9,401.60  0.00  3,053.00  960.00  960.00    0.00     0.00
                  ⑦
```

- Report Footing

FIGURE 6.9. A report summarizes captured data in a clear and organized fashion. Note the seven elements: report heading, page heading, control heading, detail line, control footing, page footing, and report footing.

Control footing The final part data item that shows the total for a specific group of data.

Page footing Appears at bottom of each page and may display page numbers, dates, or totals.

Report footing Overall and final values that appear at the end of a report.

A **control footing** is the final part of a control heading, which usually includes data totals. Certain reports do not require control footings, nor do control headings require control footings, or vice versa.

Page footings appear at the bottom of every page and may indicate page numbers or page totals.

A **report footing** occurs once, at the end of the report, indicating the report has ended and often listing report totals.

After thoroughly considering a report's eventual use, an analyst will decide which elements to include in it. For example, since upper-level management makes decisions for the company as a whole, its reports would show only overall totals, using all design elements except detail lines. On the other hand, departmental managers may need to monitor the activities of several subdepartments, and thus their reports may contain multiple levels of control breaks with headings or footings at each break. Finally, office workers' reports might include all the elements. The main point to keep in mind is the constant need to tailor a report to a user's needs.

TYPES OF REPORTS: QUERY, DETAIL, SUMMARY, EXCEPTION, AND PERIODIC

In general there are five types of reports: query, detail, summary, exception, and periodic. Although most conform to basic formatting requirements, each has a unique purpose, and hence demands its own special format considerations.

Because they allow users to obtain immediate answers to their questions, **query reports** are quite common on CRTs. For example, if an accounts receivable clerk wants to know whether a customer has sent a check for a particular invoice, the clerk can enter the customer and invoice numbers at a terminal and obtain that particular account's status at once (Figure 6.10).

A **detail report** displays all the pertinent facts about a situation. A credit card statement listing each purchase and all data relevant to it on separate lines is an example of a detail report (Figure 6.11). When printed, each line may contain some or all of the data collected about the transaction. Such reports form a handy file of information for future reference.

A **summary report** provides overall results and is especially useful when an organization needs to view historical totals or to compare a current activity with a prior one. For instance, a summary report might compare monthly and yearly sales to help management detect trends or review an employee's performance. Figure 6.12 illustrates an increase in equipment sales this year over last.

Query report Shows the details about one transaction, usually on a CRT.

Detail report Displays all the pertinent facts about a group of transactions.

Summary report Provides overall totals for specific groups of detail records.

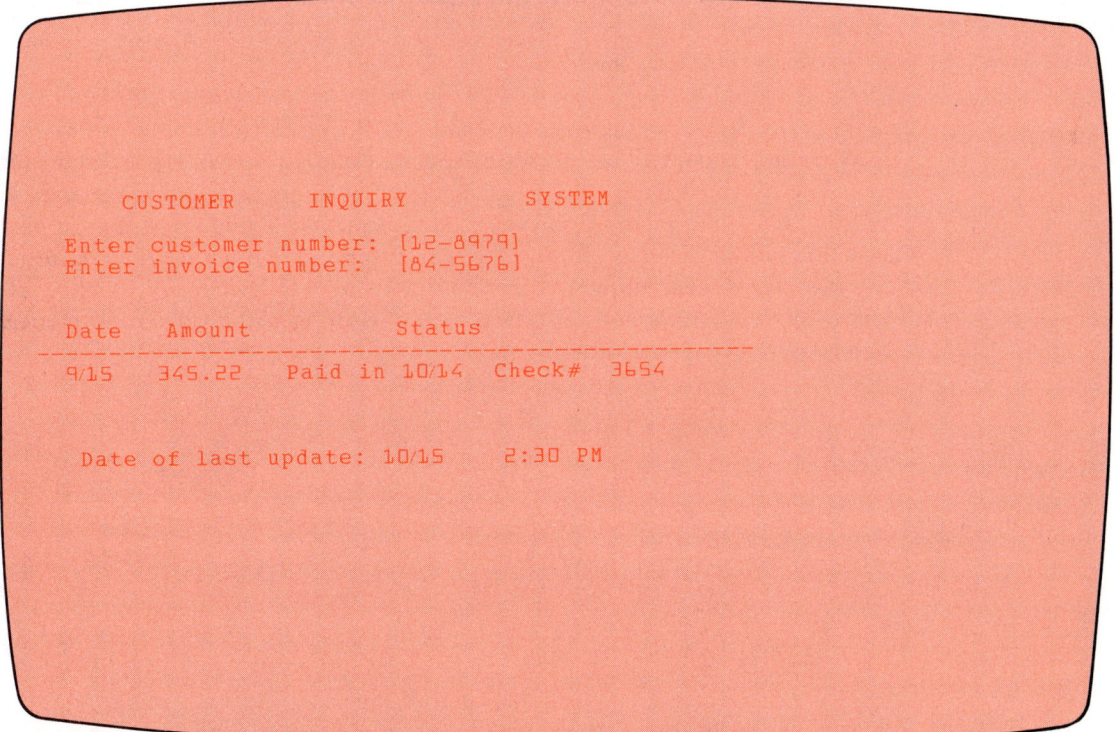

FIGURE 6.10. *Query reports require a terminal with online access to data.*

```
08/12/84    10:37 AM                         WABASH FURNITURE COMPANY
                                                     PAGE 1
PAYROLL EMPLOYEE MASTER FILE
CLOCK #           ---------- NAME ----------    BASE HRS   ST TX CD   STD  D-1   YD  HOURS   YD  FED WH   YD STD D-1   QD    GROSS
     -----        ADDRESS - 1 ------               RATE    LC TX CD   STD  D-2   YD  GROSS   YD FICA WH   YD STD D-2   QD  STATE W
SOC-SEC-#    -----  ADDRESS - 2 ------          LABOR CD   X-FED WH   STD  D-3   YD EX FICA  YD STATE W   YD STD D-3   QD   LOC WH
SORT CODE    ---CITY ---- STATE  -ZIP-          PAY FREQ   X-ST  WH   STD  D-4   YD  EIC     YD  LOC WH   YD STD D-4   QD   EIC
   DIV  DEPT                         EIC          K/H/S    FED EXMP                                      YD UNTD FD    QD EX FICA
EMPL-TYPE    START-DATE TERMN-DATE BIRTH-DATE   LOCALITY   ST  EXMP   MX U FND                           YD MSC DED
===========================================================================================================================
1001          PRESTON PRESIDENT                       40       522       0.00      160.00    1,201.95         5.00       7,500.00
                  SUITE 3-B                     3,000,000      499       0.00   27,000.00      437.15        25.00         441.00
123-45-6789   EMPTY ARKS HOTEL                      2043      0.00       0.00    1,000.00      441.00        15.00           0.00
PRESIDENT     NILES            MI    49120            12      0.00       0.00        0.00        0.00         5.00           0.00
    60   10                                            K         4       0.00                               120.00       1,000.00
     3        01/22/50   00/00/00   05/15/23           0         4     500.00                               975.00
===========================================================================================================================
1002          VINCENT V PEA                           40       514       0.00      120.00      876.51         0.00       4,500.00
              1203 SAND BAR DRIVE               1,500,000      101       0.00   27,000.00      275.85         0.00          80.76
021-42-1173   MUDDY SHORES                          2043      0.00       0.00      500.00       80.76         0.00           0.00
PEA  V        SOUTH BEND       IN    46624            12      0.00       0.00        0.00        0.00         0.00           0.00
    60   10                                            S         1       0.00                               100.00           0.00
     3        02/19/62   00/00/00   11/24/36          71         1     350.00                               875.00
===========================================================================================================================
1003          SUSAN LOGAN                             40       514       0.00      120.00      239.94         0.00       1,700.00
              62351 EDISON ROAD                   400,000      499       0.00    1,700.00      104.21         0.00          30.11
400-74-8205   SOUTH BEND       IN    46251          2043      0.00       0.00        0.00       30.11         0.00           0.00
LOGAN S                                               26      0.00       0.00        0.00        0.00       101.00           0.00
    60   10                                            S         1       0.00                                40.00           0.00
     3        06/06/79   00/00/00   01/26/49          70         1     200.00                               245.00
===========================================================================================================================
1004          CERIL CLERK                             40       514       0.00       90.00      129.98        13.50         695.60
              12045 W MEEK STREET                  6,240       101       0.00      695.60       20.32        10.00          11.54
654-40-6891   TYNER            IN    46572          2042      0.00       0.00      150.00       11.54         0.00           5.98
CLERK C                                               52      0.00       0.00        0.00        5.98         0.00           0.00
    60   10                                            S         1       0.00                                 3.75         150.00
     2        01/15/55   00/00/00   11/24/24          50         1      30.00                                19.00
===========================================================================================================================
1005          SUSIE SECRETARY                         40       514       0.00      170.00      101.26        16.50         973.45
              137 WIND ROAD                        5,150       101       0.00   26,500.00       56.05         0.00          15.15
234-55-1019   APARTMENT 22G                         2042      0.00       0.00      150.00       15.15         5.00           2.76
SECRETARY     NORTH LIBERTY    IN    46554            52      0.00       0.00        0.00        2.76         0.00         150.00
    60   10                                            M         2       0.00                                 2.00
     2        07/25/70   00/00/00   02/11/50          71         2       0.00                                18.15
===========================================================================================================================
```

FIGURE 6.11. Detail reports display all relevant data about each transaction. Such reports can provide backup information for future reference.

Exception report
Displays data about a single or selected group of transactions.

An **exception report** allows a user to specify criteria for isolating a certain set of data. The user in Figure 6.13 asked for a list of items that sold more units in the current year than in the prior year. Note that only equipment sales appear because other items, such as running shoes and clothing, did not meet the criteria. Inventory systems often employ exception reports to detect whenever stock levels fall below the reorder point. Such an exception report can save substantial clerical time and assure accurate reordering.

```
Fleet Feet
Sales by Category                                                        As of: 10/24/84

Category    Description         This           Last           This           Last
Number                          Month          Month          Year           Year

1712-22     Running shoes     $2,336.88      $2,113.55      $34,477.98     $45,607.88
5234-23     Clothing           3,856.77       2,406.66       44,712.44      56,896.07
8659-51     Equipment          2,555.91       3,550.22       89,377.12      81,652.90
```

FIGURE 6.12. A summary report stresses overall totals rather than details, thus permitting a global review of data.

```
Fleet Feet
Sales Analysis: Items selling more this year.                         As of: 10/26/84

Category      Description       This           Last           This           Last
Number                          Month          Month          Year           Year

8659-51       Equipment         $2,555.91      $3,550.22      $89,377.12     $81,652.90
```

FIGURE 6.13. *Exception reports can help users determine trends. The reader saves time by viewing only pertinent data.*

A **periodic report** gives users data at specified time intervals. At the end of a month, an accounts receivable system might produce an aging report listing all customers whose balances are above a certain amount of money and 60 days past due.

Periodic report Results printed at specific intervals, for example, every month.

DESIGNING REPORTS FOR PRINTERS

A good report design fulfills user needs. After consulting users at length, the analyst creates a rough sketch of the proposed system's output. In many cases the analyst will design the format; in others, the user may already have a good report format, perhaps an old manual one, which the analyst need merely adapt to the new system. In yet other cases, the analyst or user may borrow a format from another organization, a vendor, or even the user's competitors.

With the rough sketch in hand, the analyst uses a printer layout sheet to draw the report to the user's specifications. As Figure 6.14 shows, each horizontal line of boxes represents one printed line, with the numbers on the left identifying line numbers. The numbers across the top identify columns and specify where the data will eventually appear on the report.

When drawing the report format, the analyst uses special notations to describe pertinent facts about the design (Figure 6.15):

1. A heavy line indicates the outline of the form.
2. Dashes represent perforations.
3. X's indicate placement of alphanumeric data on the form.
4. 9's indicate placement of numeric data on the form.
5. Preprinted information, such as corporate logos, are handwritten or drawn on the form.
6. Editing, such as comma and dollar symbol insertion for numeric data items or Z's for zero suppression, provides guidelines for the eventual appearance of the data.

Any layout design displays three types of information: preprinted, constant, and variable. Preprinted information includes the words that the form supplier will

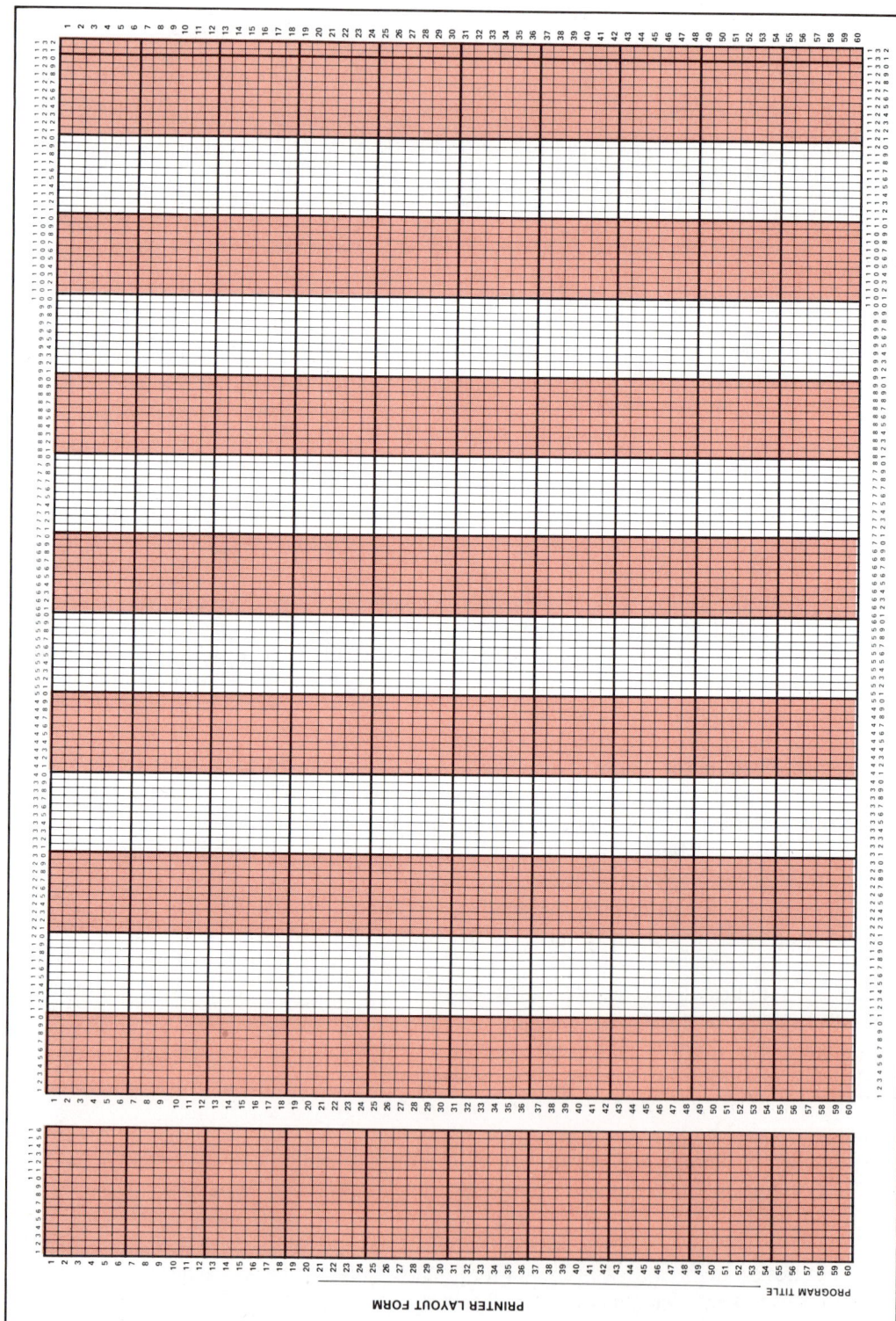

FIGURE 6.14. Printer layout sheet, or spacing chart, enables analysts to design reports.

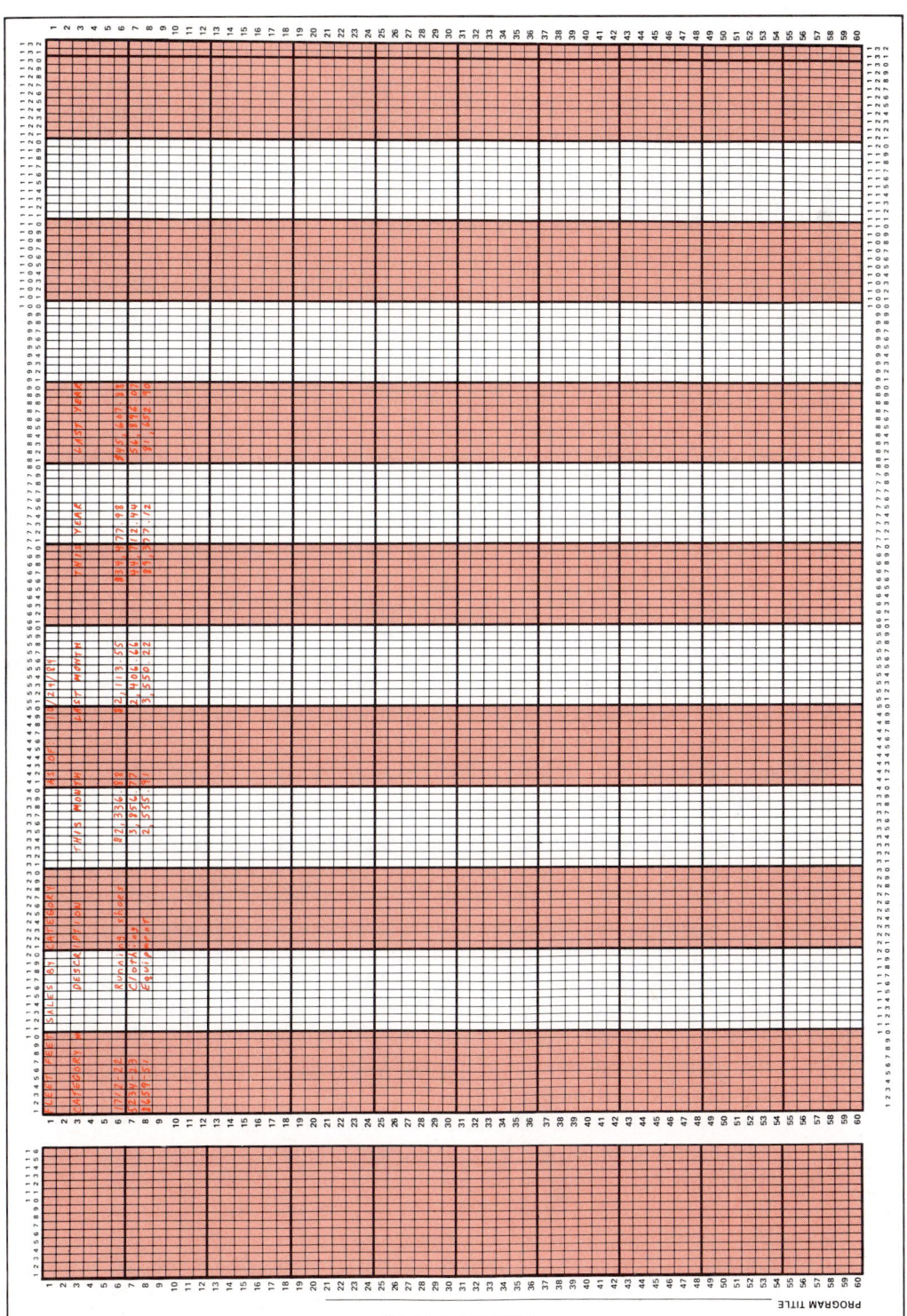

FIGURE 6.15. Spacing chart filled in with the format of the report.

print on the paper beforehand. Constant information consists of headings the computer will print or display on every form, and variable information will change for each printed line. For example, in Figure 6.11 we see a page heading, column headings identifying totals and accounts receivable aging data, detail lines with specific customer data, and a report footing giving the totals of each aging category.

The analyst must design a layout for every different type of page the system will produce. The procedure begins with discussion with the user, continues through drawing the layout on a spacing chart, and concludes with design review and user approval. Painstaking planning of reports minimizes difficulties during subsequent design stages, and report formats play a major role in the next design stage, the data-base or record layout. An incompletely designed form may cause

Guidelines for CRT Screen Designs

Terminals have become such a preferred input/output method for computer users and we see them everywhere: airline reservation desks, bookstores, dentists' offices, and as the butt of jokes in many popular cartoon strips. With so many different kinds of people using terminals, analysts must pay special attention to users' needs and capabilities when designing them. Clear, concise, and accessible screen design not only helps users enter data accurately, but it also creates an atmosphere of friendliness and ease in which users can feel comfortable and can more easily master the system.

The CRT terminal serves four functions: data collection, validation, reporting, and inquiry. In Chapter 9 we develop guidelines for designing formats for data collection and validation, and so we can restrict this discussion to data reporting and inquiry, activities directly related to systems outputs.

Because we cannot expect users to be computer experts, we focus our attention on simplicity. Despite the complexity of the system and its programs, users' screens need not be ornate or confusing. The following basic rules can help screen designers achieve good results.

1. Since users are accustomed to reading printed media in this way, screens should read from the top down, and from left to right.
2. Since users feel more comfortable with words than with numbers, screens should receive titles rather than their code numbers. Tops of columns, total lines, and footings should display captions.
3. Concise action verbs such as ENTER, SELECT, and CANCEL should direct operations. Extraneous pleasantries such as PLEASE, or WHICH DO YOU WISH? only clutter up the screen.
4. Use a standardized vocabulary throughout (e.g., do not use INPUT in one place and ENTER in another).
5. Data display should reflect cultural conventions (e.g., dates in MM/DD/YY for civilians and in YY/M/DD form for the military).
6. Always use the same termination key both for terminating a screen and for returning to menus. (Do not use both "Q" for quit and "S" for stop.)
7. A test should be provided to ensure correctness of entries. For example, provide for "Y" or "N," alternatives in a given situation, but do not allow any other possibilities.

the analyst to overlook a valuable field that should be included in the data base.

Having completed the layout, the analyst submits it to the user for approval. If it is approved, both analyst and user sign it. If the user does not approve the design, the analyst notes desired changes, redraws the form, and returns it to the user for approval. It sometimes takes several meetings between the analyst and the user before a final design is approved.

DESIGNING REPORTS FOR CRTS

The procedure used for designing terminal, or screen, reports parallels that for printed reports. After consulting the user, the analyst uses a screen spacing chart to sketch the proposed screen layout (see Figure 6.16). A screen spacing chart

8. Blinking video can attract the user's attention to crucial data.
9. Contrasting levels of intensity or colors provide visual clues for differentiating prompts from data.
10. If an application demands display of more than one screen of data during an operation, users should have to stop a given display at its end and enter a command before being allowed to continue to the next screen.
11. Signals should provide a clear pathway between multiple screens that depend on one another.
12. Pilot screens can help train users.
13. Neutral observers should review and test all designs.
14. Users must be thoroughly trained on the various screens. Their training need not include hardware or software details, but should focus on their daily interaction with the system.
15. Training manuals should describe both normal and abnormal operational procedures, and should be written in the user's vocabulary and language.
16. A review with users should occur some time after the system has been operating.
17. Help facilities can aid operators when they do not know what to do. These should be available at a single keystroke.
18. Data and captions on the screen should be:
    a. Aligned vertically.
    b. Grouped in a logical fashion.
    c. Not crowded together.
    d. Self-explanatory, not requiring the operator to refer to a reference manual.

Basically these rules reflect common sense and force analysts to keep users' needs and abilities at the top of the priority list when designing screens. No matter how sophisticated users may become as they master their systems, they will always appreciate the simplest, easiest-to-use screens.

Source: D. Dardwin, "Teleprocessing System Design," *ICP INTERFACE—Data Processing Management*, pp. 21–22, Summer 1979. Reprinted with permission from ICP INTERFACE—Data Processing Management, Copyright © 1979, International Computer Programs, Inc., Indianapolis, Ind.

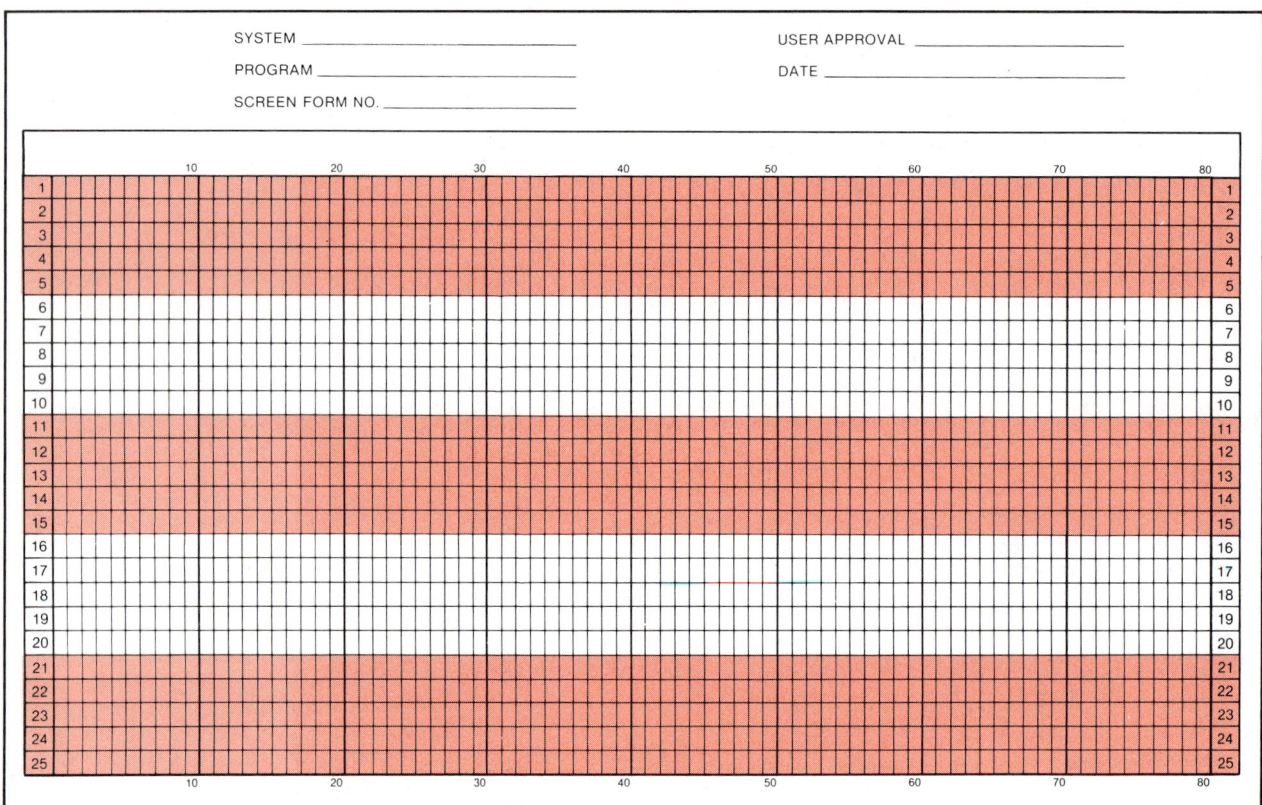

FIGURE 6.16. Screen spacing or layout charts look like those for printers but allow for 24 lines and 80 columns. Such charts may also allow a 25th, or status line.

looks like a printer spacing chart, except it depicts a CRT screen, with 80 characters across and 24 (or 25 if a status line will be used) lines down (Figure 6.17). Because a screen has certain highlighting capabilities (that is, it can provide visual cues such as reverse, blinking, secured, or bright video), the analyst must define such requirements on the screen spacing chart. As you can see in Figure 6.18, highlighting notes are written beside each field.

Terminals can accommodate input, output, or both. Though analysts design input after output, they must consider input requirements if the application will require a terminal to handle input. For example, a user may wish to use a terminal to inquire about the balance of a customer's account. A screen design must prompt the user for the customer's account number before it can display appropriate data. In other words, the terminal will ask for a little input before it releases any output.

If the analyst selects dual intensity, the screen design may show a form behind the data. The form may appear in normal video while the data are shown in bright video. The background form should resemble the form that the system will actually use.

An analyst should always avoid placing too much data on one screen. Too often, beginning analysts attempt to squeeze everything onto a single screen rather than

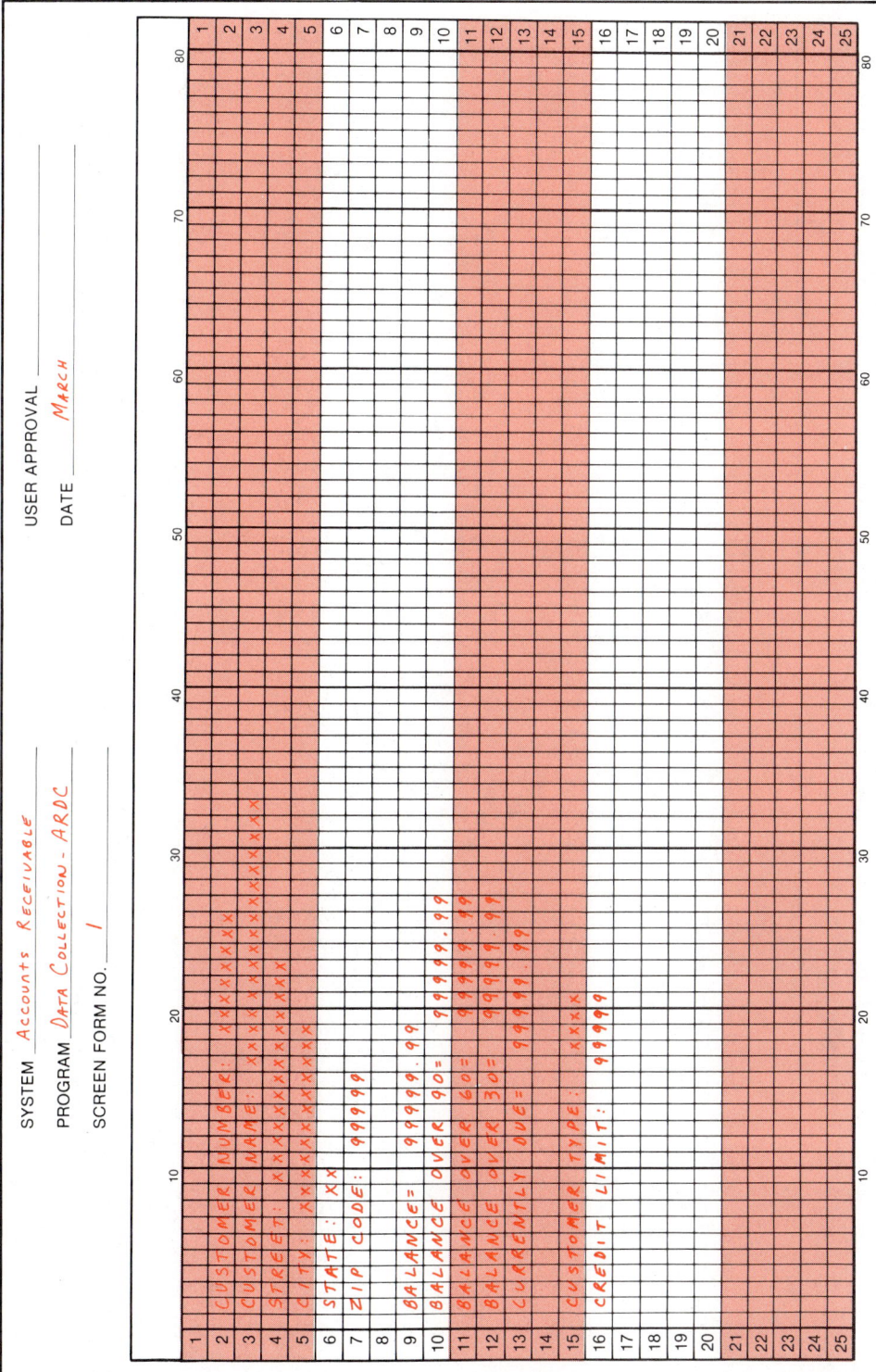

FIGURE 6.17. The analyst sketches the proposed design in the boxes on the form. Since the screen possesses physical boundaries, the design does not need to be boxed as a printer spacing chart does.

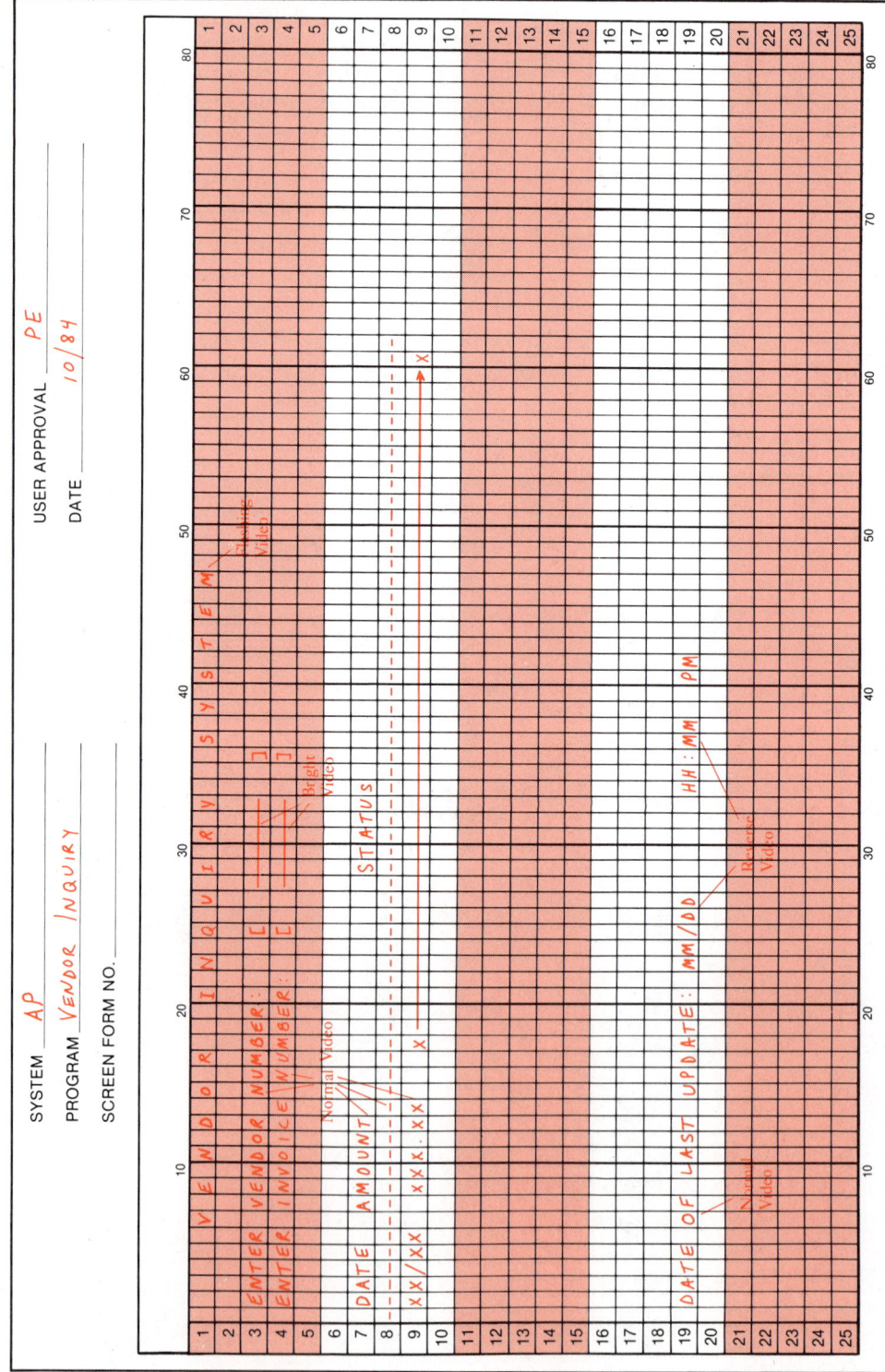

FIGURE 6.18. Highlight codes enable the analyst to specify screen requirements graphically.

use two or more for a large report. If the report requires more than one screen, the status line notifies users which they are seeing at a given time. Some terminals offer sufficient memory to store both screens, which allows the user to flip easily from one to the other. Terminals with insufficient memory require the user to store data collected on the first screen before entering data on the second.

CONTROL OF SYSTEM OUTPUTS

A **control system** is a set of procedures that ensures proper and approved system operation. You probably already know some controls: verification and validation during data collection and entry, time stamping (dating), page numbering of printed reports, and basic security provisions such as passwords, read and write access to files, and file backups.

Control system Set of procedures that help ensure proper and approved system operation.

Although a control system helps prevent computer-related fraud, the analyst uses it primarily to eliminate errors. Once an error enters a computerized system, it takes time and money to remove it. If a department store mistakenly charges one customer's purchase to another customer's account, the store eventually must perform two operations to correct the error. First it must remove the amount from the wrongly charged account, and then it must transfer that amount to the correct account. In addition to wasting time and money, the department store risks losing the goodwill of both customers. A good control system can minimize such costly errors.

An Overview of Control Systems

One can build controls into a system at three points: data collection and entry, file processing, and output distribution. Controls are costly, demanding extra effort on the part of the data-entry staff during input, and they consume additional computer time to track, count, and total data. In the long run, however, the cost of good controls outweighs the cost of errors or losses due to fraud or crime.

Control systems vary in complexity and sophistication, depending on an organization's needs and size. A retailer with few customers needs only a simple control system, whereas a large organization, such as Sears Roebuck and Company, with geographically dispersed offices and diverse points of data entry, needs an elaborate control system.

Whether simple or complex, all control systems fall into two broad categories: machine and manual. A computer-based (machine) control system uses the computer to help detect errors. To test a date, for example, the computer can determine whether a month number falls between 1 and 12, a day number between 1 and 31, and a year number within an acceptable range. With a manual control system, someone visually checks for errors. For example, the data-entry operator might enter the customer's account number from a source document and have the computer display it along with the customer's name on the terminal screen, and then visually compare the name on the screen with the one on the source document.

Quite often combinations of machine and manual control systems make sense. Before entering data from a source document, an operator can hand-total specific fields. The operator then enters the data. During the entry process the computer is programmed to total the same fields the operator hand-totaled, and at the end of the data-entry process, the operator enters the hand-calculated total, so the computer can compare the two. If the two totals match, processing may proceed. If they do not, the operator must locate the error and correct it before processing.

After an organization has collected, verified, validated, and processed its data, it wants to produce output, usually in the form of reports. The generation and distribution of reports demand additional controls. Simple output controls such as page numbers, dates, and times identify reports (Figure 6.19), and headings listing the name of the system that generated the report facilitate recognition and distribution.

Totaling and Cross-footing

Cross-footing Totaling all columns and rows of data to assure accuracy.

Analysts may also build **totaling** and **cross-footing** into their report designs (Figure 6.20). Totaling adds all numbers in a specific column, while cross-footing adds all column totals, after which the computer can compare column totals with the final total. In Figure 6.21 total gross sales ($15,760.53) minus returns ($191.79) should equal net sales ($15,768.74). Although some reports, such as an inventory report of books on hand, do not lend themselves to totaling and cross-footing, whenever possible, analysts should design this control element into report layout.

Auditing

Auditing Tracing a transaction through a system to verify accurate processing.

Another system control technique, **auditing,** traces a transaction through the system, verifying that intended processing actually occurred at the correct times and locations. Accountants frequently speak of an audit trail, which allows them easily to go back over their work to find and fix mistakes. Special software or the system itself can create such an audit trail. As a system processes a record, it can print out a copy of the old master record, the transaction record, and the new master record, noting the transaction's effect on the system (Figure 6.21).

```
SILVA BOOKSTORE              LIST OF FIXED ASSETS                AS OF 12/21/84
TRENTON NEW JERSEY                                               PAGE:      1


DEPARTMENT           DESCRIPTION         DATE          COST         LIFE         SALVAGE
NUMBER                                   ACQUIRED      $            YEARS        VALUE

    3                BOOK RACK           5/84          780          10           50
    3                LAMPS               5/84          230           5            8
    6                DESK                6/84          360           8           62
    6                FILE CABINET        7/84          210           4           25
    6                CHAIR               6/84           99           5           15



TOTAL ITEMS:     5
```

FIGURE 6.19. System outputs are usually reports distributed to users. This fixed asset report shows headings, dates, page number, and the system name.

```
SILVA BOOKSTORE              SALES BY SALESPERSON            AS OF 12/06/84
TRENTON NEW JERSEY                                            PAGE:    1


SALESPERSON     SALESPERSON              SALES        RETURNS      NET SALES
  NUMBER          NAME                    ($)           ($)           ($)
     83         LARDNER, SCOTT          1622.56        10.06         1612.50
     95         GLOYD, DOROTHY          5001.23       100.21         4901.02
    107         LUNDBERG, CAROL         2305.00         0.00         2305.00
    235         FULWEILER, JEFF          345.90        45.80          300.10
    363         OLSHEFSKY, CELIA        6009.45         9.00         6000.45
    679         CHASTAIN, CHARLENE        25.89         1.22           24.67
    890         HYATT, JAN               450.50        25.50          425.00


TOTALS FOR NOVEMBER                  $15,760.53     $(191.79)      $15,768.74
```

FIGURE 6.20. Cross-footing and totaling control procedures form part of report format and design.

Data-base management and inquiry software systems greatly assist in verifying that a system has made all desired changes. Almost every major data-base management system allows users to inquire about stored data. By entering simple English-like commands, the user or auditor can examine data-base contents to ascertain that they reflect changes (Figure 6.22). For example, a user wishing to locate a record enters LIST, which orders the data-base manager to search for the desired record. If it finds the record, the system reports so immediately. If it does not, the system reports that fact. Having found a desired record, the system displays the data record in it by responding to the user's command. Popular systems

```
SILVA BOOKSTORE              CUSTOMER UPDATE REPORT           AS OF 12/06/84
TRENTON NEW JERSEY                                            PAGE:    1
Note: Letters on the right for transaction records mean: A—New customer,
      C—Change in customer data, and D—Deletion of customer from file.

CHANGE OF ADDRESS
OLD MASTER:     12332 JOHN'S MOVING      445 SUTTER ST.
NEW MASTER:     12332 JOHN'S MOVING      122 J ST.
TRANSACTION:    12332                    122 J St.                          C

DELETION OF A CUSTOMER
OLD MASTER:     47789 GRUMPY'S BURGERS   5671 HWY. 1
NEW MASTER:
TRANSACTION:    47789                                                       D

NEW CUSTOMER
OLD MASTER
NEW MASTER:     70956 THE OFFICE         1 GREENBACK BLVD.
TRANSACTION:    70956 THE OFFICE         1 GREENBACK BLVD.                  A
```

FIGURE 6.21. Audit trails follow each transaction through the system, showing its effect on the files. Blanks indicate that the item either did not exist on the old master file or will not appear on the new master file.

```
LIST CUSTOMER#, CUST-NAME, CUST-ADDRESS FOR
     CUSTOMER# = 12332

CUSTOMER#      CUST-NAME       CUST-ADDRESS
12332          JOHN'S MOVING   122 J ST.

LIST CUSTOMER#, CUST-NAME, CUST-ADDRESS FOR
     CUSTOMER# = 47789

0 ENTRIES QUALIFIED

LIST CUSTOMER#, CUST-NAME, CUST-ADDRESS FOR
     CUSTOMER# = 70956 OR CUSTOMER# = 12332

CUSTOMER#      CUST-NAME       CUST-ADDRESS
70956          THE OFFICE      1 GREENBACK BLVD.
12332          JOHN'S MOVING   122 J ST.
```

FIGURE 6.22. *Hewlett-Packard's Image data-base manager allows novice users to interrogate the data base. The user enters an English-like command, LIST, and the system displays its findings.*

that permit users to interrogate and audit a system's outputs to ensure that desired events actually do happen include Hewlett-Packard's Image, Burroughs' Inquiry, and Adabase's Adascript.

Security Checks

Since a company's management wants a system physically secure from unauthorized use or tampering, analysts must establish physical control procedures. Several good security measures are available. The obvious first step involves locking important, negotiable documents, such as blank payroll checks, in a vault located somewhere other than the computer center.

A second measure is to print identification numbers along the top or bottom edge of documents. Such numbers facilitate easy identification of a given form, including the system from which it came and whether it has fallen into unauthorized hands. Furthermore, printing dates on all documents helps users distinguish old forms from newly revised versions.

As a third means of control, the analyst should give explicit directions about proper disposal of reports. Since some reports may hold confidential data, such as gross pay or personal telephone numbers, they should be shredded rather than tossed into the trash or recycled when the user is finished with them. Analysts should be aware of the legal aspects of form control: social security numbers, drivers' license numbers, gross income, telephone numbers, and so on, are privileged and private information. Federal law defines exactly what kinds of data need to be protected. For example, the Tax Reform Act of 1976 requires the Internal Revenue Service to obtain a court order before it can gain access to bank records. Also, the Electronic Funds Transfer Act of 1979 requires financial institutions to notify consumers of any release of customer information to a third party.

Finally, valuable forms, such as blank checks or stock certificates, should be serially numbered by the forms supplier so users can record beginning and ending numbers before and after use. This also enables users to extract and destroy forms ruined during printer adjustment. One individual usually assumes responsibility for maintaining form security, recording serial numbers, and reporting violations to an appropriate manager.

Control design is an integral part of output design. As government enacts more laws to protect the privacy of data, this aspect of output design will become even more important.

BUILDING A DATA DICTIONARY

As Figure 6.1 indicates, one result of output design is a data dictionary. After designing report forms and obtaining user approval, the analyst must determine what data the system will need in order to print the reports. A data dictionary helps state such requirements clearly and efficiently. Since an information system aims to provide information to users, the analyst creates a data dictionary to ensure that this goal is met.

To see how this is done, review the exception report in Figure 6.13, which lists sales statistics for items that sold better this year than last year at Fleet Feet. By carefully considering each element of this report, we can quickly create its data dictionary:

```
SALES BY CATEGORY = Category Number +
                    Description +
                    Sales This Month +
                    Sales Last Month +
                    Sales This Year +
                    Sales Last Year +
                    Report Date
Category Number  = General Ledger Account Number +
                    Store Number
Report Date      = Month Number +
                    Day Number +
                    Year Number
```

Leveling stops when a term or field becomes self-defining. Three fields form the report date: month, day, and year numbers. Since none of these fields requires further clarification, leveling stops here.

The analyst can complete the data dictionary by adding further details about each item in the report. For example:

Field Name	Type	Length	Editing or Comments
Category number	Numeric	6 characters	Should be edited XXXX-XX
Description	Alphanumeric	20 characters	Left justified
Sales this month	Numeric	10 digits	Edited $$$,$$9.99
Sales last month	Numeric	10 digits	Edited $$$,$$9.99
Sales this year	Numeric	10 digits	Edited $$$,$$9.99
Sales last year	Numeric	10 digits	Edited $$$,$$9.99

Note that we have not yet dealt with the processing required to produce the final report. That processing description comes later, during the detailed design of program modules.

Besides defining and naming fields, the analyst must describe other report properties.

1. *Order*—sequence of data (for example, ascending by category number).
2. *Totals*—items or categories to be totaled over a specific length of time (for example, sales totals for a specific month).
3. *Frequency*—schedule for generating reports (for example, monthly).
4. *Volume*—estimated number of pages or number of screens required by the report.
5. *Paper type*—(applicable only to printed reports) weight and bond of paper desired.
6. *Distribution*—who receives the report.
7. *Security*—report's audience (internal, external, confidential, public) and method of disposal.

The analyst can collect all such descriptive information in a single document, and some computer services departments may insist on using a standardized one they developed that has all the properties listed.

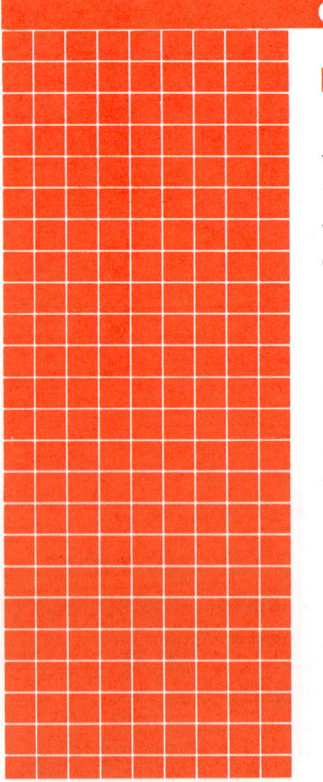

CASE STUDY

Designing the Reports for Fleet Feet's Accounts Payable System

You will remember from the preceding chapter that analyst Frank Pisciotta's feasibility study identified needed reports only by title and function. Now Frank must discuss each report with its users, Sally Edwards and Mark Stensaas, so he can design the reports' formats and determine what data the system will need in order to print them.

In addition to the check to vendors, Fleet Feet's accounts payable system should produce 10 other reports (Figure 6.23). Frank sketches those reports destined for the Fleet Feet's dot-matrix 400-line-per-minute printer on a printer spacing chart. Reports intended for display on a terminal are shown in screen layout form.

When the sketches have been approved by the users, Frank writes a data dictionary for each. As a part of each report's data dictionary definition, Frank lists the control factors: the report's order; any subtotals, final totals, or counts; the report's frequency; expected length; type of paper; distribution; and security. It is Frank's standard practice to date and page number all reports.

Vendor lists (alphabetic or numeric) allow users to locate vendor data when they only know one of the vendor's identifiers, name or number. These lists aid users in many ways: looking up a vendor number when there is doubt as to its accuracy, checking amounts owed a particular vendor, or counting the number of vendors to whom the company owes money. Figure 6.24 shows Fleet Feet's alphabetic vendor list and its accompanying data dictionary.

The numeric list of vendors (Figure 6.25) displays much of the same data as its alphabetic equivalent, except that it includes purchases and discounts for present and previous years.

CASE STUDY

Report Name	Function or Brief Description of Report
1. Alphabetic vendor list	An alphabetic listing of each vendor, addresses, discount terms, last activity date, purchases, and discounts taken this year and last.
2. Numeric vendor list	A numeric listing like the alphabetic one except that it is in order by vendor number.
3. AP transaction register	Detail list of every transaction: vendors, amounts, dates, discounts, and distributions.
4. AP open item report	Detailed list of transactions that need to be paid. Shows balances, discounts, and age of the debts.
5. AP account inquiry	A query report to show the debts owed a vendor, including total of all debts.
6. Cash requirements report	A report showing the amounts of money due on particular days and the vendors due for payment.
7. Stub-over check	The check used to pay vendors for items purchased. Top half shows invoice data and lower half is the actual check.
8. AP check register	List of checks written to vendors and the associated amounts. Totals are shown at report's end.
9. Distribution to GL	Breakdown of amounts from various invoices as they are applied to specific general ledger accounts.
10. Vendor analysis	List of vendors and a comparison of this year's versus last year's purchases.
11. Clear vendors	A screen report prompting the user to set all vendors' totals to zero. This initializing is a once-a-year report.

FIGURE 6.23. Eleven reports will flow from Fleet Feet's AP system. After discussions with users, appraisal of new user needs, and examination of other AP systems, the analyst may expand the number of reports beyond those proposed in the feasibility study.

These two data elements permit users to estimate the importance of a given vendor to Fleet Feet's success.

An important detail report for Fleet Feet's new system will be the AP transaction register (Figure 6.26), which records, in voucher number order, details about each transaction posted to the accounts payable system. It gives the name of the vendor, the date and amount of

OCT 24, 1984
2:59 PM

FLEET FEET

ALPHABETICAL VENDOR LIST

VENDOR STATUSES BLANK = NORMAL A = ALWAYS TAKE DISCOUNT H = HOLD PAYMENT N = NOT TO BE PURCHASED FROM

VENDOR #	NAME ADDRESS 1	ADDRESS 2 ADDRESS 3	TYPE STATUS	DUE DAYS	---TERMS--- DISC DAYS	DISC %	LAST ACT-DATE
000001	ACME OFFICE FURNITURE 456 BROADWAY	BIGTOWN, TEXAS 99999	CAP	2% 10 30	NET 30 10	2.00	03/15/84
000003	BIGTOWN OFFICE SUPPLIES 123 BIGTOWN AVENUE	BIGTOWN, TEXAS 99999	EXP	2% 10 30	NET 30 10	2.00	06/26/84
000005	BLUE FLASH REPAIR SERVICE 150 COMMERCIAL BLVD	BIGTOWN, TEXAS 99996	OVH	NET 15 15	0	.00	02/24/84
000200	FUTURE ELECTRONICS, INC. 300 HOLCROFT AVENUE	NEWTON, VERMONT 12345	MTL A	2% 10 30	NET 30 10	2.00	05/17/84
123ABC	HARRIS, COHEN AND SMITH 1200 DAVY CROCKETT BLVD	SUITE 3000 BIGTOWN, TEXAS 99997	LEG H	2% 30	NET 30 10	2.00	03/20/84
000004	JONES PROPERTY MANAGEMENT 250 FRIENDSHIP CIRCLE	DALLAS, TEXAS 90000	OVH H	MONTHLY PAYMENT 0	0	.00	06/01/84
PROSPY	PROSPERITY INVESTMENT BANKERS 100 GENERAL GRANT CIRCLE	PENTHOUSE SUITE CHICAGO, ILLINOIS 11111	CAP N	MONTHLY PAYMENT 0	0	.00	06/30/84
000002	RED LINE FREIGHT 3504 COMMERCIAL ROAD	BIGTOWN, TEXAS 99996	OVH	NET 30 30	0	.00	06/11/84
000100	SOUTHERN GAS & POWER 1500 FREEDOM AVENUE	EAST BIGTOWN, TEXAS 99991	UTL	NET 25 25	0	.00	06/20/84
200000	STAR ELECTRONIC COMPONENTS 8000 FRANKLIN BOULEVARD	SANTA CLARA, CALIFORNIA 94120	MTL A	5% 30 45	NET 45 30	5.00	04/12/84
000050	TEXAS TELEPHONE 555 NETWORK LANE	BIGTOWN, TEXAS 99997	OVH	NET 20 20	0	.00	06/25/84
ABC	WILLIAMS & FREGOSI 1260 BROADWAY	SUITE 500 BIGTOWN, TEXAS 99999	ADV H	5% 15 30	NET 30 15	5.00	06/22/84

12 VENDORS ON FILE

(a) *Vendor lists provide users with a means of cross-referencing vendor names and numbers.*

```
NAME of REPORT: Alphabetic Vendor List                    DATE: 10/84
ANALYST: Frank Pisciotta

ALPHABETIC VENDOR LIST = Vendor number +        *  6 Alphanumeric *
                         Vendor name +          * 30 Alphanumeric *
                         Address 1 +            * 30 Alphanumeric *
                         Address 2 +            * 30 Alphanumeric *
                         Address 3 +            * 30 Alphanumeric *
                         Type +                 *  4 Alphanumeric *
                         Due days +             *  4 Numeric      *
                         Discount days +        *  4 Numeric      *
                         Discount percentage +  *  4 Numeric      *
                         Last Activity +        *  4 Numeric      *
                         Page number +          *  4 Numeric      *
                         Today's date +         * 20 Alphanumeric *
                         Time of day            *  7 Alphanumeric *
```

Order of report: Ascending by vendor name.
Subtotals: None.
Final totals: None.
Counts: Number of vendors in the file.
Frequency: On request of user.
Length: Less than 50 pages.
Type of paper: 132-column regular computer paper.
Distribution: Directly to user.
Security: Internal use only.

(b) The data dictionary for Fleet Feet's alphabetic vendor report. Numbers in the right-hand column refer to the number of characters allowed for each item.

FIGURE 6.24. The alphabetic vendor list contains vital information about purchases, including each vendor's terms for payment.

OCT 24, 1984
2:59 PM

FLEET FEET
PAGE 0001
PAGE 0001

N U M E R I C   V E N D O R   L I S T

VENDOR STATUSES BLANK = NORMAL A = ALWAYS TAKE DISCOUNT H = HOLD PAYMENT N = NOT TO BE PURCHASED FROM

VENDOR #	NAME ADDRESS 1	ADDRESS 2 ADDRESS 3	TYPE	DUE DAYS	TERMS DISC DAYS	DISC %	LST-ACTV STATUS	PURCHASES-YTD PURCHASES-LYR	DISCNTS-YTD DISCNTS-LYR
000001	ACME OFFICE FURNITURE 456 BROADWAY	 BIGTOWN, TEXAS 99999	CAP	2% 10 30	NET 30 10	2.00	03/15/84	1,594.00 4,287.50	20.00 70.00
000002	RED LINE FREIGHT 350 COMMERCIAL ROAD	 BIGTOWN, TEXAS 99996	OVH	NET 30 30	0	.00	06/11/84	1,972.00 .00	.00 .00
000003	BIGTOWN OFFICE SUPPLIES 123 BIGTOWN AVENUE	 BIGTOWN, TEXAS 99999	EXP	2% 10 30	NET 30 10	2.00	06/26/84	7,764.00 6,000.00	151.78 120.00
000004	JONES PROPERTY MANAGEMENT 250 FRIENDSHIP CIRCLE	 DALLAS, TEXAS 90000	OVH	MONTHLY PAYMENT 0	0	.00	06/01/84 N	28,000.00 30,000.00	.00 .00
000005	BLUE FLASH REPAIR SERVICE 150 COMMERCIAL BLVD	 BIGTOWN, TEXAS 99996	OVH	NET 15 15	0	.00	02/24/84	1,500.00 2,500.00	.00 .00
000050	TEXAS TELEPHONE 555 NETWORK LANE	 BIGTOWN, TEXAS 99997	OVH	NET 20 20	0	.00	06/25/84	3,450.00 3,600.00	.00 .00
000100	SOUTHERN GAS & POWER 1500 FREEDOM AVENUE	 EAST BIGTOWN, TEXAS 99991	UTL	NET 25 25	0	.00	06/20/84	990.00 1,250.00	.00 .00
000200	FUTURE ELECTRONICS, INC. 300 HOLCROFT AVENUE	 NEWTON, VERMONT 12345	MTL	2% 10 30	NET 30 10	2.00	05/17/84 A	12,500.00 45,000.00	250.00 900.00
123ABC	HARRIS, COHEN AND SMITH 1200 DAVY CROCKETT BLVD	SUITE 3000 BIGTOWN, TEXAS 99997	LEG	2% 30	NET 30 10	2.00	03/20/84 H	200.00 2,500.00	4.00 25.00
200000	STAR ELECTRONIC COMPONENTS 8000 FRANKLIN BOULEVARD	 SANTA CLARA, CALIFORNIA 94120	MTL	5% 30 45	30	5.00	04/12/84 A	25,000.00 84,000.00	8,808.00 4,200.00
ABC	WILLIAMS & FREGOSI 1260 BROADWAY	SUITE 500 BIGTOWN, TEXAS 99999	ADV	5% 15 30	NET 30 15	5.00	06/22/84 H	4,285.00 17,500.00	95.00 500.00
PROSPY	PROSPERITY INVESTMENT BANKERS 100 GENERAL GRANT CIRCLE	PENTHOUSE SUITE CHICAGO, ILLINOIS 11111	CAP	MONTHLY PAYMENT 0	0	.00	06/30/84 N	15,000.00 50,000.00	.00 .00

12 VENDORS ON FILE

(a) *The numeric list shows much of the same data as the alphabetic one. The major difference is the order of vendor data.*

```
NAME of REPORT: Numeric Vendor List              DATE: 10/84
ANALYST: Frank Pisciotta

NUMERIC VENDOR LIST = Vendor number +         *  6 Alphanumeric *
                      Vendor name +           * 30 Alphanumeric *
                      Address 1 +             * 30 Alphanumeric *
                      Address 2 +             * 30 Alphanumeric *
                      Address 3 +             * 30 Alphanumeric *
                      Type +                  *  4 Alphanumeric *
                      Due days +              *  4 Numeric      *
                      Discount days +         *  4 Numeric      *
                      Discount percent +      *  4 Numeric      *
                      Last activity +         *  4 Numeric      *
                      Purchases YTD +         * 10 Numeric      *
                      Purchases last year +   * 10 Numeric      *
                      Discounts YTD +         * 10 Numeric      *
                      Discounts last year +   * 10 Numeric      *
                      Page number +           *  4 Numeric      *
                      Today's date +          * 20 Alphanumeric *
                      Time of day             *  7 Alphanumeric *
```

Order of report: Ascending by vendor number.
Subtotals: None.
Final totals: None.
Counts: Number of vendors in the file.
Frequency: On request of user.
Length: Less than 50 pages.
Type of paper: 132-column regular computer paper.
Distribution: Directly to user.
Security: Internal use only.

(b) *The data dictionary for Fleet Feet's numeric vendor list.*

FIGURE 6.25. *The numeric vendor list is similar to its alphabetic counterpart but includes present and previous years' purchases as well as the discounts Fleet Feet has taken.*

```
OCT. 24, 1984                                           FLEET FEET                                                              PAGE 001
3:53 PM
                                            AP  TRANSACTION  REGISTER

TRX TYPES       R = REGULAR VOUCHER      C = CANCELLATION VOUCHER     A = ADJUSTMENT TO DISTRIBUTION      P = PREPAID CHECK

VOUCH#    VENDOR   NAME                       INVOICE#    INV-AMT       DUE-DAYS    DISC-DAYS    DISC-%    CHECKS       AP-ACCT#
TRX-TYP   #        TERMS            P.O.-#    INV-DATE    NON-DISC-AMT  DUE-DATE    DISC-DATE    DISC-AMT  CHK-DATE     CASH-ACCT#

001013    000002   RED LINE FREIGHT                       150.00        30                                              20100-10000
R                  NET 30                      04/25/84        .00      05/25/84    04/25/84 0

          DISTRIBUTION:   150.00   42900-10000  Freight - Purchases                500 QUALITY CONTROL - PRODUCT 1

001014    000002   RED LINE FREIGHT            HJ         263.00        30                                              20100-1000
R                  NET 30                    G 05/23/84        .00      06/22/84    05/23/84 0

          DISTRIBUTION:   105.00   42900-20000  Freight - Purchases                500 QUALITY CONTROL - PRODUCT 1
                          158.00   42900-10000  Freight - Purchases                400 ENERGY RESEARCH

001015    000002   RED LINE FREIGHT            CU          59.00        30                                              20100-1000
R                  NET 30                    JK 06/11/84        .00     07/11/84    06/11/84 0

          DISTRIBUTION:    59.00   42900-10000  Freight - Purchases                100 SPACE SHUTTLE

001016    000003   BIGTOWN OFFICE SUPPLIES     YU          89.00        30                        2.00     25           10100-10000
P                  2% 10 NET 30             153 04/27/84        .00     05/28/84    05/07/84 10   1.79     04/27/84

          DISTRIBUTION:    44.00   53700-10000  Office Supplies                    300 UNDERSEA EXPLORATION
                           45.00   53700-20000  Office Supplies                    200 MARS UNMANNED EXPLORATION

001017    000004   JONES PROPERTY MANAGEMENT   45        5,000.00        0                                              20100-10000
R                  MONTHLY PAYMENT               05/01/84       .00     05/01/84    05/01/84 0
```

```
         DISTRIBUTION:     1,000.00   53000-10000  Rent                                100 SPACE SHUTTLE
                           4,000.00   53000-20000  Rent                                400 ENERGY RESEARCH

001019   000004  JONES PROPERTY MANAGEMENT           46       5,000.00                      0                  20100-10000
R                MONTHLY PAYMENT                  06/01/84         .00                06/01/84

         DISTRIBUTION:     1,000.00   53000-10000  Rent                                100 SPACE SHUTTLE
                           4,000.00   53000-20000  Rent                                400 ENERGY RESEARCH

001020   000050  TEXAS TELEPHONE                              1,250.00        20              0               20100-10000
R                NET 20                             05/25/84        .00     06/14/84

         DISTRIBUTION:     1,000.00   54100-10000  Telephone & Telegraph              100 SPACE SHUTTLE
                             250.00   54100-20000  Telephone & Telegraph              300 UNDERSEA EXPLORATION

001021   000050  TEXAS TELEPHONE                              1,300.00        20              0               20100-10000
R                NET 20                             06/25/84        .00     07/16/84

         DISTRIBUTION:       500.00   54100-10000  Telephone & Telegraph              300 UNDERSEA EXPLORATION
                             800.00   54100-20000  Telephone & Telegraph              500 QUALITY CONTROL - PRODUCT 1

001022   000100  SOUTHERN GAS & POWER                           250.00        25              0               20100-10000
R                NET 25                             05/21/84        .00     06/14/84

         DISTRIBUTION:       250.00   54400-10000  Utilities                          100 SPACE SHUTTLE
```

(a) You can see that Fleet Feet buys from a wide variety of vendors. The purchases vary from office supplies to swimming attire.

FIGURE 6.26. The AP transaction register, or purchases register, reports all the data about a single transaction.

```
NAME of REPORT: AP Transaction Register                DATE: 10/84
ANALYST: Frank Pisciotta

TRANSACTION REGISTER = Voucher number +       *  6 Alphanumeric *
                      Vendor number +        *  6 Alphanumeric *
                      Vendor name +          * 30 Alphanumeric *
                      Type +                 *  4 Alphanumeric *
                      Purchase order no. +   * 10 Alphanumeric *
                      Invoice number +       *  8 Alphanumeric *
                      Invoice date +         *  6 Numeric      *
                      Invoice amount +       * 10 Numeric      *
                      Invoice discount +     * 10 Numeric      *
                      Due days +             *  4 Numeric      *
                      Discount days +        *  4 Numeric      *
                      Discount percent +     *  4 Numeric      *
                      Discount amount +      *  8 Numeric      *
                      Discount date +        *  6 Numeric      *
                      Check number +         *  6 Numeric      *
                      Check date +           *  6 Numeric      *
                      Account number +       * 10 Numeric      *
                      Page number +          *  4 Numeric      *
                      Today's date +         * 20 Alphanumeric *
                      Time of day            *  7 Alphanumeric *
```

Order of report: Ascending by voucher number.
Subtotals: None.
Final totals: None.
Counts: Number of vouchers in the file.
Frequency: Weekly.
Length: Less than 250 pages.
Type of paper: 132-column regular computer paper.
Distribution: Directly to business office.
Security: Internal use only.

(b) *Data dictionary for Fleet Feet's AP transaction register.*

FIGURE 6.26 (*Continued*)

CASE STUDY

the purchase, discount terms, and the distribution of that purchase order information within Fleet Feet's organization. Thus for voucher 001014, we know that $105 worth of the $263 was allocated as freight for quality control and the balance as freight for energy research. This report provides a ready reference to all purchases and their use.

Another detail report generated by the new system will be an open item report (Figure 6.27), which shows all transactions entered and posted but not paid. The report also contains the age of each transaction: current, 31–60 days, 61–90 days, over 90 days. Totals for each aging category that reveal how much is due and when help management plan long-term financial needs.

The new system will also produce a query report called, in this case, an AP inquiry report (Figure 6.28). It enables users to call up a particular vendor's account to review all the open items on file for that vendor. Shown on a terminal screen rather than on a printed page, this report provides quick access to data when the user has a vendor on the phone or is handling urgent correspondence about the account.

The cash requirements report (Figure 6.29) is a summary report listing all accounts payable due during a specified period. It includes vendor names and numbers, voucher numbers, amounts, dates, and account totals, as well as overall totals. This is the first report prepared in ascending order by two data items: The primary order, or key, is vendor number; the secondary key is voucher number. By reviewing this report, management knows how much cash it needs to meet its immediate, or short-term, requirements.

The pre-check-writing report, another detail report, lists, by vendor number, all checks to be written on a certain date (Figure 6.30). It includes all pertinent data about the invoices covered by each check, thus giving the user a chance to make certain the checks will be correct. This is the first report with control totals for each vendor as well as an overall grand total for all vendors.

As we saw earlier, the main result of an AP system is a check (Figure 6.31). The stub part of the check includes the number, date, amount, discount, and net amounts of each invoice covered by the check. The vendor name and address appear on the check portion, along with the amount of the check. This report is the only one requiring a special preprinted form. During printing an operator must carefully align the paper in the printer because a misalignment of a tenth of an inch vertically or horizontally can render the checks useless; data will not be printed in the spaces Frank has reserved.

A check register (Figure 6.32) details each check printed or written by a certain date and is another example of a detail report. When a check has been sent to a vendor, no copy remains with the company, but this report gives management all the check's details. At the bottom of the check register is a count of the number of checks produced. These counts control the output, ensuring that the program is operating properly.

A summary report is the vendor analysis report (Figure 6.33). It shows what percentage of the total each purchase represents and thus provides management with a tool for comparing this year's purchases with last year's. The data dictionary and comments for this report contain a new category, averages, to reflect this special need.

Finally, the vendor-clear report (Figure 6.34) is prepared only at the end of the company's fiscal year. It is the only exception report in the accounts payable system. It moves "current year-to-date" totals for each vendor to "last year-to-date." Because this report places the

```
Oct. 24, 1984                              FLEET FEET                                        PAGE 0001
4:22 PM
                                       AP OPEN ITEM REPORT

AGED AS OF 05/31/84 BASED UPON DUE DATES

NOTE: "O" BESIDE VOUCHER NUMBER MEANS ITEM IS PERMANENTLY DEFERRED

                                                                            ---------AGED VENDOR NET---------
VENDOR   NAME                      VENDOR      VALID        VENDOR
  #      TERMS                    BALANCE    DISCOUNTS        NET      CURRENT   31-60-DAYS  61-90-DAYS  OVER-90-DAYS

000002   RED LINE FREIGHT          472.00         .00        472.00      472.00        .00         .00         .00
         NET 30
000003   BIGTOWN OFFICE SUPPLIES   175.00        3.50        171.50      171.50        .00         .00         .00
         2% 10 NET 30
000004   JONES PROPERTY MANAGEMENT 10,000.00      .00     10,000.00   10,000.00        .00         .00         .00
         MONTHLY PAYMENT
000050   TEXAS TELEPHONE          2,550.00        .00      2,550.00    2,550.00        .00         .00         .00
         NET 20
000100   SOUTHERN GAS & POWER       590.00        .00        590.00      590.00        .00         .00         .00
         NET 25
000200   FUTURE ELECTRONICS, INC. 1,200.00       20.00      1,180.00    1,180.00        .00         .00         .00
         2% 10 NET 30
200000   STAR ELECTRONIC          8,150.00      380.00      7,770.00    7,770.00        .00         .00         .00
         COMPONENTS
         5% 30 NET 45
ABC      WILLIAMS & FREGOSI       1,325.00       66.25      1,258.75    1,258.75        .00         .00         .00
         5% 15 NET 30
PROSPY   PROSPERITY INVESTMENT   15,000.00        .00     15,000.00   10,000.00    5,000.00         .00         .00
         BANKERS
         MONTHLY PAYMENT

         GRAND TOTALS:           39,462.00      469.75    38,992.25   33,992.25    5,000.00         .00         .00
```

(a) *This report also shows the age of Fleet Feet's debts to each supplier.*

```
NAME of REPORT: Open Item Report                           DATE: 10/84
ANALYST: Frank Pisciotta

OPEN ITEM = Vendor number +            *   6 Alphanumeric *
            Vendor name +              *  30 Alphanumeric *
            Vendor balance +           *  10 Numeric      *
            Discount balance +         *  10 Numeric      *
            Vendor net +               *  10 Numeric      *
            Current balance +          *  10 Numeric      *
            Balance over 30 days +     *  10 Numeric      *
            Balance over 60 days +     *  10 Numeric      *
            Balance over 90 days +     *  10 Numeric      *
            Page number +              *   4 Numeric      *
            Today's date +             *  20 Alphanumeric *
            Time of day                *   7 Alphanumeric *

Order of report: Ascending by vendor number.

Subtotals: None.

Final totals: 1. Vendor balance
              2. Discount balance
              3. Vendor net
              4. Current balance
              5. Balance over 30 days
              6. Balance over 60 days
              7. Balance over 90 days

Counts: None.

Frequency: Weekly.

Length: Less than 100 pages.

Type of paper: 132-column regular computer paper.

Distribution: Directly to business office.

Security: Internal use only.
```

(b) *Data dictionary for Fleet Feet's open item report.*

FIGURE 6.27. *An open item report lists the totals due to vendors. Some bills may have been partially but not fully paid.*

```
VENDOR ACCOUNT INQUIRY     VENDOR NO: 200... FUTURE ELECTRONICS, INC.
                                      TERMS 2% 10 NET 30
VOUCHR INVOICE-NO INV-DATE DUE-DATE PP?   INV-BAL   DISC-BAL   NET-BAL
001001 A1234     05/02/84  06/01/84       1,200.00    20.00   1,180.00

                             VENDOR TOTAL:  1,200.00    20.00   1,180.00

END OF ACCOUNT  -  PRESS RETURN TO CONTINUE
```

(a) Inquiry reports show all data about the vendor.

```
NAME of REPORT: Vendor Account Inquiry            DATE: 10/84
ANALYST: Frank Pisciotta

VENDOR ACCOUNT INQUIRY = Vendor number +      *  6 Alphanumeric *
                         Vendor name +        * 30 Alphanumeric *
                         Voucher number +     *  6 Alphanumeric *
                         Invoice number +     *  8 Alphanumeric *
                         Invoice date +       *  6 Numeric      *
                         Invoice due date +   *  6 Numeric      *
                         Invoice balance +    * 10 Numeric      *
                         Invoice discount +   * 10 Numeric      *
                         Invoice net amount   * 10 Numeric      *

Order of reports: Selected by vendor number.
Subtotals: None.
Final totals: 1. Invoice balance
              2. Invoice discount
              3. Invoice net amount

Counts: None.
Frequency: On request of user.
Length: At least one screen but may be more.
Type of paper: CRT with block mode, blinking, dual intensity,
               and reverse video.
Distribution: To requesting user.
Security: Internal use only.
```

(b) *Data dictionary and comments for Fleet Feet's inquiry report.*

FIGURE 6.28. *Inquiry reports usually appear on terminals so users can obtain data immediately.*

```
OCT 26, 1984                                    FLEET FEET                                                    PAGE 0002
10:17 AM

                              CASH  REQUIREMENTS  REPORT

THRU 07/31/84 FOR PAYMENT ON 06/25/84 (NEXT PAYMENT DATE IS 07/25/84)

VENDOR   NAME                          PAST-DUE-AMT     PST-VALID-DISC          NET-PST-AMT      PST-DISC-LOST
         TERMS                         CURRENT-AMT      CUR-VALID-DISC          NET-CUR-AMT      CUR-DISC-LOST
                                       OPTIONAL-AMT     OPT-VALID-DISC          NET-OPT-AMT

200000   STAR ELECTRONIC COMPONENTS
         5% 30 NET 45                       6,500.00            300.00            6,200.00                 .00
                                                  .00                .00                .00                .00

VOUCH#   INVOICE#  INV-DATE  DUE-DATE  DISC-DTE   PST-DU-AMT   CURNT-AMT   OPTNAL-AMT   DISC-LOST   VALID-DISC     NET-AMT
                                                                                                    NET-PST+CUR
                                                                                                    NET-PST+CUR+OPT
001024   989009    06/11/84  07/26/84  07/11/84                 6,500.00                                6,200.00   6,200.00
                                                                                                        6,200.00

PROSPY   PROSPERITY INVESTMENT BANKERS
         MONTHLY PAYMENT                   10,000.00                .00           10,000.00                .00
                                            5,000.00                .00            5,000.00                .00
                                                  .00                              15,000.00
                                                                                   15,000.00

VOUCH#   INVOICE#  INV-DATE  DUE DATE  DISC-DTE   PST-DU-AMT   CURNT-AMT   OPTNAL-AMT   DISC-LOST   VALID-DISC     NET-AMT
001027             04/30/84  04/30/84  04/30/84    5,000.00                                  .00          .00     5,000.00
001028             05/31/84  05/31/84  05/31/84    5,000.00                                  .00          .00     5,000.00
001029             06/29/84  06/29/84  06/29/84                5,000.00                      .00                  5,000.00
```

(a) This report shows discounts available, taken, or lost for each vendor.

```
NAME of REPORT: Cash Requirements Report                        DATE: 10/84
ANALYST: Frank Pisciotta

CASH REQUIREMENTS = Vendor number +              *  6 Alphanumeric *
                   Vendor name +                 * 30 Alphanumeric *
                   Discount Days +               *  4 Numeric      *
                   Discount Percent +            *  4 Numeric      *
                   Past due amount +             * 10 Numeric      *
                   Current amount +              * 10 Numeric      *
                   Optional amount +             * 10 Numeric      *
                   Past valid discount +         * 10 Numeric      *
                   Current valid discnt +        * 10 Numeric      *
                   Optional valid dis. +         * 10 Numeric      *
                   Net past amount +             * 10 Numeric      *
                   Net current amount +          * 10 Numeric      *
                   Net optional amount +         * 10 Numeric      *
                   Past discount lost +          * 10 Numeric      *
                   Current discount lst +        * 10 Numeric      *
                   Voucher number +              *  6 Alphanumeric *
                   Invoice number +              *  8 Alphanumeric *
                   Invoice date +                *  6 Numeric      *
                   Due date +                    *  6 Numeric      *
                   Discount date +               *  6 Numeric      *
                   Past due amount +             * 10 Numeric      *
                   Current amount +              * 10 Numeric      *
                   Optional amount +             * 10 Numeric      *
                   Discount lost +               * 10 Numeric      *
                   Discount valid +              * 10 Numeric      *
                   Net amount +                  * 10 Numeric      *
                   Page number +                 *  4 Numeric      *
                   Today's date +                * 20 Alphanumeric *
                   Time of day                   *  7 Alphanumeric *
```

(b) Data dictionary and comments for Fleet Feet's cash requirements report.

FIGURE 6.29 (Continued on next page)

```
Order of report: Ascending by vendor number and voucher number.
Subtotals: None.
Final totals:  1. Past due amount
               2. Current amount
               3. Optional amount
               4. Past valid discount
               5. Current valid discount
               6. Optional valid discount
               7. Net past amount
               8. Net current amount
               9. Net optional amount
              10. Past discount lost
              11. Current discount lost

Counts: None
Frequency: Weekly.
Length: Less than 100 pages.
Type of paper: 132-column regular computer paper.
Distribution: Directly to business office.
Security: Internal use only.
```

(b) *(Continued) Data dictionary and comments for Fleet Feet's cash requirements report.*

FIGURE 6.29. *Cash requirements report helps management plan its payments to optimize cash flow.*

OCT 26, 1984
11:19 AM

FLEET FEET
PRE-CHECK-WRITING REPORT
PAGE 0001

FOR PAYMENT ON 06/25/84
NOTE: DEFERRED ITEMS ARE NOT INCLUDED IN TOTALS. VENDORS WITH ZERO OR MINUS TOTALS ARE NOT INCLUDED IN TOTALS.

VENDOR NO	NAME	VOUCHR NO	INVOICE NO	INVOICE DATE	DUE-DATE	DISCOUNT DATE	AMOUNT TO-BE-PAID	DISCOUNT TO-BE-TAKEN	NET-CASH REQUIRED	AP ACCOUNT NO
000002	RED LINE FREIGHT	001013	45	04/25/84	05/25/84	04/25/84	150.00	.00	150.00	20100-10000
		001014	HJ	05/23/84	06/22/84	05/23/84	200.00	.00	200.00	20100-10000
					VENDOR TOTALS:		350.00	.00	350.00	
000003	BIGTOWN OFFICE SUPPLIES	001030	125368	06/26/84	07/26/84	07/06/84	175.00	3.50	171.50	20100-10000
					VENDOR TOTALS:		175.00	3.50	171.50	
000004	JONES PROPERTY MANAGEMENT	001017	45	05/01/84	05/01/84	05/01/84	5,000.00	.00	5,000.00	20100-10000
		001019	46	06/01/84	06/01/84	06/01/84	5,000.00	.00	DEFERRED	20100-10000
					VENDOR TOTALS:		5,000.00	.00	5,000.00	
000050	TEXAS TELEPHONE	001020		05/25/84	06/14/84	05/25/84	1,250.00	.00	1,250.00	20100-10000
		001021		06/25/84	07/15/84	06/25/84	1,300.00	.00	1,300.00	20100-10000
					VENDOR TOTALS:		2,550.00	.00	2,550.00	

(a) *This report is run before checks are printed, permitting management to review again what is going to be paid to each vendor.*

FIGURE 6.30 *(Continued on next page)*

OCT 26, 1984 FLEET FEET PAGE 0001
11:19 AM
           P R E - C H E C K - W R I T I N G   R E P O R T

FOR PAYMENT ON 06/25/84
NOTE: DEFERRED ITEMS ARE NOT INCLUDED IN TOTALS. VENDORS WITH ZERO OR MINUS TOTALS ARE NOT INCLUDED IN TOTALS.

VENDOR NO	NAME	VOUCHR NO	INVOICE NO	INVOICE DATE	DUE-DATE	DISCOUNT DATE	AMOUNT TO-BE-PAID	DISCOUNT TO-BE-TAKEN	NET-CASH REQUIRED	AP ACCOUNT NO
000100	SOUTHERN GAS & POWER	001022		05/20/84	06/14/84	05/20/84	250.00	.00	250.00	20100-10000
		001023		06/20/84	07/15/84	06/20/84	340.00	.00	340.00	20100-10000
					VENDOR TOTALS:		590.00		590.00	
000200	FUTURE ELECTRONICS, INC.	001001	A1234	05/02/84	06/16/84	05/31/84	1,200.00	40.00	DEFERRED	20100-10000
					VENDOR TOTALS:		.00		.00	
200000	STAR ELECTRONIC COMPONENTS	001024	989889	06/11/84	07/26/84	07/11/84	6,500.00	300.00	6,200.00	20100-10000
					VENDOR TOTALS:		6,500.00	300.00	6,200.00	
PROSPY	PROSPERITY INVESTMENT BANKERS	001027		04/30/84	04/30/84	04/30/84	5,000.00	.00	5,000.00	20100-10000
		001028		05/31/84	05/31/84	05/31/84	5,000.00	.00	5,000.00	20100-10000
		001029		06/30/84	06/30/84	06/30/84	5,000.00	.00	5,000.00	20100-10000
					VENDOR TOTALS:		15,000.00	.00	15,000.00	
					GRAND TOTALS:		30,165.00	303.50	29,861.50	

7 VENDORS TO BE PAID

(a) (Continued) This report is run before checks are printed, permitting management to review again what is going to be paid to each vendor.

```
NAME OF REPORT: Pre-Check-Writing Report                      DATE: 10/84
ANALYST: Frank Pisciotta

PRE-CHECK-WRITING REPORT = Vendor number +       *  6 Alphanumeric *
                          Vendor name +          * 30 Alphanumeric *
                          Voucher number +       *  6 Alphanumeric *
                          Invoice number +       *  8 Alphanumeric *
                          Invoice date +         *  6 Numeric      *
                          Due date +             *  6 Numeric      *
                          Discount date +        *  6 Numeric      *
                          Amount to be paid +    * 10 Numeric      *
                          Discount amount +      * 10 Numeric      *
                          Net amount +           * 10 Numeric      *
                          AP account number +    * 10 Numeric      *
                          Page number +          *  4 Numeric      *
                          Today's date +         * 20 Alphanumeric *
                          Time of day +          *  7 Alphanumeric *

Order of report: Ascending by vendor number and voucher number.
Subtotals: For each vendor
           1. Amount to be paid
           2. Discount amount
           3. Net amount
Final Totals  1. Amount to be paid
              2. Discount amount
              3. Net amount
Counts: Number of vendors to be paid.
Frequency: Weekly.
Length: Less than 150 pages.
Type of paper: Regular 132-column-wide computer paper.
Distribution: Deliver to business office.
Security: Internal use only.
```

(b) Data dictionary and comments for Fleet Feet's pre-check-writing report.

FIGURE 6.30. *The pre-check-writing report shows all the items covered by each check so the user can look for errors before actually printing the checks.*

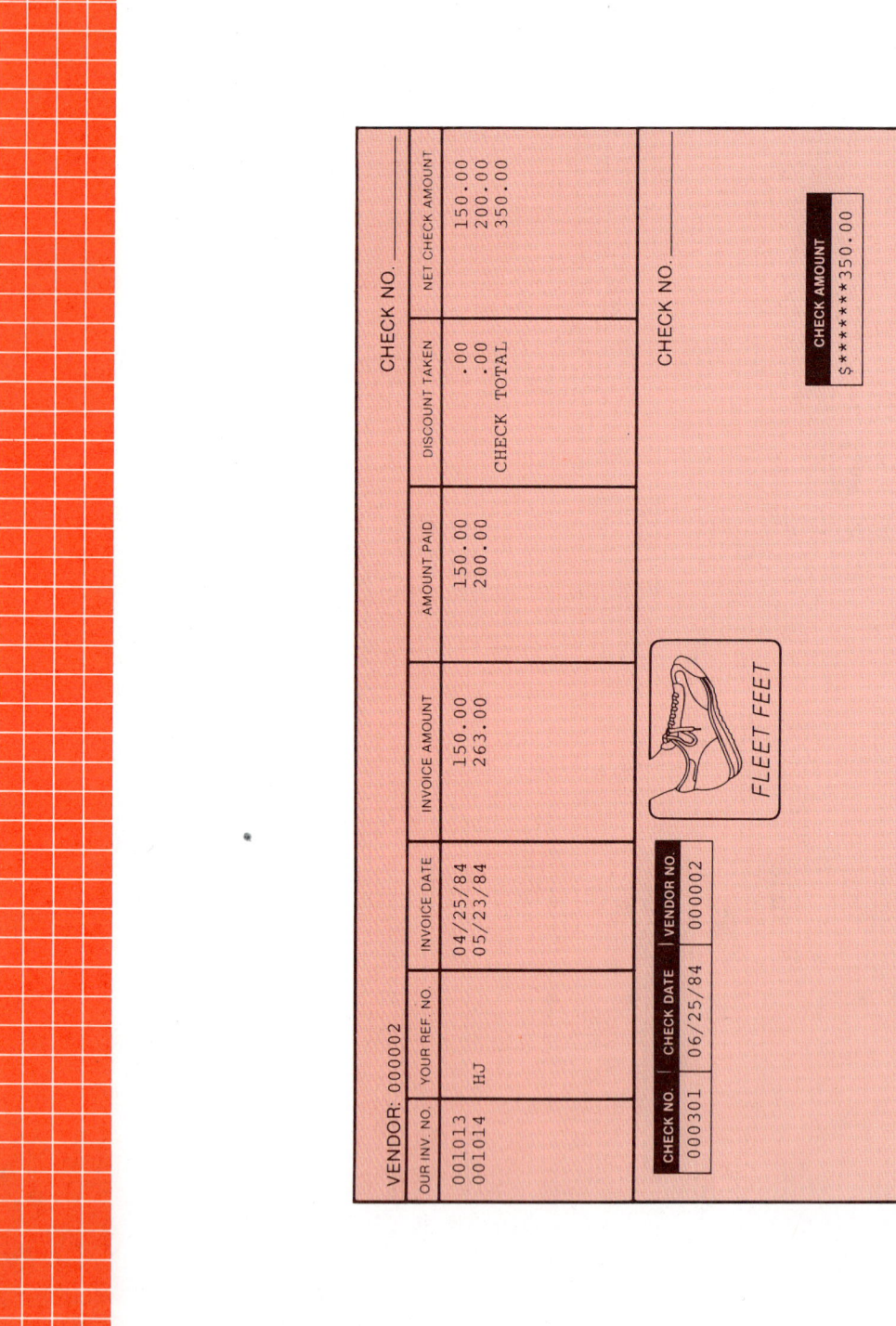

(a) Invoices being paid are listed in order by invoice number.

```
NAME of REPORT: AP Check                    DATE: 10/84
ANALYST: Frank Pisciotta

A/P CHECK = Vendor number +       *  6 Alphanumeric *
            Vendor name +         * 30 Alphanumeric *
            Address 1 +           * 30 Alphanumeric *
            Address 2 +           * 30 Alphanumeric *
            Address 3 +           * 30 Alphanumeric *
            Invoice number +      *  8 Alphanumeric *
            Purchase number +     * 10 Alphanumeric *
            Invoice date +        *  6 Numeric *
            Invoice amount +      * 10 Numeric *
            Amount paid +         * 10 Numeric *
            Discount taken +      * 10 Numeric *
            Net amount +          * 10 Numeric *
            Check total +         * 10 Numeric *
            Check number +        *  6 Numeric *
            Check date +          *  6 Numeric *

Order of report: Ascending by vendor number.
Subtotals: Check total.
Final totals: None.
Counts: None.
Frequency: Weekly.
Length: Less than 150 pages.
Type of paper: Blank check preprinted forms. Must be
               signed out and all forms accounted
               for.
Distribution: Deliver to business office for
              signature.
Security: External use only. All checks numbered by
          the paper supplier.
```

(b) *Data dictionary and comments for Fleet Feet's AP check.*

FIGURE 6.31. *The main output from the AP system is the check to vendor. The top half supplies details and the bottom is the check itself.*

Oct. 26, 1984
2:09 PM

FLEET FEET

ACCOUNTS PAYABLE CHECK REGISTER

PAGE 0002

CASH ACCOUNT: 10100-10000 Cash - First National Bank

CHECK #	CHECK DATE	VENDOR NAME	VOUCH #	P.O. #	INVOICE #	INVOICE DATE	AMOUNT PAID	DISCOUNT TAKEN	CHECK AMOUNT
307	06/25/84	PROSPY PROSPERITY INVESTMENT BANKERS	001027			04/30/84	5,000.00	.00	5,000.00
			001028			05/31/84	5,000.00	.00	5,000.00
			001029			06/29/84	5,000.00	.00	5,000.00
					CHECK TOTALS		15,000.00	.00	15,000.00
					CASH ACCT TOTALS:		32,404.00	410.28	31,993.72

7 COMPUTER CHECKS
2 PREPAID CHECKS
1 MANUAL CHECKS
0 VOID CHECKS
10 CHECKS TOTAL

(a) Check registers show the details of each check, but not on the special forms required for the checks themselves.

NAME of REPORT: Check Register   DATE: 10/84
ANALYST: Frank Pisciotta

CHECK REGISTER = Vendor number   + * 6 Alphanumeric *
                 Vendor name   + * 30 Alphanumeric *
                 Voucher number   + * 6 Alphanumeric *
                 Invoice number   + * 8 Alphanumeric *
                 Purchase number + * 10 Alphanumeric *
                 Invoice date   + * 6 Numeric   *
                 Invoice amount   + * 10 Numeric   *
                 Amount paid   + * 10 Numeric   *
                 Discount taken   + * 10 Numeric   *
                 Net amount   + * 10 Numeric   *
                 Check total   + * 10 Numeric   *
                 Check number   + * 6 Numeric   *
                 Check date   + * 6 Numeric   *
                 Page number   + * 4 Numeric   *
                 Today's date   + * 20 Alphanumeric *
                 Time of day   * 7 Alphanumeric *

Order of report: Ascending by check number and voucher number.
Subtotals: For each voucher associated with a check
           1. Amount to be paid
           2. Discount amount
           3. Check amount
Final totals: 1. Amount to be paid
              2. Discount amount
              3. Check amount
Counts: 1. Computer checks
         2. Prepaid checks
         3. Manual checks
         4. Void checks
         5. Total number of checks
Frequency: Weekly.
Length: Less than 150 pages.
Type of paper: Two-part 132-column-wide computer paper. Burst and
               decollate after run.
Distribution: Deliver to business office and to vice-president of operations.
Security: Internal use only.

(b) *Data dictionary and comments for Fleet Feet's check register.*

FIGURE 6-32. *A check register groups all checks together, including handwritten and voided checks.*

OCT 26, 1984
3:14 PM

FLEET FEET

PAGE 0001

V E N D O R   A N A L Y S I S   R E P O R T

VENDOR	NAME	LAST ACTIVITY DATE	PURCHASES	YEAR-TO-DATE %-OF TOTAL	DISCOUNTS	%-OF TOTAL	PURCHASES	LAST-YEAR %-OF TOTAL	DISCOUNTS	%-OF TOTAL
000001	ACME OFFICE FURNITURE	03/15/84	1,594.00	1.2	20.00	0.2	4,287.50	1.7	70.00	1.2
000002	RED LINE FREIGHT	06/11/84	2,444.00	1.8	.00	0.0	.00	0.0	.00	0.0
000003	BIGTOWN OFFICE SUPPLIES	06/26/84	8,028.00	6.0	157.06	1.8	6,000.00	2.4	120.00	2.1
000004	JONES PROPERTY MANAGEMENT	06/01/84	30,000.00	22.3	.00	0.0	30,000.00	12.2	.00	0.0
000005	BLUE FLASH REPAIR SERVICE	02/24/84	1,500.00	1.1	.00	0.0	2,500.00	1.0	.00	0.0
000050	TEXAS TELEPHONE	06/25/84	6,000.00	4.5	.00	0.0	3,600.00	1.5	.00	0.0
000100	SOUTHERN GAS & POWER	06/20/84	1,580.00	1.2	.00	0.0	1,250.00	0.5	.00	0.0
000200	FUTURE ELECTRONICS, INC.	05/17/84	13,700.00	10.2	250.00	2.8	45,000.00	18.2	900.00	15.5
123ABC	HARRIS, COHEN AND SMITH	03/20/84	3,200.00	1.2	4.00	0.0	2,500.00	1.0	25.00	0.4
200000	STAR ELECTRONIC COMPONENTS	06/11/84	33,650.00	25.1	8,405.00	94.1	84,200.00	34.1	4,200.00	72.2
ABC	WILLIAMS & FREGOSI	06/22/84	5,610.00	4.2	95.00	1.1	17,500.00	7.1	500.00	8.6
PROSPY	PROSPERITY INVESTMENT BANKERS	06/29/84	30,000.00	22.3	.00	0.0	50,000.00	20.3	.00	0.0
12 VENDORS ON FILE	GRAND TOTALS:		134,306.00	100.0	8,931.06	100.0	246,637.50	100.0	5,815.00	100.0
AVERAGES:			11,192.17		744.26		20,553.13		484.58	

(a) *This report also shows the last activity date for each vendor as well as a comparison of this year's versus last year's purchases.*

NAME of REPORT: Vendor Analysis Report   DATE: 10/84
ANALYST: Frank Pisciotta

VENDOR ANALYSIS REPORT = Vendor number   +   * 6 Alphanumeric *
                         Vendor name   +   * 30 Alphanumeric *
                         Last activity   +   * 4 Numeric   *
                         YTD purchases   +   * 10 Numeric   *
                         YTD discounts   +   * 10 Numeric   *
                         Last year purchases   +   * 10 Numeric   *
                         Last year discounts   +   * 10 Numeric   *
                         Percent YTD Purchase   +   * 10 Numeric   *
                         Percent YTD Discount   +   * 10 Numeric   *
                         Percent lst. purch.   +   * 10 Numeric   *
                         Percent lst. Disc.   +   * 10 Numeric   *
                         Page number   +   * 4 Numeric   *
                         Today's date   +   * 20 Alphanumeric *
                         Time of day   * 7 Alphanumeric *

Order of report: Ascending by vendor number.

Averages: 1. YTD purchases
            2. YTD discounts
            3. Last year's purchases
            4. Last year's discounts

Final totals: 1. YTD purchases
              2. YTD discounts
              3. Last year's purchases
              4. Last year's discounts
              5. Percent YTD purchases
              6. Percent YTD discounts
              7. Percent last year's purchases
              8. Percent last year's discounts

Counts: Number of vendors in the file.

Frequency: Weekly.

Length: Less than 150 pages.

Type of paper: Two-part 132-column-wide computer paper. Burst and decollate after run.

Distribution: Deliver to business office and to vice-president of operations.

Security: Internal use only.

(b) Data dictionary and comments for Fleet Feet's vendor analysis report.

FIGURE 6.33. *Management usually does not want to dig through a lot of data to learn the effects of transactions on the organization. Therefore, the vendor analysis report shows all the purchases made from all the vendors and their relative percentages of the entire amount spent.*

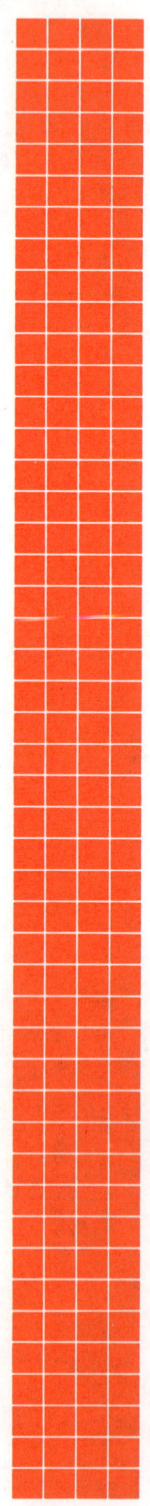
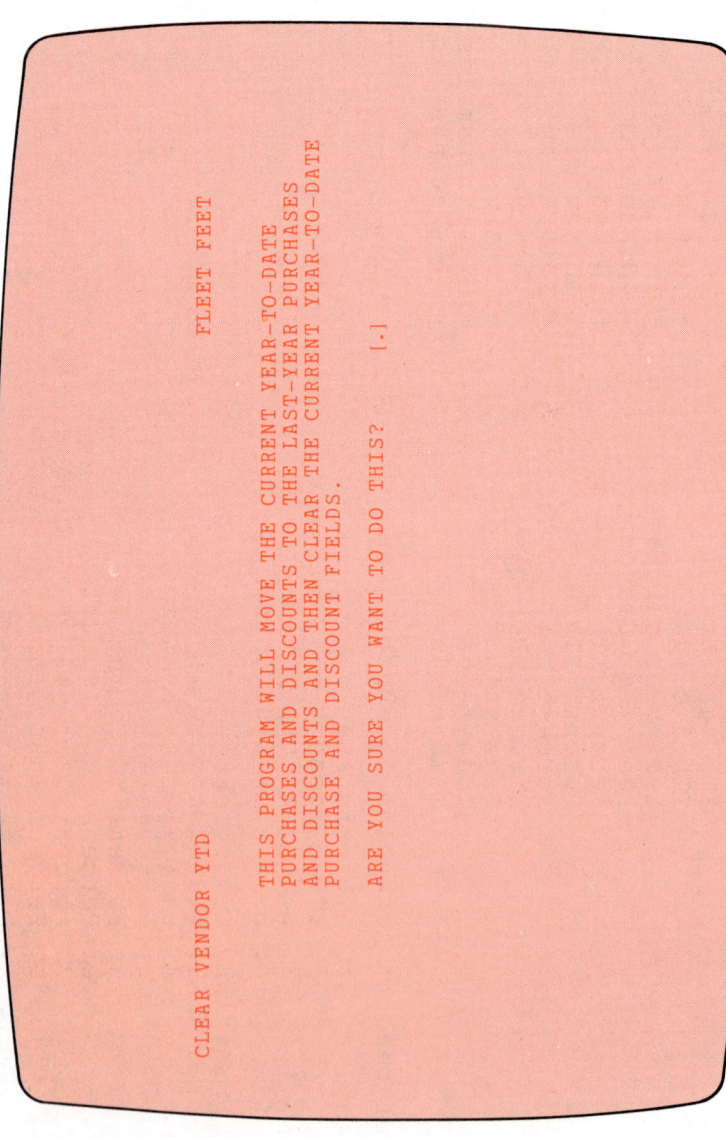

CLEAR VENDOR YTD   FLEET FEET

THIS PROGRAM WILL MOVE THE CURRENT YEAR-TO-DATE
PURCHASES AND DISCOUNTS TO THE LAST-YEAR PURCHASES
AND DISCOUNTS AND THEN CLEAR THE CURRENT YEAR-TO-DATE
PURCHASE AND DISCOUNT FIELDS.

ARE YOU SURE YOU WANT TO DO THIS?   [.]

(a) The user must verify that the clearing operation is indeed what is wanted before the computer will proceed.

```
NAME of REPORT: Clear Vendor                                    DATE: 10/84
ANALYST: Frank Pisciotta

CLEAR VENDOR YTD = User response                        * 1 Alphanumeric *



Order of report: None
Subtotals: None
Final totals: none
Counts: None.
Frequency: Yearly at beginning of business year.
Length: CRT one screen.
Type of paper: None.
Distribution: None.
Security: Can only be run with a special password and
          user code.
```

(b) *Data dictionary and comments for Fleet Feet's clear vendor report.*

FIGURE 6.34. *Because it erases all the balances owed vendors, the vendor-clear report is dangerous.*

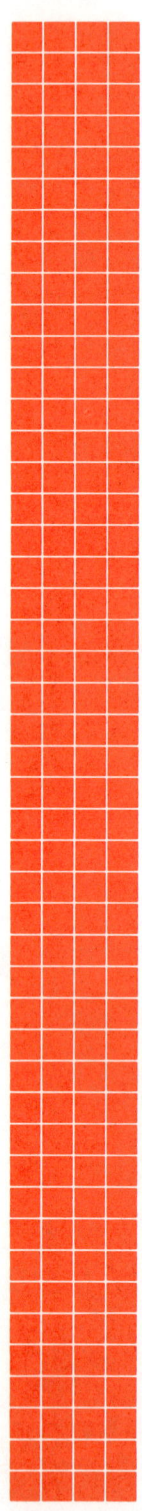

FIGURE 6.35. Fleet Feet's accounts payable check with serial numbers preprinted on both the top and bottom so the vendor can record the check before endorsing and depositing it.

CHAPTER 6 OUTPUT DESIGN

CASE STUDY

FLEET FEET Inc.

Accounts Payable Check Form

Check Number: _____

Date Taken: _____

_____   _____
Computer Operations   Business Office

Returning Check Number: _____

Date Returned: _____

_____   _____
Computer Operations   Business Office

Number of Checks Printed: _____

Number of Checks Destroyed: _____

Explain any Discrepencies: _____
_____
_____

FIGURE 6.36. *Check control form tracks blank checks. Two parties must sign the form twice, once when checks leave the safe and again when they return.*

CASE STUDY

data out of reach of further change, it should not be run until the user has decided it need make no more year-end adjustments. To be absolutely sure the report is not prepared prematurely, the computer asks users whether they want to clear.

Because Fleet Feet has used this supplier in the past, Frank chooses Moore Business Forms to print blank checks with serial numbers and Fleet Feet's name, address, and logo (Figure 6.35). Since blank checks could be stolen, filled out, and cashed, Frank must design tight controls and security to protect them before, during, and after use, even accounting for checks damaged while an operator aligns paper before printing.

To prevent theft Frank decides to store the checks in Mark Stensaas' fireproof safe, for which only Mark and the company's controller own keys. When a computer operator needs to write checks, he or she will have to ask Mark to unlock the safe, and Mark, in turn, will make sure the operator signs out for them. Before the checks are removed from Mark's office, the operator enters the beginning check number and the date on the top half of the check control form, and then signs it (Figure 6.36). After printing the checks, the operator returns any unused or damaged ones, completing the bottom half of the control form, which both the operator and Mark again sign after recording the beginning and ending check numbers. To account for them all, they attach checks damaged during printing to the control form.

Once printed, the checks go directly to Mark for his signature, after which they are mailed immediately to vendors.

To encourage vendors to cash Fleet Feet's checks promptly, Frank establishes a 60-day "stale date" policy. If a check has not been endorsed and returned to Fleet Feet within this 60-day period, the company classifies it as "stale" and will not accept it for check reconciliation within Fleet Feet's accounts payable system.

Frank has now defined all the reports the users need, including the formats and data elements required to generate them. The output design phase for Fleet Feet concludes, and the next step in the system's process, designing the data base or files to store the data, begins.

WORKING WITH PEOPLE   Involving the User from the Start

Craig Smally had just finished designing the reports for Lumberjohn's new inventory system. He thought they were terrific. After 2 weeks of uninterrupted work, and having examined five operating inventory systems, he felt he had come up with a design that would provide users (store managers) with more and better information than any other system possibly could. He called the company's seven managers and requested a meeting with them as soon as possible.

To prepare for the presentation, Craig photocopied samples of all the reports, bound them in handsome folders, and placed a folder on the table in front of each manager.

Joe Ferguson, manager of the downtown store, arrived in his usual happy frame of mind, and sat down and began leafing through the folder. Before long the smile on his face turned to a frown. Craig was curious but decided to wait for the meeting to start before questioning him. Perhaps he just was antagonistic to computers and analysts. Craig knew that some of Lumberjohn's managers thought the data processing people had forgotten that they were there to serve the organization rather than vice versa, but he thought they were just jealous.

When Lumberjohn had first installed terminals in each store, data processing had made a mistake by not properly preparing the managers for the new devices, thus causing a lot of resentment, especially when the managers had trouble with the buttons, switches, and other adjustments they had to make every time they used the terminals. Nobody likes to feel clumsy, especially an experienced manager like Joe Ferguson.

Another time, Lumberjohn had bought a new computer and notified managers through the company's monthly employee newsletter: an expenditure of over half a million dollars with no input from the people it most directly affected! Ferguson had steamed over that.

Soon the other six store managers arrived, and the data processing manager started the meeting. He began by praising his department and the job Craig had done on the inventory system. Soon all the managers' smiles turned to frowns, and they began shuffling papers and whispering to each other.

Joe Ferguson, seeming to find safety in numbers, cleared his throat and interrupted Craig's boss, complaining that data processing was telling users what they would get rather than asking them what they needed. He attacked the new computer, the terminals, and Craig's beloved inventory system. To prove his point, Joe ripped out one of Craig's reports. It didn't take a mathematical genius to calculate that this report would result in 1000 pages of printout every day, 7 days a week, for just one store! One report would weigh almost 10 pounds, and a year's worth over 3600 pounds. Joe would have to rent storage space and hire a new clerk just to deal with that ton and a half of, as Joe put it, "garbage."

"I'm sick and tired of the tail wagging the dog around here," Joe concluded.

The data processing manager became defensive. "My people know what's best; we're the experts. We understand computers."

Joe threw his folder on the table. "Well I understand the lumber business, which pays your salaries!"

Joe continued in a milder tone. "When a customer comes into my store with a problem or a complaint about a product from Lumberjohn's, the customer is always right. We exist to serve our customers. If we treat them fairly and justly, they'll continue to be our customers; if we don't, they'll go elsewhere. Why doesn't that apply to you as well?"

Craig realized he had learned an important lesson: computers serve the users, and the users' needs should overrule everything else. He promised to meet separately with each store manager and listen to their suggestions for improving his designs.

SUMMARY

Output design is the second design step; it establishes report formats and defines the data elements required to generate the reports. Users are deeply involved in this step, assisting the analyst in the designs and approving the results.

At this point the analyst selects the medium that will produce reports. Printers are most popular, followed by terminals, which can display data stored on disks or tapes.

Printer speeds vary from 10 characters per second to thousands of lines per minute and costs can run from a few hundred dollars to almost a million. Characters can be dot matrix or letter quality. Some printers offer graphics abilities; others only display letters, numerals, and special characters.

Cathode-ray tubes, like printers, have a wide variety of characteristics, including color displays and special function keys. Data displayed on a CRT can blink, and can be displayed in varying intensities, or even in inverse video.

The four basic types of reports are detail reports (which show all the pertinent facts about a transaction), summary reports (which show only overall totals or positions), exception reports (which list items that meet a predetermined condition), and query reports (which allow the users to ask the computer for specific facts).

Regardless of the type, all reports share common elements. Headings show titles and footings provide overall results. Breaks indicate subtotals or temporary halts. Detail lines show all the pertinent facts about a single transaction.

System output controls protect reports from tampering or unauthorized use. Reports should be identified by form number, page counts, and date printed. Controls also involve the disposal of sensitive documents.

After designing all report formats, the analyst lists the data elements for the report. This data dictionary is an excellent tool for listing specific report elements. Besides noting the data necessary for each report, the analyst decides on the order of the data, how often to print the report, how many pages it will probably run, what type of paper to use, and to whom the report should be distributed.

NEW TERMS

Auditing	Detail line	Page heading
Bond	Detail report	Periodic report
Column headings	Exception report	Query report
Control footing	External report	Report footing
Control headings	Internal report	Report heading
Control system	NCR	Summary report
Cross-footing	Page footing	Turnaround document

QUESTIONS FOR REVIEW AND DISCUSSION

Questions appear in three categories. You can find the answers to the A group in this chapter. The B group requires you to apply the material presented here, whereas the C group necessitates investigation or research.

A.1. List the inputs to the output design phase.
A.2. List the outputs from output design and tell what happens to each of them.
A.3. Make a list of five different output devices.
A.4. Describe the purposes of four types of reports.
A.5. What does "signing off" mean to the user and analyst?
A.6. Describe the difference between a burster and a decollator.
A.7. Explain the five characteristics of a well-designed report.
A.8. List six reports produced by an accounts payable system.

B.1. Write a data dictionary definition for output design.
B.2. If you needed to print over 100,000 pages per month, which kind of line printer would be the best choice? The worst choice?
B.3. Is a stub-over check a turnaround document?
B.4. How many lines per vertical inch does a typewriter print?
B.5. Categorize the 10 reports from the AP system as summary, detail, inquiry, or exception.
B.6. Which of the AP reports cited in Appendix A for AP systems is Fleet Feet not going to use?
B.7. Design a report similar to the AP account inquiry that would show similar data for a payroll system.
B.8. Design a report similar to the AP check register that would show similar data for a payroll system.

C.1. Obtain the names and addresses of four forms suppliers other than those given in this chapter. Where would you look for them?
C.2. From a supplier's product descriptions, obtain the cost difference between two-part and four-part carbon paper and NCR paper.

REFERENCES

Steve Eckols, *How to Design and Develop Business Systems*, Mike Murach and Associates, Fresno, Calif., 1983.

Alan L. Eliason, *Online Business Computer Applications*, Science Research Associates, Chicago, 1983.

———,Teleprocessing System Design, ICP Interface Data Processing Management, pp. 21, 22, Summer 1979.

E. R. Lawrence, "The Human Side of Software," *Datamation*, March 1982.

W. E. Moody, "How Humans Read and Understand," IBM Corporation workshop notes, 1982.

D. Verne Morland, "Human Factors Guidelines for Terminal Interface Design," *Communications of the ACM*, pp. 484–494, July 1983.

D. E. Peterson, "Screen Design Guidelines," *Small Systems World*, February 1979.

CHAPTER 7
Organizing and Designing Files

- Goals and Preview
- An Introduction to File Design
- Data-Storage Methods: ASCII, EBCDIC, Numeric Methods
- Media and Data-Storage Techniques
- Designing Disk or Tape Files
- Guidelines for Designing Files
- File and Processing Controls
- Case Study: Fleet Feet's Accounts Payable File Design
- Working with People: Professional or Personal Computer Considerations
- Summary
- New Terms
- Questions for Review and Discussion
- References

GOALS AND PREVIEW

After reading this chapter you should be able to:

1. Explain the four types of files.
2. ~~Draw the binary patterns for a character in EBCDIC, ASCII, BCD, true binary, and packed decimal.~~
3. Describe the four most common file processing techniques.
4. State the advantages and disadvantages of sequential, indexed, and direct file organizations.
5. Design record layouts for disk or tape files.
6. ~~List two control methods for files.~~

During the third stage of systems design, the analyst decides which secondary storage devices to use to store a system's data. For some analysts file design is the most enjoyable part of the systems process because it allows them fully to use their computer skills and knowledge. Not only does one determine where and in what order to store the data, but one figures out how to configure essential data and chooses identifiers users can employ to retrieve data easily. To build the description of a file, the analyst uses the report formats and the data dictionaries developed earlier. These descriptions provide input to the next design stage, input design, and they form part of the system specifications, the overall documentation for a project (Figure 7.1). In this chapter we examine the methods of storing data on disks and tapes.

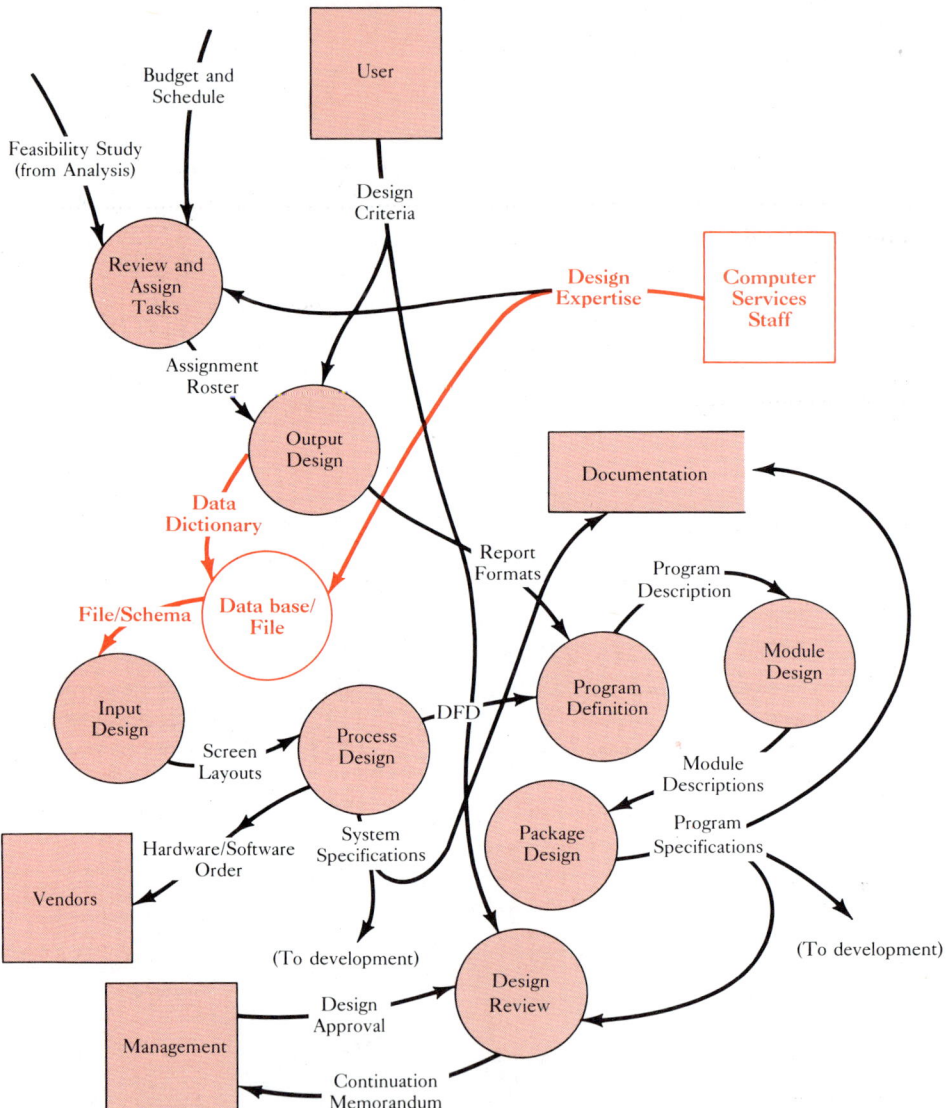

FIGURE 7.1. Data-base or file design lets analysts display their expertise. Sometimes the computer services staff hires personnel with specific skills in this area.

AN INTRODUCTION TO FILE DESIGN

Storage Capabilities

A computer system stores data in its internal memory (RAM, or random access memory), on an external reel of tape, or on a disk. The amount of memory available for data storage equals the total amount of computer memory minus the amount needed for the operating system, the application programs, and utilities. Internal machine memory is very fast (it can retrieve data in billionths of a second) but it is expensive (about 0.3 to 0.5 cent per character). Tape or disk memory is slower (operating in thousandths of a second) but less expensive (about one thousandth cent per character).

If an accounts receivable system requires 100 characters of data for each of 10,000 customers, it will need a computer with 1 million internal memory locations (100 × 10,000 = 1,000,000) to hold the entire file in memory all at once. Today many computers can store this much data.

Since most organizations retrieve only one customer's record at a time, internal memory, though faster, would be prohibitively expensive for storing data about all the organization's customers. External tapes and disks offer alternatives for massive data storage at a much lower cost. In fact companies such as Sears and Exxon, which have extremely large data bases, must rely on both tapes and disks since no computer can store the huge amounts of data they need in internal memory.

Currently tape and disk manufacturers offer a wide variety of storage devices, such as the IBM 3380 in Figure 7.2, which is capable of storing almost a billion characters or bytes on a single disk at a cost of a few thousandths of a cent per character. Of course, disks run much more slowly than internal memory, so, whenever speed is essential, they could pose problems. Still most disks retrieve data in less than 30 milliseconds (30 thousandths of a second). Tapes, although they run even more slowly than disks, are cheaper per character and can also store

FIGURE 7.2. IBM 3380 disk drives. Each drive holds almost 1000 megabytes, or a billion bytes of data. (Courtesy of International Business Machines Corporation.)

billions of characters. Because of the trade-offs among these cost, speed, and volume factors, most organizations use external memory in the form of disks, tapes, or both, to create and maintain their files.

Types of Files

Regardless of the size of an organization or the particular application in question, an analyst will set up five kinds of files: master, transaction, backup, temporary, and program.

Master file A file holding permanent or semipermanent records.

A **master file** stores all the data relevant to a particular application until the data becomes obsolete. For an accounts receivable application, an organization such as Sears would store a customer's account number, name, address, telephone number, and balance in a master file.

Transaction, or detail, file A file describing the details of a business activity.

Transaction, or **detail, files** hold data about particular events. Continuing the example above, a detail file for a charge customer at Sears would include the customer's account number, the date of a particular purchase, the amount owed, and the identifying number of the store where the purchase was made. This data would remain on file until the customer settles the account. Sears might then delete the data from the detail file or save the data for historical purposes.

Backup file A copy of another file created to protect against loss of the original file.

Backup files are copies of other files an organization can fall back on if an original is destroyed. Organizations usually store backup files on tapes (instead of the more expensive disks) somewhere other than where they store the originals.

Scratch, or temporary, file A temporary file created to hold data between processing actions.

Scratch, or **temporary, files** come into play when an organization wants to store data for a very short time. In this case a program may extract certain data from a master or transaction file and store the data on a reel of tape from which yet another program can take the data and use them for a special report. Management may analyze this report for trends, and then destroy it. In an accounts receivable system, for example, a scratch file may contain a list of customers who have not paid their debts for over 90 days. Such a scratch report could facilitate debt collection and help management improve cash flow.

Program file A file holding COBOL, Pascal, or other source-language statements.

Program files, as the name implies, store the programs needed to operate the system. Program files allow users to modify their programs without rekeying them, to transfer the program to another computer of the same or different manufacture, or to access specific portions of one program when writing a new program.

DATA-STORAGE METHODS: ASCII, EBCDIC, NUMERIC METHODS

All files contain two types of data: alphanumeric and numeric. Alphanumeric data items include letters, special characters, and numbers. Numeric data items are numbers only, such as a customer's account balance. A social security number with its dashes, 503-56-9888, is alphanumeric, whereas the social security number without dashes, 503569888, is numeric. Generally the distinction centers on how the data are used: numeric data are used for calculation; alphanumeric are not. If a retail store's customer numbers contain only numbers, they are numeric (7148), whereas ones that include letters are definitely alphanumeric (RE7148).

A system that will store massive amounts of data demands efficiency. When billions of characters are involved, the computer must store the data compactly or the disks or tapes will not be able to hold them all. With the proper storage method, a data file that might take 240 million bytes of disk storage could be compressed into 128 million bytes, a savings of almost 50 percent!

Storing Alphanumeric Data

Several methods exist for storing alphanumeric data in a computer's internal or external memory. The two most popular are EBCDIC and ASCII. The **EBCDIC** (Extended Binary Coded Decimal Interchange Code) method is common among IBM computers, and **ASCII** (American Standard Code for Information Interchange) is popular on mini- and microcomputers. Both methods use a binary format that establishes groups of bits with specific on-off patterns to represent particular characters (Figure 7.3). The capital letter L in EBCDIC is the binary pattern 11010011; its 8-bit ASCII counterpart is 11001100. The digit 5 in EBCDIC is 11110101; in ASCII it is 10110101.

While the differences between EBCDIC and ASCII may seem subtle, computers using EBCDIC will produce different results than a computer handling ASCII. For example, when an EBCDIC-based system sorts alphanumeric data, it reads alphabetic fields before numeric ones (A–Z, 0–9); an ASCII-based system does the reverse (0–9, A–Z). The order of characters is called the collating sequence. Consider each data item in a master file of automobile license plates that includes both letters and numbers:

> 219XLA, LEZ697, NCH763, 145AVE.

The EBCDIC method would sort the plate numbers as:

LEZ697   NCH763   145AVE   219XLA

The ASCII method would produce:

145AVE   219XLA   LEZ697   NCH763

The fact that ASCII and EBCDIC have different collating sequences can be critical if an organization must exchange data between ASCII- and EBCDIC-based computers. Neither EBCDIC nor ASCII is more efficient when storing data as each requires a single byte to store a character. If the analyst receives an ASCII tape, the data must be resorted for an EBCDIC type of computer. Similarly, the analyst must resort from ASCII to EBCDIC.

EBCDIC The 8-bit binary coding system most often found on IBM computers. An acronym for Extended Binary Coded Decimal Interchange Code.

ASCII The 7- or 8-bit coding system often found on micro- and minicomputers. An acronym for American Standard Code for Information Interchange.

Storing Numeric Data

Computer memories can also store numeric data in a variety of ways. ASCII and EBCDIC use digit-by-digit storage. For example, to store a 01778 Zip code for a customer in a vendor master file, both would assign 8 bits to each of the five digits, resulting in a total of 40 bits.

```
ASCII:       00110000  00110001  00110111  00110111  00111000
EBCDIC:      11110000  11110001  11110111  11110111  11111000
                 0         1         7         7         8
```

Characters	4-Bit BCD	EBCDIC	Hexadecimal Equivalent of EBCDIC	ASCII-8	Hexadecimal Equivalent of ASCII-8
(Place Values)	8421	84218421		84218421	
0	0000	11110000	F0	10110000	B0
1	0001	11110001	F1	10110001	B1
2	0010	11110010	F2	10110010	B2
3	0011	11110011	F3	10110011	B3
4	0100	11110100	F4	10110100	B4
5	0101	11110101	F5	10110101	B5
6	0110	11110110	F6	10110110	F6
7	0111	11110111	F7	10110111	B7
8	1000	11111000	F8	10111000	B8
9	1001	11111001	F9	10111001	F9
A		11000001	C1	11000001	C1
B		11000010	C2	11000010	C2
C		11000011	C3	11000011	C3
D		11000100	C4	11000100	C4
E		11000101	C5	11000101	C5
F		11000110	C6	11000110	C6
G		11000111	C7	11000111	C7
H		11001000	C8	11001000	C8
I		11001001	C9	11001001	C9
J		11010001	D1	11001010	CA
K		11010010	D2	11001011	CB
L		11010011	D3	11001100	CC
M		11010100	D4	11001101	CD
N		11010101	D5	11001110	CE
O		11010110	D6	11001111	CF
P		11010111	D7	11010000	D0
Q		11011000	D8	11010001	D1
R		11011001	D9	11010010	D2
S		11100010	E2	11010011	D3
T		11100011	E3	11010100	D4
U		11100100	E4	11010101	D5
V		11100101	E5	11010110	D6
W		11100110	E6	11010111	D7
X		11100111	E7	11011000	D8
Y		11101000	E8	11011001	D9
Z		11101001	E9	11011010	DA
+		01001110	4E	10101011	AB
−		01101101	6D	11011111	DF
'		01101011	6B	10101100	AC
.		01001011	4B	10101110	AE

FIGURE 7.3. EBCDIC and ASCII use 8 bits to represent each letter, special character, or numeral. Though their 8-bit patterns are sometimes the same, most often they are not. Remember that 1 means on and 0 means off.

Although commonly called EBCDIC or ASCII, we sometimes call this method zoned decimal since the left half of each 8 bits (called a nibble) is always 0011 in ASCII and 1111 in EBCDIC.

Another popular method for storing numeric data, called **packed decimal,** stores two digits in each 8-bit segment or byte. The last 4 bits on the right of every packed decimal number indicate sign: a negative is 1110, and a positive is 1111. If a number is entered without a sign, packed decimal assumes it to be positive. The 01778 Zip code therefore would look like this in packed decimal:

Packed decimal: 00000001 01110111 1000 1111
                       0   1   7   7 8   +

Packed decimal
Numeric storage method that places two digits in each 8-digit segment.

Packed decimal (known as COMP-3 in COBOL) needs only 24 bits of memory (3 bytes) to store the same numeric data that require 40 bits in EBCDIC and ASCII. Though packed decimal offers considerable space savings, it applies only to numeric data items. Many organizations use it to compact numeric data fields such as Zip codes, telephone numbers, and account balances.

A fourth method for storing numeric data is **straight binary,** which stores data in true binary form. Our 01778 code looks like this in binary:

Straight binary: 00000110 11110010

Straight binary
Numeric storage method using the base 2 number system.

Straight binary requires the least amount of memory—only 16 bits, or 2 bytes, for a Zip code. It is somewhat cumbersome because bits are not easily translatable into decimals before users can deal with them. Straight binary usually works best for an organization that keeps large tables of data, such as income tax tables or lists of airline departure and arrival schedules.

As you can see, these methods involve different amounts of memory and fulfill different functions. When selecting one for a given application, an analyst must weigh the hardware's coding system, volume of data, available tape or disk space, and the organization's speed requirements. If, for example, a company has a computer that requires numeric data to appear in packed decimal form for computational purposes, one should plan to store all numeric data items, such as account balances, that way. If one adopted EBCDIC, however, every time the organization needed to alter a balance, the packed-decimal-oriented computer would have to convert the EBCDIC data into packed decimal, perform the necessary operation, and then convert the new balance back into EBCDIC. Had the analyst initially stored the balance in packed decimal, the conversions would become unnecessary, and changes could be performed much faster.

As a rule, alphanumeric data are stored in ASCII or EBCDIC, and numeric in packed decimal.

MEDIA AND DATA-STORAGE TECHNIQUES

Since disks were relatively rare and very expensive in the 1960s, analysts relied on the less expensive tape, which was compatible with most computers. Disk prices have fallen, however, and today's users frequently want terminal-oriented systems that allow them direct, immediate access to data, so analysts increasingly employ disks. Almost all microcomputers use disks, and the booming popularity of such systems make disks even more attractive.

Tapes and Sequential Files

When designing a file for tape, analysts must understand the medium's physical properties. Most modern tapes are ½ inch wide, and wound on reels in lengths of 600, 1200, or 2400 feet (Figure 7-4). One may store data on a tape in EBCDIC, ASCII, or any of the numeric modes by writing the binary patterns across the tape in tracks. Tape used in the late 1960s and early 1970s used seven tracks; newer tapes use nine, with eight for data and one for parity (a built-in error-detection technique). Bits per inch (BPI) refers to the different densities available for recording data on tape: 800, 1600, or 6250 BPI, for example. A tape drive or handler reads or writes the data on the tape at speeds varying from 37.5 to 200 inches per second.

Tapes also read or write data record by record. In other words, users cannot randomly skip from one point to another at will, but must process data sequentially, record by record. This limitation dictates that tape-oriented systems store data in a certain order, so users can access the data according to unique record keys such as account, customer, part, and social security numbers.

Blocking. When designing tape-oriented systems that read or write data record by record, a space is placed by the tape drive between each record or between groups of records. Whenever two or more records are grouped together, the file is said to be blocked (Figure 7.5), and we call the number of records included in a block the **blocking factor.** For example, a file with a blocking factor of 10 is a file with an interblock gap between every 10 records.

If an inventory system stores data concerning 10,000 different items, allowing

Blocking Grouping logical records together to form a single physical record.

Blocking factor The number of logical records per physical record.

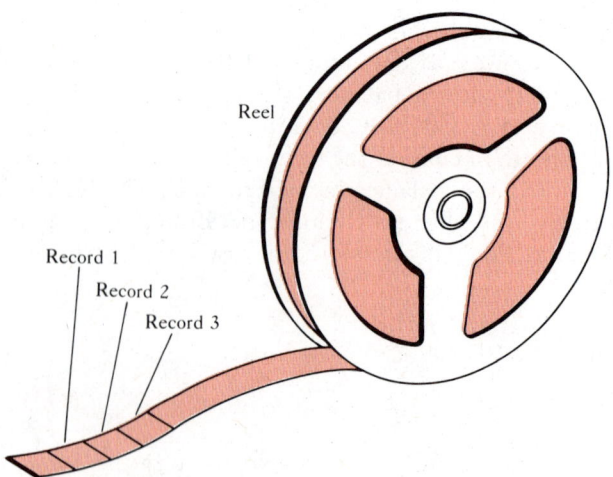

FIGURE 7.4. Tape provides a compact and inexpensive storage medium. Depending on quality and quantity ordered, a 2400-foot reel can cost between $12 and $30. The tape drive can read data in excess of 75 inches per second.

CHAPTER 7 ORGANIZING AND DESIGNING FILES 243

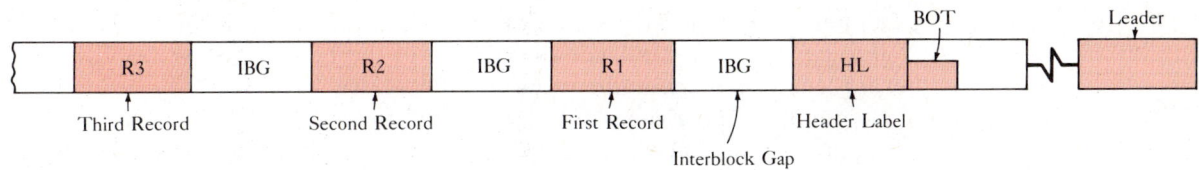

Single Blocking

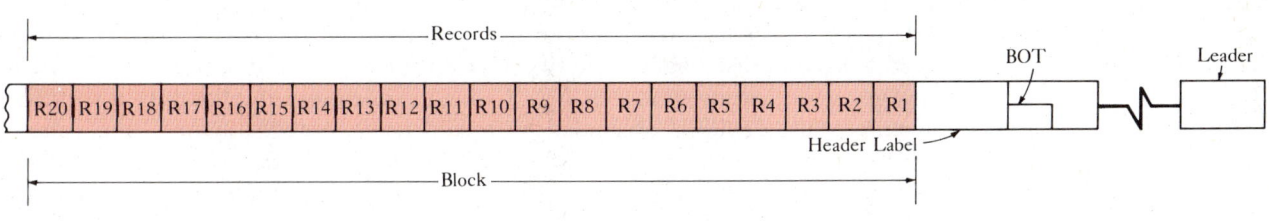

Blocked File

FIGURE 7.5. Blocking permits an organization to keep more records on a tape because it reduces the number of interblock gaps. In the top diagram, the tape holds only three records due to the space taken up by the interblock gaps. These gaps slow the speed of processing because the tape drive has to start and stop each time it encounters a gap. The bottom diagram is a blocked file with a blocking factor of 20, which means 20 records can be read or written before the tape drive stops.

80 characters for each (product number, description, quantity on hand, location in the warehouse, and number of items sold last year), the tape must hold 800,000 characters (80 × 10,000 = 800,000). Using a 1600-BPI tape to store the data record by record would involve 80/1600, or 0.05 inch of tape for each record. With a 0.6-inch interblock gap between each record, the total length of tape required to store such an inventory file would then be

Length = Number of records in file × (record length + 0.6)
       = 10,000 × (0.05 + 0.6)
       = 10,000 × 0.65
       = 6500 inches

On the other hand, we might block these records into groups of 10, resulting in 1000 groups with an interblock gap between each group. This approach would consume much less tape.

Length = Number of blocks × (Length of each block + 0.6)
       = 1000 × [(0.05 × 10) + 0.6]
       = 1000 × (0.5 + 0.6)
       = 1000 × 1.1
       = 1100 inches

Choosing an optimum blocking factor is crucial, because one that is too small will waste tape, and one that is too large may decrease available memory to hold all the data as well as the program that processes the data. Most computer manufacturers provide guidelines for blocking data on particular disk or tape drives. These guidelines will balance the density of the tape, speed of the device, and record length to make optimum use of the device.

Sequential File Processing. Not only does blocking save tape, but it also reduces computer processing time. When a tape drive reads a record from tape, it starts and stops the tape at each interblock gap, taking about 5 thousandths of a second (0.005 second). Thus our unblocked inventory file, for example, would experience 10,000 starts and stops for a total of 100 seconds [10,000 × (0.05 × 2)]. Blocked by 10s, the file contains only 1000 gaps, resulting in only 1000 stops and starts for a total of only 10 seconds [1000 × (0.005 × 2)].

If stock moves in and out of the warehouse rapidly, an inventory system may need to record 3000 changes per week. Furthermore, the system must post each change to the inventory master file. Though changes occur randomly, if the tape inventory master file is organized by record key (product number), the system must sort the changes before posting them to the master file. In other words, the computer reads 3000 changes, sorts them, then writes them out, in order, by product number (Figure 7.6), thus processing 6000 records to rearrange 3000 changes into the required order.

Once the system has sorted all changes by product number, it can match them to the correct items on the master files, and then record them. To update a **sequential file**, a system reads each record from the master file, then changes or copies it without change, thereby creating a new version of it. To update a 10,000-item inventory file, a system would have to process 23,000 records: 10,000 reads, 3000 changes, and 10,000 writes. Unaltered copies of all changes and of the original master file provide a backup in case something happens to the new master file, such as inadvertent erasure.

Conclusion. Tapes and their sequential file processing offer several advantages. First, processing automatically creates backup files. Second, tape is inexpensive and stores a large amount of data compactly. Third, since tape devices have enjoyed popularity since the 1960s, programmers are familiar with them and have developed ways to handle an array of record lengths with a variety of blocking strategies. Finally, a reel of tape can be sent through the mail, which costs less than contemporary data communications links but may not arrive for a few days.

On the other hand, sequential storage prohibits direct retrieval of records. Tape drives are expensive (as much as $22,500) and an organization must install three to achieve a system that can process changes, maintain a master file, and create a new master file (Figure 7.6). In many cases a reel of tape stores only a single file, so many reels may be needed to store all necessary records.

Though the medium of choice for storing files in the past, tape will give way to disks, just as punched cards gave way to tape. Nevertheless, tapes may continue to enjoy popularity as backup for disks.

Disks and Direct-Access Files

Commonly called direct-access storage devices (DASD), magnetic disks resemble phonograph records in diameter and thickness (Figure 7.7a). As you can see in Figure 7.7b, we classify disks as floppy (usually found on home computers or word processors) or hard.

Sequential file A file that stores records according to a key field.

CHAPTER 7   ORGANIZING AND DESIGNING FILES   **245**

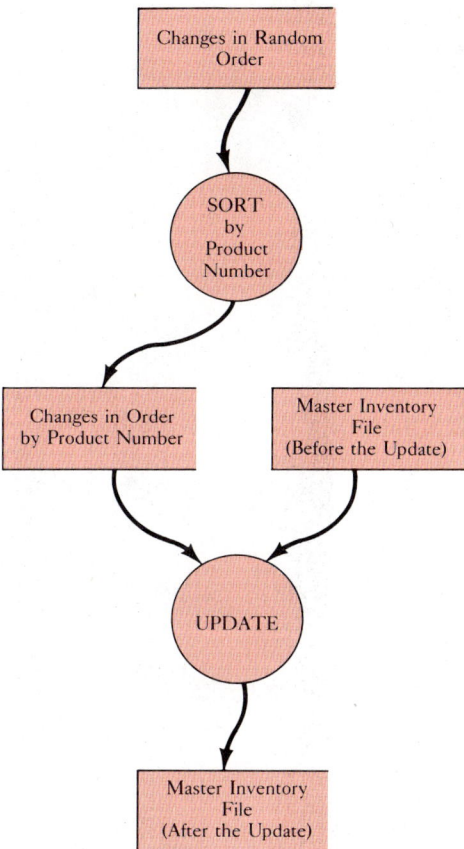

FIGURE 7.6. Data-flow diagram illustrating inventory master file update. There must be a tape for the changes, master file before the update, and the master file after the update.

Both floppy and hard disks can store data in EBCDIC, ASCII, packed decimal, or binary. Each approach involves writing binary patterns in concentric circles called tracks, which contain individually addressable sectors resembling arcs of a circle (Figure 7.8a). A read–write head attached to an access mechanism moves across the disk to access data. Thus we can retrieve data from any place on the disk by positioning the read–write mechanism over the desired location, skipping all the data in between. This ability to pick and choose the data we want gives a disk direct-access ability.

We can increase the amount of data available to the read–write mechanism by stacking 10 or more disks together. Ten disks stacked together allow data to be written on the top and bottom of each disk for a total of 20 surfaces. We call the arcs lying under the read–write access mechanism cylinders. Some disks contain as many as 500 such cylinders (Figure 7.8b).

To record or store and retrieve data, the access mechanism is positioned to the proper cylinder. Unlike a phonograph's arm and needle, which can only read sounds, the disk's access mechanism can both read and write data (not at the same time) from or to a sector. With the arm in position, the drive waits for the correct

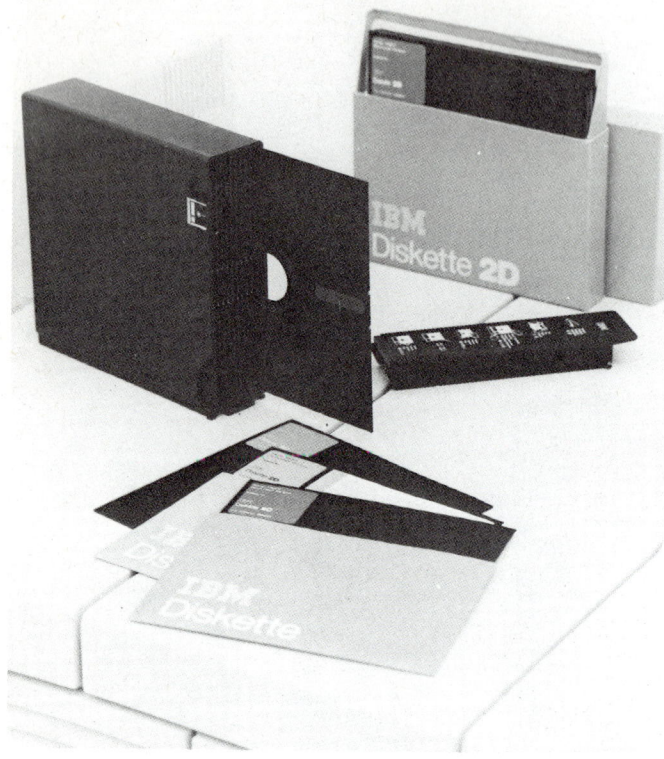

(a) Floppy disk

(b) Floppy disk drive

(c) Disk pack

(d) Disk drive

FIGURE 7.7. Hard and floppy disks are becoming extremely popular for external data storage. [(a and c) Courtesy of International Business Machines Corporation. (b and d) Courtesy of Digital Equipment Corporation.]

CHAPTER 7 ORGANIZING AND DESIGNING FILES **247**

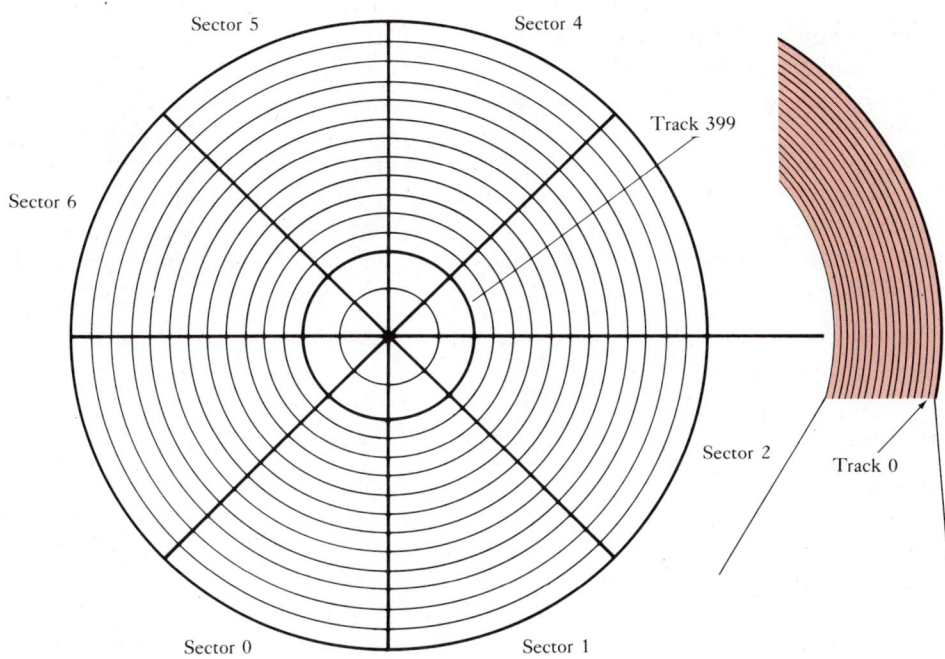

(a) A sector is a pie-shaped segment of a disk track.

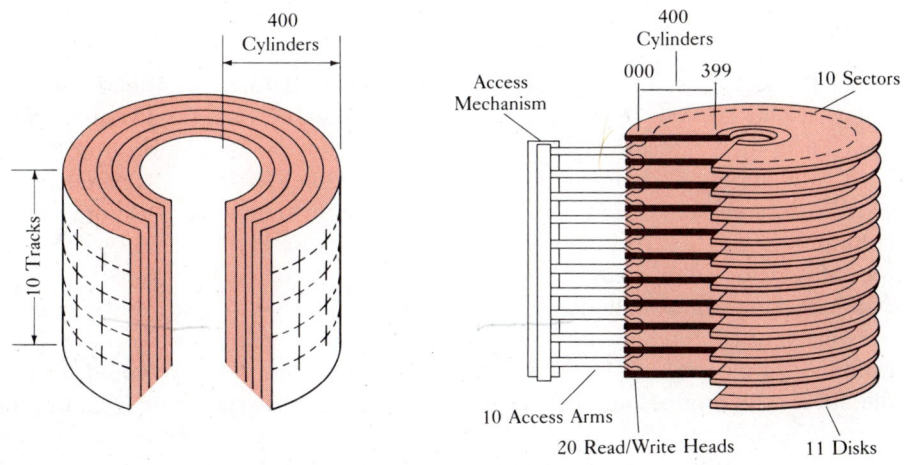

(b) Cylinders are collections of tracks on different disk surfaces. Each disk surface has its own read–write head attached to an access mechanism, which moves across the disk as a unit.

FIGURE 7.8. A disk is divided into sectors, tracks, and cylinders.

sector to spin under the access mechanism so it can read the data on that sector. The cylinder concept speeds data access since the read and write access mechanism does not have to be repositioned to read or write another track, thus saving valuable access time.

Disk capacity varies from manufacturer to manufacturer, ranging from 5 megabytes to over 1000 megabytes (a megabyte, or MB, is a million bytes; 1000 megabytes are roughly equivalent to 200,000 typed pages of text). If we stack 10 disks with 500 cylinders and 40 sectors per track, each holding 500 characters, we achieve the following capacity:

$$\begin{aligned}\text{Capacity} &= \text{Number of surfaces} \times \text{Number of cylinders} \times \text{Number of sectors} \\ &\quad \times \text{Number of characters} \\ &= 20 \times 500 \times 40 \times 500 \\ &= 200{,}000{,}000 \text{ bytes (200 MB)}\end{aligned}$$

In some cases analysts may select disks for sequential processing. For example, if one decides to offset the higher cost of disks by reducing the time computer operators spend mounting and demounting tape reels during sequential file update, one can get a disk to emulate a reel tape file.

Disks are relatively expensive: a 200-megabyte disk drive can cost $20,000 or more. Prices, though, have dropped significantly and capacities and speeds have increased dramatically since the late 1960s. With both sequential and random-access capabilities, disks are an increasingly attractive medium to store data.

VSAM and Direct-Access Files

Because the sectors on disks are individually addressable, thus accommodating both sequential and random record processing, they have a powerful advantage over tapes. To update a record, the program need merely locate the desired record, read it into memory, change it as necessary, and then write the desired record back to the same sector from which it came. Systems commonly use one of two techniques to locate a record: virtual sequential access method (VSAM) and the direct file organization method.

VSAM, sometimes called indexed sequential or **ISAM** (indexed sequential access method), locates records on a disk with a table. For example, to find data on product number C-64744 in an inventory system, the program would scan the table, locate the product number, and read its sector location (Figure 7.9). Product number C-64744 resides in sector 12228. The table giving product number and sector location is maintained by the VSAM software. Furthermore, the VSAM software "automatically" searches the table, looking for the desired product number. The program simply gives VSAM the product number and it returns the entire record.

Let us apply this procedure to our 10,000-product inventory system, where the analyst wishes to record 3000 changes. Since VSAM allows random record retrieval, we do not need to instruct our computer to sort changes. Rather, VSAM updates the master file by reading each of the 3000 changes, searching the table, reading the records at the addresses indicated by the table, altering the records in memory, and writing the records back to their original locations. To record 3000 changes,

---

VSAM A file access method that uses a table to locate a record in the file. An acronym for Virtual Sequential Access Method.

ISAM A file access method permitting both sequential and direct access to records. An acronym for Indexed Sequential Access Method.

Product Number	Location or Sector Number
A-01118	12224
A-78561	12225
A-98356	12226
B-34655	12227
C-64744	12228
G-75190	12229
K-89757	12230
M-78900	12231
Z-89222	12232

FIGURE 7.9. *The VSAM table locates a record on a disk. Thus product number C-64744 is found at sector 12228.*

VSAM processes 9000 records, as compared with 23,000 for a sequential tape system.

For large applications VSAM tables can become unwieldy, requiring enormous memory for vast numbers of records. However, one can solve this problem by dividing the main table into a number of subtables. By combining the first entries of each subtable into a "master" table, VSAM can locate a specific record by scanning the new table to find the proper subtable, then reading the subtable into memory. It then scans the subtable until the record keys match, after which the record is read, changes are made, and the altered record is written back to the disk.

The three access methods—VSAM, ISAM, and **KSAM** (keyed sequential access method)—require that files be stored sequentially by record key and the record with this key stored in a table. To add a record to the file, the system must store its key and location in an overflow table. Periodically the software rebuilds the tables and files to keep processing efficient. Some computer manufacturers include software that performs all the table operations, rebuilding them whenever necessary. In fact, one hardware vendor provides KSAM software to every purchaser of its computer system.

One advantage of VSAM is its flexibility, because it allows both random and sequential access to data. That is, we can read a specific record (product number C-64744) or we can process the entire file in order (a list of items on hand). Disadvantages include the extra work involved in maintaining the tables, the amount of memory required to store tables, and the need for several table searches before locating desired records.

If an application demands random access of records, **direct file** organization offers yet another alternative to VSAM. This method converts the record key into the disk sector number, quite often by a technique called divide and remainder. First we divide the record key by a prime number (a number only divisible by itself and 1) to determine the remainder. The prime number we use must be greater than the number of actual records plus an allotment for future file expansion. For our 10,000-product inventory system, the analyst might pick the prime

KSAM A file access method similar to ISAM. An acronym for Keyed Sequential Access Method.

Direct file A file storage method that uses the record key to locate a record in the file.

number 11,001, which allows for 1001 additional expansion positions. For product number C-64744, the division yields

$$\text{Location} = \text{Remainder (Record key/Prime number)}$$
$$= (64{,}744/11{,}001)$$
$$= 5, \text{Remainder } 9739$$

The remainder, 9739, identifies the location of the record, C-64744, in the file. With no table to search or maintain, we can locate a record by a simple act of division.

Note that if a record key has alphabetic characters, as many part numbers, license plate numbers, and drivers' license numbers do, we cannot divide until we have either stripped the record key of its nonnumeric characters or converted them to numbers. One method for performing this conversion is the Soundex system, which converts letters as follows:

B, F, P, V are assigned a 1.
C, G, J, K, Q, S, X, Z are assigned a 2.
D, T are assigned a 3.
L is assigned a 4.
M, N are assigned a 5.
R is assigned a 6.
A zero is assigned to A, E, H, I, O, U, W, Y.

Thus the word KLEGER becomes 240206 and BURNS is 1652 in the Soundex system. This system, however, is like a one-way street: you cannot convert numbers back into words.

Direct file processing works much like VSAM. It reads a change record, calculates the location, reads the disk, modifies the record, and writes the new version back to the same location. As with VSAM, it takes 9000 reads and writes to update a file with 3000 changes.

The direct file method handsomely supports applications demanding quick record retrieval, because we can locate and read the desired record into memory with a single access to the disk. The calculation for finding the record is simple, and keys can be numeric or alphanumeric. The Soundex procedure allows us to convert alphanumeric keys to numeric locations, and we can program it with a few COBOL, Ada, or Pascal instructions.

Collison When two or more record keys calculate to the same location in the file.

A disadvantage of the direct file method is that calculations on two different record keys may result in identical remainders, or **collisions** (sometimes called crashes or synonyms). For example product numbers C-64744 and F-42742 both yield a remainder of 9739 when divided by 11,001. When collisions occur, an indicator may be stored in the first record (C-64744) to warn us of the crash. The indicator tells us the sector number where F-42742 really resides, perhaps 9766, thus directing us there to determine whether it is the one we want. Because of the chance for collisions, we can reserve some of the extra disk space allotted to file expansion to hold a record that otherwise would collide with another.

Another disadvantage of direct filing is the random order of the file. Since the file does not appear in any specific order, we must add a sorting step whenever we wish to list or otherwise process the file in sequence.

Also, when a direct file becomes full, which sometimes happens, a programmer must write a special one-time program to rebuild it with expansion space. With tape, on the other hand, as the file fills up, it simply uses additional tape.

A fourth disadvantage is that disk files do not generate automatic backups. To provide an extra copy of the file, we must copy it to another disk or reel of tape. Fortunately many operating systems provided by computer manufacturers contain utility routines to copy disk files to tape.

The choice between sequential and direct or VSAM files revolves around users' needs. If users want instant access to data, direct or VSAM techniques are ideal, whereas batch processing, which collects data for later processing, perhaps at night when computer time is in low demand, historically uses the sequential access technique.

DESIGNING DISK OR TAPE FILES

When designing a file for a computer system, the analyst must choose the proper storage medium (disk or tape) and the best method for accessing data from the file: sequential, direct (random), or VSAM.

Having settled upon a medium and an access method, the analyst begins actually to design the file, consulting report formats and their accompanying data dictionaries for guidance. For example, our inventory system requires a stock status report showing a summary of inventory activity for the month. This report lists each activity, whether it is an addition or deletion of stock, as well as quantities on hand and quantities on order. Considering the nature of the report, the analyst, in this case, decides to construct two files: a master file containing data about each item, and a transaction file for changing data about items. Since the VSAM method provides both direct and sequential access to data, the analyst might prefer disk, using product numbers for record keys.

Analysts use file layout sheets to sketch the formats of records to be included in the files (Figure 7.10). Each sheet can hold several record formats and contains information on the length of the record (number of characters), the data format and size of each field, and the order in which fields will appear. Each record in our inventory master file will contain five fields and 60 characters of storage space, while each record in the transaction, or detail, file will contain five fields and 36 characters per record.

Data for both master and detail file records will come from two sources: customer orders that reduce inventory, and orders to suppliers for new items to replenish inventory (Figure 7.11). Both sources require validation to ensure the correctness of the data. Errors in dates (month number not 1–12), product numbers, or code numbers (1: reduction in inventory, 2: replenish inventory) should cause the system to reject that transaction. Data for corect orders should update the master file and should alter the "quantity on hand" field, so the organization has a record of the exact quantity that should be in the warehouse.

Since users of an inventory system demand immediate access to data, the analyst might decide to employ the VSAM storage technique for the master inventory file, using a disk for direct access. On the other hand, one might design the detail file for sequential filing on tape, because the system will need transaction details

252 PART II SYSTEMS DESIGN

DATE 10/84
SYSTEM: *Inventory-Master* PROGRAM NUMBER: _____
FILE NAME _____ FORMAT TITLE _____

1-8	9-19	20-25	26-33	34-40
Product Number	Description	Unit Price	Quantity On Hand	

(a) Master file record.

DATE 10/84
SYSTEM: *Inventory-Master* PROGRAM NUMBER: _____
FILE NAME _____ FORMAT TITLE _____

1-9	10-14	15-20	21-27	28-36	37-40
Product Number	Quantity	Date Ordered	Date Delivered	Supplier Number	

Code #

(b) Transaction or detail record.

FIGURE 7.10. File or record layout sheets allow analysts to describe a record, name it, date the design, define the fields, and identify the person who designed it. This layout sheet shows both the inventory master and detail record. A vertical line is used to separate fields.

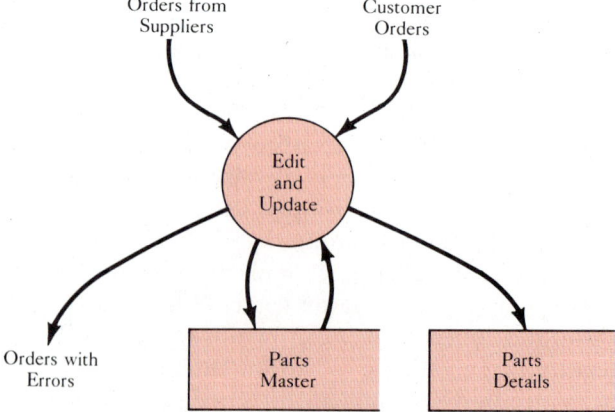

FIGURE 7.11. This data-flow diagram shows the flow of data through an inventory system. Here we see two files: Parts-Master and Parts-Details. A line above and below the file name indicates that it is a file.

Guidelines for Designing Files

Just as any storage system depends on what you wish to keep there, so do a computer system's files depend on its hardware and software. One would not rent a large storage compartment for income tax records, nor would one try to stuff athletic equipment into a filing cabinet. In the 1960s and 1970s, disks were something like two-drawer filing cabinets, with low capacity and high costs; today most computers have disk drives that are like 40-drawer filing cabinets, large storage compartments capable of holding hundreds of millions of characters of data. Many applications can take advantage of this increase in disk capacity, thereby putting an organization's data at users' fingertips.

Tape may offer the best alternative for storing data, especially if the application involves hundreds of billions of characters or does not require data to be available online via a terminal.

Whether a system uses disks or tapes; searches sequentially, randomly, or indexed sequentially; or employs a data-base management system, certain "rules of thumb" can guide us when we are designing files:

1. From the data-flow diagram (DFD), identify all data stores or files.
2. Assign each file or data store a unique and meaningful name.
3. Specify the type of file:
   a. Internal, holding data extracted from other files.
   b. Archival, requiring long-term storage.
   c. Control, containing codes (e.g., discounts for prompt payment).
4. List the data elements in the file using a data dictionary format. Include in this definition the data type (alphanumeric or numeric) as well as the length of the data element).
5. Identify the access requirements for the file, that is, ascending by vendor name, random by vendor number, Zip code for bulk mailings.
6. Determine the hierarchy among the files. Some might be subservient to others (e.g., sales invoices depend on customer lists).
7. Look for and eliminate duplication of files.
8. If files share common data elements, consider combining them.
9. Maintain the integrity of the files (e.g, last year's invoices and this year's invoices should be in two separate files).
10. Determine the medium for storage: disk or tape.
11. Estimate the number of records in the file.
12. Calculate the frequency with which records will be added, deleted, or changed.
13. Define the order of the file's data elements. Place key fields first, followed by semipermanent data, with an expansion area for additions.
14. Consider adding a data element for date of last activity for some types of records (e.g., customer payments or purchases).
15. Calculate the record length by adding together the lengths of all data elements in the file. A hardware specialist can tell you if the length will fit a particular storage device. Block the file accordingly.
16. Specify each file's need for backup.

Because input/output operations consume much more computer time than internal operations, good file design can reduce costs and improve efficiency. The fewer READ or WRITE operations the system has to make, the faster it will run and the less if will cost over time.

only once for update purposes. Once the master file has been updated, the detail file can be set aside for study, historical, and backup purposes.

Having decided to store the transaction details on tape, the analyst must block the records to pack more data per inch of tape and thereby minimize processing time. In this case, since the manager of computing services has mandated an even number of records per block plus a total block length of around 2000, 60 is an appropriate blocking factor. Each detail record contains 36 characters, so a factor of 60 results in 2160 characters ($60 \times 36 = 2160$) per block. On a 1600-byte-per-inch tape, such a block consumes 1.35 inches ($2100/1600 = 1.35$).

Let us assume that the analyst notes during the analysis phase of the systems process that the organization typically processes 12,000 detail records each month. These statistics yield a need for 200 blocks ($12,000/60 = 200$). The length of the detail or physical file is therefore 200 blocks of 1.35 inches plus 200 0.6 inch interblock gaps, or a total of 390 inches of tape [$200 \times (1.35 + 0.6) = 390$]. On a 2400-foot reel of tape, the detail file would use 390 inches, or 32.5 feet (about 2 percent of the whole tape) but the processing time would be minimized.

FILE AND PROCESSING CONTROLS

As with output controls, the analyst can choose from a wide variety of file and processing controls. After designing output controls, the analyst can consider the design of control systems for file protection and data processing.

Record Counts

Record count A control procedure that calculates the total number of records in a file.

A **record count** is one of the easiest file and processing control procedures to design. As its name implies, this type of file control simply counts the number of records read from or written to a file. If the system is set up for sequential file processing, a transaction file may undergo three basic types of updates: additions, changes, and deletions. The record count control procedure uses the following formula to monitor additions and deletions.

$$\begin{aligned}
&\text{Number of records in the new master file (NM)} = \\
&\quad \text{Number of records in the old master file (OM)} - \\
&\quad \text{Number of records dropped from the old master file (DR)} + \\
&\quad \text{Number of records added to the new master file (AD)}
\end{aligned}$$

In abbreviated form, the formula is: $NM = OM - DR + AD$.

Since existing records with changes do not affect how many records exist in either the old or new master file, the formula need not account for them.

The analyst specifies an update program that allows the computer to count these four quantities and print them at the end of the program. Then a control clerk can visually match the totals to ensure accuracy.

Backup

Backup, or file copying, offers another simple control procedure for both online and batch systems. Members of the operations staff within most computer installations routinely copy the entire contents of their disks onto tape, creating a backup in the event of a file's loss, theft, or destruction. Of course, they store these tapes in a safe place, away from the originals. Considering the time and money it would take to recreate its data, an organization can easily justify transportation and storage costs.

The best time to copy a file depends on the system. For some systems analysts may specify creating backups before and after major processing (printing of payroll checks or statements sent to customers). On the other hand, a real-time college student registration system may require that data bases be copied every day, as well as before and after all weekly or monthly closing runs.

Some systems automatically generate their own backups, Figure 7.6. For example, the data-entry staff may collect transaction data for one week on tape number 1077 (tapes are assigned numbers as they are purchased, so the staff can keep track of their contents). After collecting a week's worth of data, the processing cycle starts. The data from tape 1077 move onto tape 7634 to form a new master file on tape 8901. During the next week, the staff might record transaction data on yet another tape, 2312, which will be added to 8901 to create yet a new master file tape, 870. Since all prior data appear on individual tapes, the system has automatically built its backup. All the organization has to do is add each week's data on the preceding week's master file. Figure 7.6 illustrates weekly processing. Updating can occur daily, weekly, monthly, or whenever appropriate.

California's State Medical System routinely uses tape in this manner. The Medical Eligibility History System consists of a master file of 12 tapes storing 310 characters for each of the 12 million medical service recipients. The state updates its tapes weekly on the IBM 3081 computer, a job that requires 6 hours of processing time. The transaction file of new or dropped recipients, or changes in recipient data, takes from two to six reels, depending on the time of year. These tapes reside for 12 months in a cavern near Lake Tahoe on the California-Nevada border, while the new master stays close at hand for the next week's processing cycle.

This type of processing ideally suits tape or disk systems that use an old master to create a new one. When using disk files, the system writes any new data over the old, preventing automatic backing up, so in such applications the organization must copy each file individually on a new tape.

Backup Control procedure whereby files are copied on a routine or special basis.

CASE STUDY

Fleet Feet's Accounts Payable File Design

Fleet Feet currently uses the KSAM file system on its Hewlett-Packard 3000 computer. KSAM permits users to access data with applications programs written in COBOL, Pascal, FORTRAN, BASIC, RPG, or SPL (Systems Programming Language, similar to Algol). Furthermore, KSAM allows access to records randomly or sequentially for reporting purposes. More than one application program can retrieve KSAM data at the same time, which is a requirement in the multiterminal environment in which the accounts payable system must function. The KSAM software is an integral part of the Hewlett-Packard 3000 operating system that came with the computer.

After reviewing Fleet Feet's desired report formats, analyst Frank Pisciotta decides that the system will need six files (Figure 7.12).

1. VENDOR: To hold data about each of Fleet Feet's vendors. Stored data will include vendor name and number, address, discount terms, and total purchases this year and last.
2. AP-TRX: To hold data about each voucher paid, including date, vendor name and number, invoice and discount data, and check number.
3. CHECK: To hold data about all checks written to vendors, including the vendor and voucher numbers, and amounts.
4. AP-OPEN-ITEM: To hold data about all vouchers authorized for payment: voucher number, vendor name and number, check number, and detailed information about the purchase.
5. AP-OPEN-ADJ: Similar to AP-OPEN-ITEM, except that it also allows adjustments to invoices.
6. CHECK-RECON: To hold data, including vendor number, about issued checks that have been cashed and have cleared the bank.

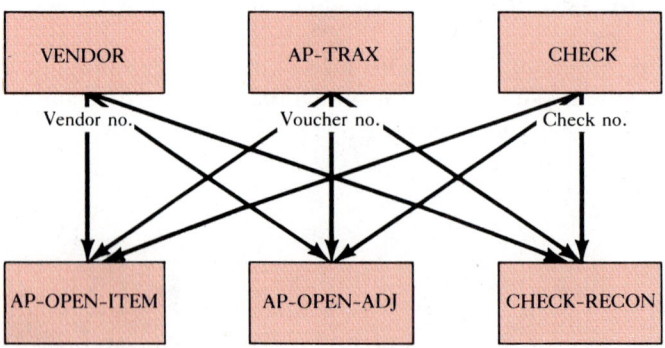

FIGURE 7.12. File relationship of Fleet Feet's accounts payable system. The master files lie above the transaction files (in uppercase letters). The record key (in lowercase letters) shows the logical links between the files.

CASE STUDY

Note that each file has a record key to allow records in various files to be logically related. For VENDOR the key is vendor number, for AP-TRX it is voucher number, and for CHECK it is check number.

After selecting the files, determining their relationships, and establishing record keys, Frank writes the data entries for each file (Figure 7.13). He gives each item a name (VOUCHER-NO, VENDOR-NO, or INVOICE-NO), determines its type (X stands for alphanumeric, 9 for numeric), then specifies its length (number of characters for alphanumeric, number of digits for numeric). The entry for vendor number in the AP-OPEN-ITEM file reads:

VENDOR-NO,   X6

All vendor numbers are alphanumeric (X) and will contain six characters.

After defining the files, Frank estimates capacities for each. The capacity statement tells the maximum number of records this file can hold. For CHECK-RECONCILE Frank decides that 4500 entires are adequate since Fleet Feet undoubtedly will not write that many for a full year.

With his file design in hand, Frank can now move to the next stage of design: input design.

```
FILE NAME: VENDOR
FILE TYPE: MASTER

FIELDS:
        VENDOR-NO,              X6        <<RECORD KEY>>
        VENDOR-NAME,            X30
        VENDOR-ADDR-1,          X30
        VENDOR-ADDR-2,          X30
        VENDOR-ADDR-3,          X30
        TYPE,                   X4
        TERMS-DESCR,            X16
        DUE-DAYS                Z4
        DISCOUNT-DAYS,          Z4
        DISCOUNT-%,             Z4
        LAST-ACTIVITY,          Z6
        STATUS-CODE,            X2
        PURCHASES-YTD,          Z10
        PURCHASES-LAST,         Z10
        DISCOUNTS-YTD,          Z10
        DISCOUNTS-LAST,         Z10

CAPACITY: 251
```

FIGURE 7.13. *File definition for Fleet Feet's accounts payable system.*

CASE STUDY

```
FILE NAME: AP-TRX
FILE TYPE: MASTER

FIELDS:
        VOUCHER-NO,              X6        <<RECORD KEY>>
        TRX-TYPE,                X2
        VENDOR-NO,               X6
        P-O-NO,                  X10
        INVOICE-NO,              X8
        INVOICE-DATE,            Z6
        INVOICE-AMOUNT,          Z10
        INVOICE-NON-DSC,         Z10
        DUE-DAYS,                Z6
        DISCOUNT-DAYS,           Z4
        DISCOUNT-DATE,           Z6
        DISCOUNT-%,              Z4
        DISCOUNT-AMOUNT,         Z8
        CHECK-NO,                Z6
        CHECK-DATE,              Z6
        ACCOUNT-NO,              Z10

CAPACITY: 6500

FILE NAME: CHECK
FILE TYPE: TRANSACTION

ITEMS:
        CHECK-NO,                Z22       <<RECORD KEY>>
        VENDOR-NO,               X6
        CHECK-TYPE,              X2
        CHECK-DATE,              Z6
        CHECK-AMOUNT,            Z10
        P-O-NO,                  X10
        INVOICE-NO,              X8
        INVOICE-DATE,            Z6
        INVOICE-AMOUNT,          Z10
        DISCOUNT-AMOUNT,         Z8

CAPACITY: 4500
```

FIGURE 7.13 (Continued)

CASE STUDY

```
FILE NAME: AP-OPEN-ITEM
FILE TYPE: TRANSACTION

ITEMS:
        VENDOR-NO,              X6      <<RECORD KEY>>
        VOUCHER-NO,             X6      <<RECORD KEY>>
        P-O-NO,                 X10
        INVOICE-NO,             X8
        INVOICE-DATE,           Z6
        DUE-DATE                Z6
        DISCOUNT-DATE           Z6
        ORIG-INVOICE-AMT,       Z10
        ORIG-DISC-AMT,          Z8
        INVOICE-BALANCE,        Z10
        DISCOUNT-BALANCE,       Z8
        PARTIAL-PAYMENT,        Z10
        CURRENT-DISCOUNT,       Z6
        CHECK-NO,               Z6
        CHECK-DATE,             Z6
        AP-ACCOUNT,             Z10
        SELECT-STATUS,          X2
        PRINT-STATUS,           X2

CAPACITY: 4500


FILE NAME: AP-OPEN-ADJ
FILE TYPE: TRANSACTION

FIELDS
        VOUCHER-NO,             X6      <<RECORD KEY>>
        TRX-TYPR,               X2
        VENDOR-NO,              X6      <<RECORD KEY>>
        CHECK-NO,               Z6      <<RECORD KEY>>
        INVOICE-NO,             X8
        INVOICE-DATE,           Z6
        INVOICE-BALANCE,        Z10
        INVOICE-DISCOUNT,       Z8
        OLD-DUE-DATE,           Z6
        DUE-DATE,               Z6
        OLD-DISC-DATE,          Z6
        DISCOUNT-DATE,          Z6
        DISCOUNT-AMOUNT,        Z8
        CHECK-DATE,             Z6
        CHECK-AMOUNT,           Z10
        DISCOUNT-TAKEN,         Z8
        CASH-ACCOUNT,           Z10

CAPACITY: 5500
```

FIGURE 7.13 *(Continued)*

260 PART II SYSTEMS DESIGN

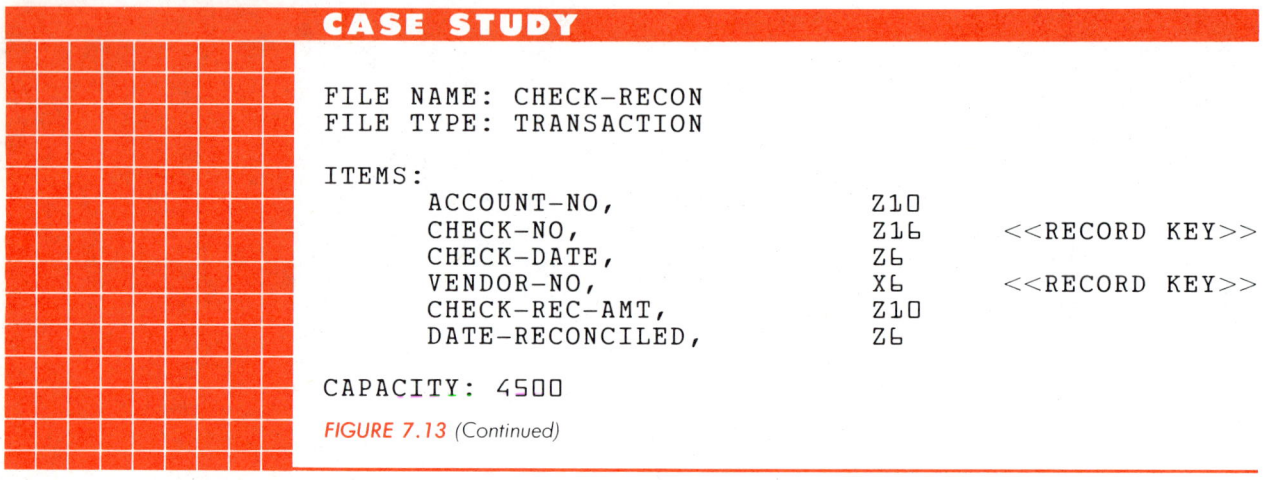

```
FILE NAME: CHECK-RECON
FILE TYPE: TRANSACTION

ITEMS:
        ACCOUNT-NO,             Z10
        CHECK-NO,               Z16       <<RECORD KEY>>
        CHECK-DATE,             Z6
        VENDOR-NO,              X6        <<RECORD KEY>>
        CHECK-REC-AMT,          Z10
        DATE-RECONCILED,        Z6

CAPACITY: 4500
```

FIGURE 7.13 (Continued)

WORKING WITH PEOPLE   Professional or Personal Computer Considerations

Bob Opie, job superintendent and estimator at Continental Construction Company, loved to work with electronic devices so he didn't surprise his co-workers when he trundled his new IBM PC microcomputer into the office. But they did wonder why Bob thought he needed an IBM PC when he had his own terminal for Continental's sophisticated time-sharing system. Continental's Digital Equipment Corporation VAX 780 ran the latest word-processing software, and data-base manager; supported many programming languages and offered a giant menu of canned programs Bob could call up from his terminal, whereas his IBM PC had a dot-matrix printer and two 320-kilobyte disk drives. Its memory was a meager 256,000 bytes, pitiful compared with the DEC's 6 megabytes.

Undismayed, Bob began explaining the software for the IBM PC: Lotus 1-2-3 (an integrated spreadsheet, file manager, and graphics system), Visi-Term, VisiPlan, and WordStar. He surprised his colleagues by showing them how his IBM PC could serve as a terminal at home (with the VisiTerm software) or on a construction job site using a modem to communicate with Continental's VAX 780.

After only 2 weeks of experimenting, Bob began using his IBM PC for job costing. Collecting data on the cost of concrete at the job site, Bob entered the data into his IBM PC and sent them back to the home office for review. Bob appreciated the IBM's lack of need for air conditioning or special power requirements, and he loved being able to transmit data directly from the site without having to return to his office.

As job superintendent, Bob schedules subcontractors, which is a very time-consuming task. Not only does he have to track costs incurred on the job, but he also has to submit weekly status reports to management. The IBM PC software, called VisiPlan, can do this too: Bob enters data on the costs and sequence of each subcontractor and projects a schedule for each part of the job. VisiPlan will display the proper sequencing, and will print a graph showing when all the subcontractors were due on and off the job; it also allows changes, thus saving valuable time in monitoring scheduling. When Bob discovered with VisiPlan that delaying by 2 days a crane scheduled for the job at $350 per day plus $65 per hour could shave 3 days off a 7-day schedule, he had saved Continental $1150. It didn't take long for Bob to convince his co-workers of the value of his IBM PC.

Bob found Lotus 1-2-3 useful for tracking each job subcontractor, comparing Continental's estimate of a particular phase of a job with actual costs incurred to date. Bob used rows in the Lotus 1-2-3 spreadsheet system for each subcontractor and columns for the various costs. The rightmost two columns displayed percentages and the difference between expected and actual costs. The bottom row totaled each column. Whenever contractors told Bob they'd finished their work, Bob inspected the work and authorized it for payment, entering the amount into the IBM PC so Lotus 1-2-3 would automatically compute totals, percentages, and differences. He was the only Continental estimator whose costs were always up to date.

The IBM word processing put the icing on the cake. With WordStar, Bob could write each week's report on the job site or at home, editing it directly on the monitor, and sending it to the office for review by his boss. There was no need for dictation, typing, editing, retyping, or duplicating.

Most of Continental's staff enjoyed Bob's success. When Bob returned to the main office for a monthly project review, his boss asked him to train two other field engineers on new IBM's the company had decided to buy. He even handed Bob a check to cover the cost of Bob's IBM PC and its software.

SUMMARY

Tapes have been popular storage devices since the dawn of computers, but today they more frequently provide backup storage for disk-based systems. Though some organizations still use tapes as the primary storage medium, new online systems will erode their popularity. Disks provide the best of both worlds—direct and sequential access to data.

Regardless of the storage medium selected, the analyst must specify a storage method: EBCDIC, ASCII, packed decimal, or true binary. The first two can store alphanumeric data, whereas the last two only permit numeric data.

Files can be classified as master, transaction, backup, temporary, or program. Each type has its own special purpose. To describe a file, the analyst assigns it a name, lists the data items for each record, chooses a data type (alphanumeric or numeric) for each field in the file, stipulates the storage method of the data (EBCDIC, ASCII, binary, or packed decimal), estimates the number of records the file is expected to hold, and classifies the file as either master, transaction, backup, temporary, or program. Finally, the analyst determines how the files will be logically tied together using record keys.

NEW TERMS

ASCII
Backup
Backup file
Blocking
Blocking factor
Collision
Direct file
EBCDIC
ISAM
KSAM
Master file
Packed decimal
Program file
Record count
Scratch, or temporary, file
Sequential file
Straight binary
Transaction, or detail, file
VSAM

QUESTIONS FOR REVIEW AND DISCUSSION

Questions appear in three categories. You can find the answers to the A group in this chapter. The B group requires you to apply the material presented here, whereas the C group necessitates investigation or research.

- A.1. List the three types of data-storage techniques.
- A.2. How long is the interblock gap?
- A.3. What do the letters in each of the following acronyms stand for?
  - a. VSAM   c. ASCII
  - b. ISAM   d. EBCDIC
- A.4. Define the five types of files.
- A.5. What is the output from the file design phase of the systems process?
- A.6. How long does it take to retrieve data from a hard and a floppy disk?
- A.7. How long does it take to retrieve data from internal memory?
- A.8. How much memory does it take to store the word "WORD" using the ASCII coding system?
- A.9. What is the storage capacity of a 2400-foot reel of tape storing data at 1600 characters per inch?
- A.10. What are the inputs to the file design phase of the systems process?

- B.1. If a record requires 180 characters, is blocked 30 records to a block, and 12,000 records are in the file, how much 1600-character-per-inch tape is required to store the file?
- B.2. Sketch the bit patterns to store the number 256 using:
  - a. EBCDIC   c. Packed decimal
  - b. ASCII   d. True binary
- B.3. How many records must a system process to update a master file of 4000 records if the transaction file contains 1000 records and the VSAM storage technique is used?
- B.4. How many records must a system process to update a master file of 4000 records if the transaction file contains 1000 records and the direct storage technique is used?
- B.5. How many records must a system process to update a master file of 4000 records if the transaction file contains 1000 records and the sequential file storage technique is used?

- C.1. Is there any mathematical relationship between uppercase and lowercase letters in the EBCDIC and ASCII systems?
- C.2. Is there any mathematical relationship between numeric characters in the EBCDIC and ASCII systems?
- C.3. From a recent issue of a computer newspaper, magazine, or other journal, find an article on streaming tape drives. Write a paragraph explaining how they differ from reel-to-reel tapes.
- C.4. Write the EBCDIC and ASCII binary patterns for
  - a. A   d. Z
  - b. a   e. z
  - c. 6

REFERENCES

Steve Eckols, *How to Design and Develop Business Systems*, Mike Murach and Associates, Fresno, Calif., 1983.

Alan L. Eliason, *Online Business Computer Applications*, Science Research Associates, Chicago, 1983.

James Martin, *Strategic Data Planning Methodologies*, Prentice-Hall, Englewood Cliffs, N.J., 1982.

Glenford J. Myers, *Software Reliability*, Wiley, New York, 1976.

CHAPTER 8 Designing a Data Base

- Goals and Preview
- Data-Base Management Systems
- Types of Data Bases
- Defining Data Bases
- Data Manipulation Languages
- Relational DBMS: What's in a Name?
- Query Languages
- Utilities
- Data-Base Controls
- Case Study: Fleet Feet's Accounts Payable Data-Base Design
- Working with People: Retraining for the Data-Base Environment
- Summary
- New Terms
- Questions for Review and Discussion
- References

GOALS AND PREVIEW

After reading this chapter you should be able to:

1. Define the three basic data-base systems.
2. Explain the schema concept.
3. List the four types of utility software that accompany most data-base management systems.
4. State the advantages and disadvantages of a data-base management system.

As an alternative to designing traditional disk and tape files, an analyst may choose a data-base manager for data storage and retrieval. Just as with the file design discussed in the previous chapter, the analyst still determines what data the computer must keep, and chooses identifiers users can employ to retrieve the data easily, but in this case one does not need to worry about where the data will actually reside on the disk. Rather one develops a description of the data files, known as a schema, from the report formats and the data dictionaries developed earlier. If a data-base manager is selected, the schema, in place of disk or tape file design, provides input to the next design stage, input design, and it forms part of the system specifications.

DATA-BASE MANAGEMENT SYSTEMS

Traditional file systems required that each application within a system retain responsibility for its own data. In other words, a data item such as an employee number or name, common to both payroll and personnel systems, had to be entered separately into both. If the employee number ever changed, that change had to be input separately into every file that used it. Because they centralize such data, data bases make that sort of duplication unnecessary (Figure 8.1).

Every computerized information system involves not only data, but also the structure, or physical location, of the data. A data base is a collection of related files in which the data are structured for easy retrieval, manipulation, and storage by many users. To achieve this goal, computer manufacturers and software developers have designed the data-base management system (DBMS). A DBMS is a software package that enables users to organize large, complex bodies of data into useful, accessible, and compact forms that avoid the awkward and inefficient file management techniques of the past. Data-base systems merge data into one pool (Figure 8.2) shared by all systems, so any change automatically affects all relevant systems. Programs to retrieve, update, add, or delete data in the data base can involve batch, online, or mixed processing.

DBMS software ties logical records together. Thus, for our inventory system, the DBMS could logically connect records with the same product number. When specifying a DBMS for an application, neither analyst nor programmer need be bothered with the time-consuming details of where or how the computer physically stores the data, because they know they can retrieve the data at will. Instead they can focus on the system itself and on users' needs.

Since DBMS organizes and manages all of an organization's information system files with a single set of software routines, it works well in an interactive environment where users with remote terminals need rapid access to data.

The DBMS concepts represent an evolutionary step from direct, sequential, and VSAM methods. The innovation lies in separating the definition and control of the data base from the specific application system, thus allowing logically con-

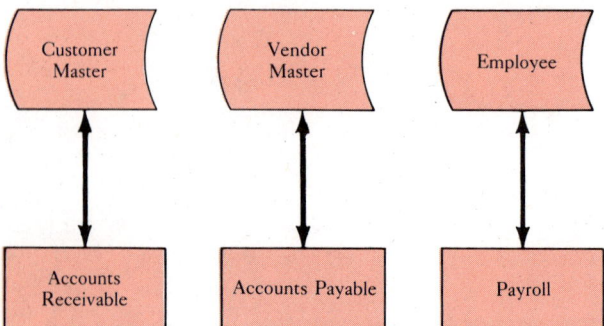

FIGURE 8.1. Traditional file systems made each system responsible for its own data. Since different systems did not share a data base, a change in one system necessitated a separate change in another.

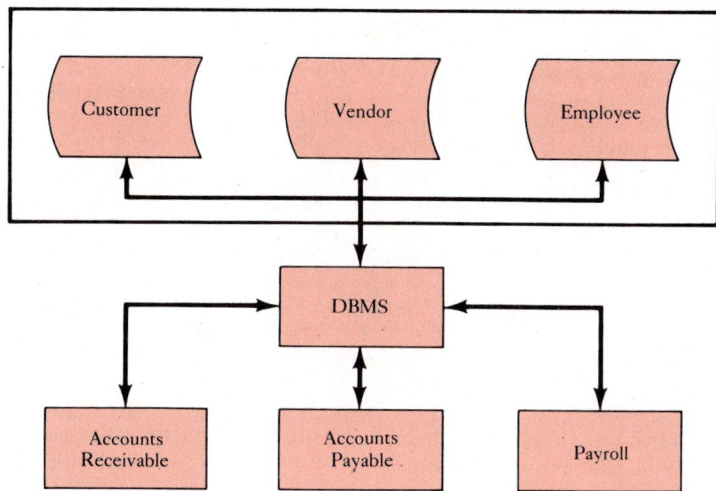

FIGURE 8.2. *Data-base management systems centralize data, merging them into a single pool so different systems can share them.*

nected files to be accessible to all programs. This results in more efficient data processing, simpler systems development, and lower programming costs.

Specific advantages of data bases are:

1. *File consolidation.* Pooling data reduces redundancy and inconsistency and promotes cooperation among different users. Since data bases link records together logically (even if they are physically separated), a data change in one system will cascade through all the other systems using the data.
2. *Program and file independence.* This feature separates the definition of the files from their programs, allowing a programmer to concentrate on the logic of the program instead of precisely how to store and retrieve data.
3. *Access versatility.* Users can retrieve data in many ways. They enjoy the best of both worlds—sequential access for reporting data in a prescribed order and random access for rapid retrieval of a specific record.
4. *Data security.* Usually a DBMS includes a password system that controls access to sensitive data. By limiting their access to read-only, write-only, or specified records, or even fields in records, passwords can prevent certain users from retrieving unauthorized data.
5. *Program development.* Programmers must use standard names for data items rather than invent their own from program to program. This lets the programmer focus on desired function.
6. *Program maintenance.* Changes and repairs to a system are relatively easy.
7. *Special information.* Special-purpose report generators can produce reports with minimum effort.

Figure 8.3 examines some of the benefits an organization derives from choosing a DBMS over traditional sequential, indexed sequential, or direct file processing.

Traditional Organization	Data-Base Organization	Benefits
Each application has its own master file, which contains data duplicated in files of related systems.	DBMS creates and maintains interfile access paths automatically so that the same data may be used by multiple applications.	Reduced data redundancy.
Application programmers supply their own data names.	DBMS standardizes names. Maintained by a data dictionary.	Encourages standardized names.
Tape handling is a major activity in production runs.	Data-base files reside on disk.	Reduces operator intervention and associated errors.
Coding of security systems is an independent task. Such systems are difficult to implement.	Control and access set by DBMS and not available to the program.	Provides data security.
Recovery procedures are not automatic and must be programmed for each system.	Recovery is an automatic feature. All transactions are logged and quick restoration is possible.	Provides automatic recoverability.
Most flat files support a single access path to data.	Multiple access paths to data are automatic.	Increases the accessibility of data.
Each file must be processed separately at different times, causing discrepancies in various reports.	Updates data in a single shared data base.	Reduces programming effort to update files.
A changed or expanded record definition must be reflected in every program.	Data description is separate from programs.	Reduces program modification.
Data communications are difficult to integrate within acceptable standards.	Most DBMSs have integrated telecommunications monitors.	Promotes data communications.

FIGURE 8.3. Traditional (sequential, VSAM, and direct) file organization versus DBMS. (Courtesy Computing Newsletter, p. 3, Feb. 1979.

CHAPTER 8 DESIGNING A DATA BASE

TYPES OF DATA BASES

In conventional file systems, groups of bytes constitute a field, one or more fields form a record, and two or more records make a file. In a data-base environment, a group of bytes constitutes a data item or segment, a collection of segments a data entry, and a series of data entries a data set. The complete collection of data sets is the data base itself.

With traditional processing of files, called flat files, records are not automatically related, so a programmer must be concerned with record relationships. Often flat files are sorted and processed by record key, just as we sorted the transaction file in our tape inventory system. Data bases relate data sets in one of three models: hierarchical, network, or relational.

Hierarchical Model

In a **hierarchical relationship,** one data set (the child or slave) is subservient to another (the parent or master). Figure 8.4a shows how we would diagram our inventory system as a hierarchical relationship: the "Parts-Master" parent data set rules for the "Parts-Detail" or child data set. You may often see rectangles representing parent and child data sets (Figure 8.4b). Parent data sets may govern more than one child set, but a child can serve only one parent (Figure 8.4c). In our inventory system, Parts-Master governs two child data sets, Parts-Detail and Sales-Detail, both of which allow for multiple data entries.

By definition the Parts-Master permits multiple data entries, each including the following data items: product number, description, unit cost, unit of measure, and quantity on hand. The Parts-Detail data set demands product number, quantity ordered, date of order, and code number. The product number appears in both sets, thereby permitting the system to relate data entries (Figure 8.4d).

The arrows in these diagrams indicate the link the DBMS has established among the children. A single-headed arrow indicates a one-to-one relationship, each par-

Hierarchical relationship Parent–child data structure in which each child belongs to one parent.

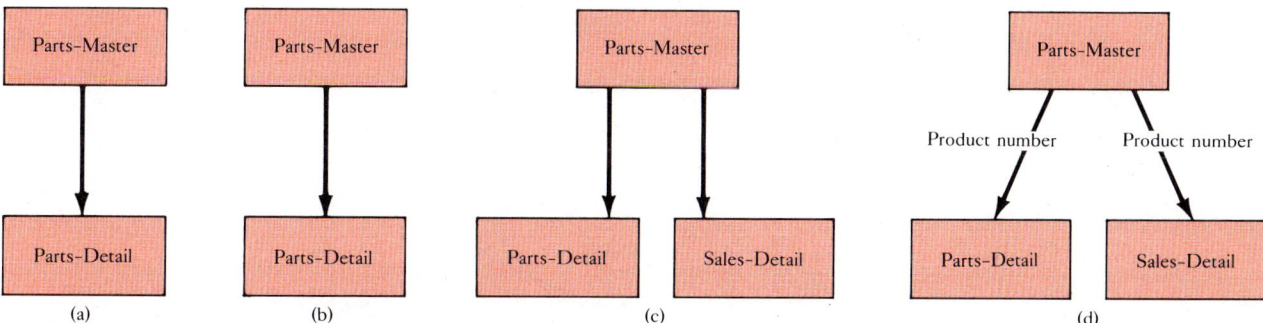

FIGURE 8.4. Hierarchical data sets employ the parent–child relationships. Each data entry in the parent data set has a key field (product number) that is repeated in each child data set. Arrowheads visually show the number of records in the child data set belonging to a record in the parent data set. A single arrow means there can be at most one child record for each parent record (a and b), two arrows mean there can be more than one child record for this parent record (c and d).

ent governing one child, while double-headed arrows indicate a one-to-many relationship, in which one parent governs many child records. Some examples of hierarchical DBMSs are IBM's IMS/VS and DL/1-DOS/VS and Intel Systems Corporation's System 2000.

Network Model

Network Parent–child structure wherein child records can have more than one parent.

Network-related data sets are similar to hierarchical ones, except that child segments can have more than one parent segment (Figure 8.5). Thus a hierarchical DBMS is a subset of network DBMSs.

Returning to our inventory system, let us suppose management wants to keep track of which staff member sold which parts. To do this, it needs two parent data sets, Parts-Master and Staff-Master. Parts-Master remains the same, but Staff-Master demands the following data segments or items: staff number, staff name, year-to-date sales, and last year's sales. Parts-Detail and Sales-Detail contain one new data item, staff number, which allows the system to link them to their second master, Staff-Master. Furthermore, if management needs a report listing how many of each part were sold, or one listing what each staff person has sold, the system can retrieve the relevant data from Parts-Detail and Sales-Detail by using the product number and the staff number. Examples of network DBMSs include Cincom System, Inc.'s Total, Burrough's DMS-II, Cullinet Software's IDMS, and Hewlett-Packard's Image.

Relational Model

Relational Data structure using two-dimensional tables to store data.

Relational data sets order data in a table of rows and columns and differ markedly from their hierarchical or network counterparts. There are no parent or child data sets (Figure 8.6). Instead, data are stored in two-dimensional data sets or tables, much as they are in a traditional file processing system.

In a relational data-base management system, we have the same concept of files, records, and fields. Files are represented by two-dimensional tables, each of which is called a "relation." Records, which can be visualized as rows in the table, are called "tuples" (rhymes with "couples"). Fields can be visualized as

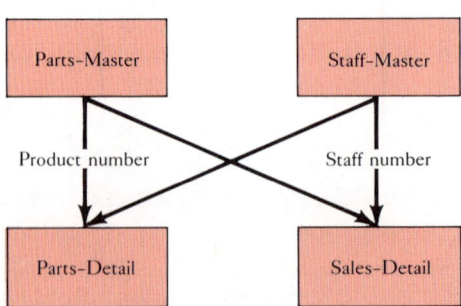

FIGURE 8.5. In a hierarchical data set a parent can govern several children. In a network children can have more than one parent. This structure handles very complex information relationships.

columns, and are called by attribute names, or domains. Note that in the supplier table in Figure 8.6 we have four tuples, or rows, and three attribute names (domains) or columns. If we need to know the name of the supplier of blue chairs, the relational DBMS searches the type and color columns of the Furniture table and finds supplier number 31, and then it scans the Supplier table for number 31, which turns out to be Campbell's. Since each "record" is a row in the table and each "field" a column, an inventory system of 1600 tuples, each with 5 attributes, would create a table of 1600 rows and 5 columns.

A relational DBMS can perform several basic operations:

1. Create or delete tables.
2. Update, insert, or delete tuples.
3. Add or delete attributes.
4. Copy data from one table into another.
5. Retrieve or query a table, tuple, or attribute.
6. Print, reorganize, or read a table or tuple (utility operations).
7. Join or combine tables based on a value in a table.

Users of a relational DBMS can create a new relation from two or more existing tables, thus permitting them to manipulate data in creative ways. For instance, we could make a new table listing furniture by type, giving us a table showing chairs, desks, lamps, clocks, and easels.

FURNITURE

Product Number	Type	Color	Quantity in Stock	Supplier Number
2763	Easel	Brown	3	48
3412	Chair	Red	4	27
3415	Chair	Blue	13	31
4002	Desk	Brown	2	48
5149	Lamp	Brown	18	101
5169	Clock	White	31	31

SUPPLIER

Supplier Number	Supplier Name	Dollar Value of Purchases This Year
27	Walkers Office	34,563.90
31	Campbell's	22,764.88
48	Associated	2,598.09
101	Rose Supply	56.81

FIGURE 8.6. *Two inventory relational tables: Furniture and Supplier. The supplier number appears in both tables for cross-referencing. (Courtesy of IBM Corporation, SQL Manual, pp. 1–6, GH24-5013-1.)*

Because of their versatility and usefulness, relational DBMSs will probably dominate the 1980s. In the past DBMSs required a powerful computer and suffered slow response times when a large number of users simultaneously tried to access the same data base, but this is changing. In fact a relational DBMS now exists for the IBM PC and Apple microcomputer, among others. With the advent of even more powerful computers during the 1980s and 1990s, relational data bases should become increasingly accessible and popular.

Hierarchical and network DBMSs require the analyst to specify particular pathways through the data base. As we see in the next section, these pathways must exist to permit the data base to locate and extract data-base records. Relational DBMSs have no built-in pathways. Instead pathways are built as required by each program. Examples of relational DBMSs include Relational Technology's Ingres, IBM's SQL/DS, Relational Software's Oracle, and Ashton Tate's dBase II.

DEFINING DATA BASES

Schema The definition or description of a data base, including fields and records.

Data definition language (DDL) Language used to define a data base.

Defining a traditional file processing or flat file involves describing the record formats: deciding how large the fields should be and their order in the record, selecting blocking factors, choosing a mode of data storage (EBCDIC, ASCII, packed decimal), and specifying a storage medium (disk or tape). Defining a data base requires the same decisions except that the designer must also describe the logic for linking records. We call such a description a **schema**, and we write it in a special format called the **data definition language (DDL)**.

Each data-base management system uses a different DDL (Figure 8.7). Burrough's DBMS is DMS-II (data management system, II) and its definition language, DASDL (data and structure definition language). Hewlett-Packard's DBMS is Image and its data definition language is DBDL (data-base definition language).

```
PARTS-MASTER DATA SET (
    PRODUCT-NUMBER               NUMBER (8);
    DESCRIPTION                  ALPHA (30);
    UNIT-PRICE                   NUMBER (9,2);
    QUANTITY-ON-HAND             NUMBER (6);
    PARTS-DETAIL DATA SET (
        QUANTITY-ORDERED         NUMBER (6);
        DATE-ORDERED             NUMBER (6);
        DATE-DELIVERED           NUMBER (6);
        AMOUNT                   NUMBER (9,2);
        SUPPLIER-NUMBER          NUMBER (8);
    );
    BY-DATE ORDERED SET OF PARTS-DETAIL KEY (DATE-ORDERED);
);
BY-PRODUCT-NUMBER SET OF PARTS-MASTER KEY (PRODUCT-NUMBER) NO
DUPLICATES; BY-DESCRIPTION SET OF PARTS-MASTER KEY (DESCRIPTION);
```

FIGURE 8.7. The data definition language used by Burroughs is DASDL. In our example the data base is hierarchical, the parent data set is PARTS-MASTER, and the child data set is PARTS-DETAIL. Although our example represents a fairly simple group of relationships, DASDL permits multiple child and parent relationships. Analysts indent and use parentheses to make the DASDL easier to read.

Before writing a schema for a data-base system, the analyst must learn the rules of the DDL just as a programmer learns the grammar rules for a particular programming language. Some of the rules in DMS-II are:

1. Data sets and data elements have names.
2. Each data element must be assigned a data type (numeric or alphanumeric) and a length.
3. A semicolon signifies the end of a data set's or data element's definition.
4. All the data elements in a set are surrounded by parentheses.
5. Slave or child data sets are defined within their parent or master data sets.
6. Sets provide logical links to records and receive names.
7. Sets include specifications for which data element will link or tie records together.

To return to our inventory system example, let us assume we are using the Burrough's DMS-II, which permits hierarchical or network data-base descriptions. Our application lends itself to hierarchical relationships. First we define the data entries and structures for two data sets, Parts-Master and Parts-Detail, listing four data items (product number, description, unit price, and quantity on hand) for each data entry (record) in Parts-Master and five items (quantity ordered, date ordered, date delivered, cost, and supplier number) for Parts-Detail (Figure 8.7). To the right of each data item, we write ALPHA or NUMBER and a number in parentheses to indicate the type of data to be stored and the number of characters or digits involved. For example, product number is an 8-digit number and unit price is an 11-digit number, nine to the left of the decimal point and two to the right.

Below the list of data entries, we describe specifications for data retrieval or pathways. A DMS-II set is similar to an indexed sequential table, in that it allows users to retrieve records only in a specific order. If, for example, one wishes to retrieve records from PARTS-MASTER in ascending order by product number, one writes the instruction BY-PRODUCT-NUMBER. By the same token, BY-DESCRIPTION will retrieve items from PARTS-MASTER in alphabetic order by description, while BY-DATE-ORDERED retrieves records according to the date the orders were placed. Once users obtain data from Parts-Master, they enjoy access to all the data in Parts-Detail, because Parts-Detail serves as a child.

With DMS-II users can retrieve data randomly (using the product number) or they can obtain the entire file sequentially by product number, description, or date ordered. No one needs to know the physical location of any record, because DMS-II handles that automatically.

DATA MANIPULATION LANGUAGES

Having written the schema to define the data base, the analyst must now compile it. The compiling process allocates space on the disk and builds all the files necessary for managing the data base.

To access the data base and add, change, or drop records, the data-base man-

Data manipulation language (DML) Commands written to retrieve or store data in a DBMS.

agement system uses special interface commands, provided by the **data manipulation language (DML),** which are built into application programming languages. If one wanted to retrieve the record for product number 3412 from Parts-Master, for example, one would write the following COBOL sentence.

LOCK PARTS-MASTER WHERE PRODUCT-NUMBER = 3412 ON EXCEPTION PERFORM 1900-NOT-FOUND.

The word LOCK tells DMS-II to examine PARTS-MASTER and locate and retrieve product number 3412. If no record exists for product number 3412, the computer must execute the COBOL paragraph 1900-NOT-FOUND. LOCK also

---

Relational DBMS: What's in a Name?

In Shakespeare's *Romeo and Juliet*, the two lovers are thwarted by a family feud, prompting Juliet to wish Romeo had a different last name, so family rivalry wouldn't keep them apart. She asks the very practical question, "What's in a name?"

Today's DP and MIS managers might well ask the same question when they regard advertisements for relational data base management systems. Does a new name mean a really new or different product? In the case of relational DBMSs, the name does mean something special.

Dr. Edgar Codd, a member of the IBM project team working on System R (a Relational Store Interface which managed devices, space allocation, transaction consistency and recovery, and more), proposed the "relational" theory in the early 1970s. System R's goal was to seek the most optimum access paths for data to travel and to offer users a logical view of data, easy data manipulation of data, and a high level of security and integrity.

To improve upon the hierarchical and network DBMSs developed in the early 1960s, Codd began searching for a simpler approach, which he ultimately found, but which required IBM's largest mainframe computer and limited use to only a few users at a time. Yes, Codd's relational theory was viable, but it demanded sophisticated hardware that lay beyond the grasp of most organizations, until IBM introduced SQL, a relational product that evolved from System R, but which could function on smaller computers.

Since Codd borrowed the term "relational" from mathematical theory, a whole new mathematic jargon sprang up around the approach, bestowing new terms upon old DP concepts: a record became called a tuple, a field an attribute name, and a file a relation.

A pure, or close to pure, relational system following Codd's theoretical constructs, should meet many criteria, among them the rule that information be represented by values in two-dimensional tables with data accessible by value only. Doing so allowed users to create new tables from existing ones by matching column values. The approach also dictated that data item definitions be centrally stored in an integrated dictionary, that extended information retrieval be available through on-line relational query facilities, and that information be accessible via

keeps one user from accessing a record on which another user is working. Once the user finishes working on the record, the computer executes the following COBOL instructions.

STORE PARTS-MASTER.
FREE PARTS-MASTER.

These two instructions place the altered record in the data base and unlock the data entry for part number 3412, so another user can access it.

We call the language enhancements that permit manipulation of the data base **host languages.** If a DBMS supports its own language (other than COBOL, PL/1, Pascal, or FORTRAN instructions), we call the DBMS "self-contained."

Host language A DBMS that permits access using high-level-language constructs.

---

distributed DP facilities. A user organization could then build applications with high-level languages in an interactive environment, with eventual on-line processing executed through a resource-conserving teleprocessing monitor.

Sound complicated? It is, but you should bear in mind the basically simple concept that relational systems allow users to dynamically change associations in files to suit their own purposes within certain authorized limits. This means that users can select fields from files to create new ones, then store data a new way. Most importantly, they can create new associations without getting bogged down in technical details.

So we see that the name relational simply refers to a better way to manage information. Users can add data to files without affecting previously existing data, and they can blend data from two different files to form what looks like a third file, all in two-dimensional, flat-file or table form.

The relational DBMS model obeyed strict mathematical consructs, and took intense effort by IBM and many universities to complete, yet it resulted in a beautifully simple result. In terms of the time it takes programming staffs to master a DBMS technique, the traditional hierarchical approach requires about a year or more of study and experimentation, the network type nine months to a year, and the relational system only about three months.

The data base management systems of the '80s must accommodate the growing needs of a broad range of users for easy learning and operation. Such systems must be flexible, providing logical access and reporting, and they must allow for continuing growth.

Already some suppliers are marketing products that offer relational capabilities. Some even claim to be almost pure implementations of Codd's theory. Despite continuing controversy about the approach, however, it should steadily gain stature and acceptance in the future.

Source: Carol T. Thiel, "Relational DBMS, What's in a Name," *Infosystems*, Sept. 1982. Reprinted from INFOSYSTEMS, September 1982. Copyright Hitchcock Publishing Company.

QUERY LANGUAGES

Query language Data manipulation language designed for use by noncomputer-trained personnel.

Data manipulation languages are designed for use by trained computer operators. To help untrained users locate and retrieve data, many DBMSs offer special-purpose languages, called **Query language** (or Inquire, Adascript, or Natural). These languages simplify data retrieval, updating, addition, deletion, or reporting because they allow casual inquiry of the data base in a language that more closely resembles English. An example of a Query reporting program to print the product number, description, and quantity on hand for Parts-Master appears in Figure 8.8.

Query systems and their host languages give the user a set of tools for referencing, reporting, and manipulating information from a data base. Users thus do

```
REPORT
LINES = 60
H1, "List of Parts on Hand by Part Number", 60
H1, "Page:", 74
H1, PAGENO, 80, SPACE A2
H2, "Product Number", 12
H2, "Description", 25
H2, "Unit Price", 50
H2, "Quantity on Hand", 68, SPACE A1
S, PART-NUMBER
D1, PRODUCT-NUMBER, 12
D1, DESCRIPTION, 40
D1, UNIT-PRICE, 50, E1
D1, QUANTITY-ON-HAND, 68
E1, "$$$,$$$,$$$.99"
END
```

(a) QUERY routine named LISTING to print a report.

```
>FIND PARTS-MASTER.PRODUCT-NUMBER > 0
>REPORT LISTING
```

(b) QUERY commands to locate and print all product numbers greater than zero in the Parts-Master data set.

```
           List of Parts on Hand by Product Number    Page:    1
Product Number  Description      Unit Price    Quantity on Hand
   2763         Easel               $45.67            48
   3412         Chair-Red          $125.98            27
   3415         Chair-Black        $456.25            31
   4002         Desk-Wood        $1,200.00            48
```

(c) Report printed by the QUERY routine for the Parts-Master data set.

FIGURE 8.8. QUERY provides a way for noncomputer personnel to access the data base. It is well suited for interactive use on a remote terminal.

not have to be proficient in a programming language. Instead they use their Query language.

Query-type languages allow users to specify:

1. Header information.
2. Sorting of data.
3. Totals and subtotals.
4. Editing of data to make the data attractive.
5. Line spacing and page skipping.
6. Averaging, counts, standard deviations, and other statistics.
7. Printing of data on one or many lines.

UTILITIES

Many DBMSs also incorporate sophisticated utility routines to help users perform common tasks. Some such utility tools provide a means of copying the data base to disk or tape, thus creating a backup in case of hardware failure. Others log database changes to tape or disk, so users can rebuild the data base if it suffers accidental damage. Still others allow users to purge (erase) the data base, move it to another disk drive, increase the data base's capacity, or reload it.

Most DBMS software also permits users to change a data entry's name, add or delete a data entry, or change the name of a data set without redoing every program accessing the data base.

Figure 8.9 compares a number of commercial DBMSs. Computer manufacturers provide some DBMSs (IMS, DMS-II, Image, SQL/DS) while software firms provide others, (Adabase, Relate/3000, Ashton-Tate).

DBMS Name	Supplier	Data Structure	Hardware
IMS	IBM	Hierarchical	IBM
Total	Cincom	Network	IBM, CDC, NCR, Prime
DMS-II	Burroughs	Hierarchical and network	Burroughs
Image	HP	Network	HP
Adabase	Software AG	Network and hierarchical	IBM, UNIVAC DEC
SQL/DS	IBM	Relational	IBM
Relate/3000	CRI	Relational	HP
dBase II	Ashton Tate	Relational	Micros
PFS	Software Pub.	Relational	Micros
1-2-3	Lotus Develop.	Integrated DBMS and spreadsheet	Micros

FIGURE 8.9. Seven popular DBMSs, each with different capabilities and data structures.

DATA-BASE CONTROLS

Analysts design controls for output, files, data bases, and data collection. For traditional file systems, we saw two control methods: record counts and backup. Both can be applied to data-base systems as well as traditional files. There are two further controls that can effectively assist controlling a system in a data-base environment: transaction logging and access security.

Transaction Logging

One of the problems associated with any online system is protecting the files from accidental destruction by the hardware (a disk failure), software, or a user. Some newer and sophisticated operating systems and data-base management systems perform **transaction logging.** With it special software, usually "built in" to the DBMS or operating system, automatically copies the old and new records plus the transaction record to tape or disk every time a record is added, altered, or deleted, creating a log file. An organization can use each log file to recreate a lost or destroyed data-base file. Log files also provide clues to the frequency and types of file changes made within a system. On a terminal-based transaction-logging system, the software can keep track of the time, the terminal, and the user making the changes.

> **Transaction logging**
> Control procedure that automatically produces copies of old and new records.

A control procedure often overlooked involves copying, or backup, of a system's programs and documentation. Some computer installations take great pains to provide backup or transaction-logging facilities for the data, but neglect to do so for the processing programs. What good would a backup serve if an organization also lost its programs? Although retrievable, the data could not be processed.

The procedure for copying programs and system documentation parallels that for files. The organization builds a duplicate disk or tape containing copies of its programs and documentation. As organizations more frequently develop programs and documentation with computer text editors or word processors, they will more easily be able to copy them to tape for storage in a safe vault or at some remote location.

Access Security

Another type of processing control restricts system access to approved users. Access security not only determines which users can enter a data base, but also how they may use it. When analysts build their schema, they may assign each user a restricted read-only or a more liberal read-and-write access status. Although access controls are generally applied to files and records, even tighter security can limit certain users to specific areas. Some data-base managers employ individual passwords to restrict access to files. For example, a bookstore clerk in Silva's bookstore might be allowed to ask the computer if the store has a specific book in stock and to alter the inventory record of books on hand after making a sale, but the same clerk might not be able to change the retail price, over which the store owner wants to maintain strict personal control.

CASE STUDY

Fleet Feet's Accounts Payable Data-Base Design

Fleet Feet currently uses the Image data-base manager on its Hewlett-Packard 3000 computer. Image is a network data-base system that permits users to access data with applications programs written in COBOL, Pascal, FORTRAN, BASIC, RPG, or SPL (Systems Programming Language, similar to Algol). Image refers to parent and child data sets as masters and detail data sets. As you will recall, Hewlett-Packard calls its data definition language DBDL (data-base definition language), and it provides the QUERY language system, enabling users to access data spontaneously. More than one application program can access data at the same time.

The network DBMS implements parent-child record relationships. A parent data set contains a record key (called the search item in the child set) that links a parent data record to one or more children (called detail data sets). A user can access parent data sets sequentially or directly.

After reviewing Fleet Feet's desired report formats, analyst Frank Pisciotta decides that the system will need three master or parent data sets and three detail or child data sets (Figure 8.10).

1. *VENDOR (Master)*. To hold data about each of Fleet Feet's vendors. Stored data will include vendor name and number, address, discount terms, and total purchases this year and last.
2. *AP-TRX (master)*. To hold data about each voucher paid, including date, vendor name and number, invoice and discount data, and check number.
3. *CHECK (master)*. To hold data about all checks written to vendors, including the vendor and voucher numbers, and amounts.
4. *AP-OPEN-ITEM (detail)*. To hold data about all vouchers authorized for payment: voucher number, vendor name and number, check number, and detailed information about the purchase.

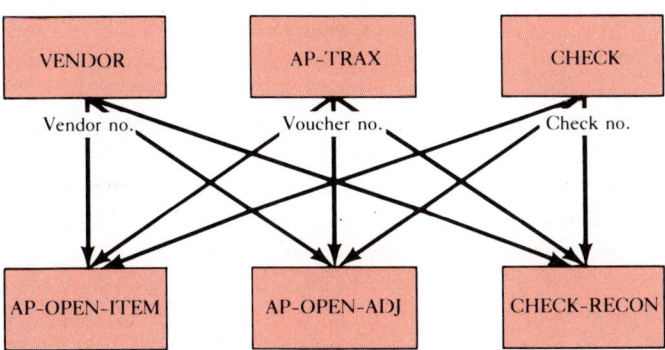

FIGURE 8.10. *The network relationship of Fleet Feet's accounts payable data base. The parent or master data sets lie above the child or detail data sets (in uppercase letters). The record key (in lowercase letters) logically links the data sets.*

CASE STUDY

5. *AP-OPEN-ADJ (detail)*. Similar to AP-OPEN-ITEM, except that it also allows adjustments to invoices.
6. *CHECK-RECON (detail)*. To hold data, including vendor number, about issued checks that have been cashed and have cleared the bank.

Note that a record key links each master set to at least one detail set. For VENDOR the key is vendor number, for AP-TRX it is voucher number, and for CHECK it is check number. Thus these record keys must appear in all linked master and slave sets.

After selecting the data sets, determining their relationships, and establishing record keys, Frank writes a schema to define the data entries for each set (Figure 8.11). He gives each item a name (VOUCHER-NO, VENDOR-NO, or INVOICE-NO), determines its type (X stands for alphanumeric, Z for numeric), then specifies its length (number of characters for alphanumeric, number of digits for numeric). The entry for vendor number in the AP-OPEN-ITEM detail data set reads:

VENDOR-NO,   X6

All vendor numbers are alphanumeric (X) and will contain six characters.

After defining the data sets, Frank sets capacities for each. The capacity statement tells Image the maximum number of data entries this data set can hold. For CHECK-RECONCILE Frank decidies that 4500 entries are adequate since Fleet Feet undoubtedly will not write that many for a full year. If any set needs more space, Image uses a utility routine to unload the data in that set, boost the capacity, and reload the data into the newly created space.

The ENTRIES phrase shows the exact number of data entries a set stores at a given time. Set at zero initially, this value will climb as the company adds data entries. Frank can monitor the difference between ENTRIES and CAPACITY to anticipate overload before it actually happens.

Data-base security and privacy features protect the data from unauthorized use. Some users have read-only clearance; others have full read-and-write access with terminals or in a batch mode. A powerful set of utilities allows users to create, purge, unload, copy, and reload the Image data base.

Since Frank wishes to create a complete system, he orders the operations staff to copy the accounts payable data base before printing checks. This backup will allow Fleet Feet to recover data quickly and inexpensively after a software or hardware failure. Like most data-base managers, Image has a utility routine that automatically unloads the data base for recovery purposes. As an added safeguard, Frank directs the HP 3000 to copy each of Fleet Feet's disks to tape at 4 o'clock each morning. During development a programmer will implement this with what HP terms a "job stream."

To ensure that only authorized personnel have access to the accounts payable data base, Frank restricts access to the data base with two sets of passwords, one for logging onto the system and one for actually accessing the data base. Frank defines the password for data-base access when he writes his Image schema, Figure 8.11. The PASSWORD statement in the Image schema defines two user classes (2 or 4) and the passwords (ACNTSPAYBLE and RDWRTAP) the computer will require to allow the user or program to read or write data

CASE STUDY

```
DATA BASE: AP

PASSWORDS
   2  ACNTSPAYBLE
   4  RDWRTAP

SET NAME: VENDOR, MANUAL (2/4)
   ITEMS:
      VENDOR-NO,                X6         <<KEY ITEM>>
      VENDOR-NAME,              X30
      VENDOR-ADDR-1,            X30
      VENDOR-ADDR-2,            X30
      VENDOR-ADDR-3,            X30
      TYPE,                     X4
      TERMS-DESCR,              X16
      DUE-DAYS,                 Z4
      DISCOUNT-DAYS             Z4
      DISCOUNT-%,               Z4
      LAST-ACTIVITY             Z6
      STATUS-CODE,              X2
      PURCHASES-YTD,            Z10
      PURCHASES-LAST,           Z10
      DISCOUNTS-YTD,            Z10
      DISCOUNTS-LAST,           Z10
   CAPACITY:   251              ENTRIES:   0

SET NAME: AP-TRX, MANUAL (2/4)
   ITEMS:
      VOUCHER-NO,               X6         <<KEY ITEM>>
      TRX-TYPE,                 X2
      VENDOR-NO,                X6
      P-O-NO,                   X10
      INVOICE-NO,               X8
      INVOICE-DATE,             Z6
      INVOICE-AMOUNT,           Z10
      INVOICE-NON-DSC,          Z10
      DUE-DAYS,                 Z6
      DISCOUNT-DAYS,            Z4
      DISCOUNT-DATE,            Z6
      DISCOUNT-%,               Z4
      DISCOUNT-AMOUNT,          Z8
      CHECK-NO,                 Z6
      CHECK-DATE,               Z6
      ACCOUNT-NO,               Z10
   CAPACITY:   6500             ENTRIES:   0
```

FIGURE 8.11. Schema for Fleet Feet's accounts payable system. The schema is written in the data definition language (DBDL) for Hewlett-Packard's Image data-base manager. Image allows two types of master or parent data sets; automatic masters contain a single data element, and manual masters contain two or more data elements. (Courtesy MCBA® MCBA is a registered trademark of MCBA, Inc.)

CASE STUDY

```
SET NAME: CHECK, MANUAL (2/4)

    ITEMS:
        CHECK-NO,                   Z22             <<KEY ITEM>>
        VENDOR-NO,                  X6
        CHECK-TYPE,                 X2
        CHECK-DATE,                 Z6
        CHECK-AMOUNT,               Z10
        P-O-NO,                     X10
        INVOICE-NO,                 X8
        INVOICE-DATE,               Z6
        INVOICE-AMOUNT,             Z10
        DISCOUNT-AMOUNT,            Z8
    CAPACITY:    4500               ENTRIES:    0

SET NAME: AP-OPEN-ITEM, DETAIL (2/4)

    ITEMS:
        VENDOR-NO,                  X6              <<SEARCH ITEM>>
        VOUCHER-NO,                 X6              <<SEARCH ITEM>>
        P-O-NO,                     X10
        INVOICE-NO,                 X8
        INVOICE-DATE,               Z6
        DUE-DATE,                   Z6
        DISCOUNT-DATE,              Z6
        ORIG-INVOICE-AMT,           Z10
        ORIG-DISC-AMT,              Z8
        INVOICE-BALANCE,            Z10
        DISCOUNT-BALANCE,           Z8
        PARTIAL-PAYMENT,            Z10
        CURRENT-DISCOUNT,           Z6
        CHECK-NO,                   Z6              <<SEARCH ITEM>>
        CHECK-DATE,                 Z6
        AP-ACCOUNT,                 Z10
        SELECT-STATUS,              X2
        PRINT-STATUS,               X2
    CAPACITY:    4500               ENTRIES:    0
```

FIGURE 8.11 *(Continued)*

CASE STUDY

```
SET NAME: AP-OPEN-ADJ, DETAIL (2/4)
    ITEMS:
        VOUCHER-NO,              X6           <<SEARCH ITEM>>
        TRX-TYPR,                X2
        VENDOR-NO,               X6           <<SEARCH ITEM>>
        CHECK-NO,                Z6           <<SEARCH ITEM>>
        INVOICE-NO,              X8
        INVOICE-DATE,            X6
        INVOICE-BALANCE,         Z10
        INVOICE-DISCOUNT,        Z8
        OLD-DUE-DATE,            Z6
        DUE-DATE,                Z6
        OLD-DISC-DATE,           Z6
        DISCOUNT-DATE,           Z6
        DISCOUNT-AMOUNT,         Z8
        CHECK-DATE,              Z6
        CHECK-AMOUNT,            Z10
        DISCOUNT-TAKEN,          Z8
        CASH-ACCOUNT,            Z10
    CAPACITY:   5500             ENTRIES:  0

SET NAME: CHECK-RECON, DETAIL (2/4)
    ITEMS:
        ACCOUNT-NO,              Z10
        CHECK-NO,                Z16          <<SEARCH ITEM>>
        CHECK-DATE,              Z6
        VENDOR-NO,               X6
        CHECK-REC-AMT,           Z10
        DATE-RECONCILED,         Z6
    CAPACITY:   4500             ENTRIES:  0
```

FIGURE 8.11 *(Continued)*

in the data base. For each set in the data base, Frank writes (2/4). Image interprets this to mean password ACNTSPAYBLE (user class 2) is given read-only access while RDWRTAP (user class 4) has both read and write capability.

With his data-base design in hand, Frank can now move to the next stage of design: input design.

WORKING WITH PEOPLE   Retraining for the Data-Base Environment

Cheryl Ruiz was excited. Her company, AAA Real Estate Brokers, had decided to purchase IBM's SQL/DS relational data-base management software. Cheryl had never used a DBMS; her experiences had always revolved around indexed sequential file systems. The IBM sales representative had recommended that the company retrain one member of its computer services staff to become the in-house data-base expert. Because she was eager to learn this powerful new technique, Cheryl hoped her boss, Salvatore Gianna, would choose her.

Though Cheryl regularly read articles in *Computerworld* about DBMSs, she only superficially understood them. She knew some of the buzzwords—hierarchical, network, relational, host language, Query language—but how they really worked was a mystery to her.

At last Sal Gianna called Cheryl into his office to discuss her career at AAA Brokers. Cheryl had earned a bachelor's degree in business 15 years earlier, had married, and then had gone back to her local community college to study computers. Since joining AAA as a programmer 4 years ago, she had won praise from Sal and her peers for establishing rapport with users and displaying eagerness to constantly update her skills. In fact it had been Cheryl who finally convinced Sal to switch from punched cards to terminals to increase the productivity of AAA's programmers and analysts. She had been right. Programmer productivity and job satisfaction had benefited and staff turnover had declined.

"I admire your enthusiasm for the new DBMS," Sal said.

"It's the wave of the future," Cheryl replied. "I only wish I knew more about it."

Sal smiled. "You're going to get your chance. The IBM training center in Boston is offering three 2-week classes on SQL/DS beginning next month and I've already enrolled you."

To prepare for the class, Cheryl asked IBM's training manager in Boston to recommend a book on data base concepts, which she quickly read.

The 6 weeks of classes were a blur of activity. In addition to reading her own books, Cheryl consumed four IBM SQL/DS manuals. When she returned to AAA, her former admirers greeted her coolly. What had she done? One day she overheard a programmer tell another analyst, "Cheryl thinks she's a wizard now. We're just a bunch of outdated hacks." Cheryl was crushed. She hadn't expected her friends to feel so jealous and insecure. But she decided to do something about it.

She invited all 10 members of AAA's staff to informal breakfast meetings to discuss SQL/DS. "You'll probably catch on much faster than I did," she said.

The first day everyone showed up early and surrounded Cheryl at a terminal to which she had attached three monitors so everyone could see. When her colleagues saw SQL/DS running, they were impressed.

At the end of the week, the group treated Cheryl to lunch. "To our new data-base manager!" toasted the formerly embittered programmers. Cheryl laughed when they all lifted their glasses. Learning new technology was fun, but sharing it with others was even better.

SUMMARY

Data-base managers assist in the managing of disk-based data, letting analysts divorce themselves from the physical aspects of data storage so they can concentrate on the logical aspects of the system and programming.

There are three types of DBMSs: hierarchical, network, and relational. Hierarchical systems have master-slave records where each slave belongs to a single master. In network DBMSs slaves have more than one master. Relational DBMS's allow users to view data as two-dimensional tables logically connected with a common key.

To describe the data base, the analyst writes a schema. This description defines the data items for each record and shows how the records will be linked.

The DBMS offers many advantages. Files are consolidated, independent of the system or program. They are versatile, fast, secure, and efficient, lessening program development time, reducing program maintenance, and permitting special needs to be met more easily than with older file storage techniques. Drawbacks include the need for additional hardware and staff training, and current high costs.

NEW TERMS

Data definition language (DDL)
Data manipulation language (DML)
Hierarchical relationship
Host language
Network
Query language
Relational
Schema
Transaction logging

QUESTIONS FOR REVIEW AND DISCUSSION

Questions appear in three categories. You can find the answers to the A group in this chapter. The B group requires you to apply the material presented here, whereas the C group necessitates investigation or research.

A.1. What are the three types of data-base systems?
A.2. Make a list of three commercial data-base managers.
A.3. What do the letters in each of the following stand for?
    a. DBMS
    b. DML
    c. DDL
A.4. Who provides data-base managers?
A.5. List three reasons why a firm would select a data-base manager over the traditional methods of data storage.
A.6. List two disadvantages of data-base managers.
A.7. Make a list of at least three utility routines commonly found with data-base management software.
A.8. What are two operations that can occur on relational DBMSs?
A.9. What are the inputs to the data-base phase of the systems process?
A.10. What is the output from the data-base phase of the systems process?

B.1. How many records must a system process to update a master file of 4000 records if the transaction file contains 1000 records and a DBMS is used to store records?

B.2. How many parent data sets can a child data set belong to in each of the following?
    a. Relational DBMS
    b. Network DBMS
    c. Hierarchical DBMS

C.1. Find the names of five other data-base management systems now in use and classify them as hierarchical, network, or relational.

C.2. Some DBMSs have a supplementary software system use called the DMCL (device media control language). What does it do?

C.3. Which type of data-base manager seems the most versatile?

C.4. Some data-base managers are classified as inverted or linked. What do these two terms mean?

REFERENCES

James Bradley, *Introduction to Data Base Management in Business*, Holt, New York, 1983.
David Kroenke, *Database Processing*, 2nd ed., Science Research Associates, Chicago, 1983.
James Martin, *Strategic Planning Methodologies*, Prentice-Hall, Englewood Cliffs, N.J., 1981.
Gio Wiederhold, *Database Design*, McGraw-Hill, New York, 1977.

CHAPTER 9: Input Design

- Goals and Preview
- Methods of Data Entry
- Controlling Data Entry
- Data-Entry Hardware
- **Guidelines for Designing Data-Entry Screens**
- Designing Data Entry for Terminals
- Designing Data Entry for Key-Entry Devices
- Case Study: Fleet Feet's Input Design
- Working with People: Ergonomics and Design
- Summary
- New Terms
- Questions for Review and Discussion
- References

GOALS AND PREVIEW

After reading this chapter you should be able to:

1. List four devices commonly used for data entry.
2. Distinguish between verification and validation.
3. List at least five types of validations.
4. Define a check digit or bit and describe appropriate applications.
5. Design a data-collection screen.
6. Discuss the advantages and disadvantages of terminals.

Having designed the files or data base (with a schema) and record formats, the analyst tackles input design (Figure 9.1). During this stage of the design process, the analyst, using the data-flow diagrams developed during output design, determines the source and method for collecting data (for example, an invoice). In addition, the analyst designs the way the organization will enter data into its computer (via online terminal or optical character reader) and selects input hardware.

The importance of data output is obvious: Reports enable users to make decisions, control activities, and monitor performance. Data input techniques are equally, if not more, important, as the output is only as good as the input used to generate it. Incorrect input, such as an erroneous account number, may generate a wrong customer billing; late input may delay billing and hurt a company's positive cash flow. As with output design, the analyst must pay strict attention to users' needs and abilities, as they, not the analyst, ultimately will be responsible for the organization's data collection.

With manufacturers offering an ever-increasing number of hardware devices, analysts must keep abreast of developing technology—weighing needs against cost. Up-to-date hardware may have all the desired features, but the cost may outweigh its worth. For example, while a new optical page reader that costs over $250,000, and which can read hundreds of pages a minute, may be ideal for the Internal Revenue Service, it would not be appropriate for a small department store.

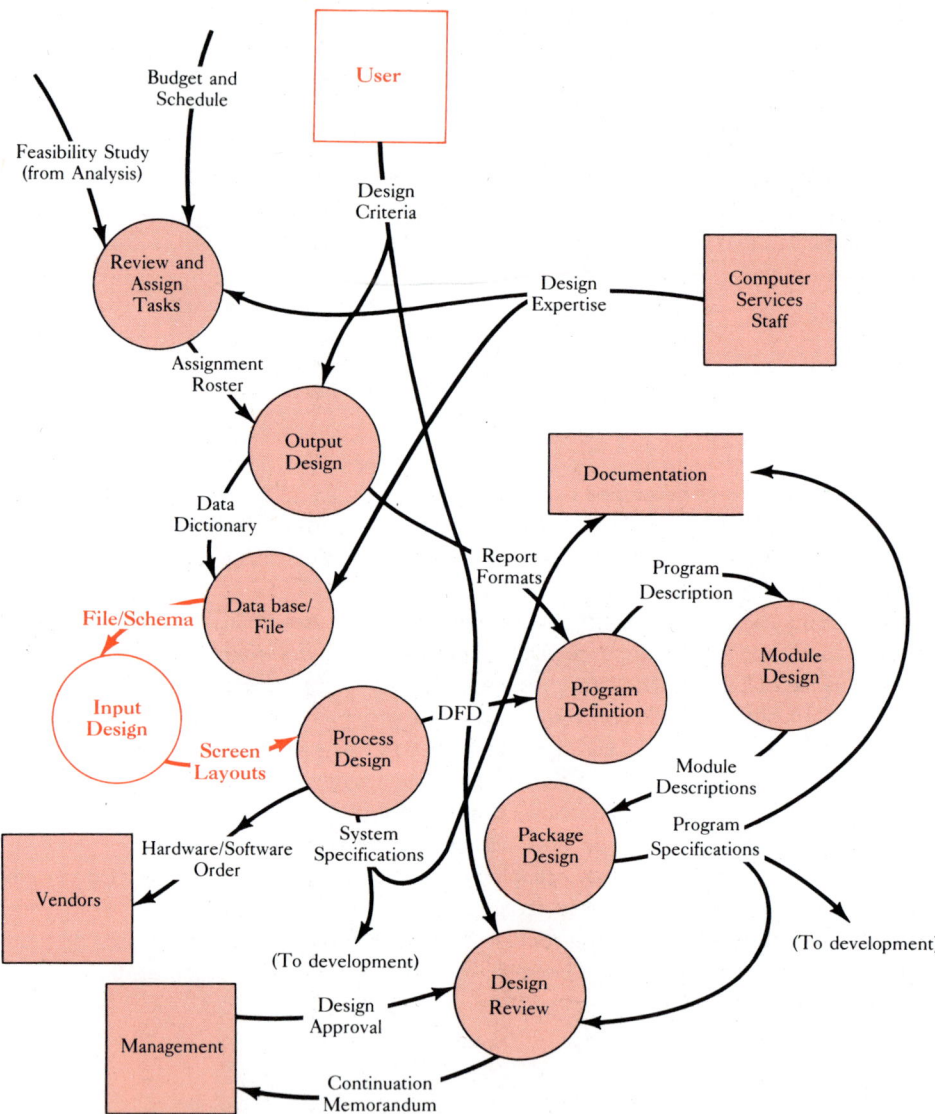

FIGURE 9.1. *Data-flow diagram for the systems design shows two inputs and a single output for the input design.*

METHODS OF DATA ENTRY

Data enter an organization's information system from two sources: internal and external. Internal data, from employees and management, can include time cards, vouchers, and checks (Figure 9.2). External data come from people and agencies outside the organization and might include credit applications, invoices, orders, payments, and packing slips (Figure 9.3). Some internal data, such as a customer's

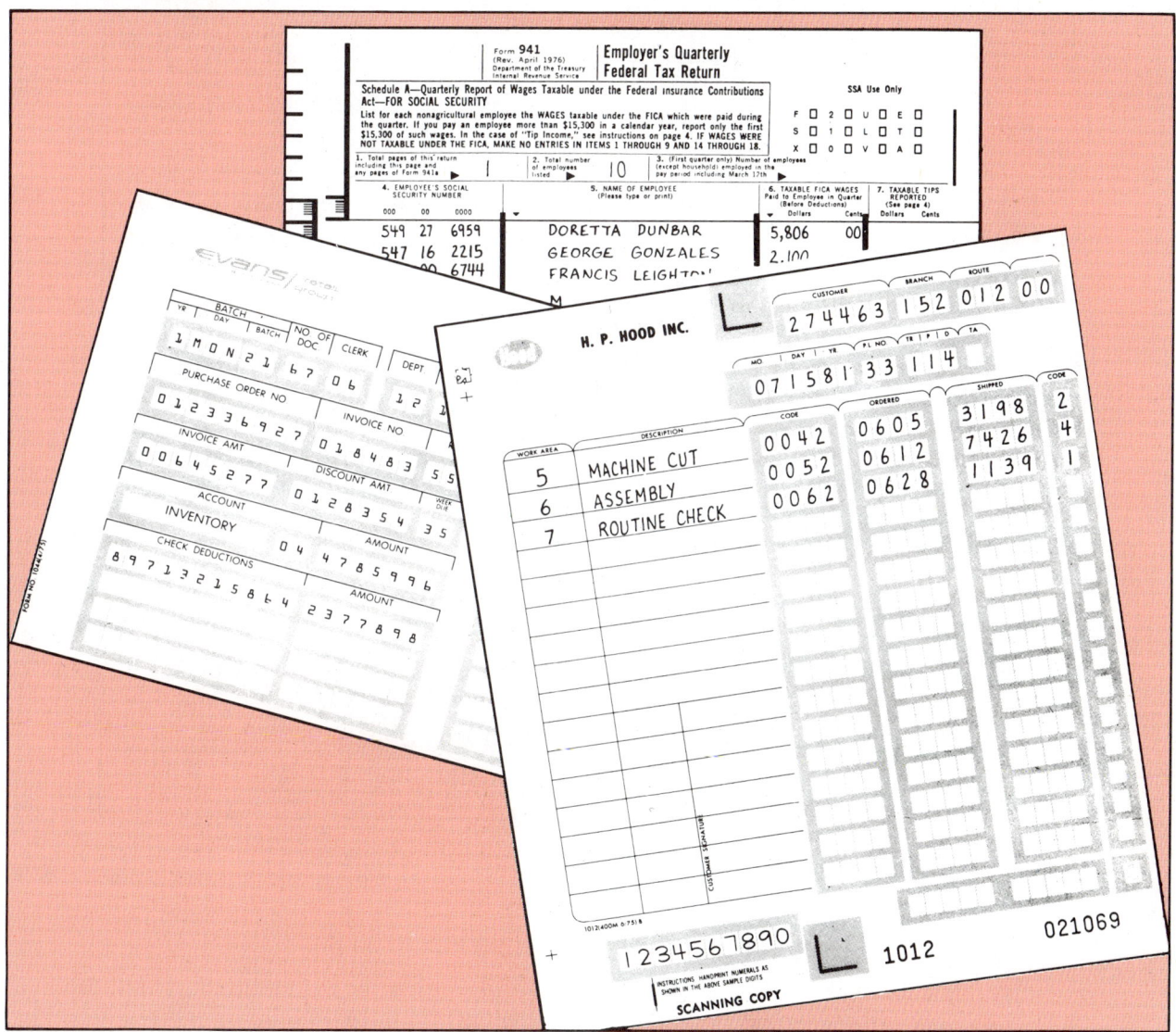

FIGURE 9.2. Internal data come from company personnel. If the data arrive in handwritten form, as often happens, reformatting is required before the data are entered. (Source: Perry Edwards and Bruce Broadwell, Data Processing: Computers in Action, 2nd ed., Wadsworth Publishing Company, Belmont, CA, p. 229. Reprinted by permission.)

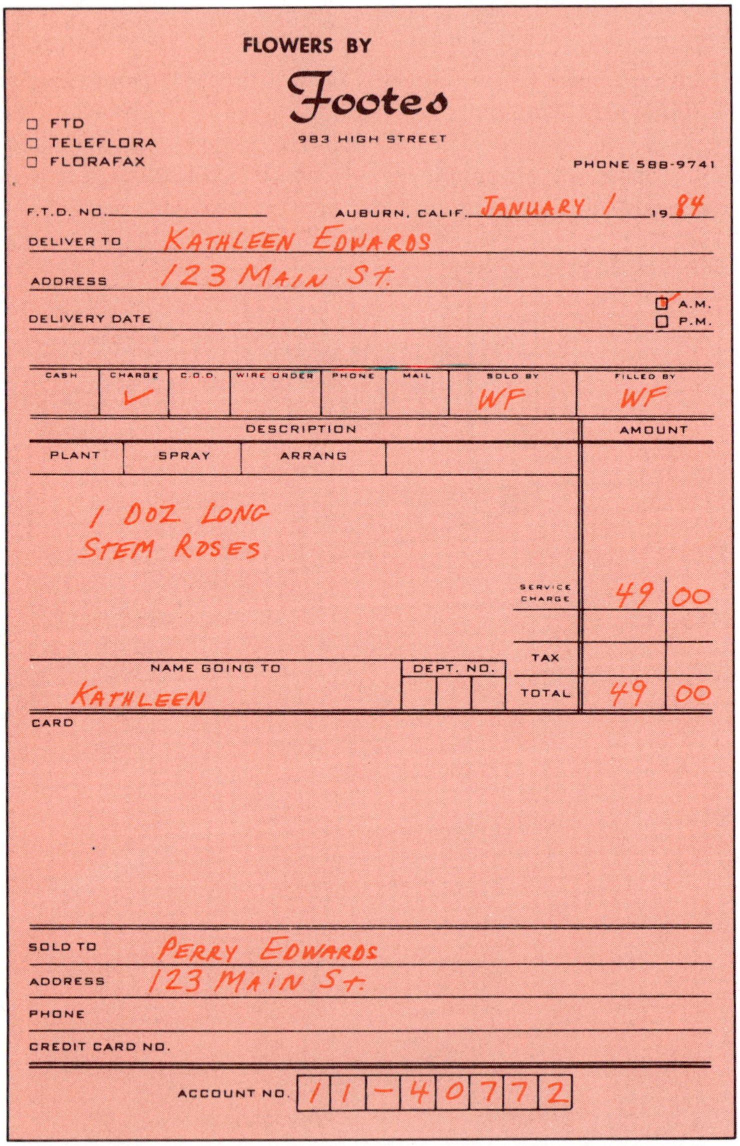

FLOWERS BY

Footes

983 HIGH STREET

☐ FTD
☐ TELEFLORA
☐ FLORAFAX

PHONE 588-9741

F.T.D. NO. ______ AUBURN, CALIF. JANUARY 1 19 84

DELIVER TO KATHLEEN EDWARDS

ADDRESS 123 MAIN St.

DELIVERY DATE ☑ A.M. ☐ P.M.

CASH	CHARGE	C.O.D.	WIRE ORDER	PHONE	MAIL	SOLD BY	FILLED BY
	✓					WF	WF

DESCRIPTION — PLANT | SPRAY | ARRANG — AMOUNT

1 DOZ LONG STEM ROSES

SERVICE CHARGE 49 00

TAX

NAME GOING TO KATHLEEN — DEPT. NO.

TOTAL 49 00

CARD

SOLD TO PERRY EDWARDS

ADDRESS 123 MAIN St.

PHONE

CREDIT CARD NO.

ACCOUNT NO. 11-40772

FIGURE 9.3. Data from customers include credit application forms, payments, and charges. Such data usually arrive in handwritten form and must be organized into a format acceptable to the computer.

previous year's total purchases, exist within the system itself. These data are, perhaps, already stored on the computer's tape or disk.

Most organizations originate their data on a printed form, such as a time card or customer order, called a source document. Such a document displays data in a form accessible to people. Users, customers, and others unfamiliar with computers may fill out a portion of the **source document,** but the organization will have to transcribe all or part of it into a form accessible to computers. Most of us have signed oil company charge card receipts that allow an attendant to imprint some data from charge cards (customer account name and number), mechanically imprint the purchase amount, and handwrite other data (car license number). Analysts must design source documents that lend themselves to both human and machine use.

After locating the data sources, an analyst determines the method of data entry: manual or direct. **Manual,** or **key, entry** requires a person to key in the data with a terminal, key-to-tape, or key-to-disk device (Figure 9.4). Each of these devices, as their names imply, store data on different media: key-to-tape employs reels of magnetic tape, and key-to-disk uses floppy or hard disks. These devices have been around for many years, often providing an organization's only data-collection method. Indeed they are particularly useful for batch applications where users do not need instant access to their data. Such methods might work well for a dentist's patient billing system that generates monthly statements.

Direct-entry machines, as their name implies, read data directly into a computer (Figure 9.5). The cash registers at many supermarkets, for example, can automatically read the universal product code (UPC) stamped on merchandise. The clerk passes each item over a glass-covered area that picks up the data. In some stores the computer actually announces each price. Although this type of equipment now costs more than a key-entry device, it reads data more quickly and reliably, and involves less processing time. As a result, if errors can be reduced by a factor of two or three with increased speed, the extra cost for a direct-entry machine may be recovered in less than a year.

Source document The paper containing the original data.

Manual, or key, entry Data collection whereby a user enters data on a terminal or other small device.

Direct entry Data collection whereby a machine reads data directly into the computer.

FIGURE 9.4. These terminals are connected to a computer. As the data-entry operators type the data, which they can check on their screens, the computer stores the data on a disk or tape. (Courtesy of Sperry Corporation.)

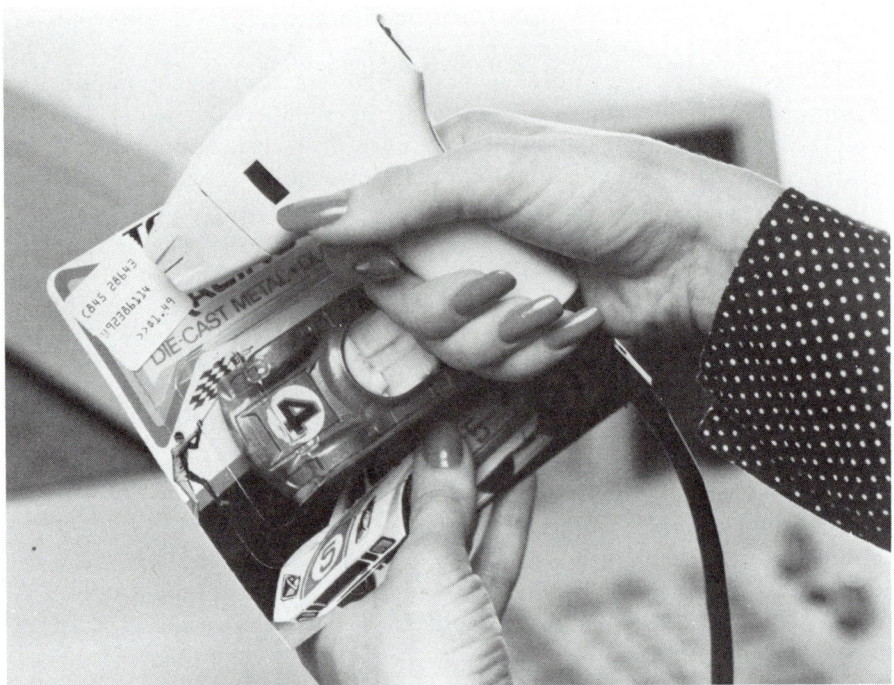

FIGURE 9.5. This optical character reader is a direct-entry device used by a retail store to read data about merchandise into the computer. (Courtesy of Recognition Equipment Corporation.)

CONTROLLING DATA ENTRY

Potential for error exists with any type of data-entry device. Clerks may write the wrong price of a purchase on a sales tag, or key-entry personnel may strike wrong keys, posting inaccurate data into an account or inadvertently posting correct data to a wrong account. Similarly, direct-entry machines occasionally malfunction, and they cannot read smeared UPC codes just as a clerk cannot read a smudged price stamp.

Regardless of how the error is made, an organization will want to develop a way to detect errors quickly. Once an error enters a system, it has a way of compounding problems: In an inventory system, for example, an item incorrectly shown as being in stock will result in the company not reordering on time, thus losing a future sale and customer goodwill, not to mention the time and money it will take to correct the error. During input design analysts spend much time designing methods to detect and eliminate errors at every step of the input process.

All input devices offer some type of error-detection techniques. Some require rekeying of data, and others use the power of the computer itself. We call these two different error-detection controls verification and validation respectively.

Verification

With **verification** a person checks data by simply looking at or rekeying the data. Reexamination or rekeying can greatly reduce errors. If an operator of an input device has a one half of a percent error rate during initial data entry, for example, the chance of making an identical mistake a second time is only 25 in a million (0.005 × 0.005 = 0.000025, or 25 millionths). Double entry effectively reduces typographic mistakes.

Verification Data-collection control procedure wherein the operator rekeys data.

Validation

Because **validation** employs the intelligence of the terminal or the computer itself, it offers much more sophisticated error-detection techniques. For example, we can program a computer to reject data that do not fall within a predetermined range. If the lower and upper boundaries for a certain drug's dosage are 10 milligrams and 30 milligrams, respectively, the computer automatically can reject 7.5 or 42, flashing an error message on the screen of a terminal to alert the operator to the error.

Eight validation tests exist.

Validation Data-collection control procedure using the intelligence of the computer itself.

1. *Class test*—determines whether the data are numeric or alphabetic (a drug number contains only digits 0 through 9).
2. *Sign test*—determines whether the algebraic value of the numeric data is less than, greater than, or equal to zero (drug number is larger than zero).
3. *Reasonableness test*—rejects data that should never occur. (For example, the program would reject more than 6 grams of streptomyacin because that dosage would be fatal.)
4. *Sequence test*—ensures that data appear in a certain order. (Date must be a two-digit month number, a two-digit day number, and a two-digit year number.)
5. *Range test*—determines that the data fall between an upper and lower limit. (For example, a month number must fall between 1 and 12.)
6. *Presence test*—ensures that the data are present to allow further processing. (For example, the operator must enter a drug number before entering anything else.)
7. *Code test*—provides that the data value must match one of several acceptable predetermined values. (The year number must be 84, 85, or 86.)
8. *Combination test*—means that a certain group of data elements must all be present at the same time and must be correct. (Drug number, dosage, date, and patient number must all be included.)

Let us assume that the Albany Veterinary Clinic enters four data elements for each prescribed drug.

Data Element	Validation Requirements
Drug number	Numeric (0001000 − 0099999)
Dosage	Alphanumeric (mg)
Date administered	MMDDYY where MM: 1–12; DD: 1–31; YY: 84, 85, 86
Patient number	AA-DDDD where AA: Alphabetic (initials of last and first name) and DDDD: Numeric (a sequential number)

If an operator enters

2345, 10 mg, 201384, AS-356S

validation should identify two errors: an incorrect month number and an improper right half of the patient number (the letter S and not the digit 8). The terminal might notify the operator of the error by flashing that data element as well as printing an error message on the terminal screen. The operator then corrects the entries so they read:

2345, 10 mg, 021384, AS-3568

Batch, or Control, Totals

Batch, or control, total Control technique whereby the computer compares a filed sum with a predetermined one.

In addition to verification and validation, an analyst may specify **batch**, or **control, totals,** which are the total amount of all the individual amounts in the batch. To illustrate this concept, let us suppose Albany uses batch totals to detect potential errors on payments received from customers. A clerk hand-totals all checks received each day, and then enters the customers' numbers and amounts received at a CRT. After posting the last check, the clerk enters the hand-calculated total to see whether it agrees with the machine-calculated one. If it does, the data are used to update the accounts. If it does not, the computer prints a list of all the entries, so the clerk can check visually for discrepancies. The clerk then returns to the terminal, reenters the correct data, and compares totals again.

Check Digits, Transpositions, and Slides

Check digit Control technique using a formula or pattern to check a field's correctness.

Check digits, transpositions, and slides help detect data-entry errors. These three detection schemes require the computer to perform mathematical calculations on the data. The schemes are easy to program and very quickly can find yet another type of error.

The **check-digit** scheme provides a formula-based error-detection routine. An operator enters a data element, and the computer analyzes it to ascertain that it follows a specific pattern or adheres to a specific formula.

The ISBN (International Standard Book Number) used to identify a copyrighted or public domain work implements the check-digit scheme. Suppose the ISBN for a particular textbook is 0-534-00615-9. The following formula has been developed to test for accuracy.

1. Multiply successive digits (starting with the second digit) by 9, 8, 7, 6, 5, 4, 3, and 2 respectively.
2. Add the products.
3. Divide by 11.
4. The result should be a number with no remainder.

In applying this formula to our hypothetical text, we obtain:

$$9 \times 5 + 8 \times 3 + 7 \times 4 + 6 \times 0 + 5 \times 0$$
$$+ 4 \times 6 + 3 \times 1 + 2 \times 5 + 1 \times 9$$
$$= 45 + 24 + 28 + 0 + 0 + 24 + 3 + 10 + 9$$
$$= 143$$

After we divide by 11 (143/11 = 13), we get a result with no remainder. If we had a remainder, the number would have been wrong because all ISBN numbers follow this formula without exception. Check-digit testing works beautifully for account numbers or other data that identify a customer, vendor, or subscriber. It works because we assign these identifiers and do not pick them at random. It cannot be used with amounts of money, customer names, or units of measure, which are random and to which we cannot apply a formula for calculation.

Another formula-based method for detecting errors concerns transposed numbers. A **transposition** occurs when the order of the numbers entered is changed. For example, a data-entry operator may enter 357, and during verification another operator enters 537. If this is merely a transposition error, the difference between the two numbers always will be evenly divisible by 9. Consider our example whose difference is 180 (537 − 357 = 180), which is evenly divisible by 9 (180/9 = 20). It thus proves itself a transposition error, but does not identify which number is actually the correct data entry. This method of detecting transposition errors only works on numeric fields and the test must be made during verification where we have two data entries.

Transposition Control procedure that checks to ensure correct order of the entries.

A third formula-based error-detection method is a **slide,** which detects an incorrectly placed decimal point. Suppose the data-entry operator enters 7125 and during verification the number 71.25 is entered for the payment for veterinary services. If the difference between the two numbers is evenly divisible by 9, we have a slide. As with transposition errors, slide error detection can be applied only to numeric fields and only during verification.

Slide Control technique that detects an improperly placed decimal point.

Visual Verification and Computer-Assisted Validation

The most complex checking procedure combines visual verification and computer-assisted validation. Let us assume that the Albany Veterinary Clinic uses a terminal. Analyst Jack Swing could program the computer to display a patient's name immediately after the clerk enters his or her account number, thereby allowing a quick visual check for accuracy. If the display on the screen of the terminal matches the name on the source document, the data go to processing. If it does not, the clerk must reenter the data.

We could repeat the process for the drug number: enter the drug number, have the computer look it up and display it on the terminal screen, and visually verify it for accuracy. If the number is correct, the computer processes the entire transaction; if it is not, the clerk must reenter the data.

This type of error detection works only on terminals linked to a computer (mainframe, mini, or micro) where we store all data on disks for rapid random retrieval. Furthermore, it currently costs more than other techniques, consuming valuable computer time and programming resources, but as computers, disks, and other prices decline over the next few years, it should become very popular.

DATA-ENTRY HARDWARE

Before designing source documents, the analyst must select the most appropriate data-entry hardware. If an organization must process a large volume of data, an optical scanner may be ideal because it provides direct input to the computer, allows for continuous file updating, and processes data at low cost (despite its high purchase price).

In another situation an application may require a turnaround document. In this instance a printed page may be designed to be torn into two pieces: one portion returned with a check and the other portion kept as a receipt. Here the analyst may find a terminal more appropriate because of low volume or handwritten data that must be keyed rather than optically read.

To illustrate the trade-offs an analyst faces when choosing input hardware, let us again consider the case of the Albany Veterinary Clinic, which uses its computer for billing and payroll. Management wants to improve the efficiency of its record keeping and to ensure accurate reporting of controlled drugs to state agencies, so it has decided to add a drug inventory application to the system. The clinic hires a consulting analyst, Jack Swing, who reviews Albany's data-entry needs, and discovers a need for four data elements: drug number, dosage, date administered, and patient number. During discussions with the clinic's eight veterinarians, Jack learns that Albany averages 240 prescriptions daily. On the basis of this information alone, he might specify punched cards, terminals, or optical scanners for this new application. However, as we shall soon see, a number of other factors will restrict his decision.

Terminals: Dumb and Intelligent

Because they allow so many users within an organization access to computerized systems, terminals are enjoying increasing popularity as key-entry devices. Terminals allow users direct access to the computer and operate like other key-entry devices. As the user depresses each key, the terminal sends the data to the computer for collection and eventual processing. Users can locate errors before the computer processes the data, thus creating the advantage of instant error correction.

Guidelines for Designing Data-Entry Screens

Users communicate with application systems running on their computer almost exclusively through the terminal or CRT. Considering the millions of people now interacting with computers on a daily basis, it behooves the analyst to design display screens as clearly, concisely, and friendly as possible. The following guidelines can help analysts design such screens.

1. Keep screens simple, having them serve a single logical user activity.
2. Make screens easy to read. They should share a common format (i.e., title, date, time, and page number in the same place on all screens).
3. At the beginning of an application, position the cursor to the first field in which users will enter data. When errors occur, the cursor should move to the first field in error.
4. Minimize operator need to use the clear key and vertical and horizontal tab keys; this speeds data entry and reduces the likelihood of errors.
5. Try to ensure that data asked for once is not requested again, except for verification.
6. Design the data format to be self-evident. For example, a screen can display hyphens for Social Security numbers or phone numbers in the proper position.
7. Insert asterisks (or some other character) to tell the operator exactly what is required in all mandatory fields.
8. Use a consistent method for signaling error messages.
9. To ensure that users see all errors, the screen should return intact. Highlight each field with blinking text or audio signals.
10. Alert users that a given data entry is acceptable by having the screen return minus the acceptable data.
11. Liberally employ menu screens, which casual users greatly appreciate. Build an accelerated screen for advanced users.
12. Include passwords to tighten security. Instruct users to record date, time, and type of screen use (inquiry, data entry, reporting) along with their passwords.
13. Write directions in a clear, logical step-by-step fashion according to the way users normally do their jobs.
14. Keep directions free of technical jargon and computer buzzwords.

Prompt, friendly, clear, and understandable input screens can boost user productivity, while slow and confusing ones can only increase frustration over computerization of any task.

Source: D. Dardwin, "Teleprocessing System Design," *ICP INTERFACE—Data Processing Management,* pp. 21–22, Summer 1979. Reprinted with permission from ICP INTERFACE—Data Processing Management, Copyright © 1979, International Computer Programs, Inc., Indianapolis, Ind.

Dumb terminal A data-entry device that cannot check data entered by the operator.

Intelligent terminal Data-entry device that can check data without support from the main computer.

Terminals fall into two categories: dumb and intelligent. **Dumb terminals** are the least expensive and offer few options or features. Most have little or no memory, have limited cursor capabilities (no flash or reverse video or direct cursor placement), and may not have 10-key numeric pads such as those found on adding machines. **Intelligent terminals,** with built-in microprocessor and memory, a keyboard that includes a numeric 10-key pad, and a variety of useful cursor capabilities, offer the analyst a wider variety of design considerations, but such machines can be more expensive. Dumb terminals are most often found where users need lowest cost and possess the least computer sophistication, while intelligent terminals are most often found in high-volume data-entry situations where users want some degree of computer sophistication during the data-collection process.

For Albany, the analyst might consider two quite different terminals: the Lear Siegler ADM-3A (a dumb terminal) and a DEC VT-220 (an intelligent terminal made by Digital Equipment Corporation), either of which could satisfy Albany's needs (Figure 9.6).

Like most intelligent terminals, the DEC VT-220 has several built-in features: function keys that the clinic can preprogram to store dates and frequently prescribed drugs, automatic cursor positioning to speed data entry, and flashing and reverse video to highlight input errors. This smart terminal, which costs around $1200, could provide quick and efficient data entry.

However, Albany could save substantial money by purchasing an ADM-3A, which costs only about $395. Although the ADM-3A is a dumb terminal, the analyst could design the input screen to compensate for the terminal's lack of memory and error-detection techniques. We examine this more thoroughly when we discuss the design of input screens. For now we only need to understand that a dumb terminal such as the ADM-3A lacks sophistication and creates a more time-consuming data-entry process, which will offset its low initial cost.

Albany's decision to buy the VT-220 was based on two factors: Jack's prediction that Albany will recoup the cost difference between the ADM 3A and the VT-220 through faster and more reliable data entry, and the fact that Albany already has three older model DEC VT-100s so users will not have to learn how to operate a completely new type of terminal.

Key-to-Tape and Key-to-Disk Devices

Key-to-tape device Data-collection device that allows data to be recorded on a reel of tape.

Key-to-disk device Data-collection device that allows data to be recorded on a hard or floppy disk.

Key-to-tape and **key-to-disk devices** are also popular input devices (Figure 9.7). With such machines users enter and verify data in much the same way as with terminals. An operator keys in data from a source document, and then another person verifies accuracy by rekeying the same data on the same machine. After making necessary corrections until both sets of data match, the operator outputs the data to a tape or to a floppy or hard disk. After all data have been recorded, the tape or disk is removed from the key-to-disk or key-to-tape device and placed on a computer for processing. Data collected on such devices are batch processed.

The first key-to-tape device was the Mohawk key-to-tape machine made in 1965. Since then many other manufacturers have entered the field. IBM, for example, sold the popular model IBM 3740 key-to-floppy-disk in the 1970s.

CHAPTER 9 INPUT DESIGN **301**

(a) A Lear Siegler ADM 3A dumb terminal.

(b) An intelligent DEC VT-220.

FIGURE 9.6. Much of today's data is entered on terminals. [(a) Courtesy of Lear Siegler, Inc. (b) Courtesy of Digital Equipment Corporation.]

302 PART II SYSTEMS DESIGN

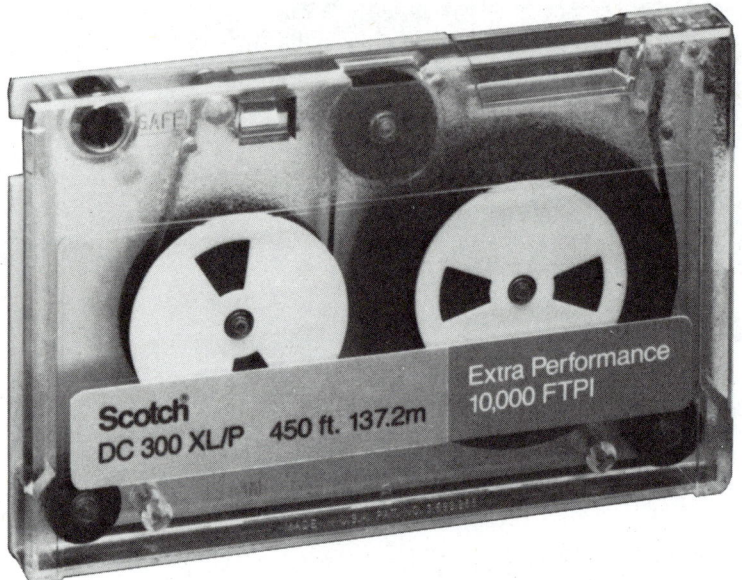

(a) Cassette tape.

(b) Floppy disks.

FIGURE 9.7. Key-to-tape or key-to-disk machines are popular input devices. Here we see the tape and diskettes they use to store data. [(a) Courtesy of International Business Machines Corporation. (b) Courtesy of 3M.]

Such input devices operate quietly, their tapes and disks are reusable, and they permit efficient data entry and verification. Organizations using them automatically can tally the number of keystrokes made by the operator, thus enabling management to pay key-entry personnel according to productivity, on a per keystroke basis.

Like a terminal, these machines operate at human speed and require two data-entry steps to get the data to the computer (data entry and reading by the computer). An operator easily can rekey data to detect typographic errors but cannot spot improper data, such as an invalid account number. However, unlike a terminal, a tape or disk does not allow people to see the data once the data have been entered.

An organization can link several key-to-disk machines into a network that shares a common disk (Figure 9.8). Since each machine does not have to have its own disk unit, this further reduces the cost per device. But sharing a common disk can increase contention for the disk by the devices and reduce data security.

Key-to-disk machines are useful for applications that involve thousands of transactions per day. Despite widespread use in the late 1960s to middle 1970s when batch processing was popular, they are gradually being phased out in favor of smart terminals or personal computers linked to minicomputers or mainframe computers.

FIGURE 9.8. An organization can link several key-to-disk devices to form a network in which they all share a single disk (in the rear of the photograph). (Courtesy of Sperry Corporation.)

Optical reader Data-collection device that scans preprinted forms to collect data.

Optical Readers

All key-entry input devices suffer a common drawback: a human operator must key in all data. Direct-entry devices such as **optical readers** (Figure 9.9) can read data directly to the computer, disk, or tape from a paper medium by scanning a code printed on the product (a can of peas), on a tag attached to the product (a coat), or on an ordinary sheet of paper (IRS 1040 forms). Optical readers cost from a few hundred dollars (hand-held wands) to hundreds of thousands of dollars for machines capable of reading 3000 sheets of handwritten data an hour.

Since we can hook optical readers directly to computers, we can effectively bypass the two-step cycle of key-to-tape and terminal devices. The error-detection technique parallels that for smart terminals. We program the machine to know where data can be found on each source document, and we give it specific error-checking rules: test for numeric value; check for specific values (A through F) or presence of a value (an amount of money). If the machine detects an error, it rejects that item or page and sends it in a different direction than that in which acceptable source documents travel. One problem with optical readers occurs when data are entered in the wrong field or are handwritten. In these instances the optical reader cannot read, and so will reject, the data.

At the Albany Clinic, Jack Swing tests a hand-held wand reader as a potential data-entry device for the new drug system. Relatively inexpensive (less than $200), the wand could be linked to the clinic's intelligent terminal, to read the bar codes

FIGURE 9.9. This optical reader "reads" data from a document and then stores the data on a reel of tape. In this case the computer then reads and stores the data on tape for later processing. (Courtesy of International Business Machines Corporation.)

on prescription bottle labels automatically. That would work for drugs administered on the premises, but in a field situation, someone without a wand would have to enter data manually later. And, of course, in neither case could the want automatically read a patient's name, the dosage, or date.

To avoid having to buy two data-entry systems (one requiring the keying of all data entry on the VT-220, and the other using a hand-held wand with the VT-220), Jack stays with his earlier decision to enter all data on a VT-220 terminal.

DESIGNING DATA ENTRY FOR TERMINALS

The use of terminals as input devices has soared for many reasons. As prices of terminals and compatible computers have decreased, their sophistication has increased. In the early 1970s, the cost of a terminal was about $2000, and a typical computer of that era could support 20 to 30 terminals, bringing the cost of such a system to nearly half a million dollars. Today, however, a computer with 20 terminals and much more capability costs well under $100,000.

Terminals can place data at a user's fingertips, thus allowing the user to call up specific data and make timely decisions on the basis of those data. A terminal-oriented system also shifts responsibility for data entry and accuracy from data-entry operators, who seldom comprehend the meaning of the data they enter, to users, who are familiar with the data. This heightens user awareness of the importance of accurate and timely data entry as well as of the entire data processing function.

To make users comfortable with this shift in responsibility, analysts have had to work hard to build user-friendly systems that noncomputer professionals can understand and use. Such users want a data input process that explains itself, prompts them to make appropriate responses, allows easy error detection, and leads them by the hand through crucial steps whenever necessary.

When designing data entry for a terminal-oriented system, the analyst "paints" the terminal's screen with identifying words and formats. Most terminal screens permit 24 horizontal lines, 80 vertical rows, and two intensities or brightnesses for painting (see terminal capabilities, Chapter 6). The less bright intensity might display directions for data entry, and the brighter one can show actual data being input. If input data survive verification and validation, the terminal might lower the intensity of the data to reflect the fact that the data have passed required tests. Data items that fail might remain brighter, or even flash an error signal.

Data-collection screens show data-entry operators headings that define their purposes. Often the analyst will utilize the heading for the system's title, date, time of day, and even the last time the program was used (Figure 9.10). By employing varying light intensities and flashing error messages, the analyst need not specify a separate place on the screen for errors. Rather the analyst may direct the computer to position the cursor automatically to the incorrect field and display an error message.

For example, Albany Clinic's new system allows an operator to enter a drug number only, and the computer automatically retrieves and fills in the drug's name

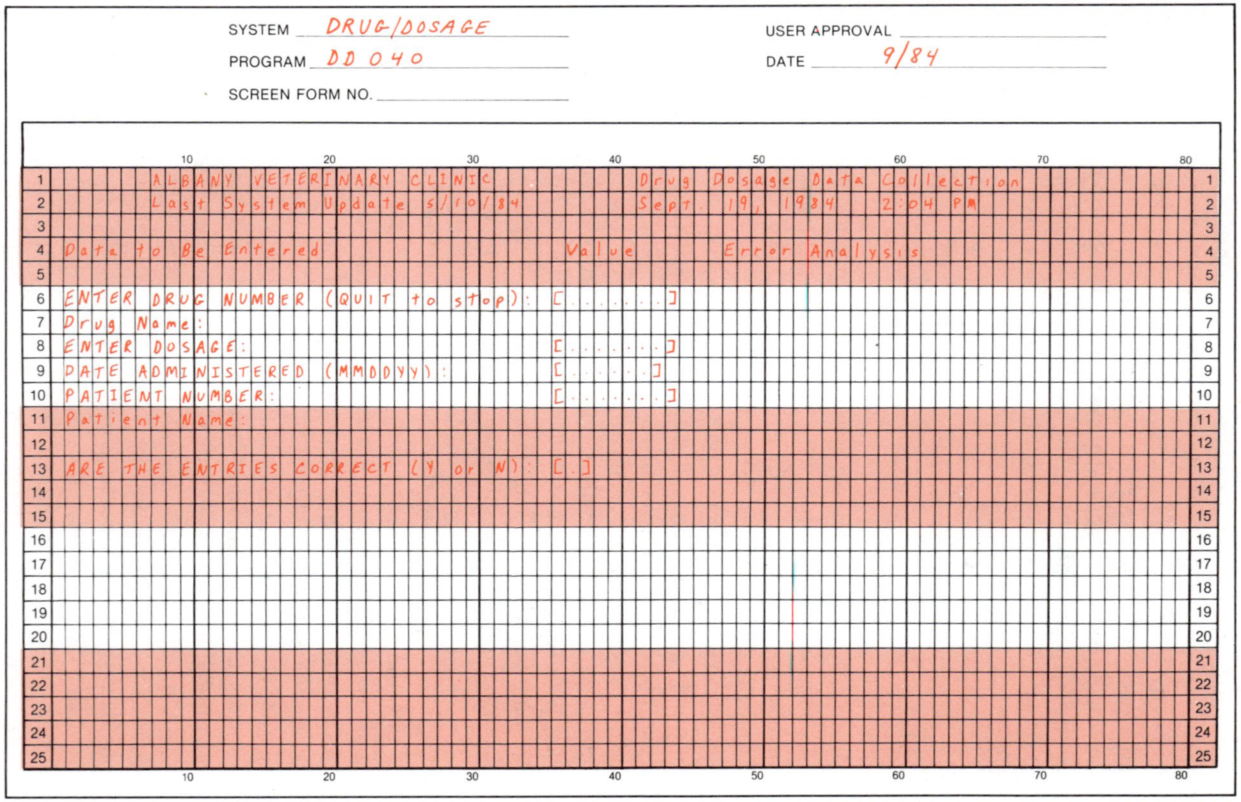

SYSTEM _DRUG/DOSAGE_
PROGRAM _DD 040_
SCREEN FORM NO. _____

USER APPROVAL _____
DATE _9/84_

```
                                        Drug Dosage Data Collection
ALBANY VETERINARY CLINIC                Sept 19, 1984    2:04 PM
Last System Update 5/10/84              Error Analysis

Data to Be Entered                      Value

ENTER DRUG NUMBER (Quit to stop):  [    ]
Drug Name:
ENTER DOSAGE:                      [    ]
DATE ADMINISTERED (MMDDYY):        [    ]
PATIENT NUMBER:
Patient Name:

ARE THE ENTRIES CORRECT (Y or N):  [ ]
```

FIGURE 9.10. A data-entry screen for the Albany Veterinary Clinic. Note the headings, prompts, and spaces for the operator responses.

for visual verification. After being assured of the proper drug, the operator can proceed to enter the dosage and other pertinent data.

The screen design in Figure 9.10 also illustrates the types of prompts that help data-entry operators enter data correctly. For example, "MMDDYY" requires double-digit month (01 to 12), day (01 to 31), and year entries. "Y or N" facilitates error detection and "Quit" permits the user to exit from the system.

Some applications require multiple screens. For example, when the Albany Clinic treats a new animal, data about the patient and its owner may be so extensive that it will take two screens to capture and verify all the details. The first screen may gather data about the owner of the animal (name, address, phone number) and the second data about the animal itself (name, type, weight, allergies). Sometimes analysts are tempted to gather all data on one screen to achieve efficiency, but it is usually wise to follow the rule that each screen be single purpose and restricted to logically related data.

The popularity of screens and terminals has stimulated software vendors to develop special systems that aid analysts and programmers in building screens. For instance, one vendor permits the screen designer to develop the desired screen, define all validation requirements for each field (including all state abbreviations, Zip codes, and telephone area codes), compile the design (just as a programmer compiles a COBOL, Ada, or Pascal program), and incorporate the screen as an operational program. The resulting screen, when placed into operation, can directly update a data base or build a batch disk file for later processing. If errors are detected during the batch update, the faulty records remain in the file, with appropriate error messages, so the next time a user performs data entry all error records are available for correction. Software with similar capabilities is available from many computer manufacturers and special software firms.

Screen designs depict the exact appearance of the screen, including:

1. Name and title of the screen.
2. List of all the entries required by the operator.
3. Positions on the screen where the data will be entered.
4. Location of all error messages.
5. Rules governing how the screen collection will end.

After designing the screens, the analyst submits them, together with screen documentation, for management approval (Figure 9.11). Screen documentation parallels screen design and stipulates:

1. Name and title of the screen.
2. Date and name of the analyst designing the screen.
3. Purpose of the screen.
4. List of data elements, their type, number of characters to be entered, and validation requirements.
5. Processing requirements giving the computer's response to data entries.

Management and users can now review the analyst's work and decide whether the design meets their needs. When errors or omissions occur, the analyst will

```
SYSTEM:   Drug Inventory                              DATE:       9/24/84
ANALYST:  Jack Swing                                  COMMENTS:   None
PURPOSE:  Collection of Drug Administration

Data Element      Type            Length    Validation Requirements

Drug Number       Alphanumeric      7       Presence, not blank
Dosage            Alphanumeric      7       Presence, not blank
Date              Numeric           6       MMDDYY where:
                                              MM: 1–12
                                              DD: 1–31
                                              YY: 84, 85, 86
                                            Presence
Patient Number    Alphanumeric      7       Presence, not blank
Correct Entry     Alphanumeric      1       Presence, "Y", "y",
                                                      "N", "n"

Processing Requirements:
  1. Validate all fields.
  2. When drug number is entered, the computer will display its name.
  3. When the patient number is entered, the computer will display the patient's
     name.
  4. When "Quit" is entered as a drug number, the system will terminate.
  5. Validation will be performed when operator enters a "Y." An "N" entry will
     cause this transaction to be discarded.
```

FIGURE 9.11. Documentation of the screen design occurs after user and management approval.

need to repair the screen design and documentation and resubmit them for approval. Screen layouts and accompanying documentation form yet another part of the output of the design phase of the systems process.

DESIGNING DATA ENTRY FOR KEY-ENTRY DEVICES

For applications involving batch processing, key-to-tape or key-to-disk devices offer a viable alternative to terminals. Let us consider one of their more effective applications at Cable Data, a cable television firm that charges each customer a flat rate of $7.75 a month and sends monthly bills requiring payment within 30 days. At the beginning of a billing cycle, a data-entry operator enters each customer's previous payment, adds and deletes customers, and updates files with new data on old customers, such as address changes. Since such an application does not need to be online, it lends itself nicely to batch-processing key-entry devices.

When designing data entry for these devices, analysts use a tool called the **multiple record layout form** (Figure 9.12). This form provides space for up to six different records and contains a place at the top for the application title, analyst's

Multiple record layout form The document used by an analyst to describe the format and fields in a file.

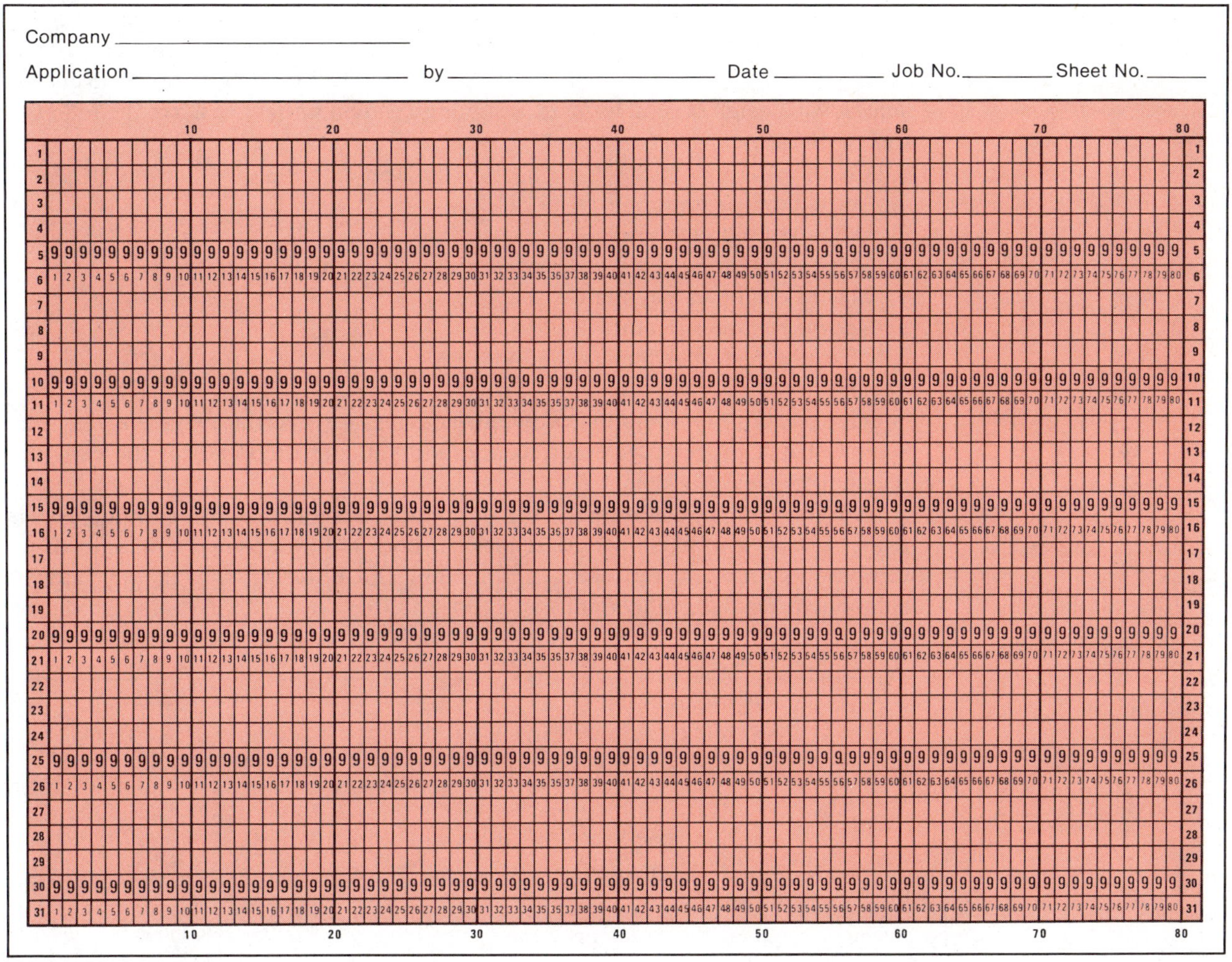

FIGURE 9.12. *Manual entry layout forms provide space for the analyst to design up to six different formats.*

name, and date. One divides the body of the form into fields, each of which will hold data entered by the data-entry operator. To avoid confusion, one crosses out unused areas.

When laying out these forms, the analyst first decides which areas to use for each data element (Figure 9.13). The data-entry instructions indicate that zeros must lead numeric fields to fill out the allotted areas. For example, to use all seven positions, customer number 1222 must be entered as 0001222. Month numbers must fall between 1 and 12, day numbers between 1 and 31, and year numbers within the predetermined range, 84–86.

If an organization such as Cable Data needs to enter large amounts of data, the analyst may specify more than one record format (Figure 9.14). Nevertheless, every record relating to a given customer will bear an identifying key (such as a customer number) in the same location. Each record will also contain a function or code number in position 1, allowing the organization to differentiate between record formats. For Cable Data, which uses key-to-disk machines exclusively, the analyst will design several formats, assigning a code number to each of the following transactions.

1. Add a new customer.
2. Delete an old customer.
3. Change customer data, such as address or telephone number.
4. Record customer payment.
5. Record customer credit.
6. Charge a customer for installation.

Once placed into operation, key-entry devices involve the two steps that characterize batch processing: enter, verify, and validate the data; let the computer read and process the data.

Like its terminal counterpart, this data-entry design results in two documents: record layouts and data-entry descriptions. Both follow a format similar to that of terminals, conveying the same information, and differing only in the hardware component for data collection.

SYSTEM: Accounts Receivable **DATE:** 9/24/84
ANALYST: Jack Swing **COMMENTS:** None
PURPOSE: Data-entry instructions for payments received from customers.

Field Name	Type	Begin Location	End Location	
Customer Number	Numeric	1	7	999999
Amount	Numeric	8	12	9.99
Date	Numeric	13	19	In form MMDDYY MM: 1–12 DD: 1–31 YY: 84, 85, 86

Entry Requirements:
1. All fields will be validated to ensure they are numeric.
2. Date must be MMDDYY.
3. Amount should not have decimal point, $4.25 should be entered 00425.
4. All numeric fields to have leading zeros.

FIGURE 9.13. Data-entry instructions tell the operator where the data should be placed and what the format should be.

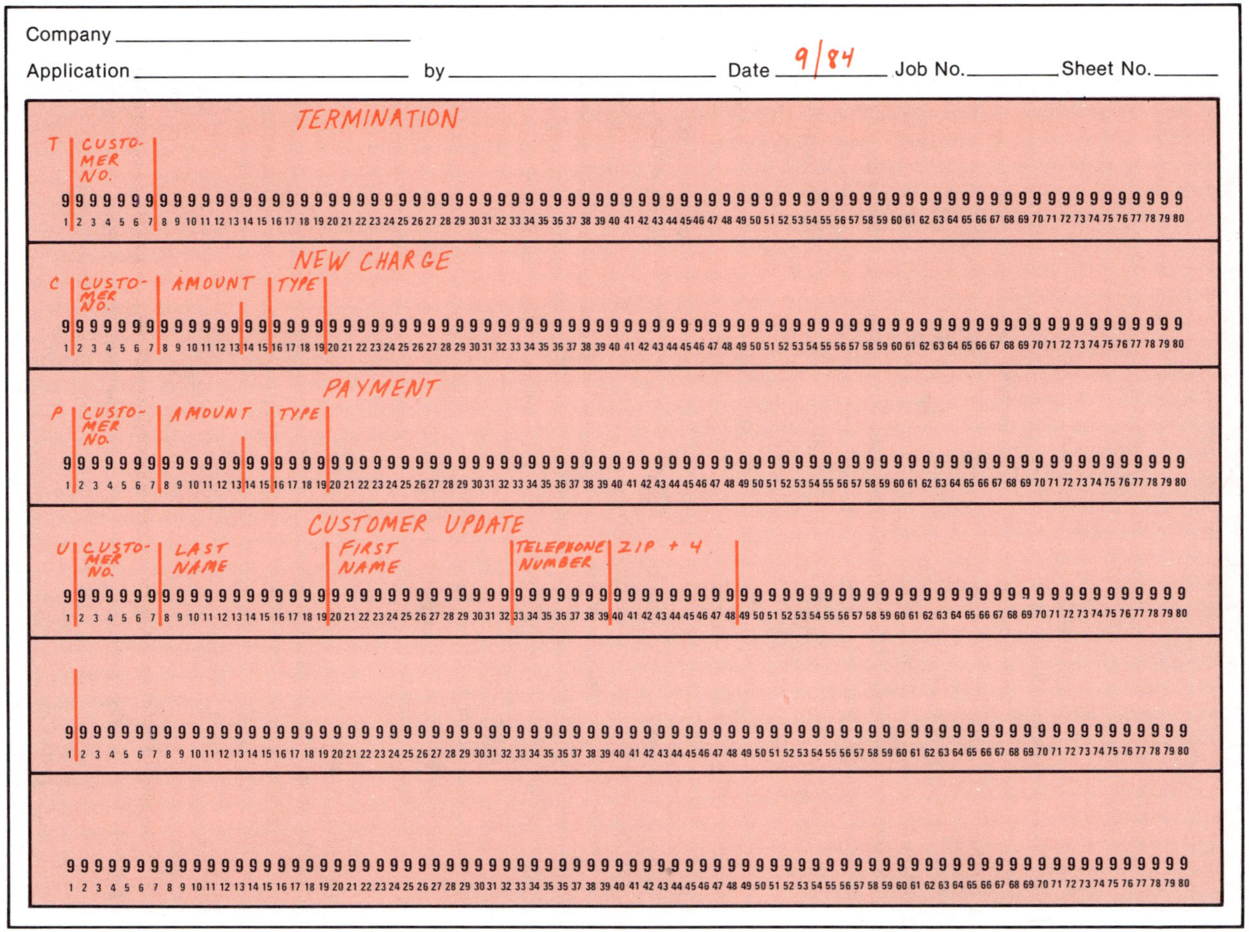

FIGURE 9.14. This key-entry design allows addition or termination of a customer, new charges and payments, and customer data updating.

CASE STUDY

Fleet Feet's Input Design

Frank Pisciotta's data-base design for Fleet Feet's accounts payable system includes three master and three detail data sets. Since the company's computer uses a network data-base management system, Frank has utilized it to define these sets. He is now ready to describe the data-entry process and techniques for verification and validation.

When Fleet Feet bought its HP 3000 computer, it also decided to buy, rather than write from scratch, the fundamental accounting software. After a long search, they decided to acquire software written by MCBA (Mini-Computer Business Applications, Inc., of Montrose, Calif.). On delivery of the HP 3000, Fleet Feet installed the MCBA software for inventory, payroll, and general ledger applications, but not for accounts payable because of the cost and conversion requirements for franchise operations (see the case study in Chapter 2). This software has been in operation for over a year and Frank chooses to pattern the new AP system after MCBA's other systems.

First he designs the selector menu (Figure 9.15a). Such menus lead users by the hand, asking what they want to do and then steer them to the desired function. The menu (and all other screens) must be accompanied by a detailed description of its use (Figure 9.15b). This documentation, coupled with the screen designs themselves, will serve two purposes: It tells programmers what to do, and it forms the basis of a packet Frank can give to users operating the new system. Notice how Frank has included data validation and verification requirements for each field.

The main selector menu offers users four types of activities: vendor file maintenance, data entry, report generation, and clearing of totals.

The first activity, vendor file maintenance, lets users add a new vendor, change a vendor's data, or delete a vendor. With this routine the system captures new or revised data about each vendor: number and name, address, phone number, payment terms. Frank decides to use a submenu to prompt users to select a specific activity involving a vendor (Figure 9.16a). As usual Frank documents this screen design explaining its purpose and function in user-friendly language (Figure 9.16b).

Figures 9.17 and 9.18 are Frank's designs for adding new vendors and changing the data about existing vendors. These screen designs and their associated documentation illustrate leveling. The main selector screen lies at the first level, vendor file maintenance at the second, and vendor additions, changes, and deletions at the third. To return to the original selector menu, the user backtracks through the screen selectors to level 1. Leveling of functions follows the structured methodology and is typical for online terminal-oriented systems.

Next comes report generation. To achieve the open item report, Frank needs to define a way to list vendors in order by vendor number on a printed form. Since this report will include all vendors in the file and not a specific subgroup, Frank need not develop a new screen for it. A prompt from the main accounts payable selector menu will tell the computer all it needs to know to print this list.

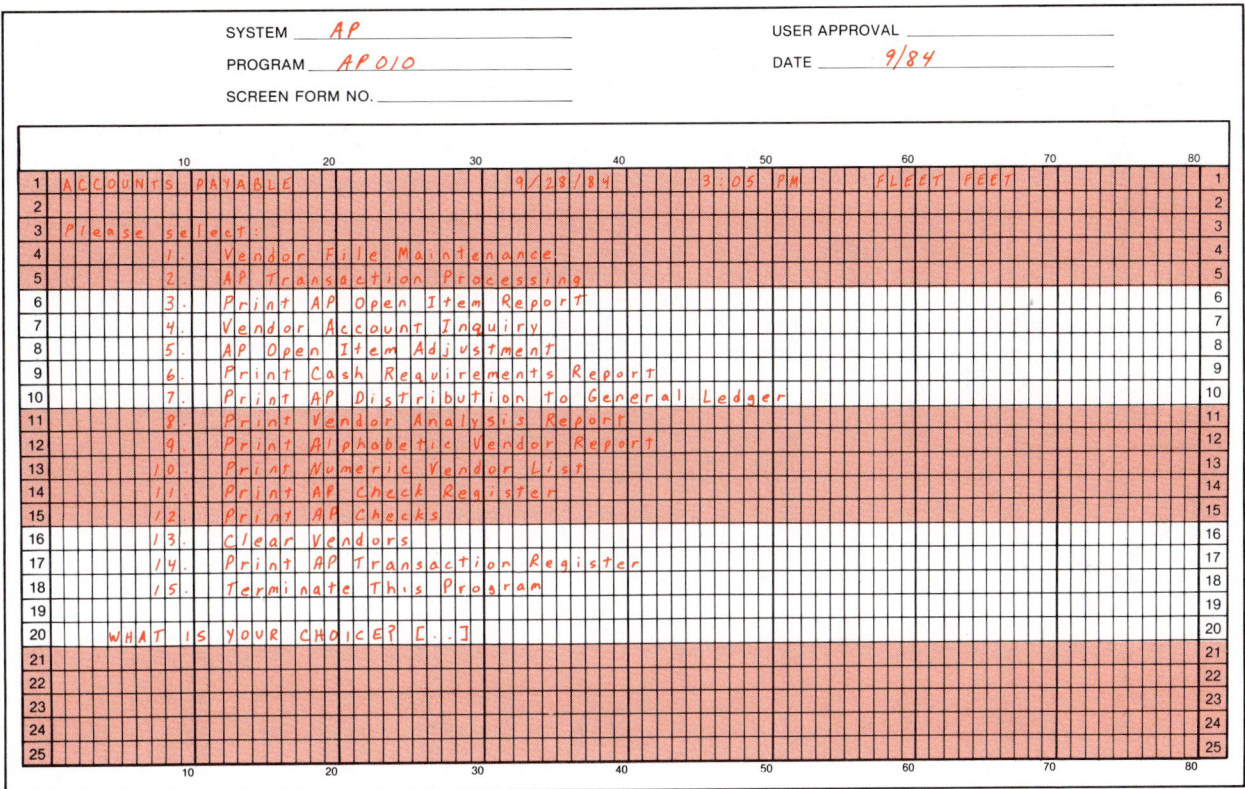

(a) The AP system consists of many programs. The user selects desired functions by choosing the appropriate entry from the main selector menu.

SYSTEM: Accounts Payable
ANALYST: Frank Pisciotta
DESCRIPTION: Main Selector Menu
DATE: 9/28/84

To use the AP main selector menu, you type the following command:

    AP (followed by striking the "RETURN" key on the terminal keyboard)

Type the number of the desired application. For example a 1 will call up the Vendor File Maintenance application.

Any entry other than a 1 through 15 will cause the screen to be displayed again.

    To terminate this selector menu, type 15 followed by "RETURN."

(b) Each screen design is accompanied by a description of its use. These descriptions of the input rules have two functions: they establish objectives for the programmer, and they show users how to enter data.

FIGURE 9.15. Frank Pisciotta's design for the main menu screen and directions for its use. (Courtesy of MCBA®. MCBA is a registered trademark of MCBA, Inc.)

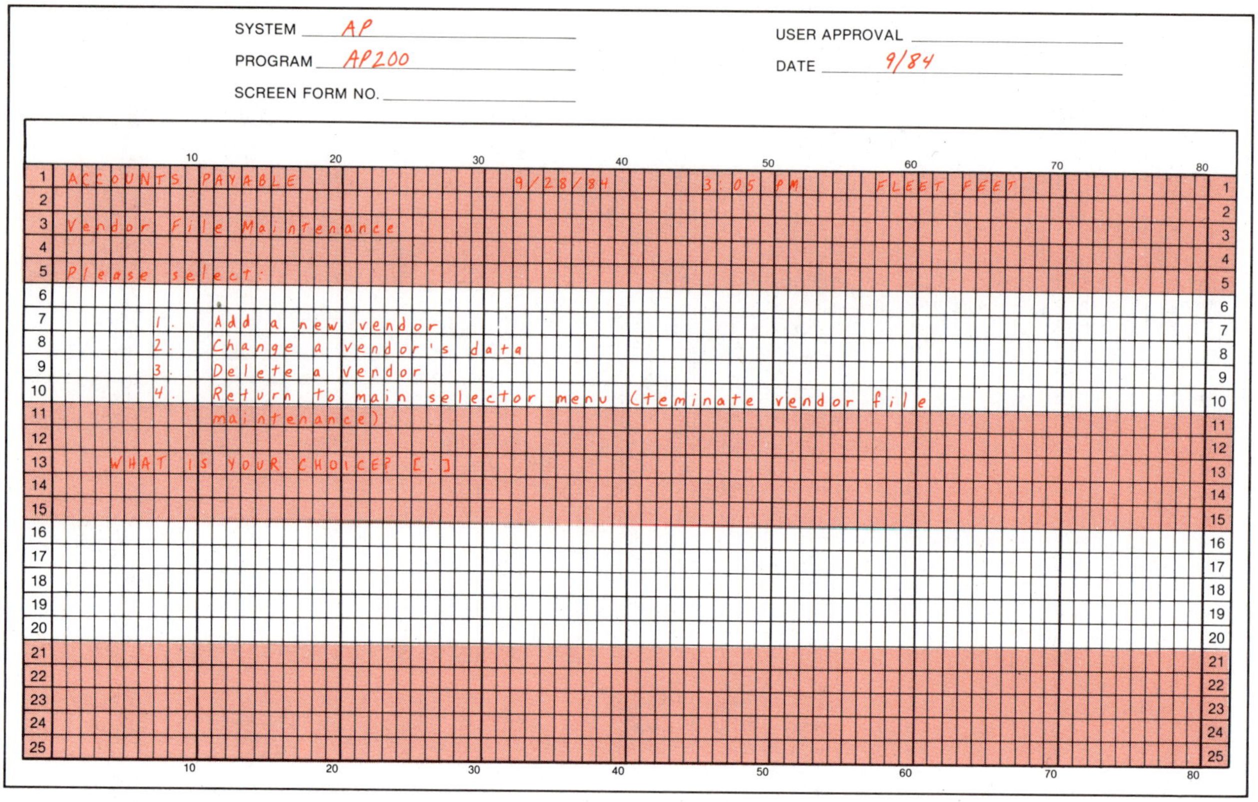

```
SYSTEM     AP                              USER APPROVAL _____
PROGRAM    AP200                           DATE          9/84
SCREEN FORM NO. _____
```

```
 1  ACCOUNTS PAYABLE              9/28/84        3:05 PM      FLEET FEET
 2
 3  Vendor File Maintenance
 4
 5  Please select:
 6
 7         1.  Add a new vendor
 8         2.  Change a vendor's data
 9         3.  Delete a vendor
10         4.  Return to main selector menu (terminate vendor file
11             maintenance)
12
13      WHAT IS YOUR CHOICE? [ ]
```

(a) The user's response dictates which screen will appear next.

SYSTEM: Accounts Payable   **DATE:** 9/28/84
ANALYST: Frank Pisciotta
DESCRIPTION: Vendor File Maintenance Selector Menu

Type the number of the desired application. For example, a "1" will call up the Add a New Vendor application.

Any entry other than a 1 through 4 will cause the screen to be displayed again.

To terminate this selector menu, type 4, followed by RETURN. This will return you to the main selector menu.

(b) This definition is similar to the earlier one, except that it does not tell users how to get to it. Rather, it tells users how to get out of it and which screen will appear when they exit this one.

FIGURE 9.16. A submenu screen for data entry for vendor file maintenance. (Courtesy of MCBA®. MCBA is a registered trademark of MCBA, Inc.)

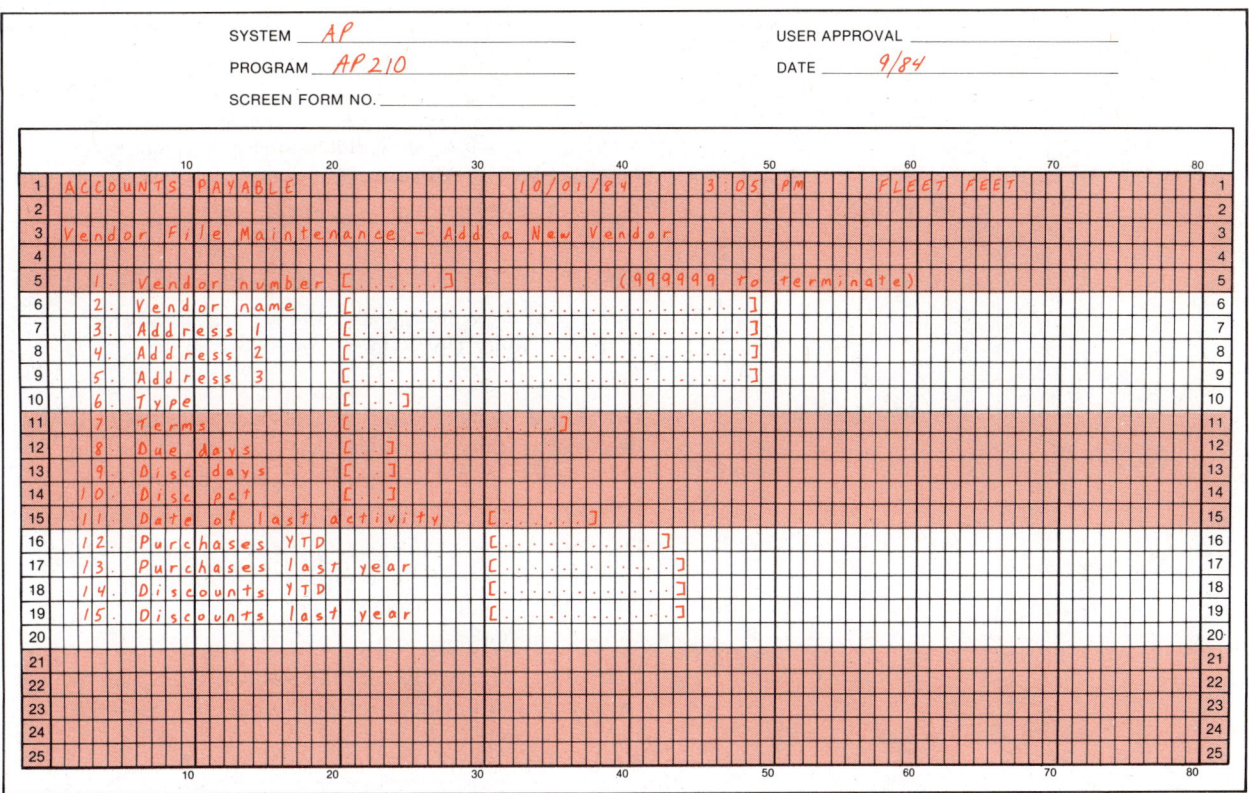

(a) The screen pops up when the user enters a "1" from the vendor file maintenance selector menu. The change vendor data pops up when the user enters a "2" from the vendor file maintenance selector menu.

SYSTEM: Accounts Payable
ANALYST: Frank Pisciotta
DESCRIPTION: Vendor File Maintenance—Add a New Vendor
DATE: 10/02/84

Enter the data as follows:

1. Vendor number — One to six alphanumeric characters. Numeric vendor numbers should be right-justified and zero filled to the left.
   Example: A-9123

2. Vendor name — Maximum of 30 alphanumeric characters. Left-justified.
   Example: Nike Shoes

3. Address 1 — First line of vendor's address. Up to 30 alphanumeric characters. Left-justified.
   Example: 444 Nevada Street

4. Address 2 — Second line of vendor's address. Up to 30 alphanumeric characters. Left-justified.
   Example: Reno Nevada 9876-0667

5. Address 3 — Third line of vendors address. Up to 30 characters. Left-justified.
   Example: Name of contact.

FIGURE 9.17. The "Add Vendor" screen. (Courtesy of MCBA®. MCBA is a registered trademark of MCBA, Inc.)

6. Type — Vendor type. Up to three alphanumeric characters.
   Example: WHL (Wholesale)

7. Terms — Description of payment terms. Up to 15 alphanumeric characters.
   Example: 2% 10 NET 30

8. Due days — Number of days allowed after the invoice date before payment is due. Up to three numeric digits.
   Example: 30

9. Disc days — Discount days: Number of days allowed after the invoice date during which a payment will qualify for a discount. Must be less than due days. Up to three numeric digits.
   Example: 10

10. Disc percent — Discount percent: Percentage allowed if payment is made within number of days specified. Up to three numeric digits. Do not enter decimal point.
    Example: 2 (for 2%)

11. Last activity date — Date of last AP activity with this vendor. Each new AP transaction will update this field. Format is MMDDYY and is validated.
    Example: 052384 (for May 23, 1984)

12. Purchases YTD — Purchases this year-to-date. One to nine numeric digits with two decimal places and optional minus sign.
    Example: 1250000 (for $12,500.00)

13. Purchases 1st yr — Amount of purchases last year from this vendor. One to nine numeric digits with two decimal places and optional minus sign.
    Example: 4500000 (for $45,000.00)

14. Discounts YTD — Amount of this year-to-date's discounts. One to nine numeric digits with two decimal places and optional minus sign.
    Example: 25000 (for $250.00)

15. Discounts last year — Amount of discounts taken last year. One to nine numeric digits with two decimal places and optional minus sign.
    Example: 90000 (for $900.00)

To change a data value on this screen, type the number of the data item. It will be blanked out and the cursor positioned at the beginning of the field. Enter the revised data.

To terminate the adding of new vendors, enter 999999 for the vendor number. You will be returned to the vendor file maintenance menu screen.

(b) Documentation of the "Add Vendor" screen. As shown, documentation for detailed screens often includes examples of how users should input and edit data.

FIGURE 9.17 *(Continued)*

```
              SYSTEM   AP                              USER APPROVAL _____
              PROGRAM  AP240                           DATE    9/84
              SCREEN FORM NO. _____
```

1	ACCOUNTS PAYABLE 10/01/84 3:05 PM FLEET FEET
2	
3	Vendor File Maintenance - Change a Vendor
4	
5	1. Vendor number [000234] (999999 to terminate)
6	2. Vendor name [NIKE SHOES..............]
7	3. Address 1 [444 Nevada Street.......]
8	4. Address 2 [RENO NEVADA 98777-0056.]
9	5. Address 3 [........................]
10	6. Type [WHL]
11	7. Terms [2% 10 net 30..]
12	8. Due days [30]
13	9. Disc days [10]
14	10. Disc pct [02]
15	11. Date of last activity [022384]
16	12. Purchases YTD [000012036 67]
17	13. Purchases last year [000010579 38]
18	14. Discounts YTD [00000025000]
19	15. Discounts last year [00000156662]
20	
21	ITEM NUMBER TO CHANGE? [..]

(a) A blank screen appears when the user enters a "2." The user then enters the vendor number and all the relevant data appear, ready for changes.

SYSTEM: Accounts Payable **DATE:** 10/02/84
ANALYST: Frank Pisciotta
DESCRIPTION: Vendor File Maintenance—Change a Vendor's Data

When the vendor file maintenance menu is displayed, enter a 2 and a blank vendor addition screen is displayed. Enter the vendor number of the desired vendor. If this is a valid vendor, all the data about this vendor will fill up the now empty screen. If there is no vendor with this number, an error message will be displayed and a blank screen generated.

Enter the item number of the field you want to change (2 through 15) and the field will be blanked, the cursor positioned at the beginning of this field, and you can now make the corrected entry.

You cannot change a vendor number. If a vendor number is incorrect, the vendor must be deleted and then added. No two vendors may have the same vendor number.

When no more changes need to be made, enter anything except a 2 through 15 and you will be returned to the vendor file maintenance selector menu.

(b) To alter data the computer must find the data about a vendor and display the data for viewing. The user then tells the computer which field to change.

FIGURE 9.18. Data entered on the "Change a Vendor" screen. (Courtesy of MCBA®. MCBA is a registered trademark of MCBA, Inc.)

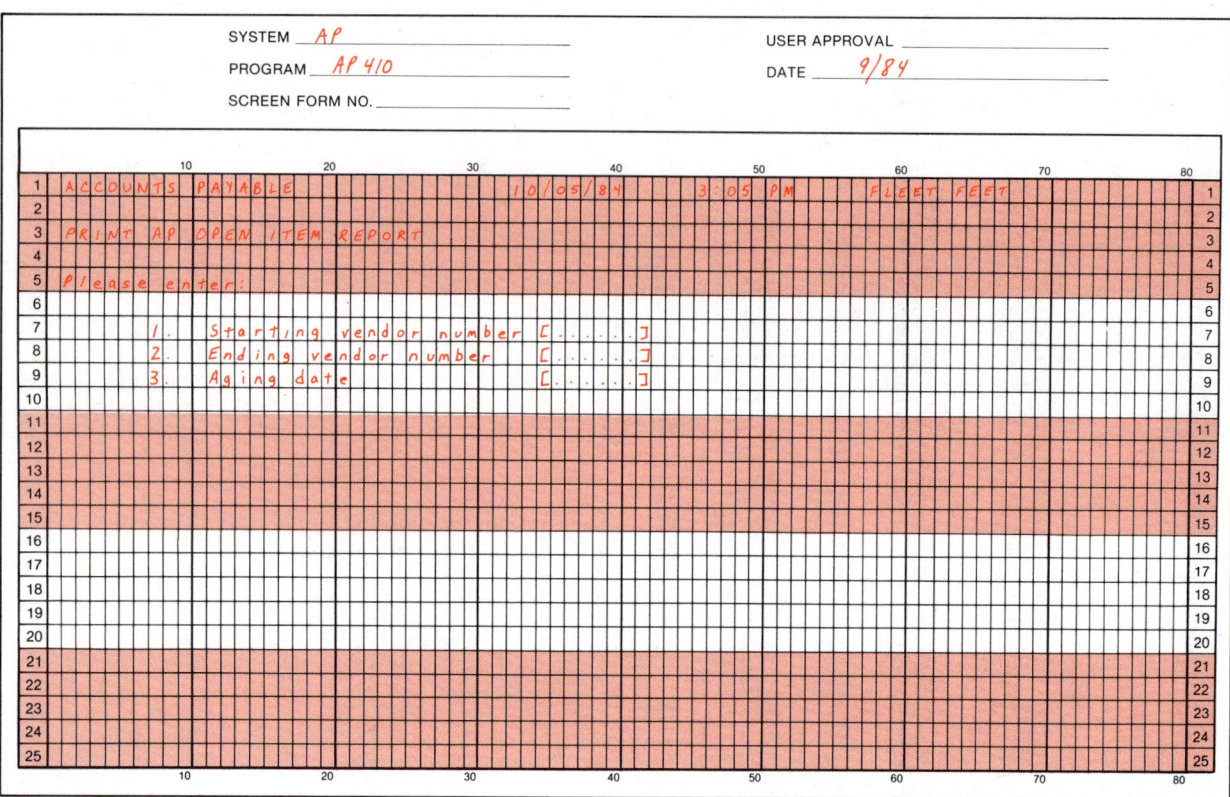

(a) This "open item" screen permits the user to specify which data are to be printed. For example, a user may want to see only open item accounts in the Midwest (if accounts are numbered according to geographic regions).

SYSTEM: Accounts Payable   **DATE:** 10/05/84
ANALYST: Frank Pisciotta
DESCRIPTION: Print AP Open Item Report

Enter the data as follows:

1. Starting vendor number — Enter the starting vendor number, a 1 to 6 alphanumeric group. Press "Return" to list all vendors.
   Example: 000100

2. Ending vendor number — Enter the ending vendor number, a 1 to 6 alphanumeric group. Press "Return" to list all vendors.
   Example: 9-A122

3. Aging date — Enter the cutoff date for posted transactions. Form is MMDDYY and must be valid.
   Example: 052384 (for May 23, 1984)

If an error is made in the aging date, an error message will be displayed and the user will be returned to the main selector menu.

(b) Like documents described earlier, this one describes use of the screen and gives examples. Unlike other documents, it does not allow the user to choose the next screen. After printing the report, the computer returns the user to the main selector menu.

FIGURE 9.19. A screen design to print a report. (Courtesy of MCBA®. MCBA is a registered trademark of MCBA, Inc.)

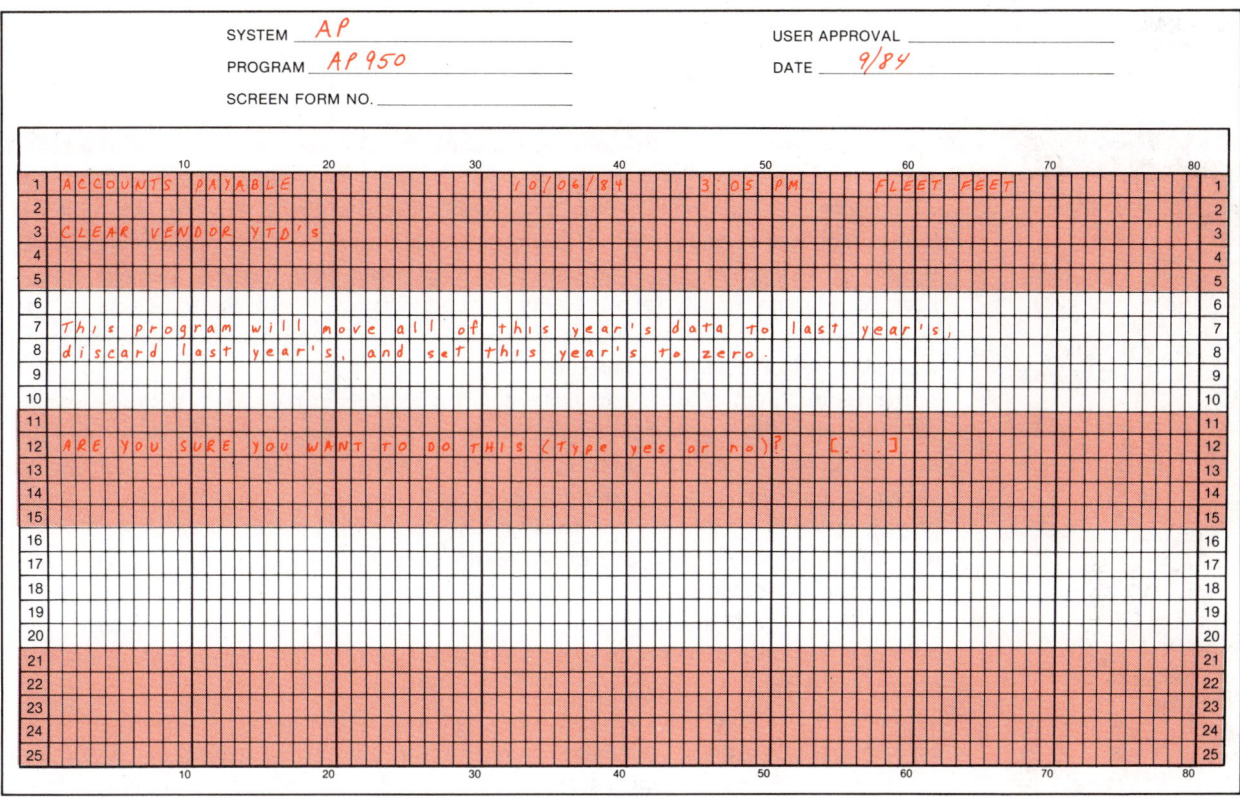

(a) Since it can be costly, time consuming, or impossible to recover purged data, only authorized users should be permitted to clear data, and only at specific times.

SYSTEM: Accounts Payable **DATE:** 10/05/84
ANALYST: Frank Pisciotta
DESCRIPTION: Clear Vendor YTD

The user enters a "Yes" or "No" to activate this program. Since this is a deletion process, the user will be given a single try to respond; if the user does not respond within 30 seconds, the program will automatically terminate.

Once the selection is made, the clearing is performed independently of the user's terminal. The user is free to perform some other activity on the computer.

(b) The definition of this screen emphasizes potential dangers.

FIGURE 9.20. This screen definition requires that the user verify a response before the computer does any further processing. (Courtesy of MCBA®. MCBA is a registered trademark of MCBA, Inc.)

CASE STUDY

The open item report differs from other reports because it must show the user all the AP transactions entered and posted up to a certain date (Figure 9.19). This screen allows users to print all vendors or a specific subgroup. It may take many minutes, perhaps over an hour, to print this lengthy report, so the print program can run at the same time the user does something else. This simultaneous running also typifies interactive online systems, thus exploiting the computer's full potential.

The menu's fourth capability facilitates clearing of totals or removal of data stored in the computer after processing (Figure 9.20). Because it is so expensive, difficult, or even impossible to recover purged data, clearing applications are dangerous and demand the utmost caution. Nevertheless, every system must have such a capability. Basic security provisions include single entry and time out. Only a specific user can make the entry, which must be correct the first time. If that person does not make the entry within a limited time, the computer will automatically interrupt the clearing process. Clearing activities are periodic, such as at the end of the month after all vendor checks have been printed and sent. Frank decides to have the computer store each purged record on tape, just in case he needs to recover the data.

Eventually Frank develops screen layouts and descriptions for every process within the accounts payable system. Though the entire AP system contains 13 screens, we have shown only seven, since they are representative of all others. Frank's collection of documents forms the basis for screen formats, which a programmer will consult to develop necessary programs during the next design stage, process and control design.

WORKING WITH PEOPLE   Ergonomics and Design

Gaylen Lewis, responsible for entering data into the payroll system at Haley's Department Store, is afraid he will not be able to input all the necessary data for this week's paychecks. The Christmas holidays, overtime, vacations, and a number of new part-time workers have made an already difficult job almost impossible. As if that were not enough to ruin Gaylen's day, the computer's response time is intolerably slow because so many other systems (such as inventory) require year-end processing.

As he had the year before, Gaylen found himself thinking how much more efficiently the payroll system could run if the company's time cards were organized for easier data entry, and he wished he had a safe place to stack them. As he had last year,

Gaylen shared his frustrations with his supervisor, Joyce Isheim. To Gaylen's amazement, this time Joyce acted on the payroll problem immediately. When they met with the payroll supervisor, Greg Goodwin, Joyce and Gaylen found him equally frustrated with the payroll system. "After all," he said, "from our employees' point of view paychecks are Haley's most important output. You can't cash an end-of-the-year inventory report." Before long Gaylen found himself working directly with Greg on the problem.

First they examined Haley's time cards. Gaylen pointed out how hard it was to locate desired data: the first data entry, employee number, appeared at the bottom of the time card; the second entry, date, appeared at the top. In addition, total hours came from the bottom of the card, sick time from the middle, and employee and supervisor signatures again from the bottom. Greg agreed that searching all over for data wasted a lot of time. It would be so much easier if one could simply read from top to bottom and from left to right. "I can't believe we've lived with this for so many years," Greg said. Gaylen sympathized. "It wasn't such a problem when we had only 50 employees."

Since Greg had been involved with the initial design, he could reasonably estimate that it would take 7 hours to repair the program, test it, and retrain Gaylen to use it. To avoid late paychecks at all costs, Greg helped Gaylen finish this period's data entry, and stayed after work to repair the system.

The following morning Greg and Gaylen reviewed the new design. Gaylen was delighted. The newly organized time cards should be a breeze to enter. But what about the other problems in the department—cramped space, poor lighting, antiquated equipment? Greg was eager to improve working conditions, but the space problem wasn't easy to solve. They experimented by moving the terminal into an adjoining typist's room with a large desk, ample lighting, and a comfortable chair. Though it was a more suitable environment, the noise level from the typing distracted Gaylen. Finally Greg suggested that Gaylen and Joyce redesign the original work space, using ideas from companies specializing in ergonomics, the science of creating work spaces with people in mind.

Gaylen and Joyce found new chairs that allowed variable heights and widths and tilted to fit each worker comfortably. New desks were adjustable, had compartments for time cards before and after data entry, and were designed to accommodate terminals. Special lighting could be positioned so as not to reflect off a terminal screen.

After the new equipment was installed, Greg looked forward to coming into work each day. Joyce summed it up for everyone. "In this computer age, it is so easy for management to forget that people still do the work."

SUMMARY

Input design is the fourth stage in the design phase of the systems process. During this stage the analyst consults users to determine the most efficient way to collect and check necessary data.

The move toward online applications lends itself to a terminal-oriented environment. Analysts must pay special attention to user-friendly systems that permit quick error detection. Verification, validation, control totals, check digits, transposition, slides, and visual verification and computer-aided validation help catch, control, and reduce errors.

The hardware to collect data can vary from dumb and smart terminals, to key-to-tape or key-to-disk devices, to optical readers. Terminals and online systems are rapidly replacing key-to-tape and key-to-disk devices for data collection. Optical readers are best suited for massive amounts of data entry.

The result of input design is a description of how the organization will collect data. This description takes two forms: screen or key-entry input form design, and details about error-detection methods (validation or verification). The description is the single output of this stage of the systems process and joins the file or data-base design and report designs in the system's documentation.

NEW TERMS

Batch, or control, total	Key-to-disk device	Slide
Check digit	Key-to-tape device	Source document
Direct entry	Manual, or key, entry	Transposition
Dumb terminal	Multiple record layout form	Validation
Intelligent terminal	Optical reader	Verification

QUESTIONS FOR REVIEW AND DISCUSSION

Questions appear in three categories. You can find the answers to the A group in this chapter. The B group requires you to apply the material presented here, whereas the C group necessitates investigation or research.

A.1. Make a list of the key-entry devices.
A.2. List the inputs to the input design phase.
A.3. List the outputs from the input design phase.
A.4. How much data can be displayed on a terminal screen, and how many rows and columns of data are there?
A.5. List the media on which key-entry devices record their data.
A.6. List all possible validation methods.

B.1. Make a chart of the number of days in each month that would be necessary properly to validate a date entered in MMDDYY form. Remember to consider leap years in your chart.
B.2. List the advantages of terminals and key-to-tape devices for data input devices.
B.3. Does an optical reader require verification of the data it reads?
B.4. What is the difference between verification and validation?
B.5. What validation methods could be performed on a field that held an amount of money?
B.6. Could Fleet Feet have a vendor in its AP system with each of the following vendor numbers?
    a. 9999   d. A
    b. 99999   e. −12345
    c. 999999   f. ZZZZZ

C.1. Is it necessary to design a screen for Fleet Feet to print an alphabetic list of vendors? Why?
C.2. Design a menu to collect the data to print AP checks.
C.3. Explain when it is necessary to have a second-level screen menu.
C.4. Why would the vendor maintenance not run simultaneously with check printing?
C.5. Does the data-base schema have any effect on input design?

REFERENCES

Steve Eckols, *How to Design and Develop Business Systems*, Mike Murach and Associates, Fresno, Calif., 1983.

Jerry Fitzgerald, *Designing Controls into Computerized Systems*, Jerry Fitzgerald and Associates, Redwood City, Calif., 1981.

Gerald Myers, *Software Reliability: Principles and Practices*, Wiley, New York, 1976.

Advanced Approaches to Systems Development, Auerbach Information Management Series, 1982.

Tyler Welburn, *Advanced Structured COBOL*, Mayfield, Palo Alto, Calif., 1983.

CHAPTER 10

Process Design and Acquisition of Hardware and Software

- Goals and Preview
- Modes of Processing Data
- Batch Processing
- Online Processing
- Real-Time Systems
- Distributed Systems: Completely Connected, Star, Ring, and Bus Networks
- LANs and Distributing the Work Load
- Hardware and Software Acquisition: RFQ and RFP
- Case Study: Fleet Feet's Online Accounts Payable System and an RFQ for New Terminals
- Working with People: Too Many Controls
- Summary
- New Terms
- Questions for Review and Discussion
- References

GOALS AND PREVIEW

After reading this chapter you should be able to:

1. Describe a batch processing system.
2. Describe an online system.
3. Describe a real-time or interactive system.
4. Describe the four types of network systems.
5. Explain the difference between an RFP and an RFQ.
6. Design an evaluation form for hardware or software acquisition.

Systems involve people, hardware, software, data, and procedures. Thus far we have concerned ourselves primarily with data, people, software, input devices, storage devices, CRTs, and printers (hardware). Machines may be useless without programs to drive them and people to operate them, but no system can deliver desired results without appropriate equipment. In this chapter we consider many factors that influence an analyst's decisions about hardware.

During the past decade, hardware prices have fallen dramatically, even for computers with a million characters of internal memory. Capabilities once restricted to the largest and most sophisticated million dollar computers are now available on a single circuit board, and organizations can acquire machines capable of servicing many terminals simultaneously for less than $15,000. The recent proliferation of powerful microcomputers further expands an analyst's resources, especially when one links them to larger machines. Having designed the data input and output and the files or data-base management system, the analyst is now ready to select a mode of system operation and specify new hardware.

How do we process the data to achieve the desired output? Should data be collected and held for future input into the computer? Should data be stored in a computer in raw form for later processing? Should data be processed when collected by the organization? In this chapter we examine the ways or modes in which computer hardware can be used to process the data, as well as the merits of the different modes in terms of cost and performance (see Figure 10.1).

Once the mode of operation is set, we also examine how the analyst specifies and acquires equipment for the new system. In some instances a new computer may be necessary; in others, perhaps a terminal, additional disk storage, or even a personal computer.

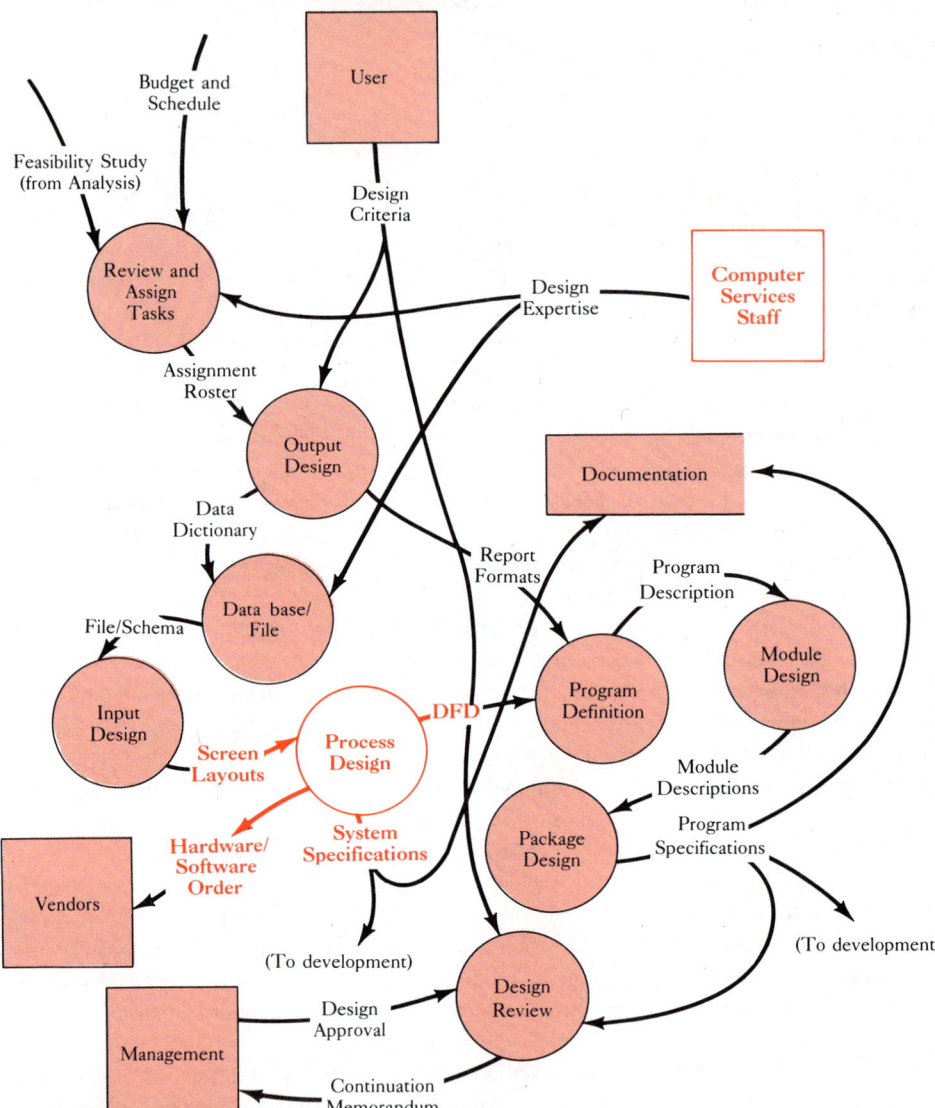

FIGURE 10.1. Design can be decomposed into nine stages. The fifth stage is process design where we decide the system's mode of operation and order hardware or software.

MODES OF PROCESSING DATA

A system can function in any of three basic modes: batch, online, or real time. A fourth mode, distributed processing, has more to do with the coordination of the subsystems than with the actual mode of processing. Thus a distributed system may employ both batch and real-time processing, or it may involve only one of the basic modes.

Depending on users' needs, budgets, and the equipment on hand or to be ordered, the analyst selects the most appropriate mode of data processing. Various applications lend themselves best to one method, whereas others demand two or more. In a bank, for example, mail deposits may be processed overnight, while deposits made in person may be processed on the spot, so the customer is told his or her account balance immediately. Similarly, a company may have several systems that require different modes of processing. The reservation system of an airline must process data from several locations immediately whereas statistical information concerning whether each flight was early or late in taking off and landing is usually processed at the end of each month.

BATCH PROCESSING

In a **batch processing** system, data are collected over a period of time, then processed in groups at specified intervals. A payroll system is usually batch: data are collected for hours worked over a specific time period (a day, a week, or a month) before paychecks are issued. An airline system gathers flight attendants' requests for flight assignments (called bids) and determines which flight attendants will work which flights. This is a once-a-month system and lends iself nicely to a batch mode of operation.

Batch processing A technique that processes a number of similar transactions in groups.

As a result of recent technological advances, the mode of operating a batch system is different from what it used to be. Formerly an operator would keypunch data onto cards, then a machine would read the cards, and the data would be processed by the computer. Today a data-entry person may collect the data with a mini- or microcomputer, then store the data on tape or a floppy or hard disk for eventual reading and processing by the computer (Figure 10.2). The computer may be located far from data-entry people who transmit data to the distant computer by means of a data communications (or telecommunications) link, telephone wire, microwave radio, satellite, or all of these.

Output devices for a batch system may be located near data entry or at a distant site (Figure 10.3). A **remote job-entry** (RJE) **device** transmits data from one location to another for processing. The results can be received immediately after processing rather than sent through the mail. For example, a large organization, such as the U.S. Bureau of Reclamation, may link many RJE devices from all over the country to a central computer. Each local office collects and enters its own data, sends the data to the home office for processing, and receives results on a local printer.

Remote job-entry device A device transmitting data to another location for further processing.

The main advantages of batch processing are its simplicity and relatively low cost. Data entry, processing, and output can each proceed separately. Furthermore

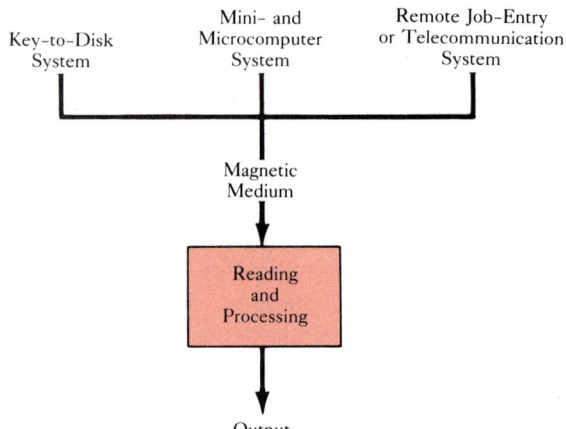

FIGURE 10.2. *An organization using batch processing may do it all at one physical location or it may send data from a remote source to a central computer for processing.*

some applications do not demand immediate feedback. For years batch processing was the only option available to the analyst; as a result many people are comfortable with the method they have used for so long.

The lack of timely data is the major drawback of batch processing. Organizations frequently need to base decisions on the most up-to-date data and cannot afford to wait a week, or even a day, for current reports. Many organizations have abandoned batch processing in favor of one of the other modes.

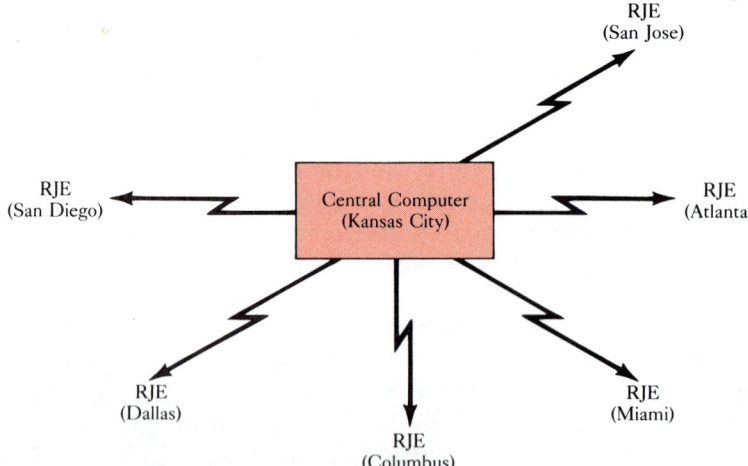

FIGURE 10.3. *Remote job-entry devices pair some type of data-entry device with an output device, usually a printer. The RJE device is tied to the computer by a telephone line or a data communications link. In a large organization, a central computer can support a number of RJE devices. Each device is operated by local staff.*

ONLINE PROCESSING

Online processing A technique that allows user access to data but does <u>not</u> permit updating.

In an **online processing** system, data can be collected and updated in a batch mode, but users with terminals can inquire about data at any time except when the system is updating their data. Users themselves cannot update data. Data are usually processed within a short time, and so are quite current. Our bank example of walk-in deposits is an online system. Online systems are more sophisticated and complex than their batch counterparts since they have the added features of terminals and inquiry.

The design of an online system is a two-step process. The analyst first designs a batch system for data collection and processing, and then designs terminal-based inquiries (Figure 10.4).

Online output can also be batch, perhaps with reports generated in the evening when computer demand is light. The Federal Bureau of Investigation's (FBI) National Crime Information Center (NCIC) implemented one of the first major online systems. Still in use, it allows local law enforcement agencies to send data concerning arrests and trial results to FBI headquarters in Washington, D.C., where the FBI enters the data into the system. In return agencies have access to the system at any time to inquire and retrieve useful information about a suspect (criminal history, identifying marks, physical description, etc.).

A county office of education on the East Coast provides an online service to each of its subscriber school districts. The county computer stores financial or budget data as well as data on students (names, addresses, telephone numbers, emergency telephone numbers). Terminals are placed in local school offices from

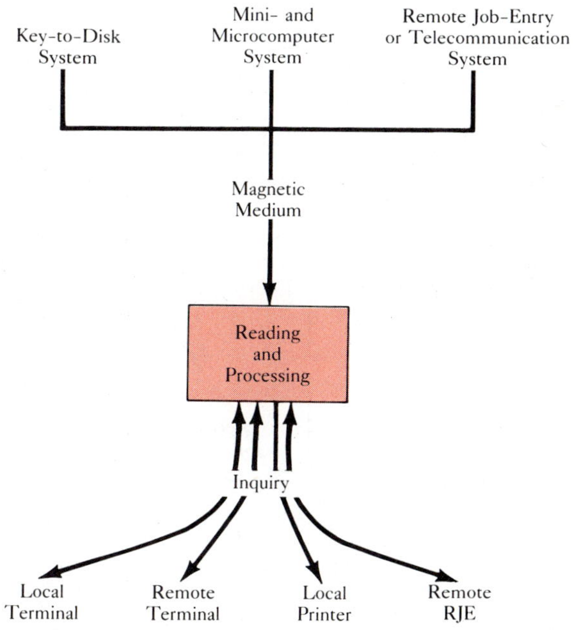

FIGURE 10.4. Online systems use batch data input but have ongoing user inquiry.

which staff can inquire about various school account balances or a student's emergency telephone number. Each school is too small to afford its own online computer system but they can pool their funds and enjoy an inquiry-based online system.

Many organizations that need timely data are now operating online systems. Online systems typically demand additional hardware and software. Hardware for online systems must be able to support nearby or remote terminals; software permits the computer to communicate with terminals or RJE devices. Online systems continue to grow in popularity, providing users with quicker access to data.

REAL-TIME SYSTEMS

Real-time processing
A technique whereby data processing occurs rapidly enough to control an activity.

Real-time processing is an interactive system. Users can enter data for immediate processing and they can access updated data any time (Figure 10.5). The system is updating data quickly enough to be useful in controlling an ongoing activity. The airline reservation system mentioned earlier is real time: passengers purchasing tickets get immediate reservations, and the system immediately updates information about available seats. Many rapid transit districts also use real-time systems for train control, as does NASA (National Aeronautics and Space Administration) for satellite launching.

Hardware for real-time systems can include multiple terminals, microcomputers, and minicomputers tied to a central computer. The computer may do all the processing and storage of data locally; some data may be stored and processed locally, and then sent to corporate headquarters for coordinated processing with data from other files. Most computers made since the middle 1970s support real-time processing. Thus organizations seldom need to purchase a new computer to support a real-time application, because analysts can design around an existing machine.

One problem analysts must address when designing real-time systems is database security. Since users can retrieve and alter data at will, the system must be

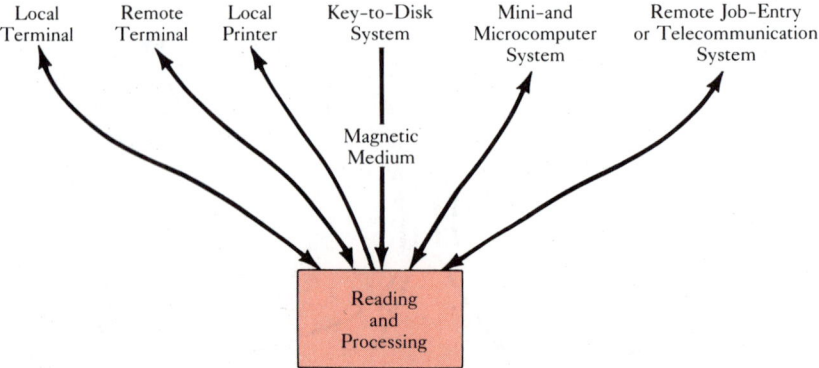

FIGURE 10.5. Real-time systems involve data input at will or in batches and allow constant updating and inquiry.

protected against unauthorized manipulation. To accomplish this, analysts may establish passwords for use of the files. Terminals also may be physically restricted from access to users and certain data. Some users may receive limited "read" capabilities; others may receive both "read" and "write" access. Yet another security control involves restricting certain portions of the entire data base to certain users. Data regarding employees' pay rates, for example, may be accessible to the payroll or personnel departments and no one else.

Another problem involves setting priorities for use of the data. Since two or more users might request identical data simultaneously, or a user might try to process a record already being processed by someone else, analysts must design a way to establish priorities among authorized users. For example, two airline ticket agents, perhaps in different cities or even adjacent terminals, may try to assign the same seat on a flight to two different passengers at the same time. In resolving these conflicts, the analyst may assign a priority to one terminal over another.

To summarize, a real-time system provides the most current data; however, it may represent a big investment. Terminals, geographically separated from the central computer, create extra costs. Additional disk drives may be required to make data interactive with the user whereas the data might be kept on cheaper tape if the system were operating in a batch mode. Software, as a result of the need for security and resolution of access rights, is more expensive to design, program, install, and maintain.

DISTRIBUTED SYSTEMS: COMPLETELY CONNECTED, STAR, RING, AND BUS NETWORKS

Most systems operating during the 1950s and early 1960s were batch, because existing hardware did not allow any other mode. As hardware and software advanced from the middle 1960s to the middle 1970s, however, online systems became feasible. Real-time systems have always been used to collect data on large mainframes for U.S. space explorations, but did not find their way into consumer and business markets until the middle 1970s, when less expensive minicomputers made real-time systems attractive.

Manual systems have always existed, but it was not until the late 1970s and early 1980s that advances in telecommunications enabled organizations to distribute data electronically to many different locations. Some of the movement to distributed systems was a direct result of inexpensive microcomputer systems, which can function as independent computers, act as terminals, or do both, collecting data and storing the data on floppy disks.

With the three basic modes (batch, online, and real time), the important consideration is the entry, updating, and storage of data relative to users. In a distributed system, data are stored in various locations where users may need access to all or part of the data base. Processing may be batch, online, or real time.

At the heart of **distributed processing** lies the notion of networking, or linking together, several computers, perhaps from different manufacturers or different models from the same manufacturer, located in different places. Each local computer provides resources for local users, whether the organization operates in batch,

Distributed processing Sharing of computerized systems among geographically dispersed computers.

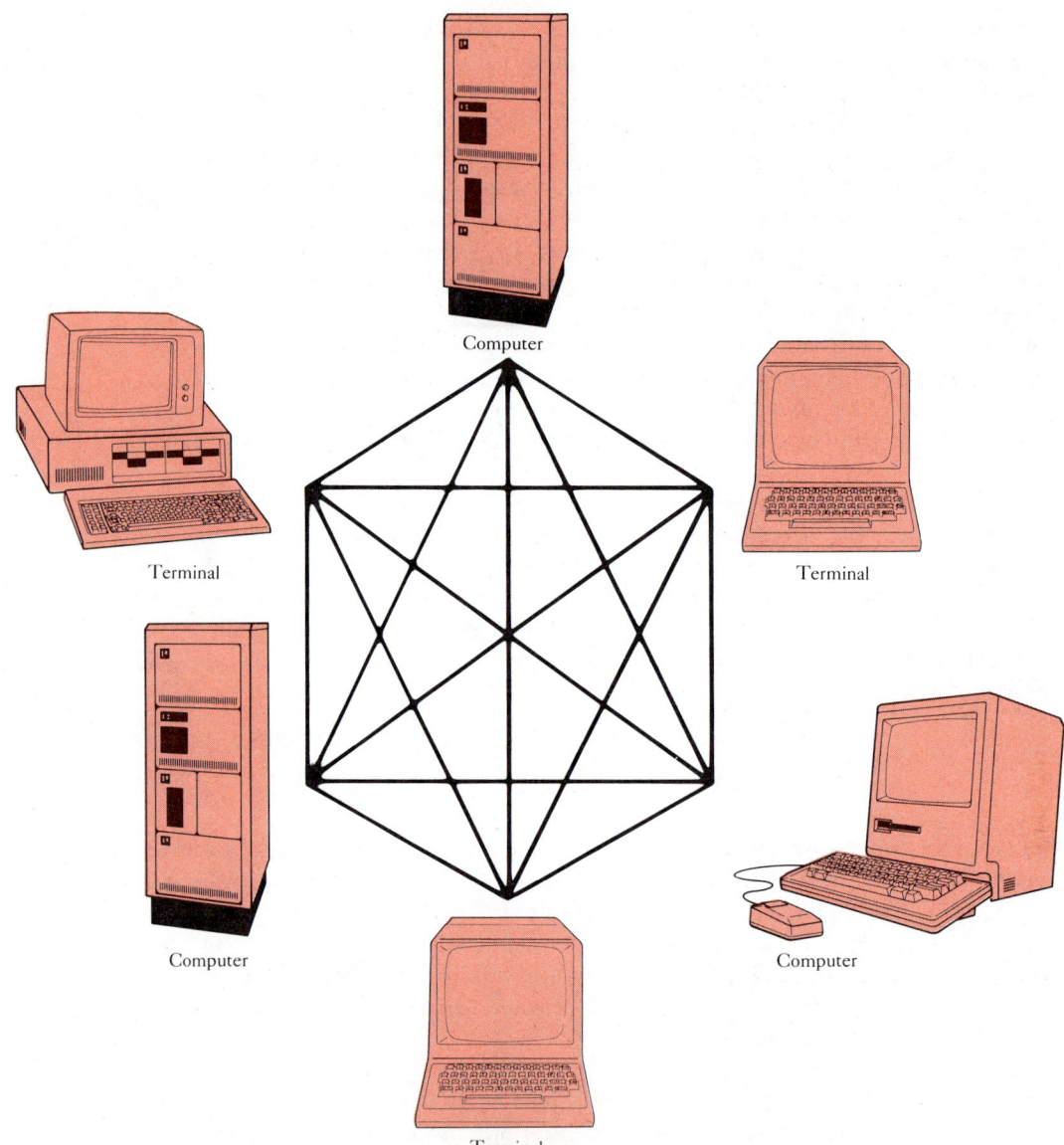

FIGURE 10.6. A completely connected network links all subsidiary terminals or computers to each other.

online, or real time. If a user at one location needs data from another location, the different machines must be able to convey the needed data. Furthermore, if data are duplicated at two or more sites, the system must be able to update the data at both locations.

Four popular methods for configuring the various computers, terminals, and personal computers are the completely connected, star, ring, and bus networks.

A **completely connected network** provides a direct path or link between any two stations (Figure 10.6). If there are eight stations, there must be 28 links. For most applications this type of network is impractical, as there are too many links when the number of stations is large.

Star networks tie one or more terminals or satellite computers to a central

Completely connected network A network providing a path or link between all stations.

Star network A network providing a path or link to one central computer.

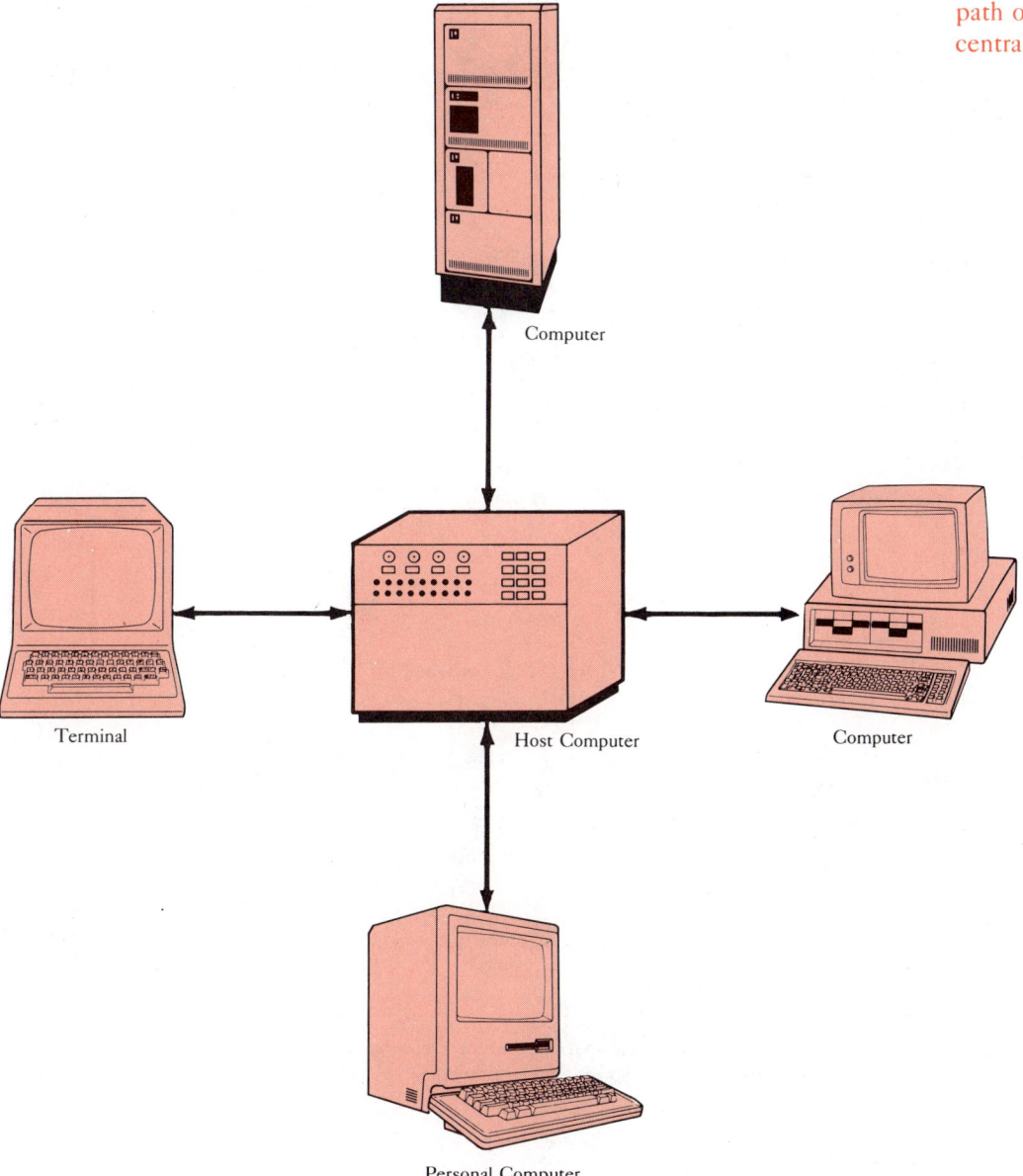

FIGURE 10.7. *A star network links all subsidiary terminals or computers to a host computer whose only job may be to tie the computers together and to route data from one to another.*

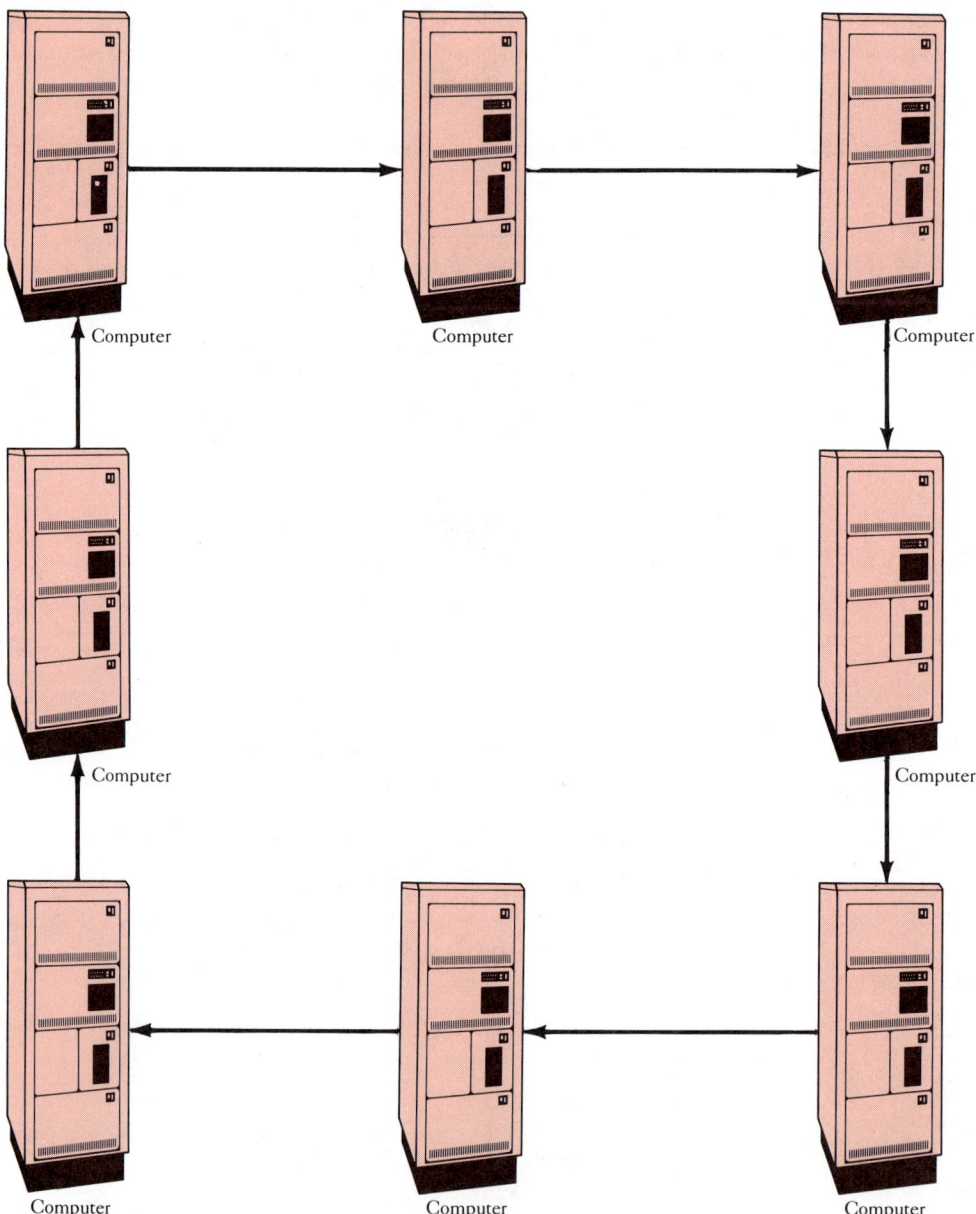

FIGURE 10.8. A ring network ties computers together in a circle. Data may travel from one machine to another through more than one machine before it reaches its destination. A host computer is not needed. Each computer in the ring may have terminals.

CHAPTER 10 PROCESS DESIGN AND ACQUISITION OF HARDWARE AND SOFTWARE **335**

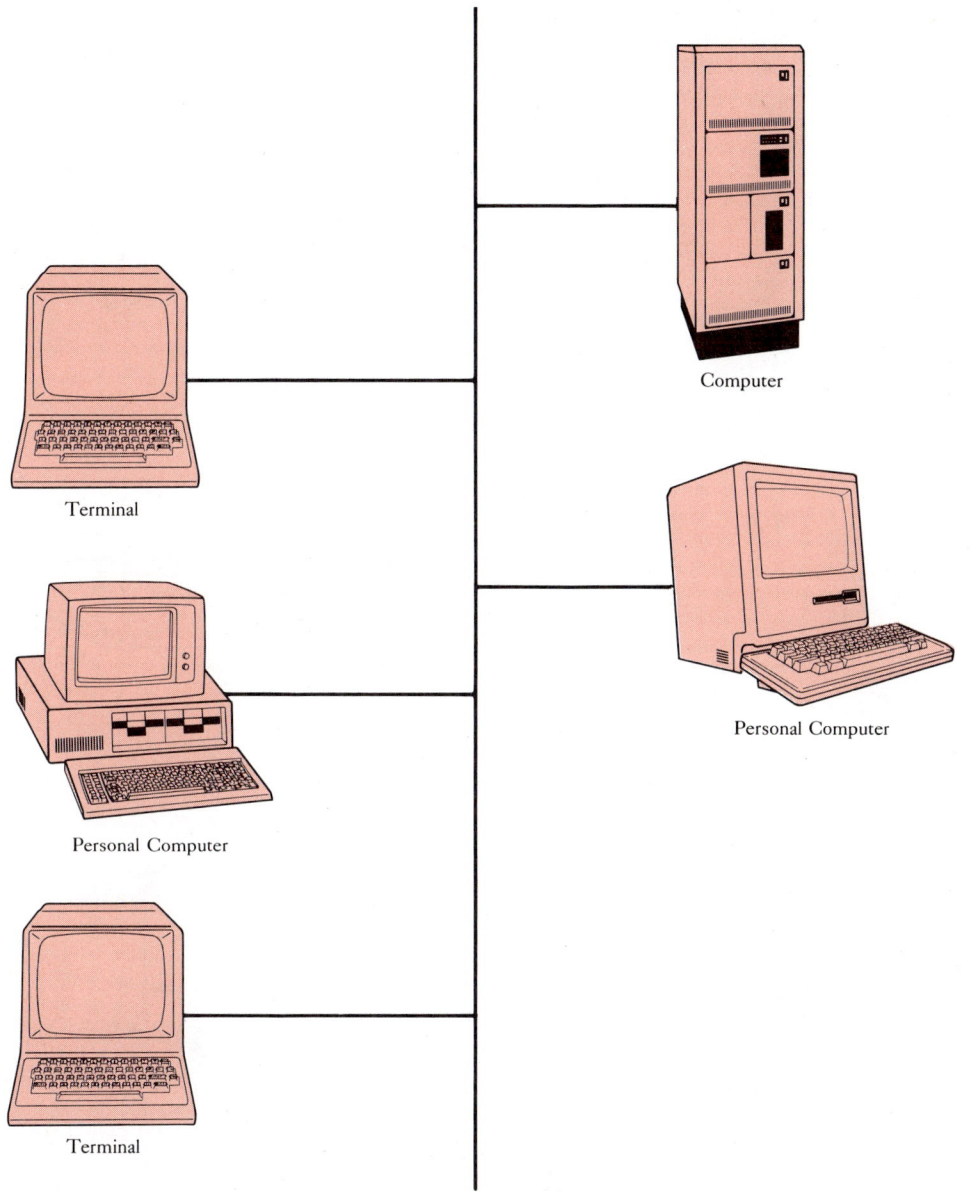

FIGURE 10.9. Bus networks have terminals or computers tapping into a central line. Each piece of equipment communicating on the line has a unique identification code.

computer—hence the name star (Figure 10.7). All data transmission or reception passes through the central computer, which routes it to the appropriate terminal or satellite computer.

Ring networks also combine multiple computers, but not to a central computer (Figure 10.8). Data in a ring network flow in one direction from station to station. When a station originates a message, it places an address with the message.

Ring network A network providing links between computers but not to a central computer.

336 PART II SYSTEMS DESIGN

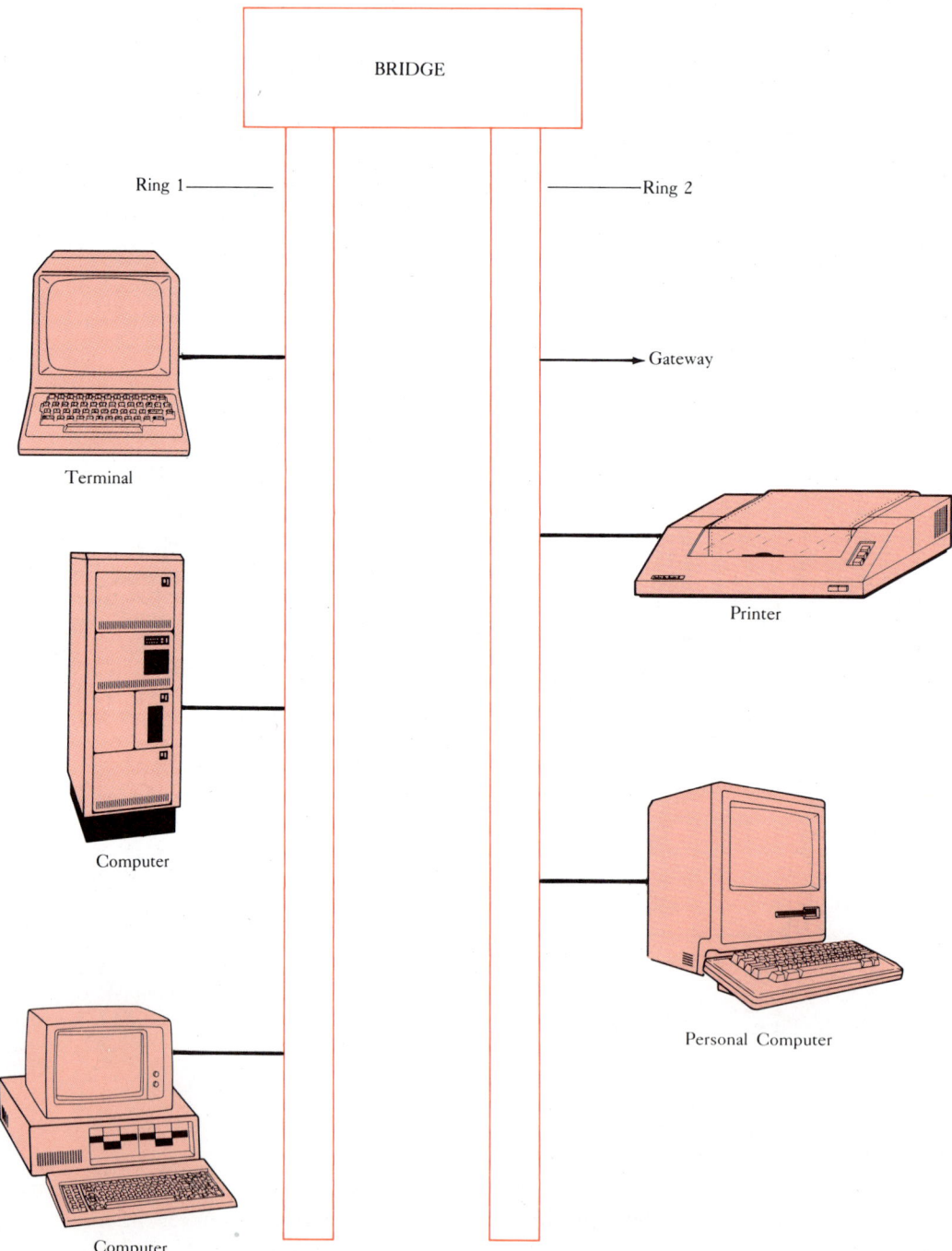

FIGURE 10.10. IBM's LAN (Local Area Network) uses the familiar ring structure. Multiple rings can be joined at bridges. Gateways permit access to outside computers or data communication links.

Incoming messages have their addresses checked, and if they match, they are accepted; if they do not match, the message is passed along to the next station. If a message progresses all the way around the ring without being accepted, it is erased when it returns to the originating station. Every station listens to and repeats every other station's message, resulting in low efficiency of the links. A failure by any station in the ring causes failure of the entire ring.

Bus networks can be likened to a water pipe projecting through an apartment building (Figure 10.9). Apartment dwellers tap on the pipe when they have a message to send. All dwellers can listen, but only one may tap at a time. If two dwellers tap at the same time, the messages become garbled. In 1980, Xerox, Digital Equipment Corporation, and Intel announced a commercial bus network, Ethernet. An Ethernet system can detect garbled data streams since devices listen as they send messages and stop sending when they detect interference.

IBM's **Local Area Network (LAN)** uses a ring network with token passing (Figure 10.10). A token is a stream of characters of fixed length and format that constantly flows around the ring. Tokens contain headers, trailer, priority, source, destination, data, and type. Headers and trailers are special characters that identify the beginning and end of the stream of characters. Source and destination give the origination and destination of the character stream. The data portion of the stream holds the specific message being sent from the originating device to the destination device. The type identifies whether the token is free or empty and a device can insert its message if the token is free and sends it on; if a token is filled, the device waits for the next free token. Rings are joined at bridges and have access to outside services via a gateway.

An analyst faced with a choice among the four approaches to networking should consider the users' needs. If all that is needed are data entry and processing at multiple sites with distribution at the end of a day to various remote users, a star network will work best. This is so because, with a star network, data for general use are sent to the host computer for later processing. On the other hand, if a number of users at different sites need access to updated data all day long, a ring will work best. This is true because, with a ring network, more than one data transmission can take place at the same time, keeping the whole system continuously current.

The analyst should consider the following six factors in measuring and evaluating networks:

1. *Capacity*—the number of users the system can support at each site simultaneously.
2. *Performance*—a measure of how quickly the system can respond to a user's request.
3. *Reliability*—the quality of the equipment in terms of how often it is likely to "crash" as a result of a hardware or software malfunction.
4. *Flexibility*—the capacity to add new components (terminals) without undue problems.
5. *Security*—protection against unauthorized access to data or interference among users.
6. *Resource utilization*—ability to keep track of time consumed by specific users and system failures.

Bus network A long line with computers or peripherals attached at various points.

Local Area Network (LAN) A medium-to-high-speed data communications system designed for intrabuilding use.

LANs and Distributing the Work Load

Since the dawn of the microcomputer revolution, unprecedented computing power has become accessible to vast numbers of office, factory, management, and data processing workers. These machines are proliferating throughout every level of virtually every organization and are creating a pressing need for users to be able to communicate with one another. Sending, receiving, and sharing information stored in computers and computerized devices has become an essential goal of analysts in all information-handling environments.

Fortunately, significant technological advances have increased our ability to link various computing resources, whether they lie across the office or across the nation. While research has long been conducted in the area of transcontinental communications, large organizations have increasingly demanded improved communications between offices in a building, between the floors of a building, or within a complex of nearby buildings; hence the current focus on what we call "LOCAL AREA NETWORKS."

A Local Area Network (or LAN) links hardware and software elements located within close geographic proximity. It may include the transmission not only of data, but of voice, video, and facsimile copies.

Unlike transcontinental hook-ups, which must rely on satellites or phone lines, a Local Area Network operates without the need to rely on external influences not under an organization's control. Furthermore, it costs less, is less complex, and does not have to accommodate the restrictions in long line communications. This independence has spawned a new technology, the Common Bus Local Area Network, where processing devices link to a common medium, such as a coaxial cable or standard multi-strand cable. While the components, or "nodes," in such a network are fully connected with one another, their transmissions do not pass from node to node. Rather, messages are broadcast to the entire "net" simultaneously without the intervention of the nodes.

The Common Bus approach eliminates the need for a central controller, thereby making it possible for communications to rely only on the nodes directly involved

Distributed data systems have several advantages. They are obviously useful in situations where local users need access to an organization's centralized data. At election time, for example, each city, county, and borough needs voting information about each precinct; the state needs summary data only. In addition, they send and receive data quickly and efficiently. Users at local sites have immediate access to their own data—they do not have to wait for processing by a central computer, and they can obtain overall company data in a timely fashion.

Distributed systems, however, have several disadvantages. They need extra hardware for data communications. The compatibility of machines from different manufacturers is another problem. For example, IBM and Burroughs computers store data in EBCDIC; HP, NCR, and Data General use ASCII. For these machines to talk to each other, software must be acquired to permit the machines to exchange data.

with a given message. Such an approach offers significant advantages over any other scheme for the configuration of a Local Area Network, because workers wishing to communicate with one another can do so independently.

In summary, an organization obtains the following benefits from the Bus Network Architecture:

Dynamic reconfiguration of nodes and functions (reconfiguration without interruption of service)
Reliable operation resistent to malfunction (no exposure to net controller failure)
Relatively low cost of installation (a fraction of the cost of other methods)
Very fast transmission of messages (200 kb to 10 megabits per second)

The Common Bus approach will undoubtedly dominate the Local Area Communications field through the 1980s and into the 1990s. The microprocessor products, whose prices will rapidly drop, will routinely offer LAN capability, thereby making the new technology available and affordable to an increasingly wide range of organizations.

The ideal LAN provides any user access to any computer resource in the LAN, regardless of location. To achieve this goal, analysts must ensure that communication among different devices which demand different protocols remains completely transparent to the user. In addition, analysts must make sure the organization can install new resources with no more effort than simple device definition and physical attachment to the network.

As organizations scramble from the Information Age to the Communications Age, they will be looking for ever greater speed, reliability, and flexibility in the Local Area Networks. For the time being, the Common Bus will provide those results.

Source: dataBASE, vol. 7, no. 3, pp. 5–8, June 1983. Used by permission.

A third drawback is that the same data may be stored at separate locations, which results in additional costs. For example, the names and addresses of registered voters are stored in county government computers as well as in state computers. When a voter moves, both computers must have their files updated.

Ever since the minicomputer arrived on the scene in the early 1970s, and now with the ever-widening use of personal computers, analysts have debated the pros and cons of centralized versus decentralized data processing: costs, efficiency of operation, responsibility, coordination, communication, security, local control, and decision making. More and more, comparisons of costs and benefits indicate that the 1980s will see increasing reliance on distributed or network systems.

Distributed systems will gain increasing favor as we solve attendant software and hardware problems: minimizing data communications costs, providing adequate security for data communications links, and developing methods to tie the computers to each other.

HARDWARE AND SOFTWARE ACQUISITION: RFP AND RFQ

By this point in the systems cycle, the analyst has chosen the system mode and, if necessary, a link joining terminals or remote job-entry stations to the computer. If the system requires some new hardware, software, or even an entire computer system, the analyst must set about the task of obtaining it.

Analysts may approach equipment acquisition in two ways: by compiling hardware lists or by specifying functional requirements. A hardware list is simply a shopping list based on exact specifications in the following areas.

1. *Memory*—measured in bytes.
2. *Disk storage*—measured in millions of bytes.
3. *Printer speeds*—measured in lines per minute.
4. *Numbers and types* of terminals.
5. *Programming languages*—COBOL, Pascal, FORTRAN, and BASIC.
6. *Operating system*—disk based with a certain number of concurrent operations, data-base managers, utility sorts or merges, file copy routines, editors.
7. *Applications software*—accounting (payroll, inventory, order entry, general ledger), electronic spreadsheet, word processing, electronic mail, and so on.

Hardware lists are distributed to vendors such as IBM, Burroughs, or Digital Equipment Corporation, which can specify the equipment in their product line that fulfills desired requirements.

If analysts elect to follow a functional requirements-specification approach, they list necessary functions and ask vendors to recommend appropriate equipment. Functional descriptions look like hardware lists, except that they omit exact measurements.

1. *Memory*—sufficient to support three active data-base systems, two concurrent COBOL compilers, 10 other program-development terminals, and 12 data-entry terminals.
2. *Disk storage*—sufficient to store all operating system programs, 1000 COBOL source programs, the data for a large accounting system, and room to expand 25 percent in each of the next 5 years.
3. *Printer*—fast enough to print all the outputs of a large accounting system within a certain number of hours.
4. *Terminals*—memory able to store at last five pages of program listing. Must operate in block or forms mode, and have a 10-key pad, tiltable screen, and separate keyboard.
5. *Programming languages*—COBOL, Pascal, FORTRAN, and BASIC.
6. *Operating system*—disk based with at least 65 active terminals, a relational data-base manager, full screen text editor for program development, data dictionary, Query language, DBMS, program generator, and spooler.
7. *Application software*—fully integrated accounting system including accounts payable, accounts receivable, payroll, inventory, and general ledger.

This second approach, while it may appear vague, nevertheless allows vendors a lot of flexibility. In using the first approach, analysts calculate system requirements based on accurate and up-to-date knowledge of the state of the art in hardware or software. However, vendors may have new and exciting products in the wings—products that even the most astute analyst may not know about. The functional requirements approach, therefore, can augment the analyst's ongoing education. Vendors know their hardware and software best and often can quickly select the right amount of memory or disk space to meet the purchaser's need. For example, one vendor's equipment may require 1 megabyte of memory for a given application, whereas another may need 2 megabytes to support it. If the analyst specifies 1 megabyte of memory, the second vendor's hardware will not suit the application.

In certain potentially costly applications, the analyst may expand either the hardware list or the functional requirements to include more detail. In any case, the lists form the basis for either a **request for quotation (RFQ)** or **request for proposal (RFP)**. RFQs usually accompany hardware lists and RFPs accompany functional requirements specifications.

To evaluate the vendors' responses to an RFP or RFQ, most analysts employ a point system, such as the following.

1. The analyst composes a list of major categories (such as software capabilities, languages, and data-base requirements) with each category weighted according to its importance to the system. For instance, a newly formed organization with no established system will not need to convert data from one form to another, so it will not assign many points to conversion. And an organization that demands only new software need not mention hardware requirements. Figure 10.11 illustrates the major categories and the weight assigned to each one.
2. Next the analyst breaks each category down into its components, with each component carrying a percentage of the entire category's weight. For example, preinstallation provisions include three components: computer time necessary, training of operations staff, and training of programming staff. The total weight of this category is 4.3, with 2.3 allotted to computer time and 2.0 to training.

Request for quotation (RFQ) A written document on which analysts write a description of functional needs.

Request for proposal (RFP) A written document on which the analyst specifies hardware needs.

Criteria	Total Weighting
1. Management summary	6.0
2. Hardware configuration	25.0
3. Software capabilities	21.5
4. Demonstration	12.5
5. Space, electricity, and air conditioning	9.0
6. Conversion	15.9
7. Support services	7.8
8. Preinstallation provisions	4.3

FIGURE 10.11. *A sample vendor evaluation point system, evaluating responses according to eight different criteria. The allocating of points occurs prior to sending the RFQ. Each major category is broken down into subcategories on the bid evaluation form.*

BID EVALUATION FORM
Software Capabilities

VENDOR NUMBER: 45
EVALUATOR: 18

Evaluation Component	Score	×	Weight	=	Points
1. 1974 ANS COBOL	5		3.5		17.5
2. FORTRAN-77	5		1.5		7.5
3. Ada	0		1.5		0.0
4. Pascal	3		1.5		4.5
5. Data-base manager	4		2.5		10.0
6. Text editor	5		2.0		10.0
7. Multiuser operating system	4		3.0		12.0
8. Data communications software	4		2.5		10.0
9. System software described in detail	4		1.0		4.0
10. Software operational for at least 1 year	5		1.0		5.0
11. Utility systems software	4		1.0		4.0

Other Considerations

12. Program generator	2		0.5		1.0

Note: Two bonus points awarded for full screen editor and not a line editor.

Total weighting	21.5	
Subtotal of points		84.5
Bonus award		2.0
Total points		86.5

Notes: Points awarded as follows; 0 = unacceptable; 1–3 = acceptable; 4–5 = outstanding. Bonus point not to exceed 15 percent of subtotal points.

FIGURE 10.12. Bid evaluation form for a software vendor's response (each vendor is numbered). Each component is rated by an evaluator (assigned an evaluator number) on a scale from 1 to 5, then multiplied by its weighting to receive the points. Points are totaled and a bonus system may be included to add extra value if the analyst feels it is appropriate.

3. The analyst completes a bid evaluation form by category for each vendor, including all components and their weights. If there are eight categories, there will be eight forms for each vendor.
4. The analyst rates each vendor's ability to provide the needed component on a scale from 1 to 5. These component ratings are multiplied by the respective weights, and their products totaled. (See Figures 10.12 and 10.13.)
5. To strengthen the objectivity of the process, several people may evaluate the vendors' responses. Individual scores can be added together to create a composite score, which, in turn, is divided by the cost of the vendor's equipment or software to obtain a point cost ratio (Figures 10.14). In this case, if we were to select the vendor solely on the basis of cost, we would pick vendor number

CHAPTER 10 PROCESS DESIGN AND ACQUISITION OF HARDWARE AND SOFTWARE   **343**

```
              BID EVALUATION FORM
              Preinstallation Provisions

              VENDOR NUMBER:    45
              EVALUATOR:        18
```

Evaluation Component	Score	×	Weight	=	Points
1. Computer time provided	3		2.0		6.0
2. Training of operations staff	3		1.0		3.0
3. Training of programming staff	2		1.0		2.0
4. Training manuals provided	3		.3		.9

Total weighting	4.3	
Subtotal of points		11.9
Bonus award		0.0
Total points		11.9

Notes: Points awarded as follows; 0 = unacceptable; 1–3 = acceptable; 4–5 = outstanding. Bonus point not to exceed 15 percent of subtotal points.

FIGURE 10.13. Bid evaluation form for preinstallation provisions.

Vendor Number	Total 5-Year Cost	Points	Point Cost Ratio
9	$630,531	313.85	2009
27	416,867	313.36	1328
45	565,999	302.47	1871
51	646,095	324.29	1992
63	539,769	338.24	1596

FIGURE 10.14. Completed evaluation of each vendor's response, showing costs, points earned, and a point cost ratio. Vendor names are confidential, so that name brands will not affect the evaluation process.

27; if we were to base our decision on points, we would pick vendor 63. However, since the point cost ratio evaluates both capabilities and costs, vendor 27 clearly wins the order.

Having chosen a vendor, the organization begins negotiating a contract. Although most vendors offer standard contracts, an organization purchasing the new equipment should have its purchasing and legal departments, controller, and the manager of computer services scrutinize any contract closely before it is signed.

CASE STUDY

Fleet Feet's Online Accounts Payable System and an RFQ for New Terminals

Frank Pisciotta has always looked forward to the process design stage because it marks a major milestone. Once design activities conclude, development can begin.

Fleet Feet's Hewlett-Packard 3000 computer will support a maximum of 32 simultaneous terminals in batch, online, or real-time modes; the system also has word processing and electronic mail capability. The company currently has nine terminals: four in the accounting office, one each in the president's office and the two vice-presidents' offices, one in the main Fleet Feet store, and one in the computer room.

All current Fleet Feet systems are real time. Data are entered at the terminals and validated and verified, then they instantly update the data base. Data are then available to management and the accounting office for inquiry purposes.

Analyst Frank Pisciotta proposes the same mode for the new accounts payable system. The bookkeeping department will enter data on terminals, and users (management, bookkeeping, and inventory) will have access to the data. The AP data base will allow management and inventory people to read the data, but only the bookkeeper can update the data.

The new AP system design provides for two terminals in the warehouse. Frank Pisciotta decides to contact terminal vendors by telephone, asking each to send sales brochures and price quotations within 2 weeks. Since the investment is fairly small, probably less than $6000, he decides not to take the time to write a formal letter asking vendors to specify terminal cost and capabilities. Besides, Frank has worked with these vendors before and feels confident their brochures and telephone responses will permit him to complete his evaluation form without any difficulties (Figure 10.15).

In the next 2 weeks, Frank receives extensive descriptive literature and fills out an evaluation form for each vendor's terminal. Many factors will influence the decision he will make after completing a bid evaluation form. Since vendors offer a variety of special enhancements, Frank includes an optional features section, where he can award additional points for capabilities exceeding his minimum standards. With the forms completed, Frank studies each terminal, rejecting the HP 2622, DEC 102, IBM 3270, and ADM 22 because they do not meet minimum standards (Figure 10.16). Frank eventually chooses an HP 2624 in spite of its higher price. It has a higher point cost ratio, which makes it the best buy.

CASE STUDY

FLEET FEET TERMINAL EVALUATION FORM

Vendor: Model Number:
Price: Monthly Maintenance:

Minimum Requirements	Status (Y or N)
1. Hewlett-Packard compatible	
2. Forms or block mode capability	
3. Programmable function keys	
4. Delivery within 60 days	
5. 8 KB of memory	
6. Separate 10-key numeric pad	
7. Dual intensity, flashing, and underline	

Optional Features	Score × Weight = Points
8. Additonal memory	2.0
9. More than eight programmable function keys	2.0
10. Printer interface built-in	1.0
11. Expanded keyboard	1.0
12. Lowercase character set	1.0
13. Detached keyboard	2.0
14. Nonreflective screen	2.5
15. Green phosphor monitor	0.5
16. Ergonomic design	2.5

Total weight for optional features   13.5
Subtotal for optional features
Minimum requirements met   300.0
Total points
Point cost ratio

Notes: Points awarded as follows; 0 = unacceptable; 1–3 = acceptable; 4–5 = outstanding.

FIGURE 10.15. Frank Pisciotta's bid evaluation form for terminals. Minimum requirements must be met, or Frank will eliminate that product from consideration.

CASE STUDY

FLEET FEET TERMINAL EVALUATION FORM

Terminal Name	Price	Minimums Met	Optional Features	Total Points	Point Cost Ratio
Visual 102	$1825	Yes	65	365	5.5
HP 2622	1907	No	—	—	—
HP 2624	1875	Yes	75	375	5.0
DEC 102	1475	No	—	—	—
IBM 3270	3500	No	—	—	—
ADM 22	875	No	—	—	—

FIGURE 10.16. Fleet Feet's responses on new terminals are evaluated by Frank, who chooses an HP 2624.

WORKING WITH PEOPLE Too Many Controls

Dick Couzens, an analyst at Solar Systems, Inc., a supplier of electronic components for missile systems, prided himself on his control designs. He loved this part of his job and had many chances to display his expertise because Solar, as a new operation, required many sophisticated new systems. For every new system, Dick designed sight verification and check digits for all key fields and validations for all other data items. When the manager of data-entry operations, Dwight Odom, scheduled an afternoon meeting to discuss data-entry routines for a new payroll system, Dick reviewed the system's data-collection, verification, and validation requirements to make sure all fields checked as entered. He expected to hear high praise for his many tight controls and eagerly anticipated the meeting.

When Dick strolled into the manager's office, however, he received a shock. Dwight barely exchanged pleasantries, and then exploded with a list of complaints: "This data-entry system is a time-consuming pain in the neck! It's too complex! Training wastes the data-entry staff's time, the prompts confuse everyone, and the system's responses make people feel stupid. Whenever someone makes a mistake, 'Try again, turkey,' flashes on the screen! And why does the screen go blank for 10 or more seconds for no apparent reason?"

Dick had tried to inject a little humor into the system but no one seemed to be laughing. What had he done wrong? Weren't his controls the best possible?

Overwhelmed by Dwight's outburst, Dick suggested they watch one of Dwight's staff enter data, but Dwight insisted that Dick enter a few transactions himself. "You play data-entry operator for 15 minutes and then come back and tell me how you like it!" Reluctantly Dick agreed. After 20 minutes on a terminal, he began to understand Dwight's complaints.

After a transaction entered the system, the new IBM computer took more than a minute to return the cursor for the next entry because it was so busy performing a multitude of class, range, check-digit, and presence tests. Such a delay could waste 50 percent or more of an operator's time, something clearly unacceptable to management. Despite the fact that he had designed the system himself, it took Dick over 10 minutes to become comfortable with it. When "Try again, turkey!" lit up his screen for the third time, he knew why the operators felt patronized.

Dick walked slowly back to Dwight's office to apologize. "I violated one of the laws of design: KISS—Keep It Simple, Stupid. I guess I got carried away with controls."

Dwight smiled. "Too many controls can be worse than none at all."

Dick agreed to overhaul the data-collection system and retreated to his office. Walking down a corridor, he stopped at analyst Susan Caldwell's office. To her door she had tacked a poster with the caption, "Keep It Simple to Make It Faster." "Well," Dick promised himself, "I'll never make that mistake again."

SUMMARY

Computerized information systems can operate in a variety of modes. This chapter examined these modes of operation (batch, online, real time, and distributed) and the linking of terminals to computers (completely connected, ring, star, or bus networks). We learned how each mode affects data input, processing and output. We also saw the capabilities of microcomputers as an alternative to terminals and centralized computers.

Batch processing systems dominated the early age of computerization and are still used today. In a batch environment, an organization collects and processes data in groups. All data are prepared, on tape or diskettes, and then read by the computer. After processing, information becomes available to users, usually in the form of printed reports. Batch processing is appropriate where the delay between data collection and processing has no significant impact on an organization's decisions.

Online systems may also collect in batches, but users have terminal access as soon as processing ends. They can examine the results, but they cannot change data. Online systems became popular in the late 1960s and are appropriate where users need immediate access to data but can wait for data entry and processing to be done at night or during off hours.

Real-time or interactive systems process the data as the data enter. Data are always as current as possible so the user sees up-to-date information. Real-time systems require a computer that can support many terminals simultaneously.

Some systems may demand a remote job-entry station. These devices permit users to collect data, transmit the data to a central computer at high speeds, and receive printed reports on their high-speed line printers.

The move from batch to online and real-time systems necessitates communication between remote devices. Organizations can link two or more computers in a ring or star network. Distributed processing will gain in popularity by the late 1980s and 1990s.

With control systems established, input and output designs completed, and the schema for the data base written, the analyst evaluates needed equipment. Descriptions of needs appear in a request for quotation (RFQ) or a request for proposal (RFP), which goes to potential vendors. The analyst weighs the vendors' responses by balancing features and costs, usually through an evaluation form.

Having established the mode of system operation, the placement of terminals and computers, and necessary data links, the analyst is ready to move to the next phase of the systems process, program definition and module design.

NEW TERMS

Batch processing
Bus network
Completely connected network
Distributed processing
Local Area Network (LAN)
Online processing
Real-time processing
Remote job-entry device
Request for Proposal (RFP)
Request for Quotation (RFQ)
Ring network
Star network

QUESTIONS FOR REVIEW AND DISCUSSION

Questions appear in three categories. You can find the answers to the A group in this chapter. The B group requires you to apply the material presented here, whereas the C group necessitates investigation or research.

A.1. List and briefly describe the four modes of system operation.
A.2. List two types of networks.
A.3. What do the following abbreviations stand for?
   a. RJE
   b. RFP
   c. RFQ
   d. LAN
A.4. List three advantages of online systems.
A.5. What are the six measures used to evalute the capabilities of a real-time system?
A.6. What are two factors to be considered when choosing a terminal?
A.7. What are the factors to be considered when choosing a new computer system?
A.8. What are the advantages of a batch processing system?
A.9. What are three disadvantages of a real-time system?

B.1. What modes have systems typically followed in their development life cycles?
B.2. What is the difference between a centralized and a decentralized data processing system?
B.3. Diagram three types of networks.

C.1. Why would a firm use a remote job-entry device?
C.2. What are the percentages of the various components of Figure 10.11? When would these percentages change?
C.3. Design a bid evaluation form similar to Figure 10.15, which would be useful in analyzing a line printer.
C.4. Design a bid evaluation form similar to Figure 10.15, which would be useful in analyzing a disk drive.

REFERENCES

Jerry Fitzgerald and Tom S. Eason, *Fundamentals of Data Communications*, Wiley, New York, 1978.

Kenneth Sherman, *Data Communications: A User's Guide*, Reston Publishing, Reston, Va., 1981.

CHAPTER 11

Program Definition, Module Design, and the Design Review

- Goals and Preview
- Program Definition
- Module Design
- **Guidelines for Planning Modules**
- Language Considerations
- Package Design and Program Specifications
- Design Walkthrough
- Design Review
- Case Study: Design Specifications for Fleet Feet's Accounts Payable Voucher-Check Printing Program
- Working with People: A Program Specifications Walkthrough
- Summary
- New Terms
- Questions for Review and Discussion
- References

GOALS AND PREVIEW

After reading this chapter you should be able to:

1. Define a module and state the rules for writing one.
2. List the three control structures.
3. Explain four criteria for good module design.
4. ~~Cite three widely used programming languages.~~
5. Describe the components of a program specification.
6. Explain the purpose of a design review.

In the earlier phases of system design, the analyst designed the outputs, data base, inputs, and processing mode, which may require new hardware (Figure 11.1). Now the analyst defines necessary programs, detailing for each its purpose, special processing requirements, and validation techniques (Figure 11.2).

During program definition one considers the priority, purpose, and function of each program, and then examines them in detail, dividing each into specific tasks or modules whose functions can be depicted with pseudocode, Warnier-Orr diagrams, or one of the other program logic tools (see Chapter 4 for a review of these techniques and tools). After identifying modules, the analyst concentrates on the details of each module, paying especially careful attention to their relationships to one another. The entire collection of program definitions and modules completes the program specifications.

With modules specified, the analyst can consider which language the system might use most effectively. For many organizations only one language may be available, even though it might not be appropriate for a new application. In a few organizations, many languages may be available, thus allowing the analyst some latitude in selecting one to suit a particular program.

Finally the entire system undergoes a walkthrough and a design review. Similar in scope and purpose to the analysis review, a design review gives management and users a final opportunity to scrutinize the proposed system, making sure it fulfills all needs and involves acceptable costs and schedules. Even at this late date, management can reject the system or request design modifications before authorizing the third phase of the systems process, development.

Program specifications and a programming language choice join output, data base, input, and process design to complete the design activities of the systems process.

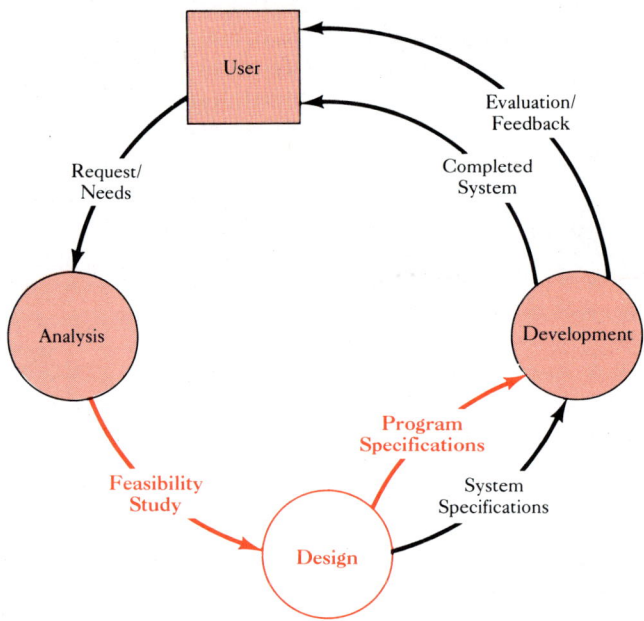

FIGURE 11.1. Part of the design phase in the systems process is concerned with completing system specifications.

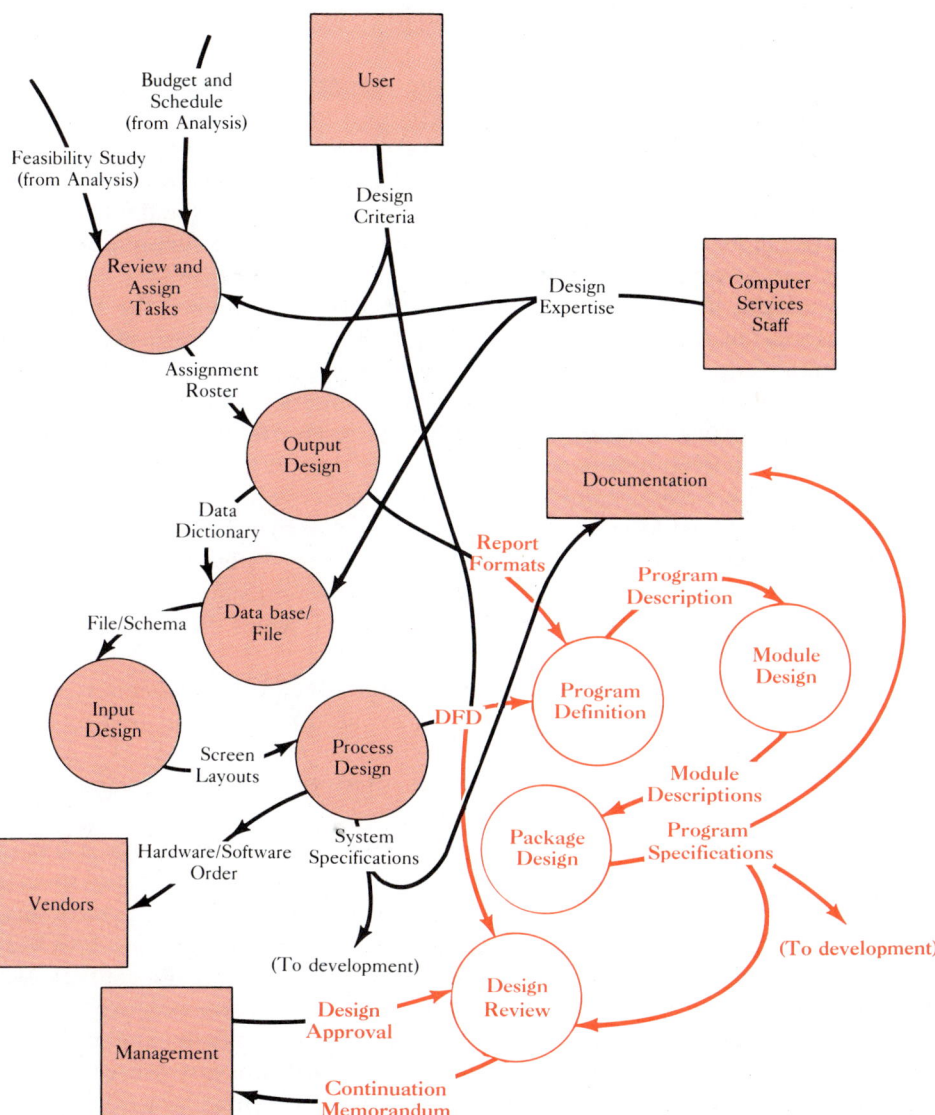

FIGURE 11.2. *Leveling of design reveals nine major tasks. The final four require the analyst to specify necessary programs that management can review before development.*

PROGRAM DEFINITION

Program definition A detailed description of each individual program in the system.

Program design A detailed description of every task a program performs.

Until a decade ago, many analysts often ignored **program definition** because their training centered on the details of advanced programming rather than on the more general aspects of program definition and the management of programmers. To complicate the situation further, the lack of structured tools made it difficult to observe the general flow and logic of programs. Now structured methodology has made it possible for modern analysts and programmers to speak a common language when it comes to **program design.**

Determining a list of programs requires the analyst to review the data-flow diagram (DFD) of the proposed system as well as the data dictionary and report formats (from output design), the schema or file layouts (from data-base or file design), and data-collection screen formats (from input design). In this review the analyst isolates the place where each report is produced by the system and examines each process activity (represented by a circle in the DFD) as a potential program(s). Since the circle or bubble in the DFD portrays a transformation of data, it tells us the potential location of a program. Just how many programs will there be? A study made in the early 1980s shows the average system produces 26 user reports and has 55 programs requiring 23,000 program source statements. However, today's users are demanding more from their systems, which could translate into more programs.

Each report may be a candidate for a program. Reporting programs typically take data from a file(s) and display the data to the user, accumulating totals or counting the number of occurrences of a certain event (Figure 11.3a). A reporting program in a DFD appears as a circle with an arrow pointing out to a named box, in this case the customer.

Often the reporting program needs to sort or merge data files before showing them to the user. Sorting and merging activities may require the analyst to specify that a set of utility routines be written using the computer's operating system sort/merge functions. Some programming languages, such as COBOL, have SORT and MERGE verbs the programmer can embed in the reporting program to achieve these functions. Whether there is a need to sort or merge will be uncovered as the analyst further defines the circle in the DFD. If we want our statements in order by Zip code (for cheaper bulk mailing postal rates), the customer record file will have to be sorted by Zip code before printing statements.

Circles may call for a data-collection program complete with verification, validation, and other data checking methods. These types of programs are easily identifiable from the DFD since we see a box (source or sink of data) and an arrow pointing to the bubble (Figure 11.3b). Here we see the need for a program to collect customer payment data, check the data for errors, and update a customer master file.

Still other circles in the DFD may call for programs that take data from a variety of origins and pass the data on in a converted form (Figure 11.3c). This DFD shows "commission-note" being created by Record payment and being used by Pay commission to produce the Commission report. This third type of potential program typically shows a named arrow between two circles.

Regardless of the types of arrow, box, and circle patterns found in the DFD, the analyst inspects each and decides on the program(s) needed to accommodate

CHAPTER 11 PROGRAM DEFINITION, MODULE DESIGN, AND THE DESIGN REVIEW **355**

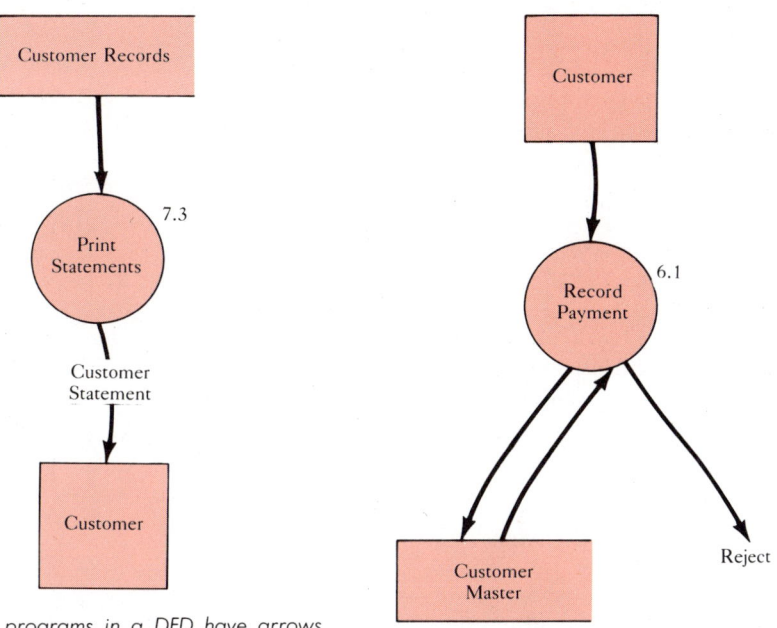

(a) Reporting programs in a DFD have arrows from circles leading to boxes or sinks.

(b) Data-collection programs in a DFD show data from a source to a circle or processing activity.

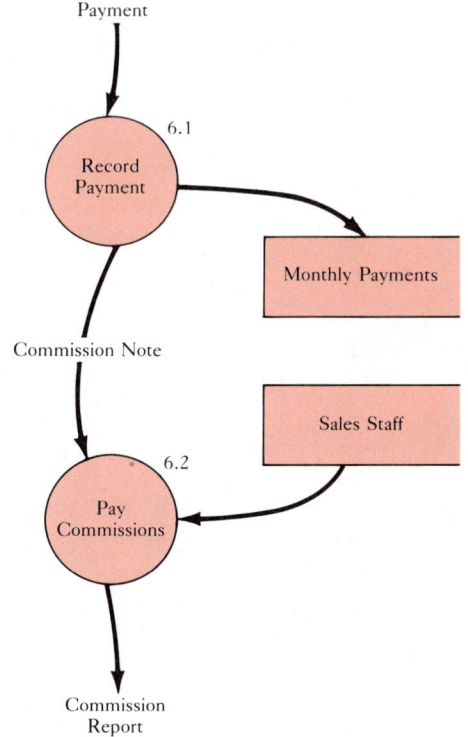

(c) Programs internal to a system exhibit themselves as data flows between circles or process symbols.

FIGURE 11.3. Data-flow diagrams show the analyst where the system's programs are likely to be found.

```
PROGRAM NAME:  PRNTSTMNT                              Date: 12/11/84
System:        Accounts Receivable                    Process Number: 7.3
Definition: For each customer in the CUSTOMER RECORDS file, print a
            statement showing the customer's balance.
```

(d) For each program in the system, the analyst defines the program, referencing its location back to the data-flow diagram.

FIGURE 11.3 *(Continued)*

the activities. The inspection yields a list of programs the system will need. Since each circle in the DFD contains an identifying number (7.3 in Figure 11.3*a*), we define each program by its number and state its purpose (Figure 11.3*d*). The collection of program definitions completes this phase of design and is input to the next phase, module design.

MODULE DESIGN

Module A discrete or identifiable single-function program unit.

After defining the overall nature of each program, the analyst divides the program into **modules**, or groups of actions with identifiable beginning and ending points. When building a house, one must allow certain events to occur in order: the foundation precedes framing, roofing, and plastering. Each event forms one distinct module, and each relates to the others in a certain way.

Modules

Just as the collection of construction modules results in a house, groups of program modules result in the ability of a computer to process data. In Figure 11.4 three

```
OPENING-MODULE.
   Open File-A, File-B, File-C.
   Read a record from File-A and File-B.

PROCESSING-MODULE.
   REPEAT until (File-A and File-B and File-C are High-Values)
      Test the keys in File-A and File-B
             Keys are equal:  Write record from File-A to File-C.
                              Write record from File-B to File-C.
                              Read a record from File-A and File-B.
            File-A < File-B:  Write a record from File-A to File-C.
                              Read a record from File-A.
                 Otherwise:   Write a record from File-B to File-C.
                              Read a record from File-B.

   CLOSING-MODULE.
      Close File-A, File-B, File-C.
      Stop program.
```

FIGURE 11.4. *Modules are groups of actions with identifiable beginning and ending points.*

modules merge files A and B into a new file, C. Notice how module names reflect their actions: Opening, Processing, and Closing.

To conform to the standards of structured methodology, each module should contain only one entry and one exit point (see Chapter 4). In Figure 11.5a you will find a correct module, and in Figure 11.5b an incorrect one. Single entry and exit points organize logic to flow smoothly from beginning to end without detours or interruptions.

The lengths of modules are important. Since most terminal screens can display only 24 lines, many applications benefit from limiting module length to that size. In cases where one does not use terminals to write programs, one may limit module size to a page, perhaps 60 lines. A service bureau in Florida limits module length to one printed page. Sizes of more than a terminal screen or a page will make it difficult for the programmer to remember what took place at an earlier point in the module.

Modularizing a program simplifies the process, making it easier for the analyst, programmer, and user to understand. Furthermore, by examining each module separately, analysts can more readily spot potential errors in logic. The earlier one can locate problems, the better.

```
Read employee's data.

REPEAT until (there is no more data)
     Calculate Gross-pay.
     Subtract Deductions from Gross-pay.
     Calculate Year-to-date totals.
     Calculate-Quarter-to-date totals.
     Print employee's paycheck.
     Read employee's data.

END {REPEAT}.
```

(a) A correct module.

```
Read-Again.
     Read employee's data.
     IF (there is no more data)
          THEN
               GO TO Print-Final-Total-Routine.
     Calculate Gross-pay.
     Subtract Deductions from Gross-pay.
     Calculate Year-to-date totals.
     Calculate-Quarter-to-date totals.
     Print employee's paycheck.
     GO TO Read-Again.
```

(b) An incorrect module in which the GO TO statement permits two exits.

FIGURE 11.5. *Modules should contain single entry and exit points.*

Control Structures

Control structure A pattern for building the logic of a computer program.

Sequence A control structure where each action follows the next in a linear fashion.

Decision A control structure where the next activity depends on a conditional test.

In 1964, Corrado Bohm and Giuseppe Jacopini demonstrated that one can build all programs from three **control structures**: sequence, decision, and repetition or iteration (Figure 11.6). Regardless of a system's complexity or the advanced techniques required to program it, all programs are combinations of only these three structures.

Sequence structure (Figure 11.6a) describes a series of actions that follow one another in a linear fashion (Figure 11.7). The payroll calculations leading to the printing of a paycheck exhibit a typical group of sequential operations.

Decision control structure describes a situation in which an action depends on which of two conditions has been met (Figure 11.6b). For example, suppose you wish to determine whether a customer deserves a discount on a purchase (Figure 11.8a). You determine two conditions, good or not good, that govern your decision about awarding the discount. If the customer falls into the "good" category (makes many purchases and pays promptly), the customer gets a 2 percent discount. If the customer falls into the "not good" category (no credit history or chronically late payments), the customer does not get a discount. We also call the decision structure IF-THEN-ELSE, because many computer programming languages use these words to implement such decision making.

Decision control structures can involve more than one action based on a given

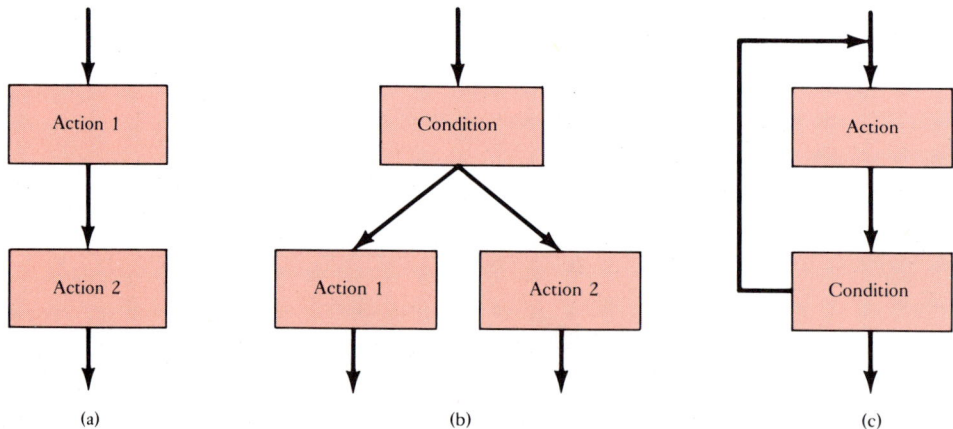

FIGURE 11.6. Bohm and Jacopini postulated three control structures at an international computer science conference in 1964.

```
Calculate Gross-pay.
Subtract Deductions from Gross-pay.
Calculate Year-to-date totals.
Calculate Quarter-to-date totals.
Print employee's paycheck.
Read employee's data.
```

FIGURE 11.7. Sequence control structure requires that each action occur in a set order.

condition. For instance, water utility customers pay one rate until they use enough water to qualify for a lower rate. The ultimate bill could include two separate totals, one for the amount consumed prior to meeting the condition, and one for the amount consumed thereafter (Figure 11.8*b*).

Since writing a long series of IF-THEN-ELSE commands can become cumbersome, we sometimes resort to a modified decision structure, **case**, which can depict complex logic. Consider newspaper classified advertising where the rates depend on the number of times an ad will run (Figure 11.8*c*). The case structure makes complex logic easy for programmers and users to follow. But although case offers an alternative to writing long or involved IF-THEN-ELSE, it is still a decision control structure.

Case Control structure supplementing IF-THEN-ELSE.

```
IF (the customer type is 'good')
    THEN
            Set the discount percent to 2%.

    ELSE
            Set the discount percent to 0%.
```

(a) In the decision control structure above, action depends on one condition.

```
IF (Amount-of-water used is less than 1500 cubic feet)
    THEN
        Amount-owed equals $2.00 plus 0.015 times the
            Amount-of-water.
        Add Amount-owed to Total- owed-by regular customers.


    ELSE
        Amount-owed equals $1 plus 0.01 times the
            Amount-of-water.
        Add Amount-owed to Total-owed-by-large-users.
```

(b) This decision control structure allows two actions based on one condition.

```
IF (Number-of-times the ad is run is)

            1 or 2:  Set Rate to  $3.75.
                 3:  Set Rate to   2.96.
            4, 5, 6: Set Rate to   2.33.
            7, 8, 9: Set Rate to   1.68.
    10 through 13:   Set Rate to   1.54.
    14 through 29:   Set Rate to   1.49.
         Otherwise:  Set Rate to   1.43.
```

(c) Case control structure allows one of many actions, depending on numerous conditions.

FIGURE 11.8. Decision control structure tests whether a predetermined condition exists, and then acts accordingly.

Repetition Control structure permitting looping or iteration.

Looping A control structure that allows a set of tasks to be repeated.

Iteration A control structure that allows repetition as long as a condition remains true.

Decomposition The continual breaking down of a task into its elementary components.

Refinement Dividing a task into more specific details.

Repetition (**looping** or **iteration**) provides the third control structure (Figure 11.6c). In certain programming languages, WHILE-DO or REPEAT-UNTIL statements allow for an action to repeat a specific number of times or until certain conditions arise. In Figure 11.9 six actions relating to each employee's data repeat until the last employee's data have been processed, at which point the repetition ceases.

Decomposition and Refinement

Sue Wurst, analyst for Class Video, decides the company's new video tape rental system needs modules to initialize the file, add a new tape, change a tape, delete a tape, print the tape file, post a rental, and terminate the program. First she draws a context data-flow diagram (Figure 11.10a) and from it develops a series of data-flow diagrams for each module the system needs (Figure 11.10b).

One can decompose most systems into modules, each of which may be decomposed into submodules, which one can sometimes level or refine even further. For example, a West Coast college's online registration system developed 130 modules from the original three, requiring over 9000 COBOL statements. We call the breaking down of a system into its elementary components **decomposition** or **refinement**.

For Class Video Sue goes on to develop a main selector menu screen (Figure 11.11) for displaying the date of the last file update and prompting the operator to select a desired operation. Sue's screen design contains the seven modules: Initialize, Add, Change, Delete, Print, Post Rental, and Terminate. Since Sue anticipates that Class Video will want to print gummed labels for selective advertising campaigns and a tape analysis report showing which tapes have earned more than their cost, she provides for two further modules, 7 and 8. Number 9 will still allow users to terminate the program.

She now takes each module in turn (in this case PRINT THE FILE), examines the report it must produce (Figure 11.12), and refines it into its most basic components (Figure 11.13), using pseudocode to show details. In some modules (ADD A TAPE, for example), Sue will refer to a data-collection screen design rather than a report design. Regardless of the module under consideration, the analyst

```
Read employee's data.

REPEAT until (there are no more data)
    Calculate Gross-pay.
    Subtract Deductions from Gross-pay.
    Calculate Year-to-date totals.
    Calculate Quarter-to-date totals.
    Print employee's paycheck.
    Read next employee's data.

END-REPEAT.
```

FIGURE 11.9. Repetition control structure repeats actions a specific number of times or until a certain condition arises.

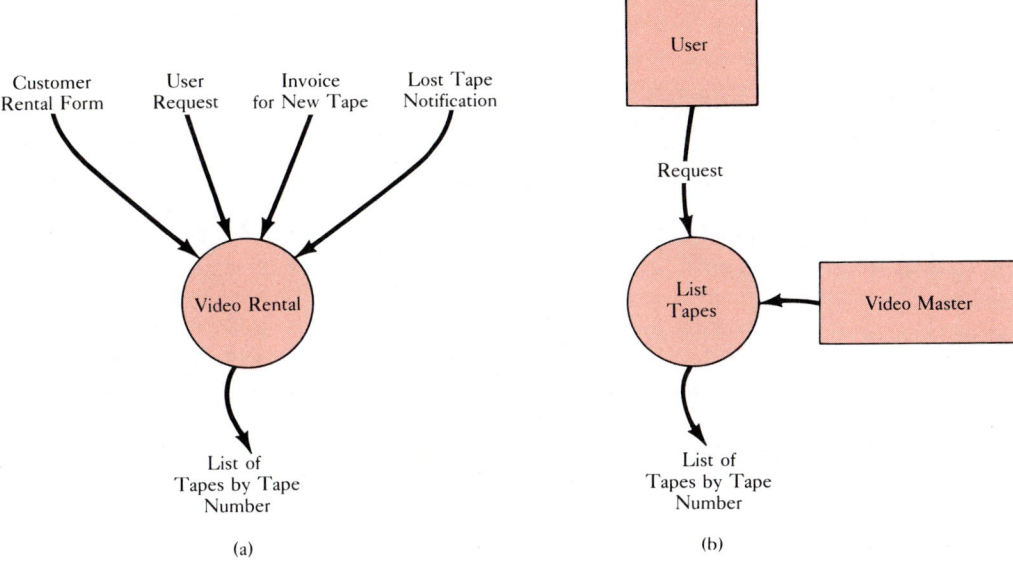

FIGURE 11.10. Data-flow diagram for Class Video's tape rental system. (a) Context data-flow diagram for the overall tape-tracking system. (b) Data-flow diagram to list a tape by tape number; this is only one of six diagrams Sue Wurst draws. (Courtesy of MCBA®. MCBA is a registered trademark of MCBA, Inc.)

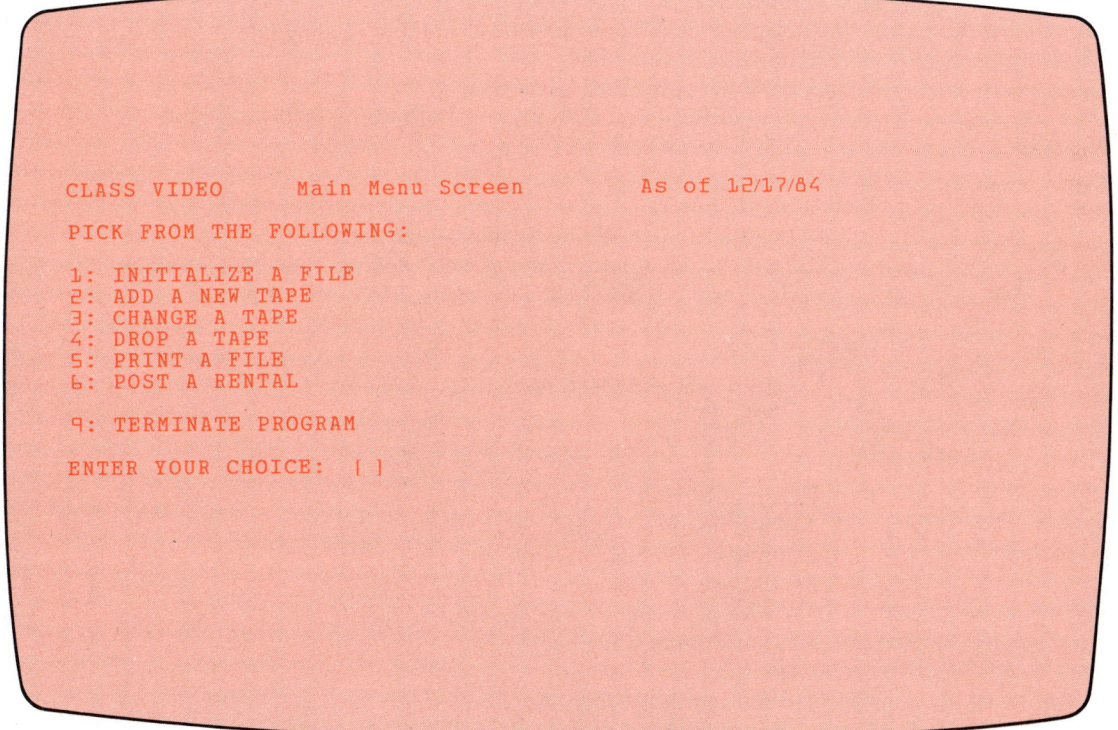

FIGURE 11.11. Main selector menu for the Class Video tape rental system which gives the terminal operator a choice of seven separate operations. (Courtesy of MCBA® MCBA is a registered trademark of MCBA, Inc.)

```
CLASS VIDEO RENTAL     List of Tapes by Tape Number    MM/DD/YY
1234 Sunrise Blvd.
Citrus Heights, CA                                     Page: 99
------------------------------------------------------------------
                                       Last    Number
Tape      Title              Date      Rental    of    Tape
Number    of Tape            Acquired  Date    Rentals Format
------------------------------------------------------------------
1         Superman - II      12/22/81  12/23/83  137     B
2         Star Wars - II     12/20/81  1/30/84   412     B
```

FIGURE 11.12. *Output report formats developed during preliminary design paid close attention to user needs.*

Opening Module.
    1. Open files.
    2. Access today's date and place it into the heading line.
    3. Initialize page number to 1.
    4. Perform Page Headings Module.
    5. Find or Read first record from VIDEO-MASTER using TAPE-NUMBER-KEY set.

Processing Module.
    1. Build up the detail line by moving the data.
    2. Print the detail line; if end of page, perform Page Headings Module.
    3. Read next VIDEO-MASTER using TAPE-NUMBER-KEY set.

Closing Module.
    1. Eject paper to a blank page.
    2. Close files.
    3. Terminate program.

Page Headings Module.
    1. Print first heading line.
    2. Print second heading line.
    3. Add one to page counter.
    4. Move page counter to third heading line.
    5. Print third heading line.
    6. Print a line of dashes.
    7. Print fourth heading line.
    8. Print fifth heading line.
    9. Print sixth heading line.
    10. Print a line of dashes.

FIGURE 11.13. *Pseudocode for modules to print a list of tapes by tape number.*

examines the data-flow diagram and output and input requirements, lists modules, and describes details of the modules.

Coupling

After all modules have been sufficiently defined and refined, the analyst begins the important job of determining their relationships (Figure 11.14). We call relationships among modules **coupling.** Independent modules are loosely coupled, and highly dependent ones are tightly coupled. Structure charts graphically depict module relationships (Figure 11.15). For Class Video's application, second-level modules (Initialize, Add, Change, Drop, Print, Post, Terminate) do not interrelate and are, therefore, loosely coupled. In contrast, the page heading functions within the opening and processing submodules of the PRINT module are tightly coupled.

> **Coupling** The measure of control and interdependence among modules.

```
MAIN DRIVER MODULE
    1. Perform Opening Module.
    2. Perform Processing Module until end of VENDOR-MASTER.
    3. Perform Closing Module.

Opening Module.
    1. Open files.
    2. Access today's date and place it into the heading line.
    3. Initialize page number to 1.
    4. Perform Page Headings Module.
    5. Find or Read first record from VIDEO-MASTER using
       TAPE-NUMBER-KEY set.

Processing Module.
    1. Build up the detail line by moving the data.
    2. Print the detail line; if end of page, perform page Headings Module.
    3. Read next VIDEO-MASTER using TAPE-NUMBER-KEY set.

Closing Module.
    1. Eject paper to a blank page.
    2. Close files.
    3. Terminate program.

Page Headings Module.
    1. Print first heading line.
    2. Print second heading line.
    3. Add one to page counter.
    4. Move page counter to third heading line.
    5. Print third heading line.
    6. Print a line of dashes.
    7. Print fourth heading line.
    8. Print fifth heading line.
    9. Print sixth heading line.
   10. Print a line of dashes.
```

FIGURE 11.14. *Coupling links the modules together in a logical manner.*

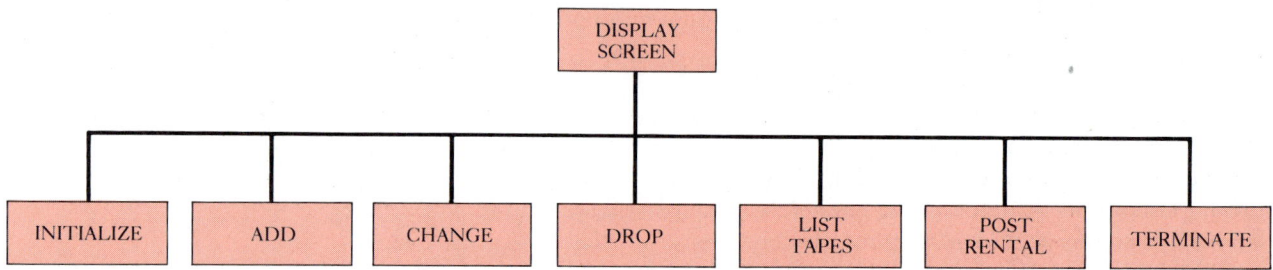

FIGURE 11.15. Like organization charts, structure charts graphically depict coupling among modules, in this case they are the program modules for Class Video's tape rental system.

```
005000 PROCEDURE DIVISION.
005100 000-MAIN-DRIVER.
005200     PERFORM 100-OPENING-MODULE.
005300     PERFORM 200-PROCESSING-MODULE UNTIL DONE.
005400     PERFORM 300-CLOSING-MODULE.
005450     STOP RUN.
005500
005600 100-OPENING-MODULE.
005700     OPEN OUTPUT PRINTER-FILE.
005800     OPEN INPUT VIDEO-DATA-BASE.
005900     MOVE CURRENT-DATE TO LINE-3-DATE.
006000     PERFORM 400-HEADINGS-MODULE.
006100     READ VIDEO-RECORD AT END MOVE "DONE" TO FINISHED.
006200
006300 200-PROCESSING MODULE.
"
"
007000 300-CLOSING-MODULE.
007100     CLOSE PRINTER-FILE.
007200     CLOSE VIDEO-DATA-BASE.
"
"
008900 400-HEADINGS-MODULE.
009000     ADD 1 TO PAGE-COUNTER.
009100     MOVE PAGE-COUNTER TO PAGE-NUMBER.
009200     WRITE PRINTER FROM HEADING-LINE AFTER ADVANCING PAGE.
```

FIGURE 11.16. A portion of a COBOL program providing access to modules for Class Video's listing of tapes. COBOL allows access to modules with its PERFORM verb.

```
PROGRAM LISTTAPES (INPUT, OUTPUT);
VAR
    PAGECOUNTER: INTEGER;
    DONE: BOOLEAN;
"
{ ------------------------------------------------------------------ }
PROCEDURE HEADINGSMODULE;
BEGIN
   WRITELN (CHR(16));
   PAGECOUNTER:=PAGECOUNTER + 1;
   WRITELN ('PAGE NUMBER; ':10, PAGECOUNTER:4);
END; { HEADINGS MODULE }
{ ------------------------------------------------------------------ }
PROCEDURE OPENINGMODULE;
BEGIN
   REWRITE (INFILE, 'PRINTER:');
   RESET (INFILE, '#5:VIDEO.DATA');
   WRITE ('ENTER DATE:');
   READLN (TODAYSDATE);
   DONE:=FALSE;
   PAGECOUNTER:=0;
   HEADINGSMODULE;
   READLN (INFILE, VIDEORECORD)
END; { OPENING MODULE }
{ ------------------------------------------------------------------ }
PROCEDURE PROCESSINGMODULE;
BEGIN
"
"
"
"
END; { PROCESSING MODULE }
{ ------------------------------------------------------------------ }
PROCEDURE CLOSINGMODULE;
BEGIN
"
"
    CLOSE (INFILE, LOCK);
"
END; { CLOSING MODULE }
{ ------------------------------------------------------------------ }
BEGIN { MAIN PROGRAM }
   OPENINGMODULE;
   REPEAT
        PROCESSINGMODULE
   UNTIL DONE;
   CLOSINGMODULE;
END; { MAIN PROGRAM }
```

FIGURE 11.17. A portion of a Pascal program providing access to modules for Class Video's listing of tapes. Unlike COBOL's use of a PERFORM verb, Pascal accesses a module by using the module's name.

The fact that changes within loosely coupled modules do not always require changes in other modules eases programming. Programming tightly coupled modules can become complicated because any change within one module may dramatically affect others. In Chapter 4 we saw other types of coupling: data, stamp, control, common, and content. Ideally modules should be as independent as possible, but in actual practice very few modules are entirely loosely coupled.

As an example of a loosely coupled module, consider our page heading, which gives the effective date of three different reports. In other words, this one submodule does the same job for all three modules: Opening, Processing, and Closing.

Now let us suppose the user wants to show the date in two of the three possible places (Opening and Processing, but not in Closing). To do so, the programmer will have to write a new page heading module or reconstruct the existing one to print the date only for two rather than three modules. In such a case, we have introduced a decision within the module and must include some sort of indicator or flag that identifies when and where a heading should be printed. The new module grows more complex, now requiring two data items, the report date and the flag, and it is no longer loosely but control coupled (see Chapter 4).

Guidelines for Planning Modules

Technically, a module is "a bounded contiguous group of statements having a single name by which it can be referred to as a unit." In plain language that means a single block within a pile of blocks. Though we must identify a module's inputs and outputs when designing it, we need not worry about how it actually works. Later, of course, that becomes a concern, but at this point we can treat modules more or less as blocks or "black boxes."

Second, module design implements the structured methodology and requires both intuitive and practical skills from an analyst. The following guidelines help analysts design good modules:

1. Remember the purpose of any module: to receive data, perform one or more operations on the data, and return output data.
2. Make sure modules obey the single task, single entry, and single exit rules of structured methodology. They should read from the top down, following a simple algorithm with clear and understandable and appropriate names.
3. A module should seldom contain more than 40 lines of code or less than 7 lines. For fewer lines which one may only invoke once, write the instructions into the main line of the calling module.
4. Try to minimize the amount of data that any module can reference. Instead of having the system pass an entire record, have it pass only the individual fields needed for a given operation. The less data passed, the less likelihood of errors.
5. Consider writing an initialization module to establish key values for a system.
6. Isolate a system's input/output operations into a small number of modules.

Each programming language has its own technique for depicting coupling. In COBOL the verb PERFORM calls up a paragraph (Figure 11.16), whereas Pascal uses the name of a procedure or function: OPENINGMODULE, PROCESSINGMODULE, and CLOSINGMODULE (Figure 11.17). Both COBOL and Pascal enable users to call up other modules while still working within a module. For example, a user working within OPENINGMODULE can call up HEADINGSMODULE. Structured methodology allows programs to call up a module, execute it, and then return to the statement that accessed it. Notice that accessing another module from within a module does not violate the structured methodology of single entry and exit because the accessed module will return the user to the one that accessed it.

Nonstructured programs use GO TO to access and return from specific statements, from any point in a program (Figure 11.18). However, the GO TO does not obey the single-entry/single-exit and single-purpose program module logic rules. GO TOs result in a program that looks like a plate of cooked spaghetti: no modules, no control structures, no decomposition, and no coupling.

Such isolation enhances the portability of the system. If an organization adapts a different data structure, it can easily convert the system by replacing existing I/O modules with new ones.

7. Identify system dependent functions, such as acquiring today's date, and isolate them to a small number of modules so that changes can be easily made if the system is moved to another computer.
8. Avoid content coupling which leads a module back to itself or to the middle of another module. This violates the single entry and exit concept fundamental to the structured methodology.
9. Try to write loosely or functionally coupled modules. Such modules accomplish a single goal (e.g., load a table, look up a table element, build a master file record, or calculate FICA or hours worked).
10. To test the strength, binding, or cohesion of a module, write a sentence describing what the module does. Does the sentence:
    a. contain multiple verbs? If so, the module is probably performing more than one function.
    b. contain words relating to time (first, next, then, after, start)? If so, the module is probably procedurally bound.
    c. contain words such as initialize and clean-up? Such jargon implies classically bound modules.
    d. have a predicate without a single specific object following a strong verb? If so, the module is probably logically bound.

Source: Seminar notes provided by Gopal Kapur on "Structured Design and Structured Programming," Kapur and Associates.

```
005000    PROCEDURE DIVISION.
005100
005150    FIRST-PARAGRAPH.
005200        OPEN OUTPUT PRINTER-FILE.
005300        OPEN INPUT VIDEO-DATA-BASE.
005400        MOVE CURRENT-DATE TO LINE-3-DATE.
005500        ADD 1 TO PAGE-COUNTER.
005600        MOVE PAGE-COUNTER TO PAGE-NUMBER.
005700        WRITE PRINTER FROM HEADING-LINE AFTER ADVANCING PAGE.
005800        GO TO READ-AND-PROCESS.
005900
006000    CLOSE-ROUTINE.
007100        CLOSE PRINTER-FILE.
007200        CLOSE VIDEO-DATA-BASE.
007300        STOP RUN.
007400
007500    READ-AND-PROCESS.
007600        READ VIDEO-RECORD AT END GO TO CLOSE-ROUTINE.
007700        WRITE PRINTER FROM DETAIL-LINE AFTER ADVANCING 1 LINE AT
007800            END-OF-PAGE
007900              ADD 1 TO PAGE-COUNTER
008000              MOVE PAGE-COUNTER TO PAGE-NUMBER
008100              WRITE PRINTER FROM HEADING-LINE AFTER ADVANCING PAGE.
  "
  "
  "
008900        GO TO READ-AND-PROCESS.
```

FIGURE 11.18. A portion of an unstructured COBOL program.

LANGUAGE CONSIDERATIONS

After management approves the system design, the analyst selects an appropriate programming language. In most situations analysts find themselves restricted to a certain language because the organization relies on that particular one. Perhaps an organization's programming staff feels familiar with only one language, or may have only one language translator (compiler or interpreter) at its disposal. If no such restriction exists, the analyst can select the language best suited for the application. For example, whereas COBOL might work best for a system that processes long lists of data containing nonnumeric data (addresses, descriptions, etc.), Pascal or FORTRAN may offer more benefits in a scientific setting where the system must handle complex mathematical formulas.

Languages vary in terms of capabilities, syntax structure, and ability to process data efficiently. Certain language capabilities will benefit any system. However, those that employ structured concepts lend themselves nicely to a system designed with a structured approach. Pascal, Ada, and to some extent COBOL allow structured programming more handily than BASIC or FORTRAN. The former support modules, the three control structures, and top-down program composition. Studies reveal that structured programming facilitates quicker program development, introduces fewer errors, and permits easier program maintenance than does nonstructured programming. One county government reports that its programmers spend over 80 percent of their time maintaining nonstructured systems. In stark contrast a 2-year community college using structured concepts estimates its programmers spend only 30 percent of their time on maintenance.

One cannot follow precise rules when comparing or choosing among these languages, and personal preferences often outweigh other factors. Though most business application systems currently rely on COBOL, Pascal, Modula-II, and Ada will become popular as more people learn them and they become available on a wider variety of computer systems.

PACKAGE DESIGN AND PROGRAM SPECIFICATIONS

Having defined the programs, planned the modules, selected control structure, decomposed and refined modules, established the coupling, and chosen a programming language, the analyst gathers together the outputs from all phases of the design activities. This collection of materials includes system specifications (output from process design) and all the program definitions and module descriptions forming the entire design package.

1. System specifications
    a. System overview
    b. System data-flow diagrams
    c. Output report format or design
    d. Data base schema or file design
    e. Screen layout or input formats
2. Program specifications
    a. Program definitions
    b. Module descriptions

Package design pulls together and summarizes all the materials developed during design. Analysts understand data-flow diagrams, report formats, data-base specifications, and input/output layouts, but programmers may not. Instead they rely on the program modules for keys to coding the desired programs.

The system overview provides background information as well as system goals and objectives (Figure 11.19). Overviews often include descriptions of data-collection methods and systems operations, explanations of each program's purpose within the system, a schedule for programming, and the names of the analyst, programmer, and programs.

> **PROGRAM SPECIFICATION**
>
> System: VTR—VIDEO TAPE RENTAL
> Program Name: VTR04—Printing the List of Tapes by Tape Number
>
> Analyst: Sue Wurst   Programmer: Allen Boucher
> Date Assigned: 10/22/84   Completion Date: 11/9/84
>
> Background: Class Video rents video tapes to customers on a short-term basis. The owner wants to keep track of the number of times each tape is rented as well as specific data about the tape (date of purchase, date last rented, title, purchase price). Tapes are numbered (starting at 1) to identify them. The tape number is affixed to the cartridge and is recorded each time a tape is rented by a customer.
>
> System Data-Flow Diagram: The data-flow diagram for the video rental system is found in Appendix A. The leveled data-flow diagram for the printing of the list of tapes by tape number is in Appendix B.
>
> Output Reports: Appendix C shows the report format desired. All pages are numbered as well as dated. A maximum of 50 tape titles is printed on a page.
>
> Date-Base Layout: See Appendix D for a listing of the schema for the video rental data base.
>
> Input Date Layout: All data to produce the report are in the VIDEO-MASTER data set. The set TAPE-NUMBER-KEY will permit retrieval sequentially by tape number.
>
> Processing Requirements: Appendix E shows the pseudocode to print the list of tapes by tape number.

FIGURE 11.19. *System specifications include goals and objectives for the system as well as relevant background information.*

System data-flow diagrams show program inputs and outputs and how a single program fits within the entire system (Figure 11.10a). Developed at the beginning of the design phase, output report formats also form part of the program specifications (Figure 11.12). Analysts often describe system designs on printer layout sheets, although many analysts now use computerized text editors to create page designs.

The fourth component of program specifications is the data-base or file layouts (Figure 11.20). If a program accesses only a portion of the data base, the person programming it need only worry about relevant portions of the data-base design.

A screen design specifies data input requirements that allow programs to collect, validate, and store data on disk or tape (Figure 11.11). Screen designs should specify data type (numeric or alphanumeric), editing requirements, and any nec-

```
VIDEO-MASTER DATA SET
      TAPE-NUMBER              NUMBER (8);
      TAPE-TITLE               ALPHA (40);
      DATE-ACQUIRED            NUMBER (6);
      LAST-RENTAL-DATE         NUMBER (6);
      NUMBER-OF-RENTALS        NUMBER (6);
      TAPE-FORMAT              ALPHA (1);
      TAPE-COST                NUMBER (9,2);
      SUPPLIER                 ALPHA (40);
      );
TAPE-NUMBER-KEY SET OF VIDEO-MASTER KEY (TAPE-NUMBER) NO DUPLICATES;
TAPE-TITLE-KEY SET OF VIDEO-MASTER KEY (TAPE-TITLE) DUPLICATED PERMITTED;
```

FIGURE 11.20. *Data-base or file layouts are developed during the design phase of the systems process. This schema uses the Burroughs' DMS-II definition process. A set provides a logical path for retrieving records from the data set.*

essary order in which entries should occur. For example, before accepting a date entry, the screen might demand that the entry be a 1, 2, 3, 4, 5, 6, or 9.

Depending on personal experience or preference, an analyst proceeds to write **program specifications** that define each program the system requires and can describe the program in pseudocode, with decision tables, IPO charts, Warnier-Orr diagrams, or any of the other tools (Figure 11.14). In many cases the computer services manager will dictate the tool and the analyst will have no choice in this matter. Because of its wordiness, some analysts avoid pseudocode, preferring to use a VTOC, IPO, Nassi-Shneidermann chart, or Warnier-Orr diagram.

Program specifications Functional descriptions of each program the system needs.

DESIGN WALKTHROUGH

We conclude the design phase with a crucial structured walkthrough (discussed in Chapter 4). A walkthrough supplements earlier formal analysis and design reviews by providing a careful scrutiny of system and program specifications. During this phase the analyst, the programmer (or leader of a programming team), another analyst, and sometimes a user critique specifications to locate module errors, to check for completeness, and to ensure clarity for eventual programming.

Module errors may arise from improper linkage, incomplete refinement, or violation of the single-entry, single-purpose, single-exit, and control structure rules. No matter how insignificant a detail may seem, it can create nagging problems in a sensitive and complex computer system.

Remember that a walkthrough aims at detecting, not correcting, problems. At the conclusion of the walkthrough, the analyst may study problems, make necessary corrections, and then release revised system and program specifications for further review. If the team finds a lot of errors, it may schedule a second walkthrough. Since a walkthrough should not attempt to evaluate an analyst's performance but to assure quality, no formal summary goes to management. It is important that a walkthrough remain as free from emotion as possible.

DESIGN REVIEW

Design review A decision-making meeting to determine whether the system meets organizational needs.

Once the analyst has corrected any mistakes or added any missing materials, then management, users, and the analyst conduct a **design review** (Figure 11.21). During a formal presentation of the system, everyone studies the specifications and all materials developed during design: detailed data-flow diagrams, data dictionaries, report or output formats, data-base or file requirements, input layouts, and program definitions. The review process occurs during a formal presentation of the system. This is the last chance for all concerned to verify that the system will meet the organization's needs within acceptable cost and schedule limits.

In some cases last-minute changes may delay the system's progress to development, but making changes now will save the greater time and money involved in making them after programming begins.

If everyone agrees that the system should proceed to development, a memo formally authorizes the next step (Figure 11.22). The sooner management circulates it, the better. Otherwise false rumors can begin, alarming staff and stimulating resistance rather than cooperation.

If the organization has suffered financial setbacks since it first requested analysis, has altered priorities, or has more pressing needs, it may still (although very rarely) abandon a new system at this point. In that case an immediate memorandum should explain why the organization has decided against the system. Again, false rumors can do a lot of damage to staff confidence and morale.

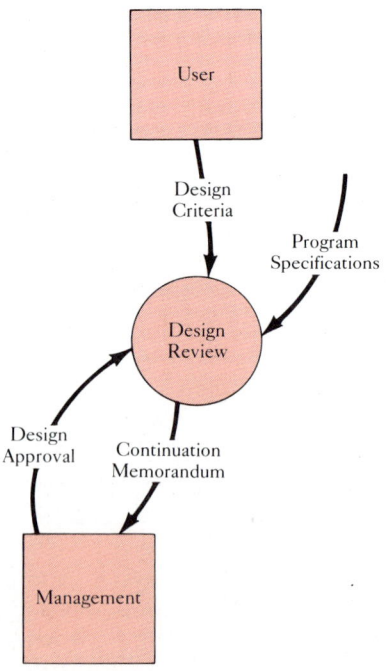

FIGURE 11.21. The design review is the last chance to catch errors before system development begins.

CHAPTER 11 PROGRAM DEFINITION, MODULE DESIGN, AND THE DESIGN REVIEW **373**

MEMORANDUM Oct. 14, 1984

TO: Sue Wurst and Allen Boucher, Analysts

FROM: Connie and Harry Allen, Owners of Class Video

SUBJECT: Decision on a New Video Tape Tracking System

For the past 5 months, our video rental system has been undergoing an extensive analysis and design. Yesterday we approved the new system design. This approval carries with it direction and authority for you to proceed immediately to development. We are ahead of our original schedule by 10 days and plan to have the new system operating by March 15.

FIGURE 11.22. *A memorandum authorizes system development.*

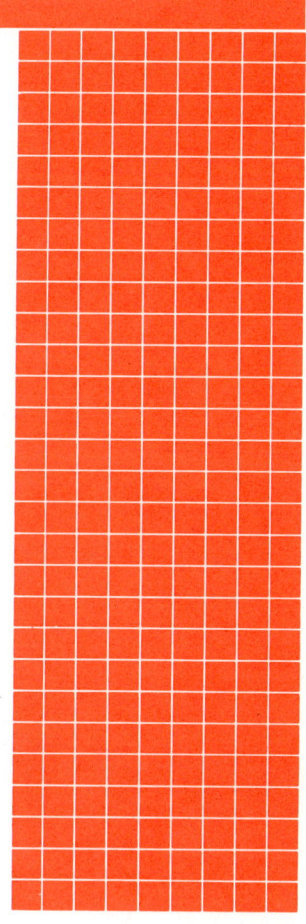

CASE STUDY

Design Specifications for Fleet Feet's Accounts Payable Voucher-Check Printing Program

An accounts payable system's primary output is a check to a vendor. Checks contain two sections: the upper stub details the payment so vendors can reconcile their own accounts receivable systems, and the lower stub provides actual payment.

The upper stub of Fleet Feet's vendor check indicates Fleet Feet's voucher number (a voucher is an authorization for payment of an invoice), the vendor's invoice number, the gross amount of the check, deductible discounts, and the net amount due. The lower stub is the check itself, listing vendor number, name, address, date, Fleet Feet's check number, and the amount of the check (Figure 11.23). This check folds in half to fit inside a window envelope.

The voucher-check printing program requires the printing of data on specific lines and in specific columns. Payment details can only appear on lines 4 through 18. On the bottom portion, the 20th line contains the check number and amount; line 29 shows date and check number, line 35 shows vendor name and amount, and 36 and 37 show address, city, state, and Zip code. Column locations precisely position data items horizontally.

Frank Pisciotta, Fleet Feet's analyst, completes the specifications for the voucher-check program (Figure 11.24), providing programmers with a background narrative, a system data-flow diagram, output report column and line numbers, data-base schema, location of input data (data sets), and pseudocode for the check printing program. Frank's pseudocode divides the check printing program into modules with linkages connecting them.

With these specifications in hand, Frank conducts a design review by management, users, and other members of Fleet Feet's staff. After Fleet Feet's staff carefully examines the specifications, management gives Frank the green light to proceed to development (Figure 11.25). A memorandum notifies all parties that the third and final phase of the new AP system has begun.

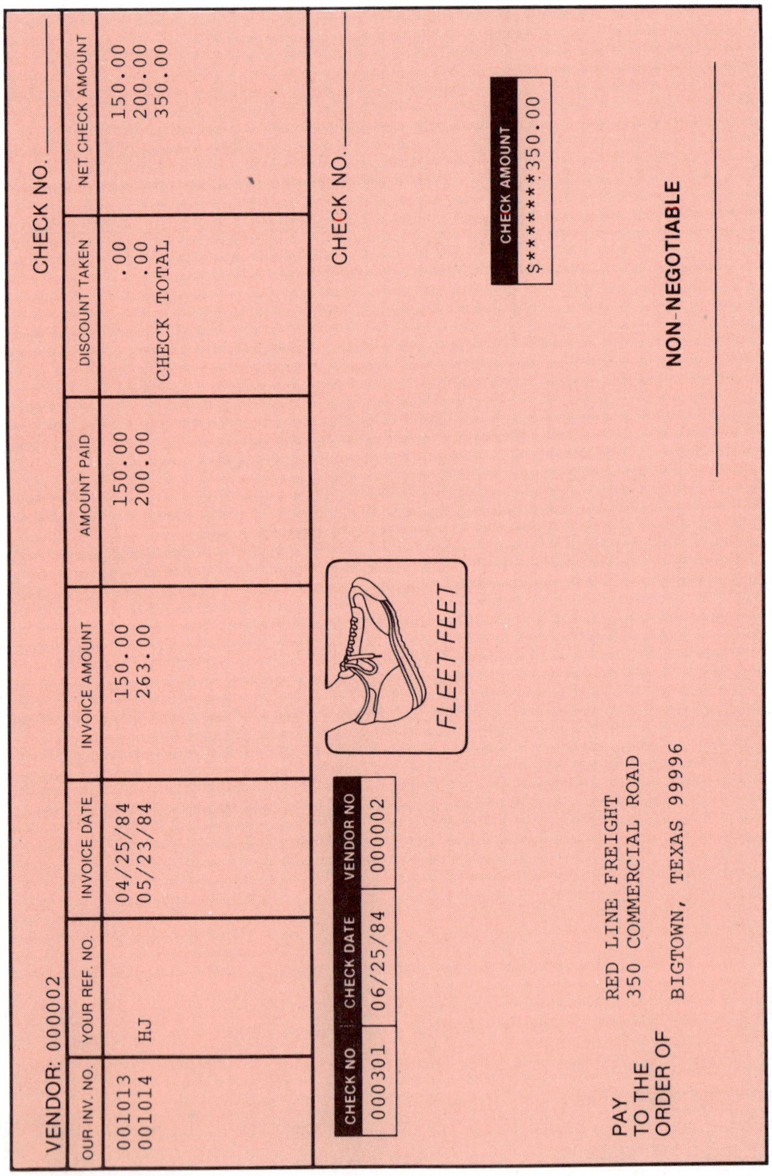

VENDOR: 000002

OUR INV. NO.	YOUR REF. NO.	INVOICE DATE	INVOICE AMOUNT	AMOUNT PAID	DISCOUNT TAKEN	NET CHECK AMOUNT
001013	HJ	04/25/84	150.00	150.00	.00	150.00
001014		05/23/84	263.00	200.00	.00	200.00
					CHECK TOTAL	350.00

CHECK NO.

CHECK NO.	CHECK DATE	VENDOR NO
000301	06/25/84	000002

FLEET FEET

PAY TO THE ORDER OF RED LINE FREIGHT
350 COMMERCIAL ROAD
BIGTOWN, TEXAS 99996

CHECK NO. _____

CHECK AMOUNT
$*******350.00

NON-NEGOTIABLE

FIGURE 11.23. Voucher checks for an accounts payable system usually have two parts, whether the stub appears above or below the check.

CASE STUDY

PROGRAM SPECIFICATION

System: AP
Program Name: AP075–Printing the AP Voucher Check

Analyst: Frank Pisciotta
Date Assigned: 10/31/84

Programmer: Steve Osterday
Completion Date: 12/14/84

Background: The voucher check is a principal output from our AP system. To encourage prompt payment, our vendors allow discounts to be taken if we pay within a specified time period. Freight charges are usually excluded from the discount since the vendor must pay this amount in full. Some vendors charge us interest on unpaid or long overdue amounts. We will not pay interest on past due amounts, but prefer to deal with another vendor.

System Data-Flow Diagram: The data-flow diagram for the AP075 program is found in Attachment 1.

Output Reports: A copy of the AP check we will use is also a part of this specification (Figure 11.23). The table below specifies where each data item should be printed on the check.

Data Item Name	Line Number	Column Number
Voucher Number	4–18	2–11
Vendor Number	4–18	14–21
Gross Amount	4–18	19–28
Discount	4–18	35–44
Net Amount	4–18	50–61
Check Date	20	2–9
Check Number	20	19–24
Check Amount	20	50–61
Check Date	29	8–15
Check Number	29	19–24
Vendor Name	35	11–40
Address1 (Street)	35	50–61
Address2 (City, State)	36	11–40
Address3	37	11–40

Date-Base Layout: See Attachment 2 for a listing of the schema for the data sets required to print the voucher checks. The entire schema for the AP data base is not shown.

Input Data Layout: All data to produce the report are found in the AP data base. The data sets needed are VENDOR, CHECK-MASTER, AP-TRX, AP-OPEN-ITEM, and AP-OPEN-ADJ.

Processing Requirements: Attachment 3 shows the pseudocode to be followed to print the check in the format required.

FIGURE 11.24. *Structured specifications for a program contain all the information the programmer will need to write machine instructions.*

CASE STUDY

Attachment 1. Data-flow diagram for AP075.

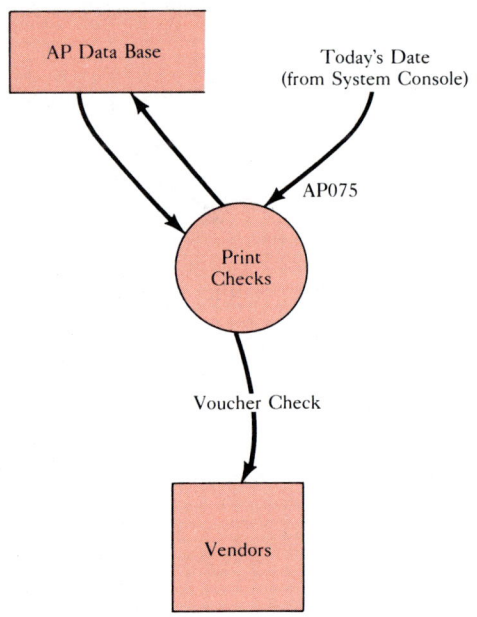

Attachment 2. Accounts Payable data base for the AP075 program.

```
SET NAME:
    VENDOR, MANUAL

    ITEMS:
        VENDOR-NO,              X6          <<KEY ITEM>>
        VENDOR-NAME,            X30
        VENDOR-ADDR-1,          X30
        VENDOR-ADDRR-2,         X30
        VENDOR-ADDR-3,          X30
        TYPE,                   X4
        TERMS-DESCR,            X16
        DUE-DAYS,               Z4
        DISCOUNT-DAYS,          Z4
        DISCOUNT-%,             Z4
        LAST-ACTIVITY,          Z6
        STATUS-CODE,            X2
        PURCHASES,YTD,          Z10
        PURCHASES-LAST,         Z10
        DISCOUNTS-YTD,          Z10
        DISCOUNTS-LAST,         Z10

    CAPACITY:   251             ENTRIES:    0
```

FIGURE 11.24 *(Continued)*

CASE STUDY

Attachment 2. Accounts Payable data base for the AP075 program.

```
SET NAME:
    AP-TRX,MANUAL

    ITEMS:
        VOUCHER-NO,              X6         <<KEY ITEM>>
        TRX-TYPE,                X2
        VENDOR-NO,               X6
        P-O-NO,                  X10
        INVOICE-NO,              X8
        INVOICE-DATE,            Z6
        INVOICE-AMOUNT,          Z10
        INVOICE-NON-DSC,         Z10
        DUE-DAYS,                Z6
        DISCOUNT-DAYS,           Z4
        DISCOUNT-DATE,           Z6
        DISCOUNT-%,              Z4
        DISCOUNT-AMOUNT,         Z8
        CHECK-NO,                Z6
        CHECK-DATE,              Z6
        ACCOUNT-NO,              Z10

    CAPACITY:   6500             ENTRIES:   0

SET NAME:
    CHECK-MASTER,MANUAL

    ITEMS:
        ACCT-CHECK-VCHR,         Z22        <<KEY ITEM>>
        VENDOR-NO,               X6
        CHECK-TYPE,              X2
        CHECK-DATE,              Z6
        CHECK-AMOUNT,            Z10
        P-O-NO,                  X10
        INVOICE-NO,              X8
        INVOICE-DATE,            Z6
        INVOICE-AMOUNT,          Z10
        DISCOUNT-AMOUNT,         Z8

    CAPACITY:   4500             ENTRIES:   0
```

FIGURE 11.24 *(Continued)*

CASE STUDY

Attachment 2. Accounts Payable data base for the AP075 program.

```
SET NAME:
    AP-OPEN-ITEM, DETAIL

    ITEMS:
        VENDOR-NO,              X6      <<SEARCH ITEM>>
        VOUCHER-NO,             X6      <<SEARCH ITEM>>
        P-O-NO,                 X10
        INVOICE-NO,             X8
        INVOICE-DATE,           Z6
        DUE-DATE,               Z6
        DISCOUNT-DATE,          Z6
        ORIG-INVOICE-AMT,       Z10
        ORIG-DISC-AMT,          Z8
        INVOICE-BALANCE,        Z10
        DISCOUNT-BALANCE,       Z8
        PARTIAL-PAYMENT,        Z10
        CURRENT-DISCOUNT,       Z6
        CHECK-NO,               Z6      <<SEARCH ITEM>>
        CHECK-DATE,             Z6
        AP-ACCOUNT,             Z10
        SELECT-STATUS,          X2
        PRINT-STATUS,           X2

    CAPACITY:   4500            ENTRIES:    0

SET NAME:
    AP-OPEN-ADJ, DETAIL

    ITEMS:
        VOUCHER-NO,             X6      <<SEARCH ITEM>>
        TRX-TYPR,               X2
        VENDOR-NO,              X6      <<SEARCH ITEM>>
        CHECK-NO,               Z6      <<SEARCH ITEM>>
        INVOICE-NO,             X8
        INVOICE-DATE,           Z6
        INVOICE-BALANCE,        Z10
        INVOICE-DISCOUNT,       Z8
        OLD-DUE-DATE,           Z6
        DUE-DATE,               Z6
        OLD-DISC-DATE,          Z6
        DISCOUNT-DATE,          Z6
        DISCOUNT-AMOUNT,        Z8
        CHECK-DATE,             Z6
        CHECK-AMOUNT,           Z10
        DISCOUNT-TAKEN,         Z8
        CASH-ACCOUNT,           Z10

    CAPACITY:   5500            ENTRIES:    0
```

FIGURE 11.24 *(Continued)*

CASE STUDY

Attachment 3. Pseudocode to print the AP voucher check.

MAIN DRIVER

1. Perform Opening Module.
2. Perform Processing Module until end of VENDOR.
3. Perform Closing Module.

Opening Module.
1. Open data base and files.
2. Access today's date and place it into Check Date.
3. Find or Read first record from VENDOR.

Processing Module.
1. Try to read a record from AP-OPEN-ITEM.
2. If found:
    Space paper down two lines.
    Perform Print Detail Line Module until no more AP-OPEN-ITEM records.
    Perform Print Bottom Module.
    Build a record for CHECK-MASTER.
    Write a record to CHECK-MASTER.
3. Read next VENDOR record.

Closing Module.
1. Eject paper top of next page.
2. Close data base and files.
3. Terminate program.

Print Detail Line Module.
1. Read a record from AP-TRX for this AP-OPEN-ITEM.
2. Calculate Discount.
3. Build a detail line.
4. Print detail line before advancing one line.
5. Write updated AP-TRX record back.
6. If line number is greater than 18, skip paper to top of the next page.
7. Add net amount to total.
8. Read next record from AP-OPEN-ITEM.

Print Bottom Module.
1. Skip paper to line number 29.
2. Print Check Date and Check Number.
3. Skip paper to line number 35 and print Vendor Name and Check Amount.
4. Skip paper to next line and print ADDRESS1.
5. Skip paper to next line and print ADDRESS2.
6. Skip paper to next line and print ADDRESS3.
7. Eject paper to top of next page.

FIGURE 11.24 *(Continued)*

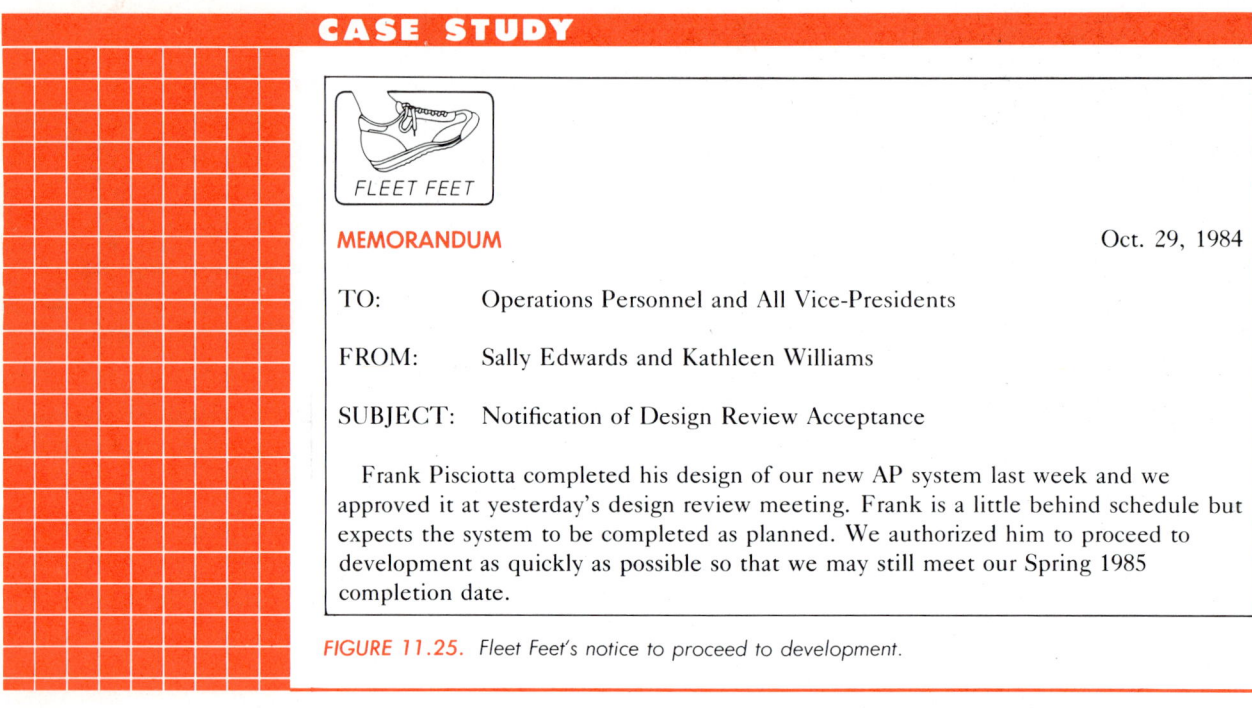

FIGURE 11.25. Fleet Feet's notice to proceed to development.

WORKING WITH PEOPLE A Program Specifications Walkthrough

Steve Osterday glanced at the clock as he walked back to his desk—2:47. It had taken only 47 minutes to walk through his specifications for Class Video's program for posting rentals to the data base. After this 16th walkthrough in 2 years, he felt comfortable with the process; his first walkthrough had dragged on for more than 3 hours. Returning to his cubicle, Steve glanced up at his old wall plaque: "In God we trust; everything else we walkthrough."

Steve shook his head over that first ordeal, his design for a sales analysis system that was supposed to print a report comparing the current year's sales with last year's, including subtotals for a variety of categories and computations of averages and standard deviations. Determined not to get caught with a glaring error on his first major project, he had spent 2 days preparing for the walkthrough. What a wonderful design it had seemed at the time.

(Steve had provided for six digits, whereas seven were occasionally needed). However, what had really alarmed him was the number of typographical mis-

And what a shock it had been when the walkthrough team tore it to shreds! His data-base design did not permit the system to capture last year's sales statistics for comparison with the current year's, and some fields were too small to store sales amounts takes people found in his specifications. To top it off, some sentences that were perfectly clear to Steve made no sense to the review team.

When the first walkthrough had concluded, no one had said, "Good job, Steve," or even, "Not bad for a beginner." Although Steve had anticipated praise, he felt he could have handled criticism—but silence depressed him. It took over a year and a half dozen more design walkthroughs for him to appreciate that silence. Walkthroughs are supposed to improve systems, not to evaluate analysts. Evaluations tend to be clouded by emotional judgments that can detract from the pure issue of quality. He had gradually learned not to take objective criticism personally.

Reflecting on 2 years of walkthrough experience, Steve marveled at how much he had gained from them professionally. He'd been able to expand his bag of tricks by picking up useful design and programming techniques from other analysts and programmers. His writing style had improved too. A short business communications course had taught him how to replace clumsy passive sentences with crisp active ones. He found that English-language skills benefited his career as much as his mastery of FORTRAN and COBOL.

Steve's reputation had grown as he eliminated most major errors from his designs. No longer did users worry that his lack of business experience would result in complex, hard-to-understand systems. Suddenly the new analyst, Sue Aimes, barged into Steve's cubicle, interrupting his reverie. "Steve," she cried. "I just went though my first walkthrough on the new accounts payable system, and the review team shot it full of holes."

"That usually happens," Steve observed.

"Maybe, but that's not what bothered me."

"What bothered you?" Steve thought he knew.

"No one said anything. It was weird. They must think I'm an idiot. Why are you laughing?"

"Let me tell you about my first walkthrough," began Steve. "You think you fouled up!"

SUMMARY

The last phase of design involves program definition, which allows programmers to write programs for the new system.

When developing program specifications, the analyst must divide the program into single-entry, single-purpose, and single-exit modules that reflect the three control structures: sequence, selection, and iteration.

Having divided programs into modules, the analyst establishes their relationships with linkage. Modules may be further refined into basic components, provided those components obey the rules of single entry, single purpose, and single exit and support the three control structures.

If no programming language has been adopted, the analyst selects an appropriate one. Some languages, such as Pascal and Ada, lend themselves to structured methodology, whereas others, such as FORTRAN and BASIC, do not. COBOL is the most widely used business programming language.

Before program specifications go to programming, the system undergoes a structured walkthrough to locate errors, assure completeness, and make sure specifications are clear. After a successful walkthrough, the specifications move to a design review, and then, if approved, into the hands of programmers.

A memo authorizes development or explains why development will not proceed.

NEW TERMS

Case	Design review	Program definition
Control structure	Iteration	Program specifications
Coupling	Looping	Refinement
Decision	Module	Repetition
Decomposition	Program design	Sequence

QUESTIONS FOR REVIEW AND DISCUSSION

Questions appear in three categories. You can find the answers to the A group in this chapter. The B group requires you to apply the material presented here, whereas the C group necessitates investigation or research.

- A.1. What are the three control structures?
- A.2. List two synonyms for each of the three control structures.
- A.3. Draw a flowchart to represent each of the three control structures.
- A.4. List three popular programming languages.
- A.5. List the events that occur during the module design phase.
- A.6. List the three terms associated with modules.
- A.7. List the major parts of the system specifications.
- A.8. Who are the participants in a structured walkthrough of the program specifications?
- A.9. Who are the two individuals first associated with the idea of three control structures?
- A.10. During which phases of the systems process were the following developed?
   - a. Program narrative.
   - b. Data-base design.
   - c. Input design.
   - d. Output report design.
- A.11. What is the newest part of the program specifications?
- A.12. What are the outputs from the design phase of the systems process?
- A.13. What are the inputs to the design phase of the systems process?

- B.1. Convert the decision table of Figure 11.8c into pseudocode.
- B.2. List the goals and objectives of a structured walkthrough.
- B.3. Who is responsible for scheduling and conducting the structured walkthrough of the program specifications?

- C.1. Interview someone in a management position for a computer installation. Determine whether the person uses structured methodology, and if so, what benefits have been observed.
- C.2. The "*New York Times* Project" is an often-cited example of the success of structured methodology. Write a short paper describing the project and its results.
- C.3. Find the names and authors of three books on structured analysis or design.

REFERENCES

Tom DeMarco, *Structured Analysis and System Specification*, Yourdon Press, New York, 1979.

Jean Dominique, *Logical Construction of Programs*, Van Nostrand Reinhold, New York, 1974.

Daniel P. Freedman and Gerald M. Weinberg, *Handbook of Walkthroughs, Inspections, and Technical Reviews*, Little Brown, Boston, 1982.

Robert T. Grauer, *Structured Methods Through COBOL*, Prentice-Hall, Englewood Cliffs, N.J., 1983.

G. J. Myers, *Composite Structured Design*, Van Nostrand Reinhold, New York, 1978.

PART III

Systems Development

CHAPTER 12 Programming and Conversion

- Goals and Preview
- Overview of Development
- Scheduling and Assignment of Tasks
- Programming a Structured System
- Guidelines for Writing IF and GO TO Statements
- Conversion
- Case Study: Developing Fleet Feet's Accounts Payable Voucher-Check System
- Working with People: Coping with Change
- Summary
- New Terms
- Questions for Review and Discussion
- References

GOALS AND PREVIEW

After reading this chapter you should be able to:

1. Draw a data-flow diagram for the development phase of the systems process.
2. Describe the major activities involved in programming a system.
3. Develop three examples of programming standards.
4. Cite the purpose of a program walkthrough.
5. Define an implementation plan.
6. List the three types of conversion.

Most analysts feel a great sense of accomplishment when they enter the development phase. After so much earlier general planning and organization, they look forward to seeing the system actually begin to come to life. However, as exciting as this prospect may be, one must finish a few remaining tasks before the job is complete.

Having obtained management's approval of the new system during the design review, the analyst moves into the third and final phase of the systems process, development, during which programmers write necessary programs and writers develop documentation (users' and operation staff manuals). Also, a thorough training program must be devised and everyone affected by the new system must be properly prepared for it. If the organization will be adopting a new system, the analyst must decide how to convert data from the old system to the new.

During the two earlier phases of the process, we focused our attention on the organization's goals for the system, much as an architect drafts plans for a building with its future occupants in mind. But now we change our point of view to one similar to that of a building contractor. Using all his or her skills and tools and coordinating the efforts of programmers, users, and specialists, the analyst finally begins to craft the system.

OVERVIEW OF DEVELOPMENT

Development Final phase of systems cycle: programs are written, tested, and installed.

Development moves the system closer to realization (Figure 12.1). Since an organization has invested a lot of time and effort in getting to this point in the systems process, it hesitates to halt it at this late date; but there are occasions when that makes sense. For example, after a lengthy analysis and design of a computerized payroll system, a small contractor might decide to continue with a manual system in the light of recent changes in workers' benefit plans negotiated

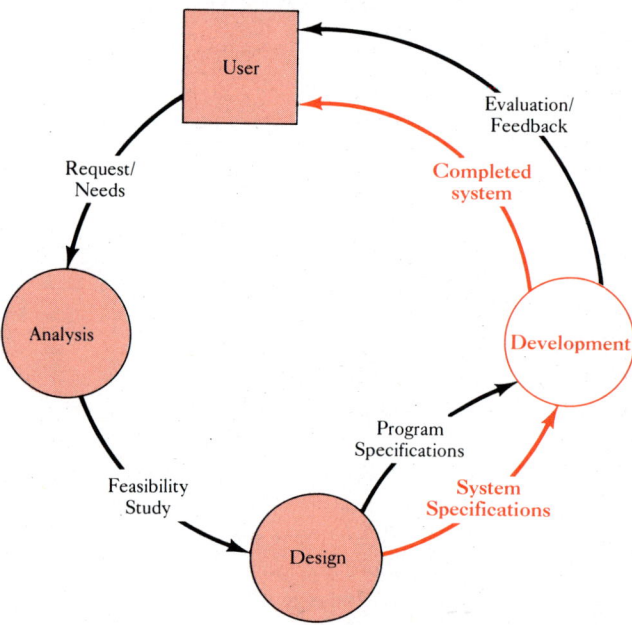

FIGURE 12.1. *Development is the third phase in the systems process. The system specifications provide inputs; outputs include the completed system and its documentation.*

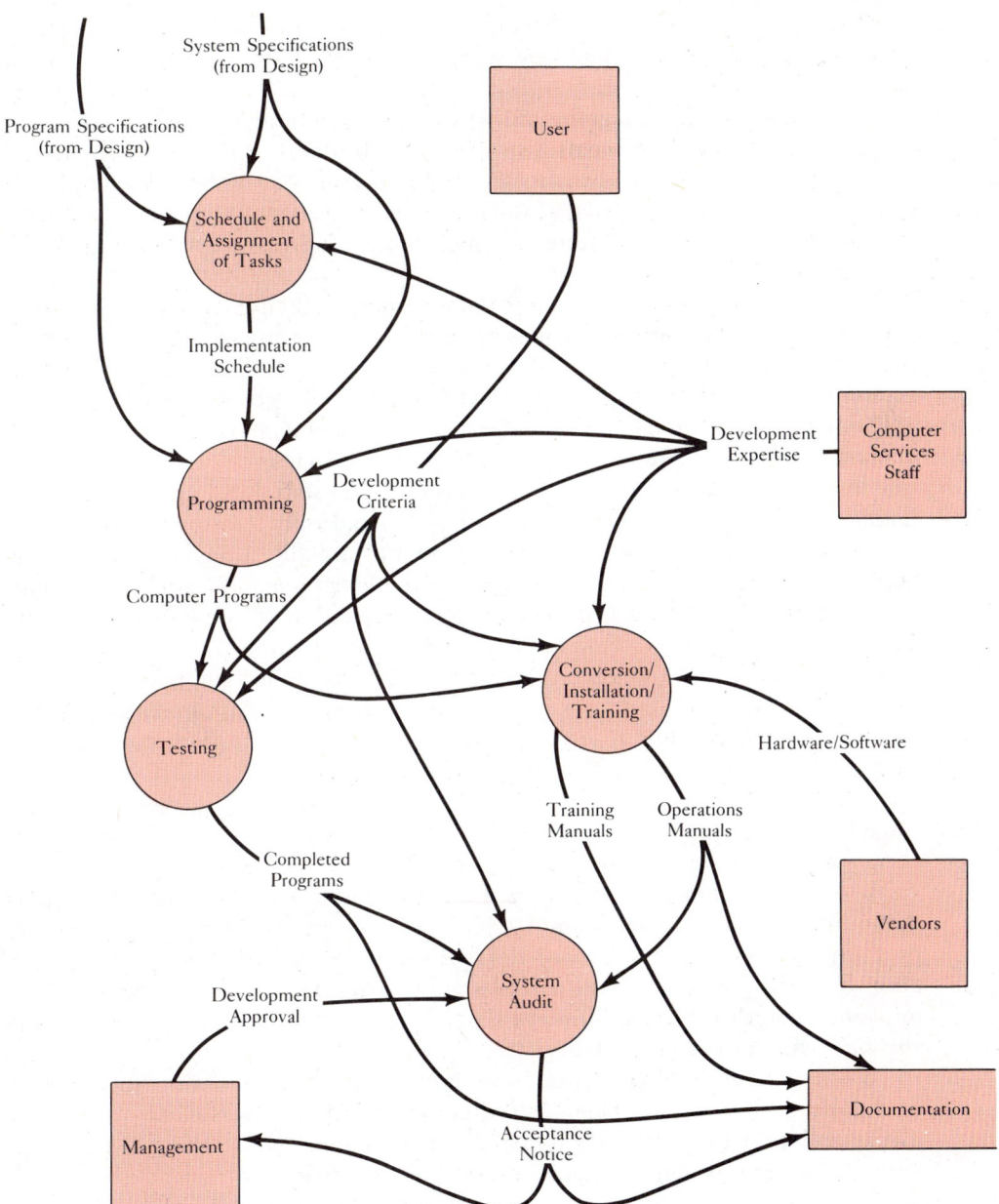

FIGURE 12.2. *Leveling of the data-flow diagram for development reveals six major tasks. Scheduling comes first; conversion, training, and installation occur later.*

by a labor union. Perhaps the proposed payroll system will not be able to accommodate constant and unforeseen changes.

As we have done throughout analysis and design, we can level or decompose the data-flow diagram for development (Figure 12.2). The first subtask is scheduling and assignment of tasks. How much time will the entire development phase take? With a schedule settled, the analyst can choose programmers, determine in what order the programs should be written, and select a means for testing their accuracy.

After scheduling and assigning tasks, actual programming proceeds with coding the system according to the analyst's written specifications. This task may consume from only a few weeks to several months, depending on the number and complexity of the programs. For example, it might take only a month to program a billing system for a small hardware store but might take a full year to program an online college registration system.

Program testing occurs next. Here programmers might use real data pulled from a prior system to "test drive" the new system. While program testing is in progress, the organization may begin to convert from the old to the new system. If the system will replace an existing one, data must transfer between them.

Hardware installation and training of users, management, and computer center operations staff also take place. Outputs from these concurrent tasks include user training manuals, instructions to the computer center staff detailing its anticipated interaction with the system, and the completed and fully tested programs.

Finally the system must pass a system review or audit, a final evaluation that occurs only after the system has been operating long enough for users to become familiar with it. At this point users, management, and the computer services staff carefully critique the system, citing strengths and weaknesses and comparing anticipated and actual budgets and schedules. Results of the audit not only benefit the system in question, but form an important part of the organization's experience with computers, enabling it to make even better decisions in the future.

> **Program testing**
> Checking the correctness of an entire program.

SCHEDULING AND ASSIGNMENT OF TASKS

> **Implementation plan**
> An outline showing activities, times, and events of development.

Analysts begin development (Figure 12.3) by writing an **implementation plan** that outlines tasks to be performed (Figure 12.4). A review of the system and its program specifications helps those unfamiliar with the new system more fully to grasp its goals and how it will achieve those goals. If the analyst charged with implementation did not perform the earlier analysis and design, this review will also bring him or her up to date.

To complete an implementation schedule or plan, the analyst fits together all of the pieces of a puzzle: facilities, equipment, technical personnel, users, and management. If the system calls for new equipment, the analyst must arrange with the computer center operations staff for delivery and installation by the supplying vendor. Quite often the analyst must indicate to the computer center operations staff any alterations that need to be made within the computer center or a user's work area. When making program assignments, one must allow for adequate computer time and test data, and begin users' training. Since training sessions usually disrupt normal work schedules, management must remain aware of all schedules and costs.

Two tools are used to schedule and plan all these activities. A Gantt chart depicts overall schedules, personnel needs, and each task's schedule, and PERT charts describe these events (Figure 12.5). You will remember from Chapter 4

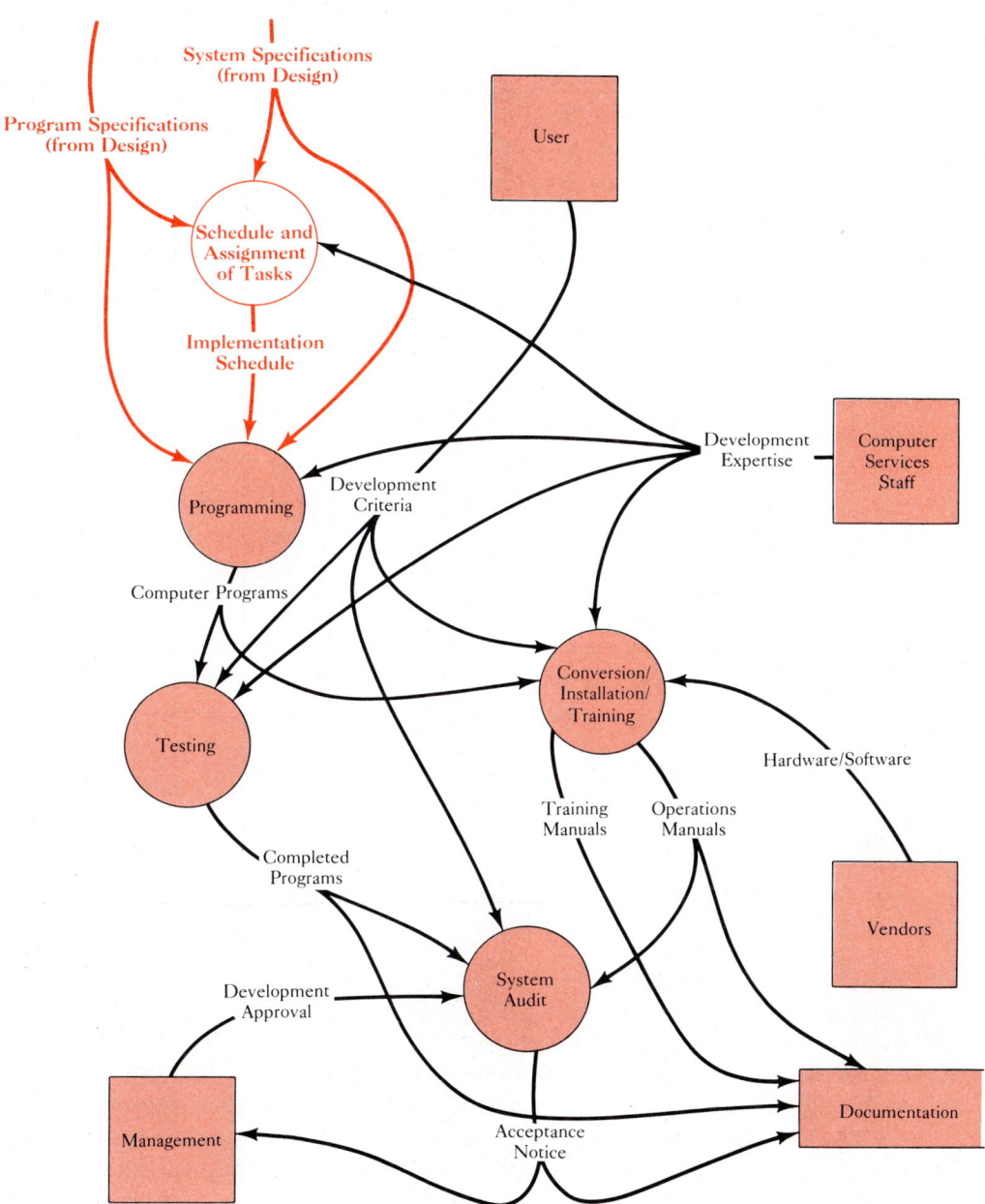

FIGURE 12.3. *Planning precedes all other development tasks.*

that such a chart emphasizes tasks critical to maintaining a schedule. If a task on the critical path falls behind schedule, the analyst might allocate additional resources to succeeding tasks to get implementation back on schedule. The PERT chart also reveals "slack" times during which certain events can begin earlier than anticipated, provided a preceding event has concluded. For example, the PERT

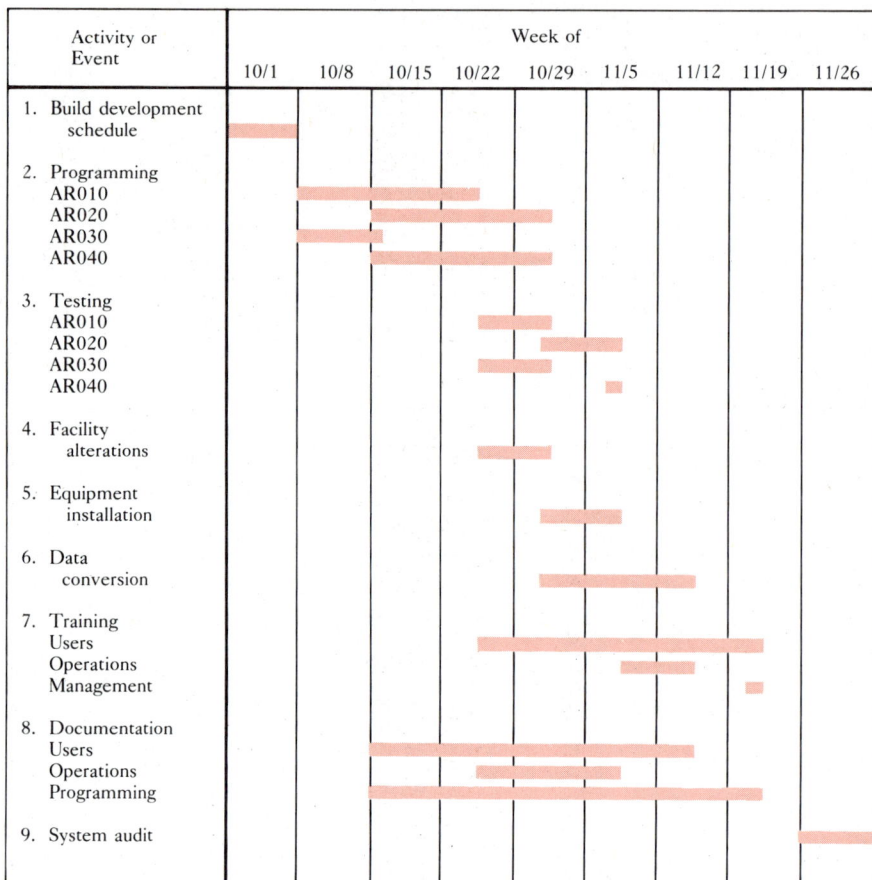

FIGURE 12.4. *Gantt charts depict implementation activities, showing a schedule of weekly activities.*

chart in Figure 12.5*b* shows that events 1, 4, 3, 7, 8, and 9 are on the critical path. Thus event 2 can start at any time so long as it is completed by day 11. Likewise, event 5 cannot start until day 6 and must be completed by day 15; therefore, since it takes 2 days, the earliest it can start is day 7 and the latest is day 13. Software, such as VisiSchedule and Apple's Lisa Plan, can automatically analyze activities, their relationships, and time factors, and calculate the critical path that will ensure completion of the system on schedule. In fact this software will show all this information to you, using its graphics capability.

Within the implementation schedule, the analyst includes a personnel chart detailing people assigned to each event (Figure 12.6). The personnel chart for Class Video (our video tape rental business from Chapter 11) involves vendors (Holcenberg, a systems engineer from IBM, and Fane of Easters Electric), technical staff (Overmiller, Robinson, McAdams, and Sprenger), users (Fenolio and Darlington), and management (Valentine and Baker). The lead analyst (Wurst) formally contacts people early to make sure they understand how and when they fit into the implementation schedule.

Event Number	Predecessor Event	Successor Events	Time Required (Days)
1. Build implementation plan	None	2	4
2. Programming			
a. AR010	1	3a	12
b. AR020	1	3b	14
c. AR030	1	3c	8
d. AR040	1	3d	13
3. Testing			
a. AR010	2a	5	9
b. AR020	2b	5	10
c. AR030	2c	5	7
d. AR040	2d	5	4
4. Facility alterations	1	5	9
5. Equipment installation	4	5	8
6. Conversion data	3a	7a	9
7. Training			
a. Users	3b	9	24
b. Operations	3a,3b,3c,3d	8	9
c. Management	6	8	3
8. Documentation			
a. Users	3	9	22
b. Operations	6	9	15
c. Programming	6	9	29
9. System audit	8	None	9

(a) Precedence relationships of the implementation plan and times required to complete each task.

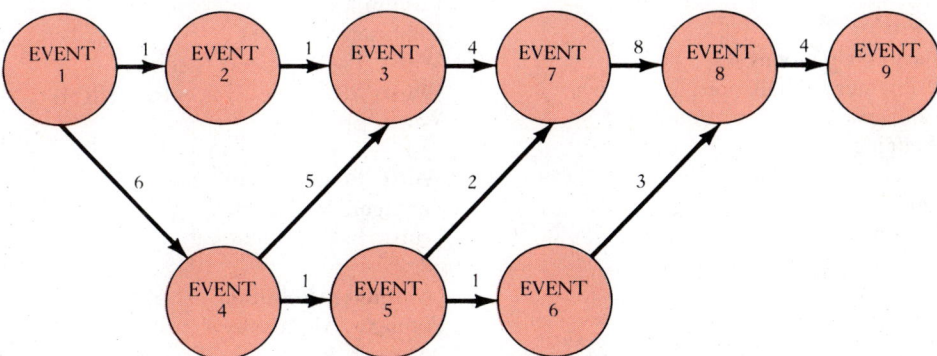

(b) A PERT chart for the implementation phase illustrates interrelationships between events.

FIGURE 12.5. PERT (Program Evaluation Research Technique) charts graphically represent an implementation schedule.

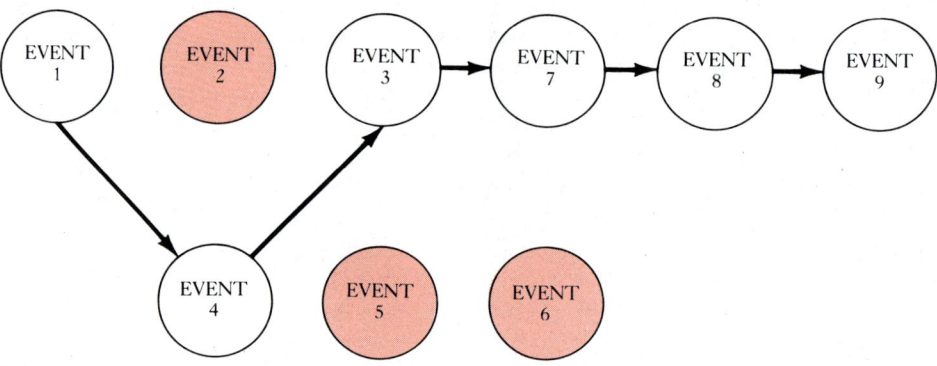

(c) Critical path (highlighted) shows events which, if they occur in sequence from EVENT 1 to EVENT 9, will provide the shortest possible schedule.

FIGURE 12.5 (Continued)

CLASS VIDEO
Personnel Roster

SYSTEM: Video Tape Tracking System   Date: 12/4/84

Event	Personnel Assigned
1. Build implementation plan	Wurst
2. Programming	
AR010	Robinson
AR020	Overmiller
AR030	Overmiller
AR040	McAdams
3. Testing	
AR010	Robinson
AR020	Overmiller
AR030	Overmiller
AR040	Robinson
4. Facility alterations	Wurst, Fane (Easters Electric)
5. Equipment installation	Wurst, McAdams, Holcenberg (IBM)
6. Conversion data	Wurst, Sprenger
7. Training	
Users	Wurst, Fenolio, Darlington
Operations	Sprenger, McAdams
Management	Wurst, Baker, Valentine
8. Documentation	
Users	Overmiller, Robinson
Operations	Sprenger, McAdams
Programming	Sprenger, Robinson
9. System audit	Wurst, Fenolio, Baker, Valentine

FIGURE 12.6. Personnel chart for system development. Some of the personnel are technical (programming) staff and others are users.

Wise analysts invest a lot of time planning for development, because planning can reap many rewards, forestalling hurt user feelings, improving communication, and smoothing the transition to the new system.

PROGRAMMING A STRUCTURED SYSTEM

After completing the implementation plan, the analyst turns attention to programming, inputs to which include the implementation plan, the computer service staff's expertise, and system and program specifications (Figure 12.7). Outputs will be the completed programs along with their test results.

Standards

Standards, a set of rules programmers must follow when writing programs, have been adopted by most computer services departments (Figure 12.8). They promote a consistent programming style within a department, thereby making it easier for new personnel to maintain programs. Many departments demand consistency for file names, record names, variable or data names, and module names, and most dictate strict rules for writing modules.

Actual **program writing**, or **coding**, begins with a review of the program specifications, which should reveal program logic (Figure 12.9). During this review the programmer identifies and numbers modules if analysts have not already done so. The VTOC in Figure 12.9 displays four modules: opening, processing, closing, and a main driver module that causes the system to perform each module in the correct sequence.

Stubs

With the organization's program standards and specifications in mind, the programmers begin coding in the specified language. In our Class Video example, the programmers first write the main driver, 000-MAIN-DRIVER, and then, in a top-down fashion, 100-OPENING, 200-PROCESSING, and 300-CLOSING. If they wish to concentrate on one specific module at a time, programmers can **stub** other modules, setting them aside to be written later (Figure 12.10*a*). For example, while writing the 100-OPENING module, a programmer might stub the others by inserting a temporary statement such as DISPLAY "300-PROCESSING" at the point where the fully developed module will eventually appear. Stubbing unwritten modules allows programmers to test related modules even though they have not completed the whole program. In Figure 12.10*b* a programmer has replaced the 100-OPENING and 200-PROCESSING temporary statements with full program statements. Stubbing also permits programmers to write modules in any order, a benefit when a particular individual (user or data-base administrator) is unavailable to assist the programmer.

Because stubbing frees programmers to write the most critical modules first, it strengthens program development and permits testing at intermediate stages. This allows users to "test drive" early versions of programs, thereby enhancing their

Standards A collection of rules analysts and programmers must follow.

Program writing Conversion of the program design into computer language form.

Coding Translating module descriptions into a set of computer instructions.

Stub An abbreviated version of a program module written to facilitate programming.

394 PART III SYSTEMS DEVELOPMENT

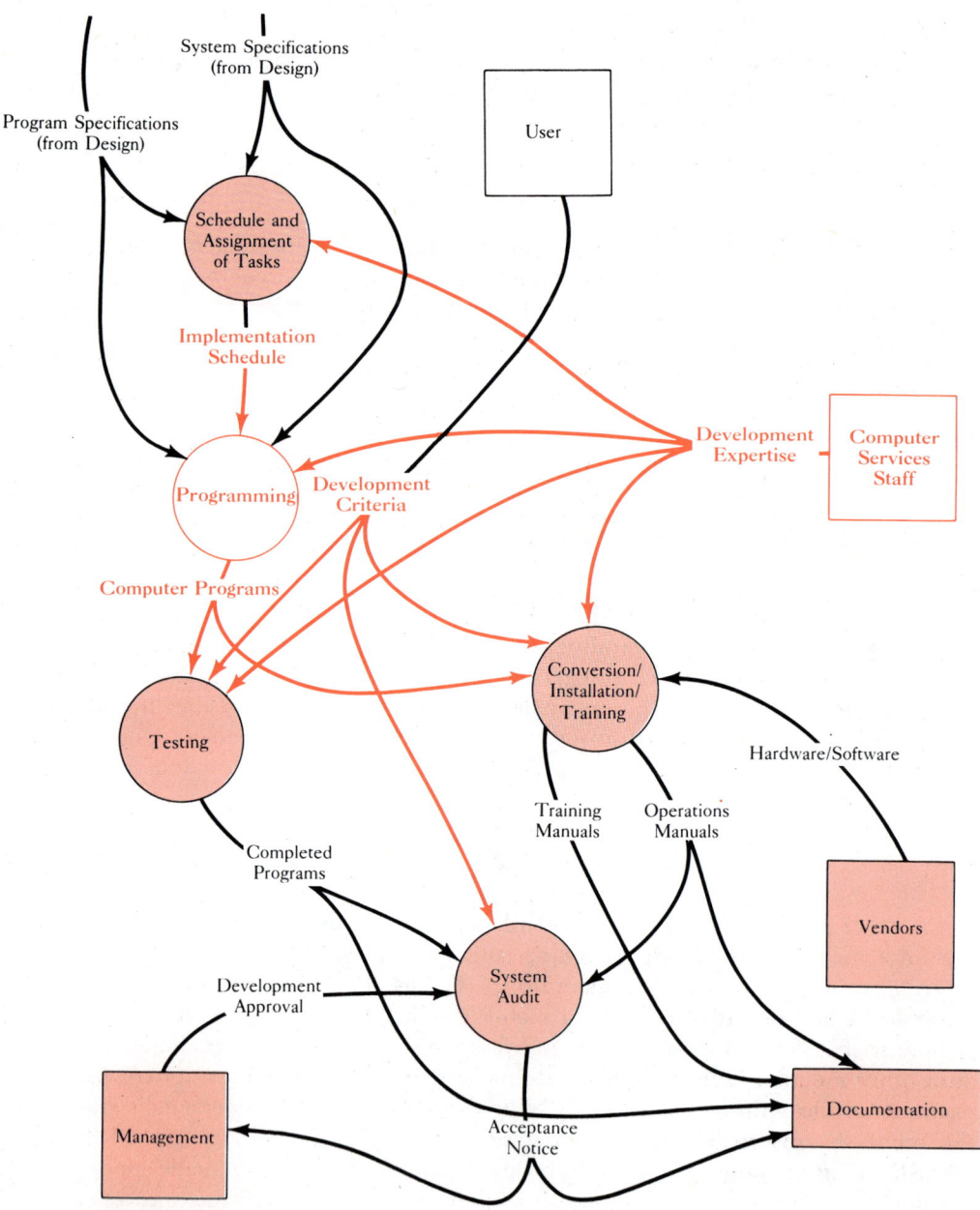

FIGURE 12.7. Programming converts program specifications developed during design.

Bottom-up approach
Testing approach that checks lowest level routines first, and then each module.

morale as they watch their system begin to materialize. This can especially boost morale when schedules suffer setbacks. Stubbing also forces programmers to follow the top-down structured methodology that insists that they consider the most important aspects of the program before dealing with details.

The **bottom-up approach** offers an alternative to top-down program construc-

> **CLASS VIDEO**
> **Programming Standards**
>
> Adopted 12/03/84
>
> 1. File names: All file names will be plural words. Examples are MASTERS, TRANSACTIONS, VENDORS, and CUSTOMERS. File names shall be meaningful as to the data they represent.
>
> 2. Record names: All record names will be the singular version of their file names. Examples: MASTER, TRANSACTION, VENDOR, and CUSTOMER.
>
> 3. Data names: Data names shall have prefixes of their file names. Examples are VENDOR-NUMBER, CUSTOMER-LAST-NAME. Prefixes for working storage will be prefixed with WS. Examples are WS-PAGE-NUMBER, WS-CUSTOMER-TOTAL. Data names must be unique and meaningful. Programs should not have such names as PIZZA, RAVIOLI, or HERE.
>
> 4. Module names: All module names will be prefixed with a module number and all modules will be in ascending order. Examples are 100-OPENING, 900-END-OF-PAGE.
>
> 5. Module rules: All modules will be single entry, single exit. Module size should be less than 24 lines unless it is simple sequence in structure. Modules must follow one of the three control structures: sequence, iteration, or IF-THEN-ELSE.
>
> 6. Statements: Every program statement will begin on its own line. No multiple statements on one line.
>
> 7. Indentation: Statements must be indented to show their hierarchy. Indentation is especially important when writing IF-THEN-ELSE statements.

FIGURE 12.8. Programming standards vary from one computer installation to another.

tion. With this approach the programmer writes all of lowest level routines first, progressing up the hierarchy of modules toward the main driver (paragraph 000-MAIN-DRIVER in Figure 12.10). Bottom-up programming requires the programmer to build a test driver, sometimes called a test harness, test monitor, or exerciser. Such routines, though ultimately replaced by the actual drivers, exercise the lower level modules. Bear in mind, however, that bottom-up construction violates the structured methodology hierarchy of most important to least important.

Program Walkthroughs

After the programmers have written and tested all program modules, the analyst schedules a **program walkthrough** attended by the programmer, analyst, other programmers, and perhaps a member of the operations staff if the program requires

Program walkthrough Peer review to find errors, omissions, faulty logic, or improper language use.

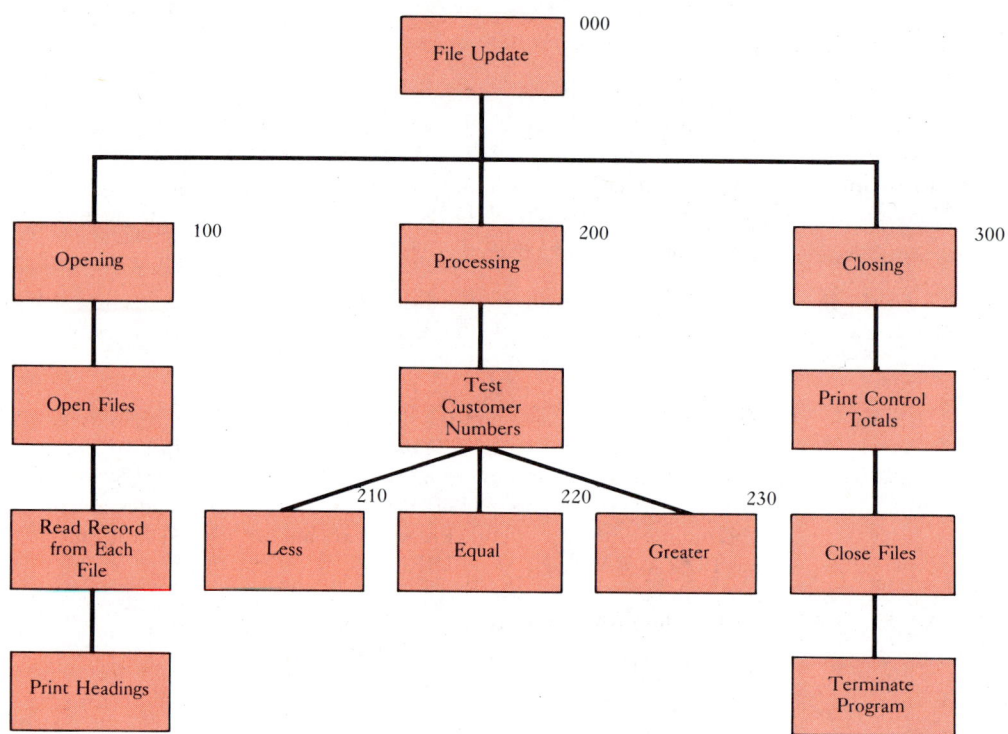

FIGURE 12.9. Programming specifications should reveal program logic. One can depict program logic with a flowchart, pseudocode, Nassi-Shneidermann diagram, or VTOC chart.

operator intervention (entering dates, changing types of paper in the printer). Similar in format, scope, and purpose to the walkthroughs encountered earlier in the systems process, program walkthroughs focus on finding omissions, errors, faulty logic, improper language usage, or otherwise faulty programs.

Whenever a programmer forgets to take into account certain key factors, omissions may result. However, if something small but critical such as omitting an identifying number on the customer statements occurs, the walkthrough will undoubtedly locate the error.

Faulty logic may even appear in syntactically correct programs that do not perform as intended. For example, if a programmer neglects to couple an IF statement with an ELSE statement to accommodate the negative case, the program will fail whenever it encounters a negative situation. In an accounts receivable system, a late arriving payment may determine that a finance charge should be applied to the customer's account, but a test for late payment date without an ELSE statement may cause all payments to be considered late, resulting in incorrect finance charges to all customers' accounts.

We refer to an otherwise syntactically correct program that does not perform correctly all the time as improper language usage. Suppose a programmer has indicated a MOVE statement, but the receiving field is too small to receive the result. In such a situation, truncation occurs. A large city on the West Coast apparently lost over a million dollars in traffic fines when a COBOL program's PICTURE clause could only hold a number up to 999,999.99 and the programmer did not use COBOL's "ON SIZE ERROR" clause to catch field overflows. When the collected amount of fines exceeded this amount, the million portion disappeared, leaving only the hundred thousands on the report. Unfortunately the cross-footing control on the report, which might have helped spot the error, also could only show numbers up to 999,999.99. County auditors discovered the discrepancy a few days later when hand-kept totals did not correspond with computer totals.

Users do not usually participate in program walkthroughs because they lack sufficient technical knowledge to be very helpful. Therefore, the team generally includes only technical members of the computer services staff, such as the analyst, programmers, and a member of the operations staff. Copies of the programs should go to each person involved well in advance, and if team members are not familiar with the data-base design or input or output requirements, they should also receive copies of these materials.

One member of the team takes responsibility for recording all the errors discovered by the team and reports them to the programmer. The team does not correct errors, but leaves that job for the programmer. Some organizations hold walkthroughs before the program is compiled, and others do so afterward. If the team spots few errors, there is no need for a second walkthrough, but it if finds more than an acceptable number, then the analyst may wish to conduct a second one.

```
006000 IDENTIFICATION DIVISION.
006100
006200 000-MAIN-DRIVER.
006300     PERFORM 100-OPENING.
006400     PERFORM 200-PROCESSING UNTIL DONE.
006500     PERFORM 300-CLOSING.
006600
006700 100-OPENING.
006710     DISPLAY "100-OPENING STUB SENTENCE".
007500
007600 200-PROCESSING.
007710     IF (TRANSACTION-NUMBER IS LESS THAN CUSTOMER-NUMBER)
007720         PERFORM 210-LESS-THAN
007730     ELSE
007735         IF (TRANSACTION-NUMBER EQUALS CUSTOMER-NUMBER)
007740             PERFORM 220-EQUAL-TO
007750         ELSE
007755             PERFORM 230-GREATER-THAN.
007760
007900 300-CLOSING.
008000     DISPLAY "300-CLOSING STUB SENTENCE".
009000
009100 900-TOP-OF-PAGE.
009200     DISPLAY "900-TOP-OF-PAGE STUB SENTENCE".
009300
```

(a) Update program with the main driver paragraph and the opening module fully developed.

FIGURE 12.10. Program stubs allow programmers to set aside modules to be written while they focus on the one at hand.

```
006000     IDENTIFICATION DIVISION.
006100
006200     000-MAIN-DRIVER.
006300         PERFORM 100-OPENING.
006400         PERFORM 200-PROCESSING UNTIL DONE.
006500         PERFORM 300-CLOSING.
006600
006700     100-OPENING.
006800         OPEN INPUT TRANSACTIONS.
006900         OPEN INPUT CUSTOMERS.
007000         OPEN OUTPUT NEW-CUSTOMERS.
007100         OPEN OUTPUT PRINTERS.
007200         READ CUSTOMERS AT END MOVE HIGH-VALUES TO CUSTOMER.
007300         READ TRANSACTIONS AT END MOVE HIGH-VALUES TO TRANSACTION.
007400         PERFORM 900-TOP-OF-PAGE.
007500
007600     200-PROCESSING.
007710         IF (TRANSACTION-NUMBER IS LESS THAN CUSTOMER-NUMBER)
007720             PERFORM 210-LESS-THAN
007730         ELSE
007735             IF (TRANSACTION-NUMBER EQUALS CUSTOMER-NUMBER)
007740                 PERFORM 220-EQUAL-TO
007750             ELSE
007755                 PERFORM 230-GREATER THAN.
007760
007770     210-LESS-THAN.
007772         DISPLAY "210-LESS-THAN STUB SENTENCE".
007774
007776     220-EQUAL-TO.
007778         DISPLAY "220-EQUAL-TO STUB SENTENCE".
007780
007790     230-GREATER-THAN.
007792         DISPLAY "230-GREATER-THAN STUB SENTENCE".
007800
007900     300-CLOSING.
008000         WRITE PRINTER FROM WS-TOTAL-LINE BEFORE ADVANCING 3 LINES
008100             AT END-OF-PAGE PERFORM 900-TOP-OF-PAGE.
008200         CLOSE TRANSACTIONS.
008300         CLOSE CUSTOMERS.
008400         CLOSE NEW-CUSTOMERS WITH LOCK.
008500         CLOSE PRINTERS.
008600         STOP RUN.
009000
009100     900-TOP-OF-PAGE.
009200         DISPLAY "900-TOP-OF-PAGE STUB SENTENCE".
009300
```

(b) Update program with the main driver paragraph, opening module, and closing module fully developed.

FIGURE 12.10 (Continued)

Guidelines for Writing IF and GO TO Statements

Two of the most dangerous, yet often used, commands in a programming language are the If and GO TO. Although many programmers know how to correctly write them, these statements can, as we've seen earlier, violate the rules of the structured methodology. In fact, purists insist that we abolish the GO TO altogether. However, IF and GO TO can offer valuable alternatives in certain situations where they can improve the ease with which an organization can read and maintain its programs.

When writing IF statements:

1. Align and indent the IF with its associated ELSE. Place the ELSE on a separate line by itself.
2. Pair every IF with an ELSE to make the logic of the program apparent.
3. Write pseudocode before writing multiple levels of nested IF's, especially when there are null statements within the nested logic. The pseudocode will help you spot faulty logic.
4. Avoid over five levels of nested IFs; they are too hard to read.
5. If you can't avoid more than five levels, consider performing the inner IF statements in another module.
6. To improve readability, place parentheses around conditions being tested.
7. In the situation where the physical span of the IF is too large (if too many statements appear before the ELSE), move these statements to a separate module.

GO TO statements may be needed occasionally. On these rare occasions:

1. Don't use GO TO to move from one module to another.
2. Permit GO TO to move downwards to the end of the module.
3. GO TO downwards, but not to the end of the module, should not be permitted.

Though the structured methodology should provide the meat and potatoes of the programmer's diet, a few old fashioned techniques can make the dish tastier, provided one uses them as sparingly as tabasco sauce.

Source: Seminar notes provided by Gopal Kapur, Kapur and Associates.

Testing: Modules, Module Integration, and Programs

After the programs have been written, walked through, and compiled to remove all syntax errors, they are ready for testing with actual or hypothetical data (Figure 12.11*a*). During **testing** one tries to locate errors that might hamper the system in the future. This step is so important that some programming groups spend 30 percent of their development time and budgets on software testing. They may borrow data from the old system, use artificially constructed data, or build test data files with special software.

Testing (1) Verifying that a system performs as expected. (2) Checking a program to see that it performs as expected.

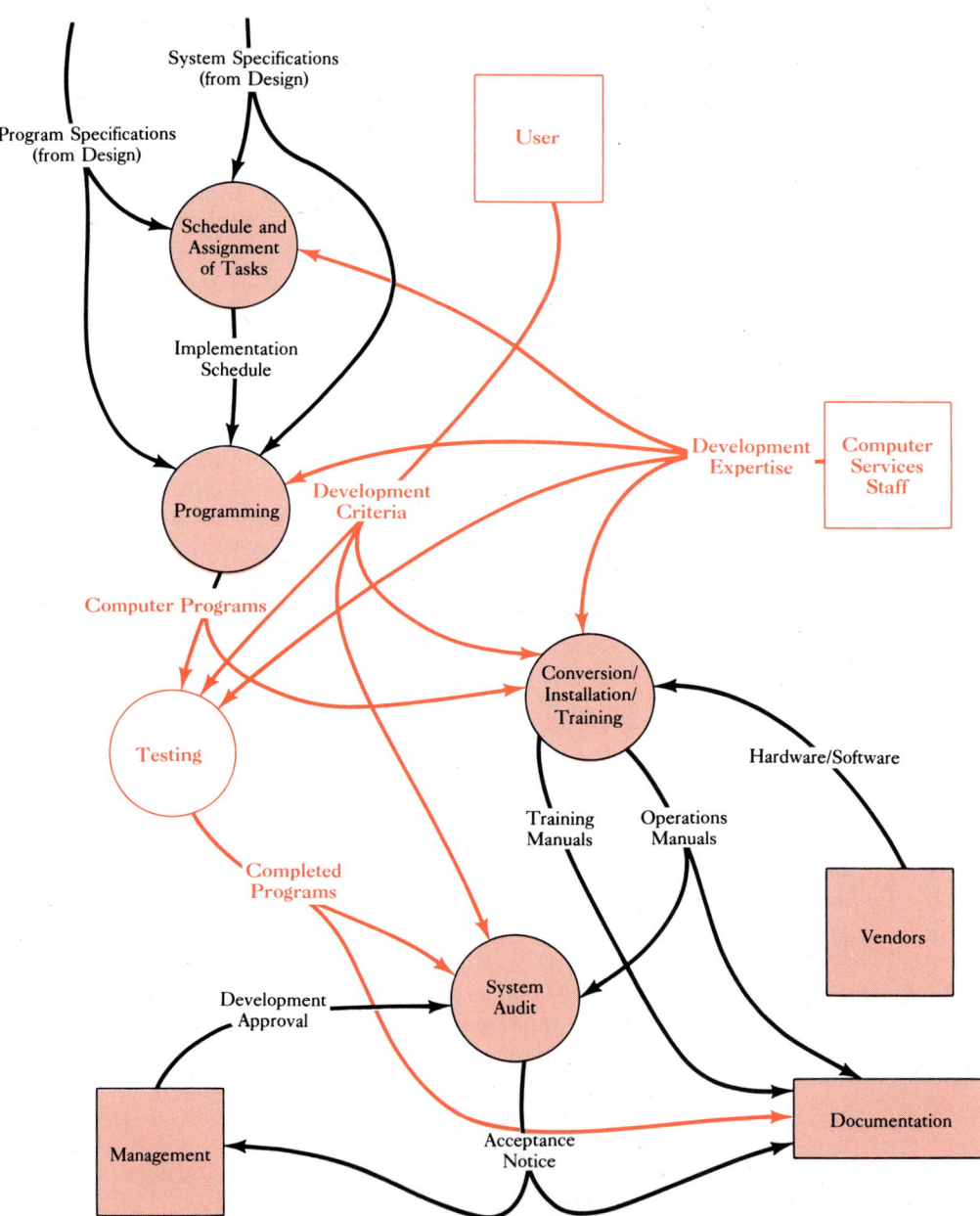

(a) Testing can be decomposed into six components. In Chapter 12 we look at the first three phases. Chapter 13 covers the last three.

FIGURE 12.11. Testing involves six phases leading to a completed system.

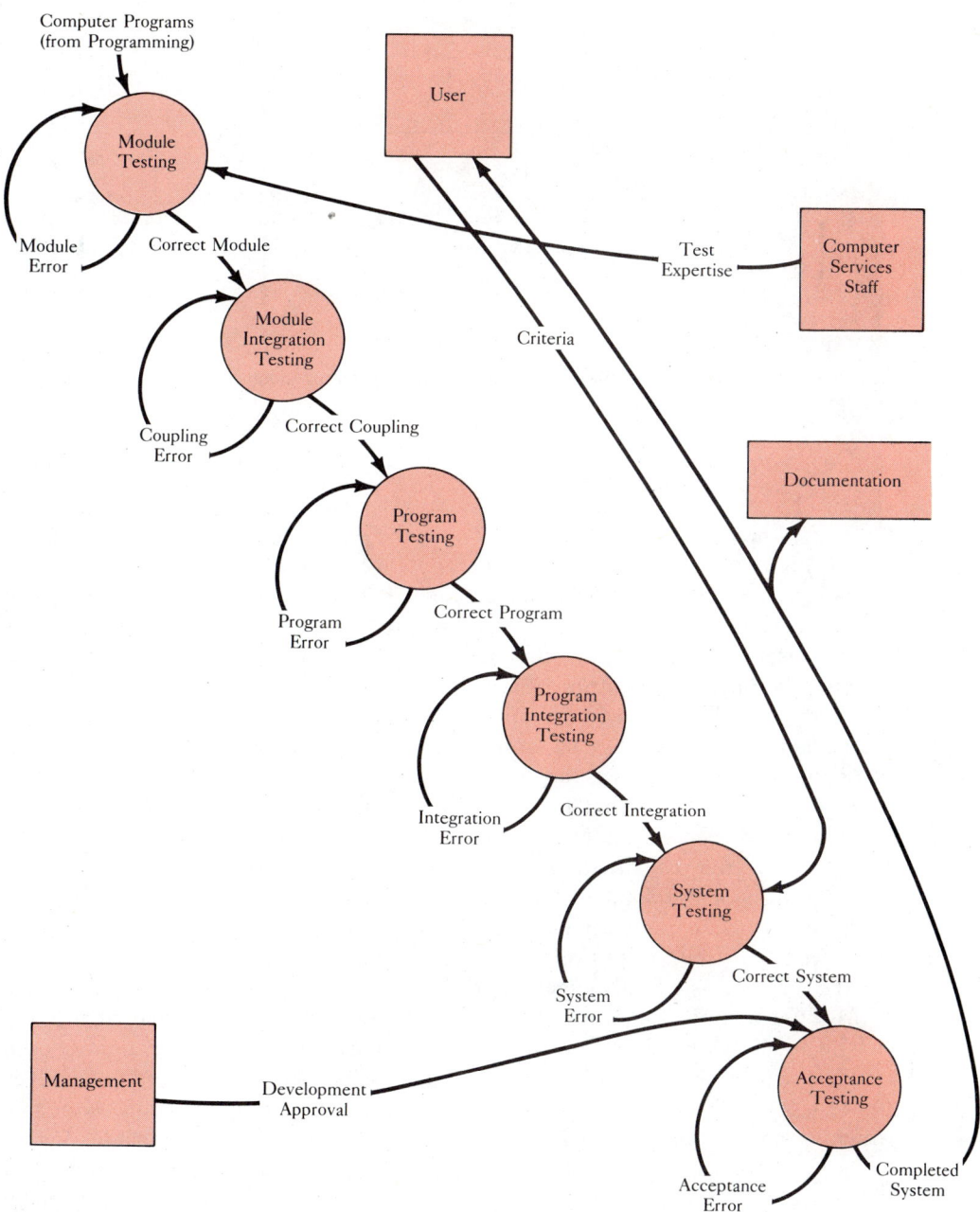

(b) The first three phases of testing find errors in modules, module coupling, and programs.

FIGURE 12.11 *(Continued)*

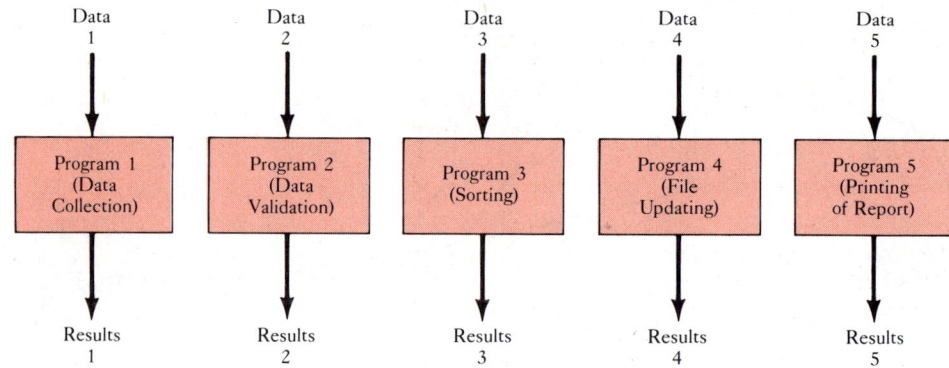

(c) During program testing the analyst and programmer run data through each program and examine the results.

FIGURE 12.11 (Continued)

When a new system replaces an old one, the organization can extract data from the old system to test on the new. Such data usually exist in sufficient volume to provide ample testing, and can create a realistic running environment that ensures eventual system success. However, most students and beginning analysts use artificial data. In such a case, the analyst builds a data file specifically designed to test all conceivable future situations the system might encounter. Although artificial, such data can more carefully test each module in every program. On the other hand, the data may lack sufficient volume, and thus not test the system adequately under normal load.

Finally, a computer program or special software can produce test data. Though such programs can generate sufficient volume to test normal load, the data may not mirror all the real situations the system will encounter later.

Regardless of the source of test data, one must conduct module integration, program, program integration, and acceptance tests (Figure 12.11b). The first three can be carried out at once, by the programmers, and the last one later on by the analyst, user, and programmer (Chapter 13).

Module testing, sometimes called unit testing, centers on validating the correctness of each module. During this type of test, the analyst examines each module separately, and actually tries to make it fail. Once each module survives its test or has been corrected, **module integration** is evaluated to ensure that modules are coupled together properly. Faulty module coupling includes failure to access another module, such as the one that prints captions across the top of the page. Most vendors supply software to assist in module integration. Such software can tell how many times a module will be accessed and the amount of computer time that will be spent in the module.

The third test concentrates on the programs: Do the programs work as they should? With test data available, programmers can test each program individually. The responsibility for conducting these tests falls on programmers' shoulders. If the programs do not produce intended results, programmers must repair them and continue testing until they do.

Module testing
Validating the correctness of each module.

Module integration
Validating the coupling of modules in a program.

In Figure 12.11c Program 1 will run Data File 1 to produce results 1, and so on for each program. When Class Video tested its new tape tracking system, it collected data for five programs:

Program 1 collects raw data
Program 2 validates the data.
Program 3 sorts the data into order by tape number.
Program 4 updates tape master files.
Program 5 prints a report stating tape rental statistics.

After all errors have been corrected, tests cease until the time comes to evaluate program integration, run the completed system, and recommend acceptance. However, in a very real sense, testing never ends because errors and bugs continue to crop up even after an organization has been using a system for a long time.

CONVERSION

While programming is under way the analyst considers the steps to be taken when converting from the old system to the new (Figure 12.12). Smooth **conversion** from one system to another depends a great deal on thorough preparation. The new system may involved installing new equipment, altering facilities to accommodate new hardware, and preparing data—an operation that may demand special programs. Furthermore, the analyst must formulate rules for using the new system, including appropriate user responses to errors or exceptions.

Most organizations adopt one of three standard methods for converting to new systems: parallel, phased, or direct. Each method suits certain situations and has its own advantages and disadvantages.

Conversion The changeover from one system to another.

Parallel

Parallel conversion requires simultaneous operation of both the old and new systems. An operator enters data into both systems for processing so someone can compare the results. If both systems produce the same results, then the new system replaces the old one (Figure 12.13). If results do not match, the analyst must repair the new system before conversion can take place.

Raintree Florists used parallel conversion when computerizing a manual billing system. During August transactions were posted to both systems and management was delighted to see that on September 1 the customer bills matched. It was decided to drop the manual system immediately in favor of a microcomputer system for September's billing cycle.

One minor problem arose at once. Raintree discovered that when a bill was placed in a window envelope, the Zip code did not show with the address; it was printed too far to the right. This rather simple error was rapidly corrected for the October billing by moving the location of the Zip to the left.

Parallel conversion works best when a new system is replacing a similar old one.

Parallel conversion Simultaneous operation of two systems crosschecking the results of each.

404 PART III SYSTEMS DEVELOPMENT

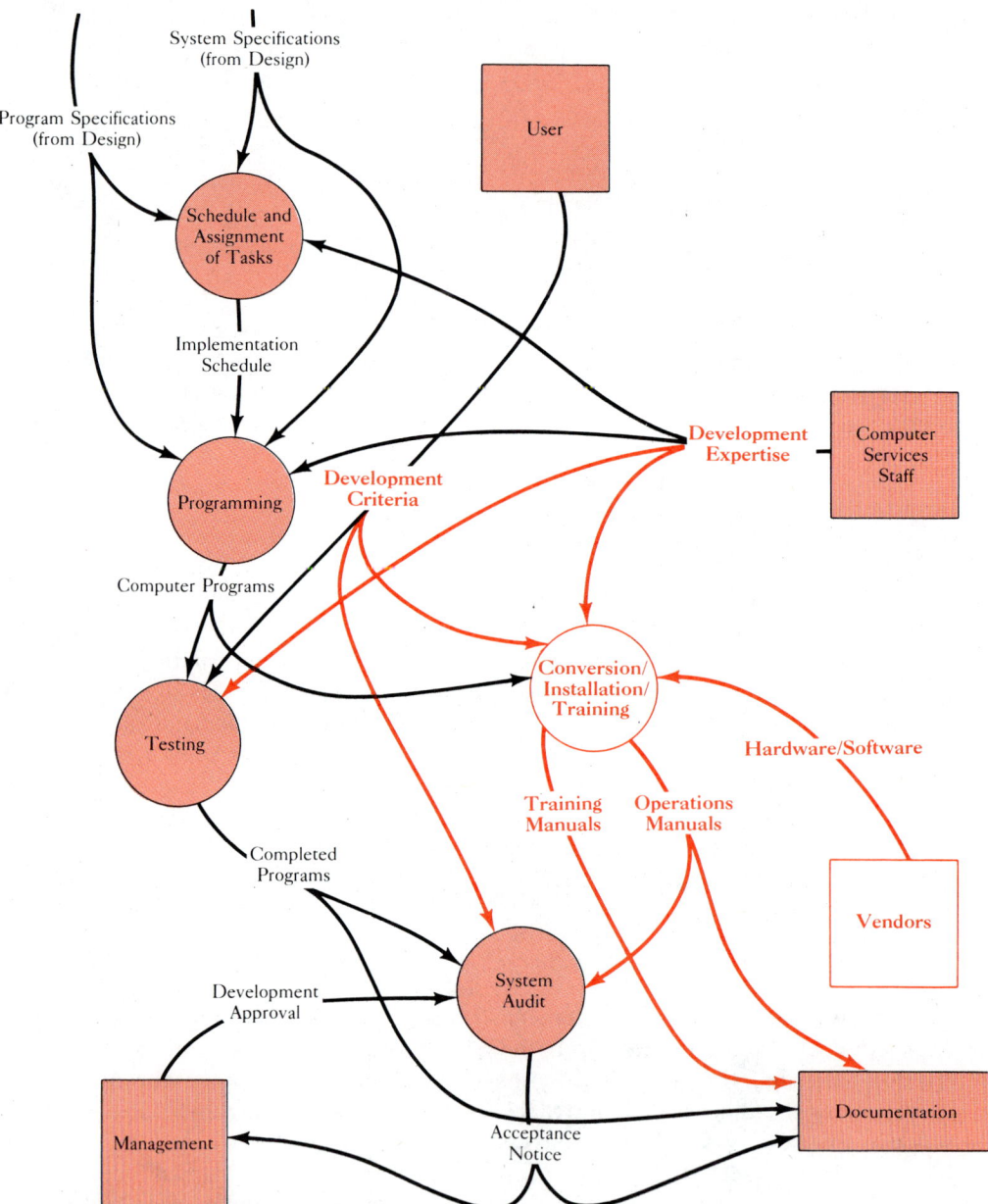

FIGURE 12.12. Conversion can begin during programming.

It also offers some security. If the new system fails, the old system can keep on working. However, when an organization feels comfortable with an old backup, it may take longer for it fully to accept a new system. Furthermore, parallel conversion is expensive because the organization must, in effect, do everything twice.

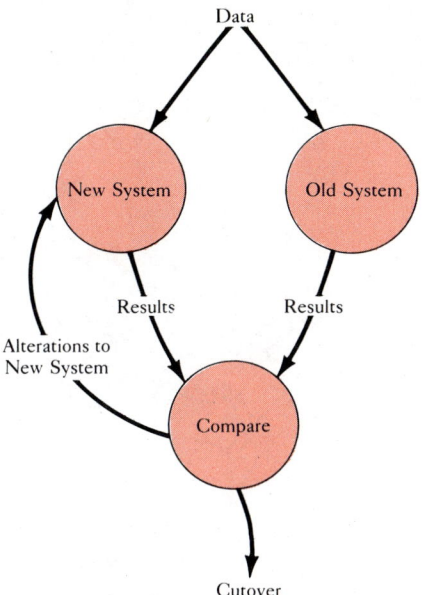

FIGURE 12.13. Parallel conversion requires the old and new systems to operate simultaneously.

Phased

Phased conversion gradually replaces the old system with the new (Figure 12.14). As users become familiar with specific, manageable portions or functions of the new system, they can discard corresponding portions of the old one. In Figure 12.14 we see a system with three distinct components: Function 1, Function 2, and Function 3. During Time Frame 1, the organization tests all three functions but finds only Function 1 satisfactory. Therefore, the organization converts Function 1 to the new system, while retaining Functions 2 and 3 on the old one. In Time Frame 2, the organization tests Functions 2 and 3, which results in the conversion of Function 3 to the new system, and so on until the whole new system is fully operational by the end of Time Frame 4.

The Smiles Dental Clinic used phased conversion to change its accounts receivable system from batch to online. It continued its batch system's courier service for picking up charge slips but began to use the new online system terminal to generate patient reminder notices. Since this function worked well, Smiles converted it after the first month. The second month the clinic concentrated on insurance billing, which also worked well, resulting in computerized printing of insurance bills at the end of the second month. In the third month the clinic cut over to the complete patient billing system, and finally canceled the unnecessary batch backup.

Phased conversion is less expensive than parallel because the organization does not duplicate all data entry, does not process everything twice, and distributes work loads evenly. Since it evaluates the system on a module-by-module basis, it reflects the spirit of structured methodology.

Phased conversion
Gradual replacement of one system with another.

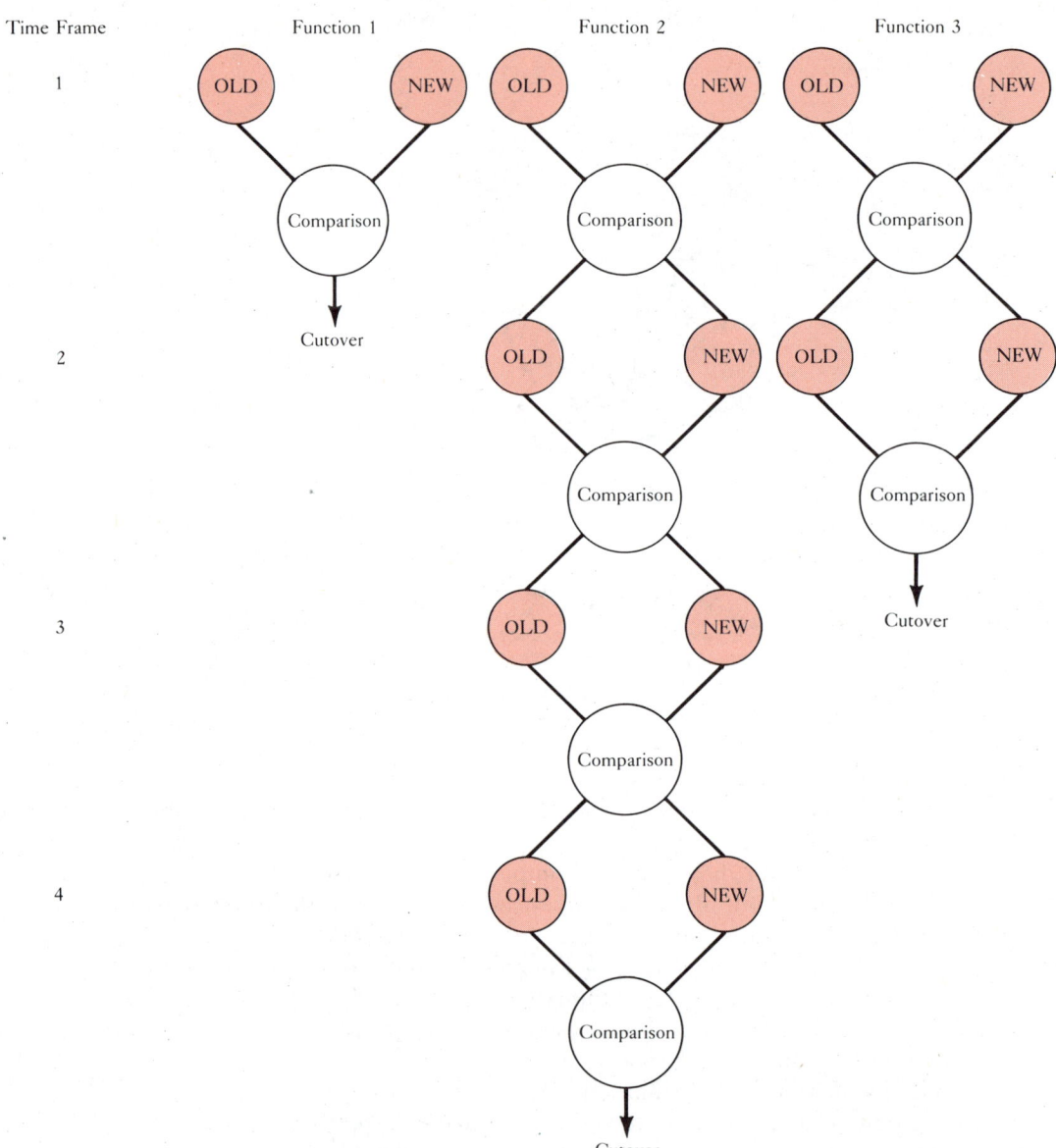

FIGURE 12.14. Phased conversion gradually replaces the old system with the new one. In this case an organization gradually converts its functions over four time frames.

However, phased conversion may confuse users or customers if they see results of both systems simultaneously. Phased conversion also stalls management's appraisal of overall business performance because it cannot easily pull together data spread over two systems. For example, if a company cannot compile its monthly list of customers who have not paid their bills for over 90 days, it can imperil its positive cash flow, thereby placing its own good credit with vendors in jeopardy.

A delay in spotting certain trends or events that could have profound impact on the system itself could also create a problem.

Direct

Direct conversion (also called "cold turkey," "slash," "cut," or "inventory") involves immediate conversion from an old to a new system (Figure 12.15). Because it eliminates any backup, this method requires a thoroughly tested new system and is the most hazardous of all conversion methods.

Direct conversion The immediate changeover from one system to another.

Mountain Motor Supply, an auto parts retailer, adopted direct conversion when it computerized its manual inventory system. On July 1 (the beginning of the company's fiscal year), all manual write-up of customer orders ceased, and clerks began entering customer orders directly on terminals located at each parts counter. Sales staff simply entered part numbers, to which the computer immediately responded by printing a sales receipt via online printers positioned next to the terminals. When one of the online printers suffered a mechanical failure, the spare

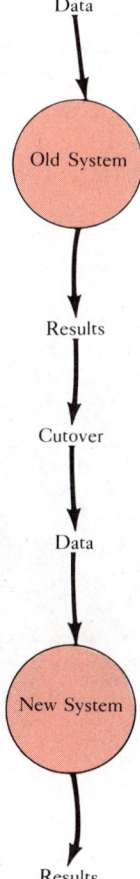

FIGURE 12.15. Direct conversion calls for immediate conversion from the old to the new system.

printer the analyst had in reserve for just such an occasion replaced the broken one.

Direct conversion is the least expensive but riskiest of the three conversion methods. Though it eliminates customer or user confusion over which system produced the results, the lack of sufficient backup can preclude its use.

Programs, Facilities, and Procedures

Regardless of the conversion method an organization uses, analysts must convert data files, programs, facilities, and procedures to any new system. Data file conversion becomes necessary whenever new systems transfer data from tape to disk, change from sequential to random file access, or adopt a data-base environment, and new data files must implement the file design developed earlier in the systems process. The organization must collect and enter new data and drop data elements no longer needed. Sometimes a special program can be written to remove data automatically from the old system and enter the data into the new one, thereby speeding the file conversion process.

The programmers for a small building supply firm changing from a batch magnetic tape to an online disk data-base system wrote programs that could read data from the batch tapes directly to magnetic tapes. Other special programs transferred the data from the tapes into the data base, while yet another one collected new data required by the new system (veteran status, high school attended, and birth date), entering the data directly into the data base. Once all the new data were successfully stored, the supplier discarded these special-purpose programs.

FIGURE 12.16. Computer vendors will provide plans to help the analyst position the computer in a room. They can also specify the electrical and cooling needs of their equipment. (Courtesy of Sperry Corporation.)

If an application requires a new computer, the organization usually must alter its old programs to get them to run properly. If the new computer fully supports old COBOL programs, then program conversions may be minimal, but if old assembly-language programs are not compatible with the new system, conversion will be a major undertaking. Some packaged software can assist the process by letting the computer help in the laborious conversion from one machine to another.

An organization may have to alter existing facilities to accommodate a new system. Such alterations might include raised floors, provision for additional electricity and air conditioning, dust and humidity controls, and new lighting. Hardware vendors will readily supply power, space, and air conditioning requirements and specifications for their equipment, and may even draw floor plans showing the optimal layout (Figure 12.16).

An analyst also accepts responsibility for equipment removal. If an organization

CLASS VIDEO	ERROR REPORT FORM
ATTENTION:_____	REPORT SUBMITTED BY:_____
	DATA SUBMITTED:_____
DESCRIPTION of ERROR:_____	
_____	
_____	
_____	
_____	
SPECIFY URGENCY:_____	
SAMPLE or EXAMPLE ATTACHED:_____	
M. I. S. Use Only	
REPORT RECEIVED ON:_____	REPORT RECEIVED BY:_____
	REPORT ASSIGNED TO:_____
CAUSE of ERROR:_____	
_____	
SOLUTION to ERROR:_____	
_____	
_____	

FIGURE 12.17. A preprinted form allows users to report problems or errors they encounter with a new system.

buys its new computer from a vendor other than the one with which it has done business in the past, the old vendor may not cooperate in this endeavor. Therefore, an analyst must pay special attention to rapport. For example, one might assure the old vendor that it will receive consideration for future equipment, even if it failed to land this particular order.

If the organization finds itself with outdated equipment on hand, it might sell it to a used-computer vendor specializing in older hardware. One organization sold its 10-year-old computer to a junk dealer who melted it for the gold, silver, and other metals it contained. A used-computer vendor had offered $15,000 for the same machine but the junk dealer paid over $25,000 for what had once cost a million dollars.

Regardless of how thoroughly a new system has been tested, problems inevitably arise. Preprinted forms can help users report such problems (Figure 12.17). Forms designed to report errors must allow space for dates, descriptions, personnel involved, and response. When some user spots errors or omissions during this stage of the systems process, the analyst must respond rapidly, correcting them before they proliferate and strengthening user confidence that the analyst cares about user needs.

Regardless of the reporting system, weekly meetings among the analyst, vendors, and users can help prevent problems as well as solve them. For example, the same organization that sold its old computer to the junk dealer initially met weekly with its new vendor. During these meetings new problems were analyzed, potential future problems aired, personnel training discussed, and future wants defined. Weekly meetings became monthly during the second year of the new system's operation, and eventually ceased altogether.

CASE STUDY

Developing Fleet Feet's Accounts Payable Voucher-Check System

Analyst Frank Pisciotta looked forward to the development phase of Fleet Feet's accounts payable system. Formally authorized to proceed by management's October 24 memorandum, he began by reviewing the design phase documentation and building an implementation schedule (Figure 12.18). After scheduling all vital tasks, he assigned them to staff and an outside vendor (Figure 12.19).

Considering the relatively high number of programs required and the fact that they must be ready in 2 months, Frank elects to contract some programming to a software development firm, Data Management (DM). Having used DM in the past, Frank knows its work will meet his high standards. Data Management has its own HP 3000 computer, so its development efforts will be compatible with Fleet Feet's HP 3000. At $60 per hour, including machine time, disk storage, and paper costs, DM agrees to deliver the software in 4 weeks at an estimated cost of $2200, well within Frank's cost and schedule requirements.

Hewlett-Packard will deliver the new terminals Frank ordered after evaluating the company's response to his terminal evaluation form (see Chapter 10, Case Study). Since Maasta Construction handles all of Fleet Feet's remodeling, electrical, plumbing, heating, and air

CASE STUDY

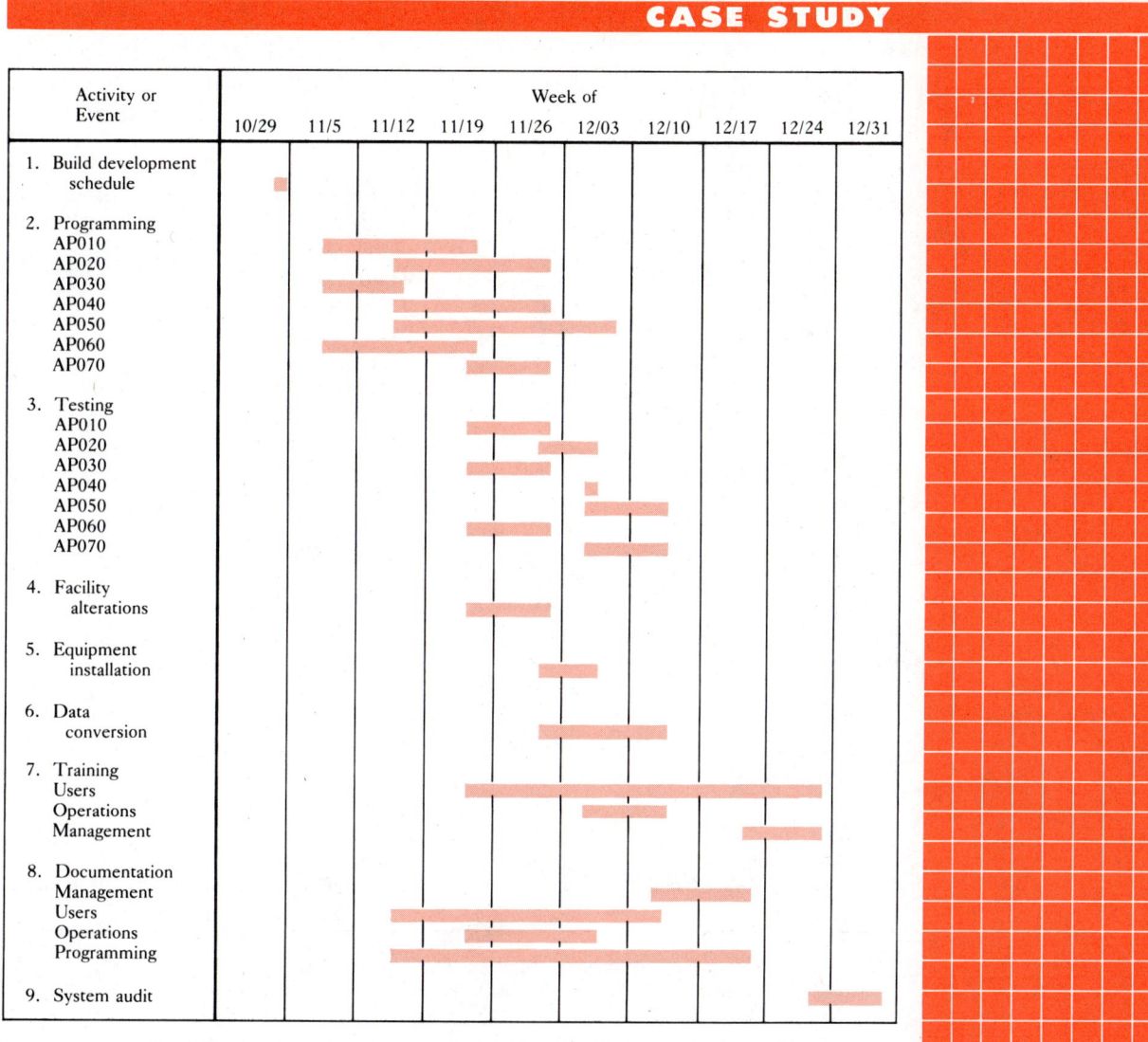

FIGURE 12.18. Frank Pisciotta's implementation schedule for Fleet Feet's accounts payable system.

conditioning work, Frank asks Maasta to estimate costs for minor electrical alterations to the warehouse and bookkeeping areas.

Fleet Feet's current AP system is manual, so Frank can select any of the three data conversion methods: parallel, phased, or direct. Since Fleet Feet is not replacing an existing computerized system, and since Frank cannot break the manual AP system into components, he decides against phased conversion. Considering the risks direct conversion poses for such a small business, Frank avoids it in favor of parallel conversion. An examination of the schedule reveals that the APO10 main data-collection program for vendor data should be completed by November 23 and tested by December 3, so Frank plans to commence the massive job

CASE STUDY

Event	Personnel Assigned
1. Build implementation schedule	Pisciotta
2. Programming	
AP010	Keachie (DM)
AP020	Williams
AP030	Williams
AP040	Hardt (DM)
AP050	Hardt (DM)
AP060	Williams
AP070	Keachie (DM)
3. Testing	Keachie (DM)
AP010	Williams
AP020	Williams
AP030	Keachie (DM)
AP040	Hardt (DM)
AP050	Hardt (DM)
AP060	Keachie (DM)
AP070	
4. Facility alterations	Pisciotta, Brown (Maasta Construction)
5. Equipment installation	Pisciotta, Neill, Engberg (HP)
6. Conversion data	Pisciotta, Hardt (DM)
7. Training	
Users	Pisciotta, Smith, Fane, Darlington
Operations	Hardt (DM), Neill
Management	Pisciotta, Edwards
8. Documentation	
Users	Pisciotta, Keachie (DM)
Operations	Hardt (DM), Neill
Programming	Hardt (DM), Keachie (DM), Williams
9. System audit	Pisciotta, Smith, Edwards

FIGURE 12.19. *Personnel assignments for Fleet Feet's AP system include tasks and schedules for individuals from Fleet Feet, Data Management, Hewlett-Packard, and Maasta.*

of entering data to the vendor master data set immediately after the data-entry program passes its tests.

Aware of Thanksgiving and Christmas holidays, Frank chooses to wait until early December to convert the rest of the data, figuring he can build other data sets during the first 2 weeks of December, thereby allowing a parallel run in January. Since the Christmas and New Year's holiday season is a peak selling time for Fleet Feet, data for this period should provide an adequate check of the new system. If it fails, the company can fall back on its old manual system. The first 2 weeks in January are slow sales periods, so Frank and his people will have time to run both the manual and computerized systems to compare results.

WORKING WITH PEOPLE   Coping with Change

Franklin Pest Control's customer service representative, Margot Sue Lyon, sat glumly at her desk, worrying over impending job changes. Next Wednesday the computer department would install a terminal on her desk and put her "online," whatever that meant. With only 3 years until retirement, Margot Sue wished the company could have waited until after she had left to enter the computer age. After all, she hadn't needed any new-fangled equipment for 17 years. Why mess with a good thing?

Margot Sue liked her job. Whenever anxious customers whose names began with A through J called to ask when Franklin would spray their property for gypsy moth caterpillars, she simply looked up the answer in her manual delivery schedule. It took a day or so for her to call back with the answer, but no one had ever complained. Since she cherished the friends she had made as a result of her warm friendly telephone manner, she was afraid the new terminal inquiry system would ruin the best part of her job. Her boss claimed the new computer would allow her to work more efficiently, but she didn't think she could handle additional customers K through M. And what about poor Helen Leslie, one of the three existing service representatives who would have to move to a new job? Was Helen going to be fired?

Even more than the loss of her friend Helen and personal contact with customers, Margot Sue dreaded the ungainly terminal that would dominate her desk. It looked like something out of a science fiction movie.

Fidgeting with a paperclip and stewing about all this, Margot Sue didn't notice Stan Munoz, Franklin's systems analyst, approach with two cups of coffee.

Stan smiled brightly, "How about some black coffee? You don't take cream and sugar, do you?"

Margot Sue's eyes brightened, "How did you know that?"

"Helen told me. You know, she's working in my office now."

"No, I didn't know. What does she do?"

"She's learning all about computers." This amazed Margot Sue, because Helen had been as upset about the new machines as she had.

"Computers. Ugh!"

Stan laughed. "If it weren't for computers, I'd be out of a job."

"And because of computers," moaned Margot, "we're *all* going to be out of jobs!"

Her response seemed to bother Stan. "Where did you get that idea?" he asked.

"I'm too old for all these changes—Helen leaving, new customers I don't know, new faces in the computer department, a stupid machine with a lot of buttons. I won't even be able to set a coffee cup on my desk any more, my parking space has been blocked all week by that truck from IBM, and my granddaughter is getting married on Saturday."

Stan shook his head. "I see I haven't been doing my job. Sorry about your parking space, but you'll have it back next week. Helen? She's changed jobs. When I began designing the new order-entry system, I knew one person would have to go. I've spent a lot of time with Helen, so I chose Helen as our new data-entry control clerk. She loves it. She's been taking a night class in computers, with the company paying for tuition, books, and expenses."

"Really!" exclaimed Margot. "Next thing you know I'll become a robot."

Stan frowned. "I should have talked to you sooner about all this. I assumed you'd be grateful to see your job streamlined."

"Streamlined so much I won't be able to make friends on the phone any more. Everyone's just a number."

"Not at all," said Stan. "Your terminal will help you answer customer questions faster and more accurately. People will appreciate that."

Margot mulled this over. "I guess that's good," she said grudgingly. "But what if I push the wrong button and blow up the computer? Then we'll both be out of job."

"You can't damage the computer, Margot. We'll label all the buttons so you'll know what they're for."

"That doesn't solve my space problem. I don't even have room for a flower vase on this desk."

"The terminal will sit on its own table. And you'll be able to throw these away." Stan pointed to several thick binders that held 3 years' worth of customer data.

"I didn't know that."

"As I said, it's my fault. I've been so preoccupied with the new system, I've neglected the people it will affect the most. I'm sorry."

Margot laughed. "Don't apologize. I just might figure out these computers some day."

SUMMARY

System development is the third and final phase of the systems process. It includes actual construction of the system on the basis of specifications and plans developed during analysis and design.

Analysts begin development by framing implementation plans, which outline a schedule and assign tasks to specific individuals: users, managers, computer services staff members, and hardware or software vendors.

With the plan in place, the analyst assigns programs to the programming staff. Programmers write the code necessary for generating desired reports, collecting necessary data, and manipulating data files. Technical walkthroughs of programs occur before the programs are actually run on a computer.

While the programmers are writing code, the analyst decides how to convert from the old system to the new one. Conversion can be parallel, phased, or direct. Parallel conversion requires an existing system to remain in operation while the replacement operates beside it. Phase conversion involves gradually replacing components of the old with corresponding components of the new system, and direct conversion demands total and immediate change to a new system.

NEW TERMS

Bottom-up approach	Module integration	Program walkthrough
Coding	Module testing	Program writing
Conversion	Parallel conversion	Standards
Development	Phased conversion	Stub
Direct conversion	Program testing	Testing
Implementation plan		

QUESTIONS FOR REVIEW AND DISCUSSION

Questions appear in three categories. You can find the answers to the A group in this chapter. The B group requires you to apply the material presented here, whereas the C group necessitates investigation or research.

A.1. List the activities that take place during the development phase of the systems process.
A.2. What items does the analyst have to juggle and schedule when the implementation plan is built?
A.3. What are the inputs and outputs in programming a system?
A.4. What three factors do program walkthroughs try to find?
A.5. List the three types of conversions.
A.6. List the items that must be converted for the new system.
A.7. List three common standards a systems department may choose to employ.
A.8. In building the implementation schedule, what groups of people should the analyst contact?
A.9. What are two synonyms for direct conversion?
A.10. List the two types of program construction.
A.11. Define a program stub.

B.1. Who are the participants in a program walkthrough?
B.2. Could Mountain Motor Supply have used the parallel method of conversion? Explain.
B.3. Write a data dictionary definition of systems documentation.

C.1. What are the differences between top-down and bottom-up development of a program or system?
C.2. What did Stan Munoz learn as a result of his talk with Margot Sue Lyon?

REFERENCES

F. T. Baker, "Chief Programmer Team Management of Production Programming," *IBM Systems Journal*, Jan. 1972.
B. Dickenson, *Developing Structured Systems*, Yourdon Press, New York, 1980.
Robert T. Grauer, *Structured Methods Through COBOL*, Prentice-Hall, Englewood Cliffs, N.J., 1983.
G. J. Myers, *Software Reliability, Principles and Practices*, Wiley, New York, 1976.
Gary L. Richardson, Charles L. Butler, and John D. Tomlinson, *A Primer on Structured Program Design*, Petrocelli, New York, 1980.
E. Yourdon, *Techniques of Program Structure and Design*, Prentice-Hall, Englewood Cliffs, N.J., 1975.

CHAPTER 13

Testing, Training, Documentation, and System Maintenance

- Goals and Preview
- Testing
- Training
- Guidelines for Terminal Users
- Documentation
- System Maintenance
- Case Study: Testing, Training, and Documentation for Fleet Feet's Accounts Payable Voucher-Check System
- Working with People: Writing with Style
- Summary
- New Terms
- Questions for Review and Discussion
- References

GOALS AND PREVIEW

After reading this chapter you should be able to:

1. Describe the three types of tests a new system must pass.
2. Identify the individuals who should be trained to use the new system.
3. Describe system documentation and its use.
4. Define system maintenance.
5. List the components of a training manual.

By this point in the development phase of the systems process, the analyst has put together an implementation plan, has ordered programs to be written and tested, and has chosen a conversion method. Now the analyst concentrates on system testing, training, documentation, and maintenance.

Testing helps assure that the system will achieve its goals. Earlier, each module underwent a variety of tests, using sample data, until all modules appeared correct, their linkages proper, and each program worked as intended. If problems and errors arose, programmers worked to correct them until the programs ran smoothly. At this juncture three other tests will occur: program integration, system, and acceptance.

Before, during, and after testing, the analyst or a technical author writes documentation and training materials. Users, management, and operations personnel need specific instructions, usually in the form of reference and user manuals, that will enable them to use the system and the information it will generate.

Provisions for maintenance of the system conclude this phase of development. Maintenance involves not only the upkeep of the system, but a means to update it if additional problems develop or users demand improvements or modifications.

TESTING

Testing (1) Verifying that a system performs as expected. (2) Checking a program to see that it performs as expected.

Automobile buyers not only examine the specifications and appearance of their intended purchase, but they actually take the car out for a drive to see how it runs on the road. Similarly, when all components of a computer system have been assembled, the analyst begins **testing** them (Figure 13.1). Module, module integration or coupling, and entire program testing occurred earlier in development (see Chapter 12). Now the tests center on program integration, system proving, and, finally, an acceptance test (Figure 13.2). Tests run the gamut from something as elementary as whether or not work stations have been arranged most conveniently, to something as sophisticated as full processing of millions of records.

Program Integration

Program integration test Verification that individual programs work together as expected.

String test Evaluation of programs where output from one program is input to the next.

Link test Verifying that programs work as intended.

Test data The collection of data the analyst plans to use in proving the system's accuracy.

After each program passes its test, it must be linked by the analyst to the other programs in a **program integration (string** or **link) test** (Figure 13.3). This fourth test assures that the programs work together as intended. (Chapter 12 shows module, module integration, and program tests.) If an error appears, the analyst and programmer isolate and correct it. For example, even though each of Class Video's individual programs ran perfectly, Program 1 (which collects raw data) did not place the tape rental date in the file Program 2 used for validation purposes. The program integration test caught this error, which the module and module integration tests missed.

The analyst may devise **test data** that actually try to force the system to fail. In an accounts payable system, one should be able to force failures by having users enter improper account numbers or purchase amounts, credit amounts they should debit, or enter alphabetic data when the system requires numeric. Such failures help teach programmers procedures for error correction, and they allow analysts to observe certain system behaviors: the time it takes to process a transaction, terminal response time, and the impact of a new system on the other demands the organization makes of its computer.

Class Video conducted a system test on its new video tape tracking system and found a 3-minute terminal response time between data entries. Since the organization could not tolerate such a lengthy response time, the analyst investigated the situation and discovered that the problem stemmed from the fact that the data-base design required sequential processing of over 1.5 million entries. By replacing the sequential requirement with end-of-the-month file sorting, the analyst was able to reduce terminal response time to 3 seconds, well within the company's requirements.

System Test

System test Users verify that a system performs as expected.

Though the programmers, analyst and occasionally managers of the computer services department conduct a variety of tests before a system is running, it ultimately must satisfy users and their managers. After all users, not the computing services department, will interact with the system daily. During **system tests** users enter data (Figure 13.4) and observe the results. Unlike program integration

CHAPTER 13 TESTING, TRAINING, DOCUMENTATION, AND SYSTEM MAINTENANCE 419

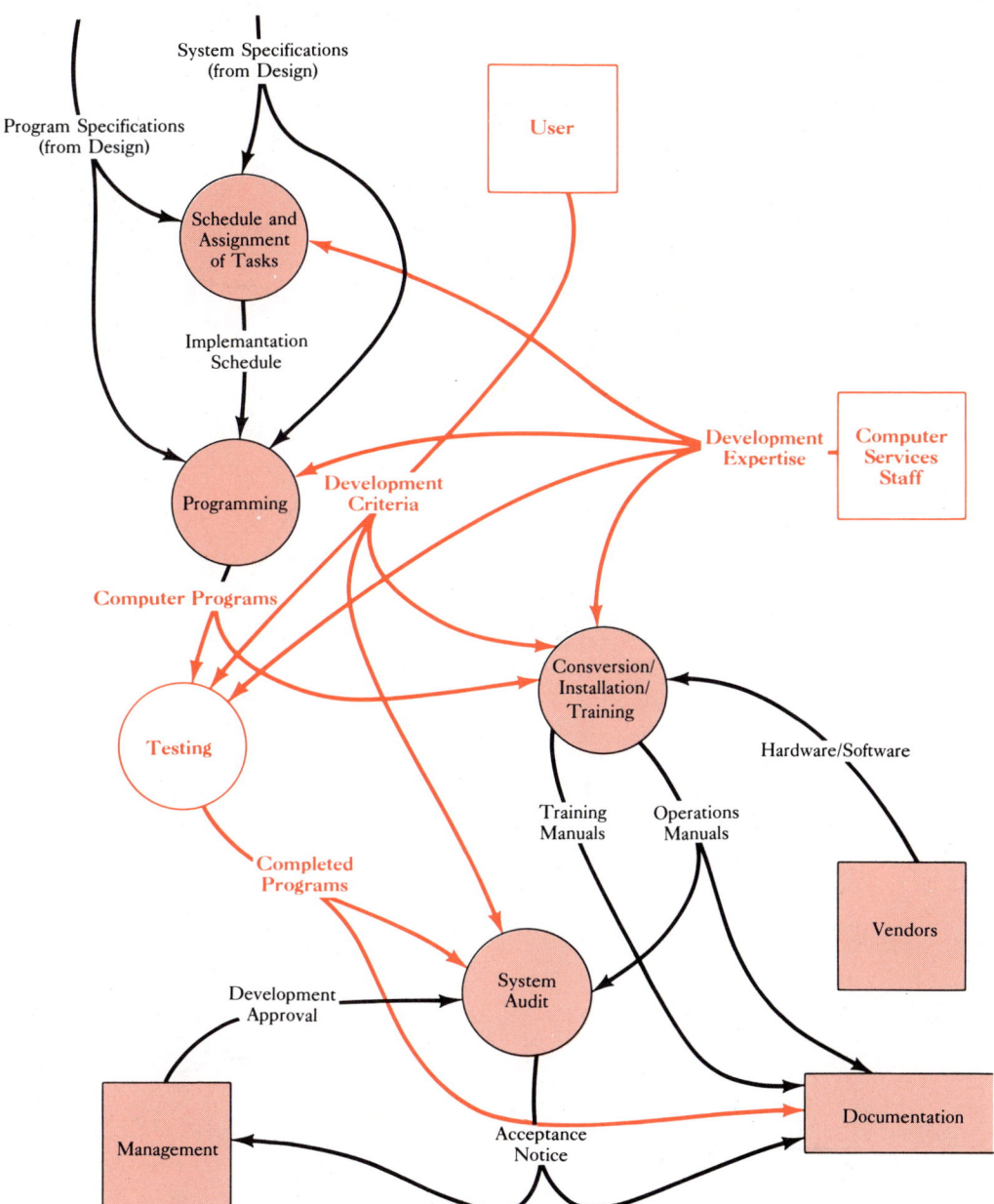

FIGURE 13.1. Testing involves all personnel and components of the system.

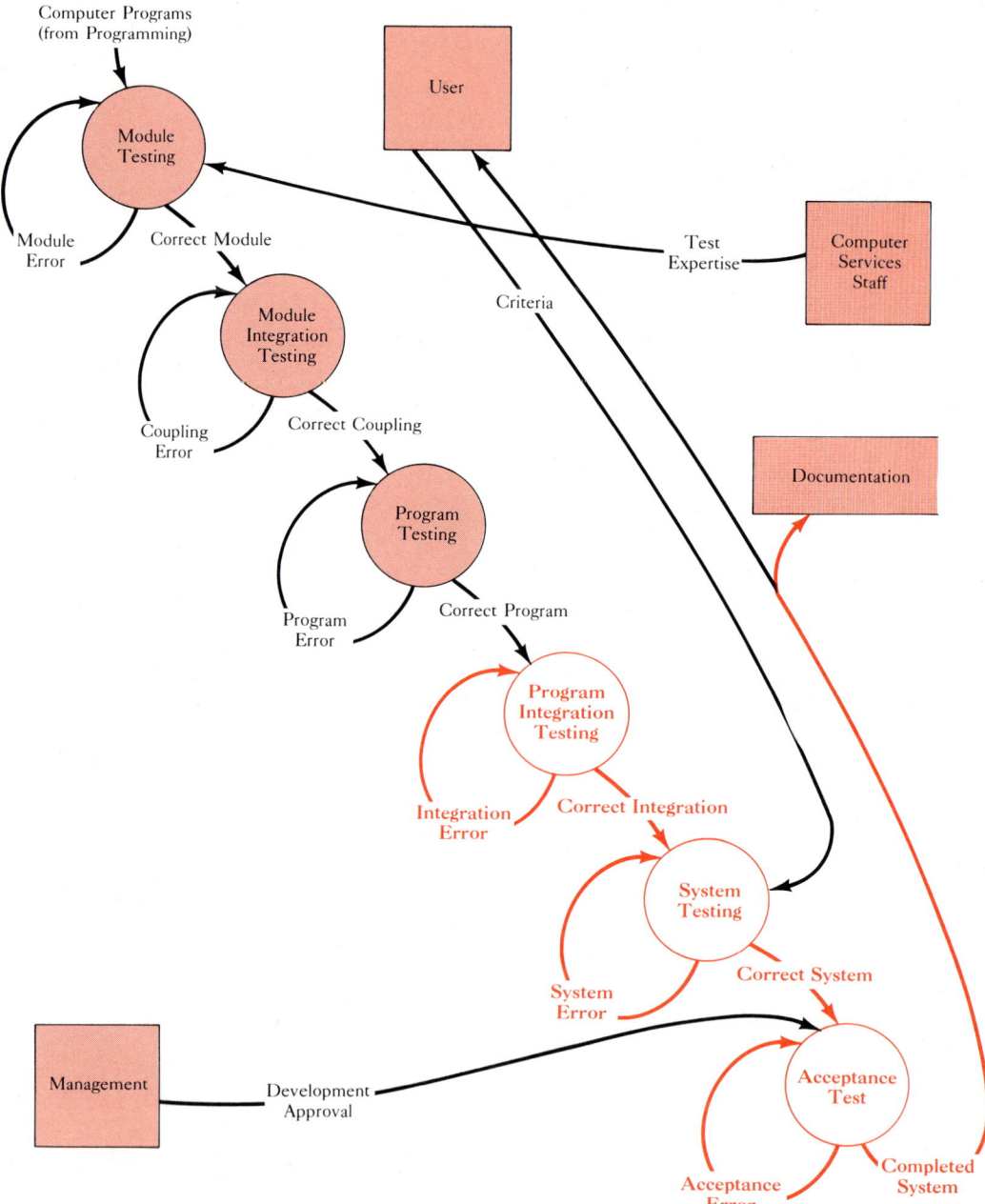

FIGURE 13.2. Testing can be decomposed into six components. In Chapter 12 we followed testing through the first three phases. Now the analyst looks at the last three leading to a completed system.

CHAPTER 13 TESTING, TRAINING, DOCUMENTATION, AND SYSTEM MAINTENANCE **421**

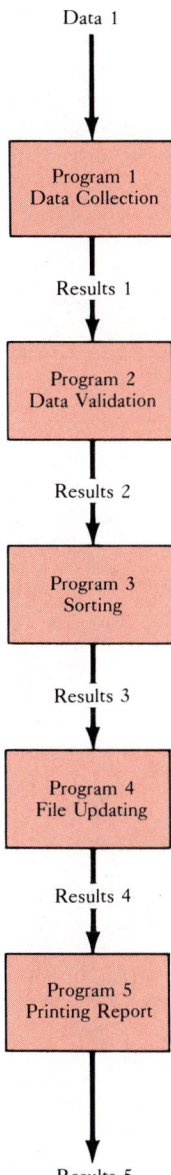

FIGURE 13.3. Program integration, string, or link tests unite all the programs to ensure that data files created by one program are compatible with the next.

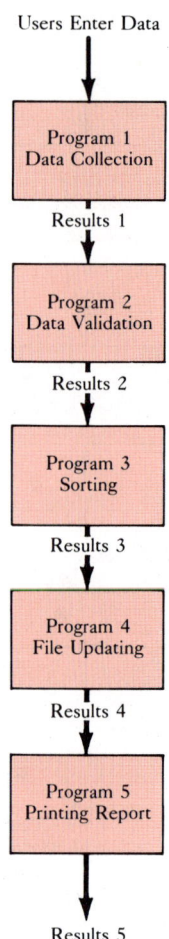

FIGURE 13.4. A systems test occurs after individual program and string tests. This test makes sure all components of the system function properly as a unit by actually trying to force the system to fail.

tests where the analyst or programmer provided data, responsibility for data now falls to users, who must carefully study the reports the system produces, matching them against any known results to authenticate accuracy.

Acceptance Tests

Acceptance test
Verification that the system meets its original goals and objectives.

Assuming users find no major problems with the system's accuracy, it undergoes a final **acceptance test** to confirm that it meets the original requirements, which were established during the analysis phase of the systems process. Like the system test, acceptance tests are the responsibility of users and management. You will recall that testing tries to make a system fail. If the system fulfills all requirements, it is finally acceptable and ready for operation.

TRAINING

Training Explaining to all parties how a system operates.

As the new system nears completion, the analyst must finish materials for **training** users and the computer center's staff. The new system may bring major changes to users' jobs, introducing new colleagues and equipment, altering work hours, and affecting even the most minute aspects of the workplace. As we saw earlier, changes can cause anger, resistance, and even sabotage of the new system. Therefore, analysts should exploit training as one means of minimizing problems associated with change.

Tools for Training

Diverse training methods exist, but with the most popular, users, their managers, and the computer center staff are trained separately. No one enjoys sitting through a training program of which only one third directly concerns one's job. Class Video's owners chose to hold two separate training sessions, one for the owners, Harry and Connie Allen, and the other for the clerical staff. All learned how to incorporate the new system into their daily jobs, but the owners' training focused on inputs required of them and the outputs they would see, whereas the clerks learned the mechanics of actually operating the new system.

The analyst may choose to train one user thoroughly, who will, in turn, train others. Delegating training responsibilities to users helps build confidence because it makes them "experts," and nothing stimulates success as confidence does. Users are more familiar than analysts with ongoing staff needs, capabilities, problems, and fears, and they often respond more quickly and cheerfully to one of their own.

Some large organizations maintain special training departments. For example, the State of California employs an entire separate department to train state employees on IBM hardware and software; it not only offers basic instruction to programmers, operators, analysts, and management, but follows up with special classes to upgrade skills. Such in-house programs have become quite popular because they can be tailored to the specific needs of an organization.

Since vendors routinely provide a certain amount of training, sometimes free of charge, analysts should consider inserting a training clause in the original purchase agreement for hardware or software. Though vendors may not be familiar with another organization, they do more fully understand their own equipment and how it relates to equipment from rival manufacturers. Vendors may send representatives to the organization, or they may invite the organization to send people to regularly scheduled classes. Large cities throughout the country support many such vendor schools.

Colleges and vocational training programs can supplement training with courses that closely parallel vendor and large in-house training courses. Colleges also offer more generalized courses, which students take to broaden their education.

Specialists and consultants have begun to offer seminars on data-base management, structured methodology, testing, electronic spreadsheets, data communications, time management, and other up-to-date topics. New video, audio, and computer-assisted instruction training materials have also begun to proliferate. For example, Boeing Computer Services offers a video-taped series of classes ranging from introductory material to an in-depth look at data communications.

Management Training

Regardless of the training mix an analyst selects, he or she must anticipate the need for different levels of training for each of three groups of people: users, their managers, and the computer services staff. Managers need a broad view of how the new system will help them more productively fulfill their responsibilities. Therefore, management training focuses on the information each report will contain. For example, if the new system has an inquiry facility, managers may want to know how to interrogate the data base with such pressing questions as "Which customers' debts within area code 305 are more than 90 days old?" Of course, these managers do not need to know technical details about entering data to the system.

Small group seminars usually work best for the training of supervisors and managers. Placing managers in a general training session with subordinates who possess a high level of technical knowledge may cause the managers to avoid asking questions they fear will make them appear ignorant. In some instances management training should be conducted away from the office, thus eliminating interruptions resulting from the daily crisis situations so common in any active organization.

Users and Operations Staff

User training differs somewhat from management training. Not only do users need to know how to interrogate the data base, but they must know in detail how to enter data, how to respond to error messages, and how to call up routines to print reports.

Since the computer center's operations staff concerns itself with the system's operation, operators must learn procedures to run various programs, how to back up the system, and such mundane matters as which paper the printer will use for various programs, on which disk drive the data base resides, and how to increase or decrease data-base file sizes.

Intensive classes, frequently lasting many days, often work best for users. If the system is particularly complex, it may take months for them to feel fully comfortable with it.

Any user training program should begin with a system overview, which includes the system's goals and objectives, its mode of operation (batch, online, or distributed), and the kinds of data the system will collect, store, and report. Then, since effective training depends on the analyst's ability to determine and fill individual needs, one can customize more specific training to reflect these.

Once the system is in full operation, the analyst should conduct a follow-up training session aimed at more seasoned users, management, and operators who have become familiar enough with the system to pose more sophisticated questions and needs. While such follow-up training may not fall within the analyst's realm

Guidelines for Terminal Users

The computer revolution has introduced new health hazards into the home and workplace. Families and workers who spend long hours every day playing games or processing data can suffer from strained backs and eyes. Already furniture manufacturers have begun offering posture-sensitive chairs, but what about eyestrain? Despite recent government studies which claim constant computer use does not permanently damage eyes, symptoms of eye irritation, if unchecked, can develop into serious health problems, according to Dr. Steven Groubert, an optometrist and partner in Personal Support Computers in Los Angeles.

Headaches, blurred vision, eye fatigue, shooting pains, double vision—all are beginning to manifest themselves in most computer-oriented professions and businesses, he said.

As users strain their bodies to compensate for poor visual adjustment to their terminals, they also suffer physical discomfort, including lower back ailments, shoulder pains, and neck muscle tension.

Dr. Groubert offers some interesting suggestions to help overcome many such symptoms:

1. Use a 9- to 14-inch screen. Anything smaller strains the eyes. Anything larger requires excessive eye movement.
2. Try to maintain a 24-inch viewing distance from your screen.
3. Position yourself before your terminal so your eye-level remains 10 to 15 degrees, or about 4 inches, below the middle of the screen.
4. If you normally wear bifocals, consider ordering special computer glasses if the screen is not close enough for your reading lenses, yet not far away enough for your distance lenses.
5. Install overhead lighting. Light from behind causes glare and reflections on the screen. For this same reason you should face your screen away from windows and doors.
6. Both green-toned screens, popular in the United States, and amber ones, popular in Europe, are better than white ones. Amber and green fall near the center of the visible light spectrum, placing minimal strain on the focusing mechanism of the eyes. Choose the color that feels most comfortable to you.
7. Try out different chairs that provide good back support and allow the feet to rest flat on the floor.
8. Rest your eyes occasionally by closing them for a few minutes. Exercise them gently by looking into the distance now and then. Use a cold compress at the outset of even slight eyestrain.

Whether computer-related symptoms cause any long range problems or not, they can not only cause very real discomfort, they can reduce your productivity and cause you to make more than the usual number of mistakes.

Source: Vonne Godfrey, "Take Care of Your Eyes at the Computer," distributed by *Los Angeles Times* Syndicate, Jan. 23, 1984.

of responsibility, it can reap huge rewards. For example, after members of Class Video's staff had worked with the new system for a few months, they learned that they could enter abbreviations of codes (B for Betamax, V for VHS), which sped data entry by 5 percent, quite a savings over a full year.

DOCUMENTATION

Documentation Written materials describing a new system or changes mode to an existing system.

To support training, analysts or technical writers develop written materials that describe the new system both generally and specifically. We call such a written description **documentation**, and, like training, it may include a variety of manuals tailored to individual user needs.

Throughout analysis, design, and development, the analyst has accumulated a great deal of documentation: report formats, cost estimates, screen formats, database and file layouts, program and module descriptions, and schedules, all of which form the basis for final documentation. When one prepares for college exams, steady ongoing study produces better and more lasting results than last-minute all-night cramming. Therefore, analysts should compose all early documentation with its ultimate use in mind. Since so many analysts have neglected writing skills in favor of technical expertise, they can find documentation quite frustrating if they save it until the last minute, when what could have been a dream turns into a nightmare.

Traditionally analysts have treated documentation almost as an afterthought because they are preoccupied with developing expensive and profitable hardware or software systems. This tendency has filtered all the way down to users. Though management often realizes its value, documentation sometimes lacks quality and completeness because people forget that writing good manuals takes skill, time, and money. Modern structured methodology emphasizes strong documentation, insisting that it be developed in an ongoing fashion. Good documentation:

1. Encourages clearer communication between all parties involved with the system.
2. Protects the system when personnel are promoted, transferred, or leave.
3. Represents long-term money savings because it reduces the cost of training.
4. Eases system maintenance by centralizing materials describing the system.
5. Provides a permanent reference on the system.

Of course, good documentation is not easy to write. Some of the major documentation problems include:

1. Incomplete or badly written documents.
2. Out-of-date or inaccurate documents that do not reflect the evolution of the system.
3. Unnecessarily technical and jargon-filled manuals that display no sensitivity to readers.
4. The reluctance of technically trained analysts and programmers to invest time in writing something accessible to users.

Fortunately, modern technology has eased the task. Word processors and text editors can speed schedules and save money by allowing analysts or technical writers to store and update manuals as the system progresses toward completion.

Management Documentation

The amount of detail contained in a given document depends on the intended audience. **Management documentation** needs the least detail, but should include:

1. An overview of the system.
2. System goals and objectives.
3. Examples of key reports that enhance decision making.
4. Final versus budgeted costs.
5. Final versus proposed schedule.

Such documents should be especially businesslike and jargonfree, stressing the value of the new system in increasing the productivity or profitability of the organization.

Management documentation Written materials for supervisors, including overview, goals, costs, and schedules.

User Documentation

User documentation should contain little more technical detail than management documentation, and it should employ the same clear and concise writing style that avoids data processing jargon. The user manual must offer all the information users need to perform their jobs satisfactorily. The screen menu selector in Figure 13.5 shows the sequence of steps a user must take to activate the payroll system, process a transaction, and terminate the program. The explanation gives a brief overview of the system and describes exactly what the user can expect to see on the terminal screen. In this application similar user documentation would be necessary for each of the 19 possible entries the payroll system permits.

Though the analyst or specially trained writers actually write the user's manual, users should participate in developing and testing it by reviewing preliminary drafts. User manuals must be clear, accessible, and jargonfree.

In some situations the analyst may even design online documentation with "help" screens. These screens can be called upon by users when they need guidance, telling them what is wanted or expected.

User documentation Written materials that teach people how to use a system.

Program Documentation

Because programmers need detailed information about the system, **program documentation** demands greater technical detail (Figure 13.6). Strong program documentation permits other programmers to learn the system quickly, a critical need in an environment with a high rate of programming staff turnover or in which a program has been operating successfully but eventually needs modification.

Furthermore, the body of every program should contain comments explaining the purpose of each module (Figure 13.7). Comments should begin every pro-

Program documentation Written materials explaining the details of each module in a system.

CLASS VIDEO USER'S INSTRUCTIONS

System: PR (Payroll)   Date: Sept. 1984   Page 1 of 2

Application: PAYROLL MENU

Comments: The payroll system consists of many programs that provide a range of useful payroll and related accounting functions. You select which program you want to use from the "menu" that is displayed on the screen.

RUN INSTRUCTIONS

1. To use the payroll menu, you type the following:
   PR (followed by striking the RETURN key on your terminal)
2. The terminal screen will show:

PAYROLL MENU   Class Video
1. EMPLOYEE FILE MAINTENANCE
2. PAYROLL TIME PROCESSING
3. CALCULATE PAYROLL
4. PRINT PAYROLL CHECKS
5. MANUAL PAYROLL TRX PROCESSING
6. PRINT PR DISTRIBUTION TO G/L REPORT
7. PRINT PAYROLL HISTORY REPORT
8. PRINT EMPLOYEE REPORTS
9. PRINT QUARTERLY PAYROLL REPORT
10. PRINT YEAR-END W-2 FORMS
11. CHECK RECONCILIATION
12. JOB FILE MAINTENANCE
13. PRINT JOB DISTRIBUTION REPORT
14. COMPANY FILE MAINTENANCE
15. PAYROLL CONTROL FILE MAINTENANCE
16. PR VALID G/L ACCOUNT FILE MAINTENANCE
17. DEDUCTIONS/EARNINGS CODES FILE MAINTENANCE
18. STATE/CITY TAX CODES FILE MAINTENANCE
19. CLEAR EMPLOYEE TOTALS

PLEASE SELECT [..]

---

3. Type the number of the application you want followed by a RETURN. For example, if you print year-end W-2s, type 10 and press RETURN. If you enter anything other than a 1 through a 19, nothing will happen and the screen will wait for a correct entry.
4. To terminate the program, press the END key on your terminal, which will terminate the Payroll system and log you off the computer.

FIGURE 13.5. User documentation consists of a manual that guides the user through the system. A terminal screen selection menu program for Class Video appears here. (Courtesy of MCBA®, Payroll Section 4.12. MCBA is a registered trademark of MCBA, Inc.)

ITEM	DESCRIPTION
Program title	Most programs are named. Some systems adopt naming and numbering systems, AR040. This pattern identifies the program as belonging to the accounts receivable system and that it is the 40th program in the system.
Abstract	Background information about the program: 1. General description or narrative of the program stating its purpose, and where it fits into the system. 2. Date written. 3. Author of the program. 4. Hardware requirements: printers, disks, tapes, or terminals.
Revisions	Similar to the abstract but a list of the changes made to the program: 1. Date of change. 2. Name of programmer making the change. 3. Description of the change. 4. Person authorizing or requesting the change.
System Logic	A HIPO, structure, or flowchart showing a visual perspective of the location of this program relative to the entire system.
Layouts	Definitions of the reports produced, data-base requirements, and input screen designs. This portion of the program documentation comes from the design phase of the system.
Module definitions	Expanded system logic describing each module, its function, data inputs to the module, and data outputs from the module. If the module validates data, the validation requirements are listed. If the module selects another module, the rules for the selection are listed.
Program listing	A copy of the program showing all program statements, a list of variables, cross-reference listing of data names, and module names. A copy of the original as well as the most recent version of the program should be kept. This listing should be produced by the language compiler.
Test data	A copy of the data used to test the original and revised versions of the program. Notes showing expected results of the test are also a portion of this portion of a program's documentation.

FIGURE 13.6. Program documentation provides technical detail about the system.

gram's procedural section, explaining the function of the module, any subtle programming practices used, or complex logic involved. They assist maintenance programmers because they reduce the need to read the program line by line when making enhancements. A software house in New England goes so far as to try to average one comment line per computer instruction.

```
008000************************************************************
008010*                                                           *
008020* MAIN SELECTOR MODULE                                       *
008030*                                                           *
008040* THIS MODULE COMPARES THE INCOMING CODE NUMBER TO DETERMINE *
008050* WHAT SHOULD HAPPEN TO THIS TRANSACTION. THE CODE NUMBER    *
008060* HAS ALREADY BEEN VALIDATED TO BE A1, 2, 3, OR 4.           *
008070*                                                           *
008080************************************************************
008090    500-SELECTION.
008100        IF (CODE-NUMBER = 1)
008105            PERFORM 510-ADD
008110        ELSE
008115            IF (CODE-NUMBER = 2)
008117                PERFORM 520-CHANGE-TELEPHONE-NUMBER
008120            ELSE
008125                IF (CODE-NUMBER = 3)
008127                    PERFORM 530-CHANGE-CREDIT-LIMIT
008130                ELSE
008135                    PERFORM 540-CHANGE-STATUS.
```

FIGURE 13.7. *Comments explaining module purpose, data requirements, and module results should accompany every program.*

Operations Documentation

Operations documentation tells the computer center staff how to run the programs. Without such directions the computer center staff can only guess about such requirements as disk space, backup, frequency of operation, and disposition of printed reports. Newer online terminal-based systems reduce the amount of operations documentation because they allow each user to control relevant programs.

Computer center personnel need a **run manual** for each system. Such a manual contains:

1. System function and purpose.
2. System flowchart (or similar tool) detailing each program in the system.
3. All error conditions and operator responses.
4. Program run information:
   a. Special forms requirements for the printer.
   b. Names of data base or files required by the program.
   c. Hardware assignments, including which disks or tapes should be used and where these tapes or disks should reside.
   d. Disposition instructions: who should receive the printed reports, where tapes should go after the programs has been run, bursting requirements of printed reports.
5. Security: who can use the system when, including log-on procedures and passwords.

For easy reference the run manual should fit into a three-ring binder, which is placed next to the computer operator's console.

Operator instructions range from lengthy ones for a complex program to short ones for a simpler program. Note that the payroll W-2 operator instructions in

---

Operations documentation Written materials for the operations staff explaining how to run a system.

Run manual Operator manual showing how to handle errors, special forms, files, and security.

CHAPTER 13 TESTING, TRAINING, DOCUMENTATION, AND SYSTEM MAINTENANCE   **431**

Figure 13.8 lead the operator step by step through the running of the program, including such major items as terminating the program, and such minor ones as what to do if the printer runs out of paper. Using a system flowchart format, the system logic flow diagram graphically shows operators the sequence of steps they should perform.

CLASS   VIDEO
OPERATOR   INSTRUCTIONS

System: PR (Payroll)   Date: Sept. 1984   Page 1 of 2

Application: PRINT YEAR-END W-2 FORMS

Paper Requirements: W-2 Forms

Comments: This program should be run at the end of the year. It provides all the necessary income and withholding information on a standard W-2 form.

RUN INSTRUCTIONS

1. Log onto the system with the master payroll user code and select option 10 from the Payroll menu. All data files will be present under this user code.

2. Mount the forms in the printer. The form must be mounted so that printing begins with the first form on the page. (A subtotal will be printed every 42nd form. The IRS requires that this form be at the bottom of the page. The program will comply with this requirement if the forms are mounted correctly.)

3. Print as many alignment forms as necessary for the forms to be properly aligned. One full page will print for each alignment; this is three W-2 forms.

4. Enter the starting and ending employee numbers.

5. You may exit this program by pressing the END key while positioned for entry of the starting employee number. The program returns to the Payroll menu.

Data-Entry Specifications

Item No.	Description	Required	Format
1	Starting employee number	Yes	9(6)
2	Ending employee number	No	9(6)

6. If the printer runs out of forms while printing, put some plain paper on the printer and let it finish the run. Then go back and find the last subtotal form that was successfully printed and reprint all W-2 forms from that point on. (The W-2 forms are printed in employee number order.)

7. W-2 forms will not be printed for nonemployees.

8. When the W-2s are finished, they should be burst and delivered to the controller's office.

FIGURE 13.8. *Operator instructions tell the computer operator how to start up and run a specific program. (Courtesy of MCBA®, Payroll Section 4.12. MCBA is a registered trademark of MCBA, Inc.)*

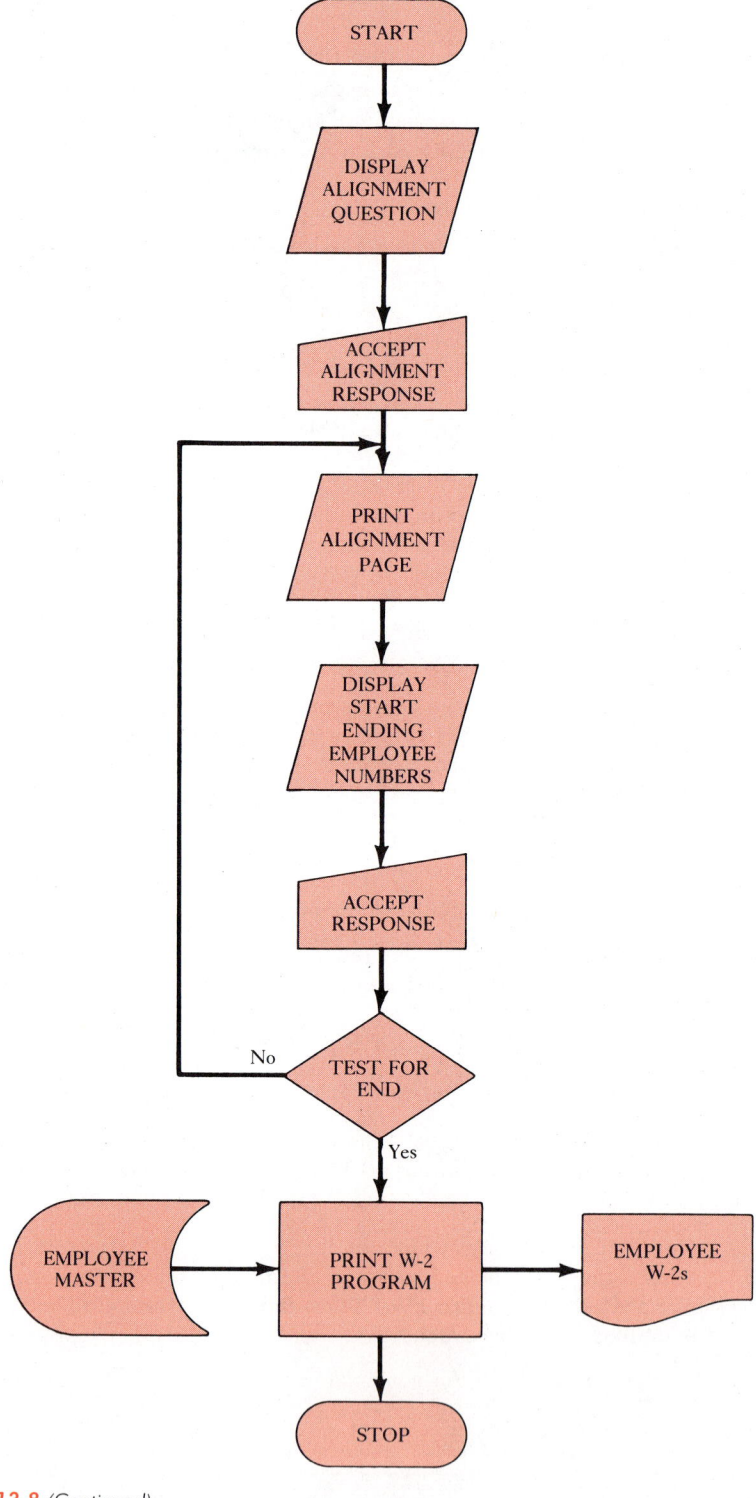

FIGURE 13.8 *(Continued)*

SYSTEM MAINTENANCE

Systems change: users find errors, develop new needs, or want enhancements. Though structured methodology may lead to more error-free systems, any good system should be able to evolve under use. **Maintenance** is the name of this kind of evolution. Because Class Video wanted a report showing the last two tapes various customers rented (to identify potential breakage), its system needed a modification 2 months after installation. Fortunately the system was already collecting the customer number of the last two rentals, so the analyst could quickly respond by having only a single program written to print the new information.

To permit and encourage changes, the analyst can develop a form (Figure 13.9). Such a form provides space to describe an error or desired change and its anticipated effect on the organization. The bottom portion of the form indicates what action, if any, the computer services department takes, including personnel assigned to the task.

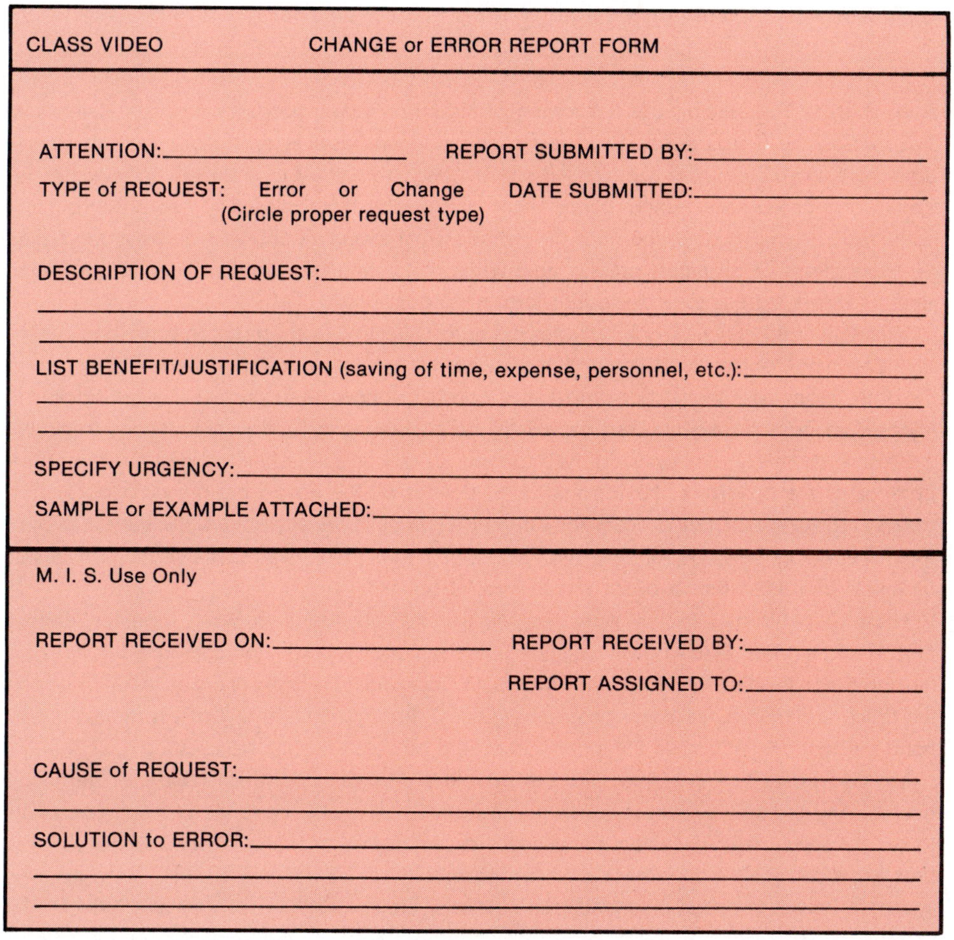

Maintenance Changing or modifying an existing system to fix errors or provide enhancements.

CLASS VIDEO	CHANGE or ERROR REPORT FORM

ATTENTION:_____ REPORT SUBMITTED BY:_____
TYPE of REQUEST: Error or Change   DATE SUBMITTED:_____
(Circle proper request type)

DESCRIPTION OF REQUEST:_____
_____
_____

LIST BENEFIT/JUSTIFICATION (saving of time, expense, personnel, etc.):_____
_____
_____

SPECIFY URGENCY:_____
SAMPLE or EXAMPLE ATTACHED:_____

M. I. S. Use Only

REPORT RECEIVED ON:_____   REPORT RECEIVED BY:_____
                                     REPORT ASSIGNED TO:_____

CAUSE of REQUEST:_____
_____

SOLUTION to ERROR:_____
_____
_____

FIGURE 13.9. Change and error reporting form for Class Video.

Such a form helps encourage necessary changes. When users have to fill out a form, they think about improvements that will really benefit the organization's productivity or profitability. They are also forced to define the change in specific language, listing benefits and urgency.

The completed form is designed to be forwarded to the computer services department for action. Most departments have a system already in place to manage changes. This change system operates like a small systems study: analysis, design, and development. Analysis targets how much the change will cost, its benefits, and the impact on the existing system. Design focuses on how to make the change. Development concentrates on implementing the change. Some simple changes may take a few days, and complex ones may take months.

CASE STUDY

Testing, Training, and Documentation for Fleet Feet's Accounts Payable Voucher-Check System

Analyst Frank Pisciotta divides the testing of Fleet Feet's new accounts payable system into two phases; the first is designed to test each individual module and program with sample data from Mark Stensaas's manual AP system, and the second to link them all together for system and acceptance testing. Since Frank will do the input himself to see how the computerized system processes the data, he hands his colleague a handwritten memo requesting 40 transactions of all types: adding new vendors, posting charges and vouchers, deleting vendors. Mark agrees to provide the data in 10 days.

When the data arrive, Frank activates the data-collection and validation program, AP010 on his terminal, enters the data, and notes what the system does to the data. After entering the data, Frank uses the Query-data-base reporting language (a part of Hewlett-Packard's Image data-base manager) to request results. Before calling up Query, he manually calculates the results so he can compare the two for any discrepancies. After locating the few faulty portions of the programs, he repairs them, deletes all the data from the data base, then retests. Frank repeats this process for all the programs in the system until he feels it is error free. The errors he finds fall into two categories: those that derived from his own design mistakes, and those that a programmer had introduced.

Phase 2 of Frank's testing plan requires all the programs to work together. Before performing this test, Frank again calculates results manually so he can compare results. Then he again purges the data from the data base, activates the first program (AP010), enters the data, and allows the other programs (AP020 through AP070) to process the data. The hand calculations match the computer-generated ones, so he can stop testing.

Training poses few problems. All the people in Mark's department have used terminals before, and the new system requires no new staff. All Frank has to do is teach users how to run the seven new AP system programs. To prepare for training, Frank decides to write user and operations manuals for each program (Figures 13.10 and 13.11). Since the new system is online, it needs only brief run manuals. However, the user manuals, which must provide a good deal of detail about data entry, updates, and report printing, will be longer.

CASE STUDY

OPERATOR INSTRUCTIONS

System: AP (Accounts Payable)   Date: Dec. 1984   Page 1 of 2

Application: PRINTING THE AP VOUCHER CHECK

Paper Requirements: AP CHECKS

Comments: AP checks are one of the primary outputs from the AP system. Check forms are kept by the controller and should be retrieved from that office before the run is begun.

RUN INSTRUCTIONS:

1. This application may be run on request and will generate checks as well as various reports and associated check registers.

2. Check printing requires exclusive use of the AP data base. It will only run if no other users are running AP applications at the same time. If other users are accessing the AP data base, notify them to log off or wait until a later date to print checks.

3. Mount the check forms when requested. Type "DONE" when check forms are all in position.

4. If you wish to print a check with X's for alignment purposes, answer "Y" to the question "PRINT ALIGNMENT ?". You may continue to request an alignment form until you are satisfied alignment is correct. When you wish to proceed with check printing, answer "N" to the alignment form question.

5. Enter the payment and check-printing data requested on the screen. The "END" key is allowed on any screen in case you wish to abort the check run.

6. Checks are now printed.

7. When printing is complete, you will be asked to mount the standard paper again. Next the check register prints for all checks printed and voided.

8. The system will now update all files for the payments just made. This update will be run as a separate job.

9. Burst the checks and the two check register reports and deliver them to the controller's office. All voided checks should be separated from the good checks.

FIGURE 13.10. Accounts payable operator instructions tell the computer operator how to start up and run the accounts payable voucher-check printing program. (Courtesy of MCBA® AP Manual. MCBA is a registered trademark of MCBA, Inc.)

CASE STUDY

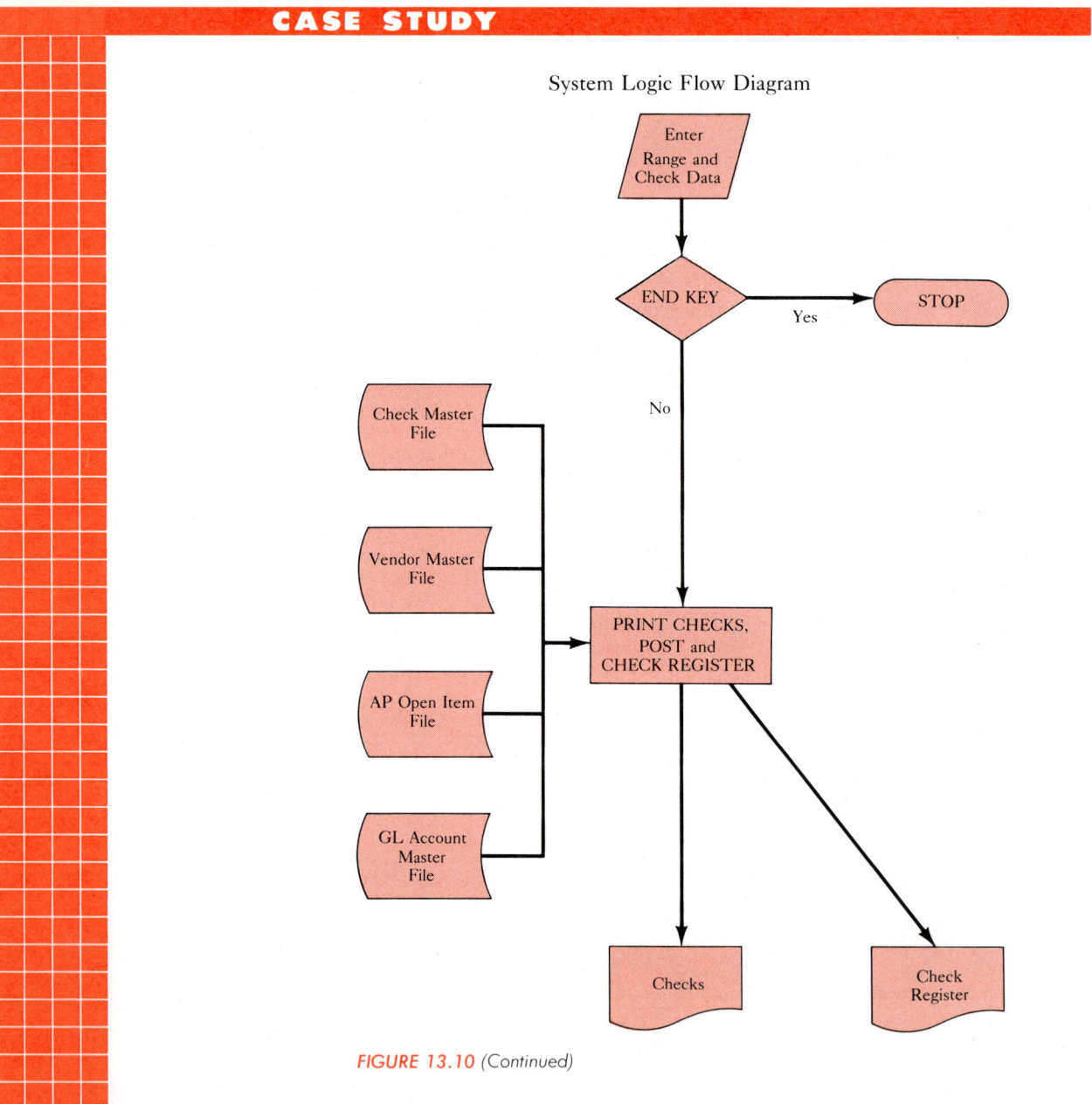

FIGURE 13.10 (Continued)

CASE STUDY

USER'S INSTRUCTIONS

System: AP (Accounts Payable)   Date: Dec. 1984   Page 1 of 3

Application: CHECK PRINTING

Comments: You may use this application to print the computer checks and post these payments to the AP files.

RUN INSTRUCTIONS

1. To run the application, select No. 7 from the AP main selector menu.
2. The terminal screen will show:

```
ACCOUNTS PAYABLE                                      FLEET FEET
PRINT CHECKS AND POST
    PLEASE ENTER
       1. PAYMENT DATE        [MM/DD/YY]
       2. CHECK DATE          [MM/DD/YY]
       3. STARTING CHECK NO   [        ]
       4. STARTING VENDOR NO  [        ]
       5. ENDING VENDOR NO    [        ]

    ITEM NUMBER TO CHANGE ?   [ ]
```

3. Enter the data as follows:

   1. PAYMENT DATE — MM/DD/YY format

   2. CHECK DATE — Enter check data (the date to be printed on the checks) in MM/DD/YY format. These data do not have to be the same as the payment date used by the system to compute discounts. Press RETURN to default to the payment date.

   3. STARTING CHECK NO — Enter the starting check number, from one to six numeric digits. The starting check number is compared with the numbers of the other computer printed checks (valid and void) to ensure the starting check number is higher than the number of the last check printed. If the starting check number is the same or lower than that of the last check printed, a warning is displayed and you are asked to verify that what you entered is what you want and not an error. Type Y and press RETURN in response to "IS THIS CORRECT?" to verify the starting check number you entered.

FIGURE 13.11. User documentation for the accounts payable voucher-check program guides users through the system. Typical terminal screen selection menu programs appear here. (Courtesy of MCBA®, AP. MCBA is a registered trademark of MCBA, Inc.)

CASE STUDY

    4. STARTING VENDOR NO — Enter the starting vendor number of the range of vendors for whom checks are to be printed. Press RETURN to default to "ALL" vendors.

    5. ENDING VENDOR NO — Enter the ending vendor number of the range of vendors for whom checks are to be printed. Press RETURN to default to the starting vendor number.
If "ALL" was entered for the starting vendor number, the ending vendor number field is skipped.

4. Enter the number of any data item to be changed. The data in that field will be blanked and the cursor positioned at that field. Enter the revised data. When no more changes are required, press RETURN in response to "ITEM NUMBER TO CHANGE?"
5. Mount blank checks in the printer after the message:
"PLEASE MOUNT CHECK FORMS ON PRINTER"
Load the blank check forms into the printer, and when they are positioned properly, type "DONE". If you attempt to do anything else at the keyboard, it will not be accepted.
6. After you enter "DONE", you are asked: "PRINT ALIGNMENT?" Answer Y to print a dummy check form filled out with 9's and X's to verify the alignment of the forms in the printer. Answer N to bypass printing of a dummy check form. When the dummy form has been printed (if you request one), the alignment question will be asked again. Adjust the form in the printer. You may align as many times as is necessary to get the checks in the proper position.
7. Checks are now printed for the range of vendors requested.
8. Each check contains a remittance advice detailing each voucher paid by the check. Up to 16 vouchers are detailed. If more than 16 vouchers are to be paid to a vendor, 16 vouchers are detailed on the first stub, the corresponding check is voided, and the detailing of vouchers is continued on the next check stub. After the last voucher is detailed, the check is then printed for the total amount due the vendor. Only one check is printed per vendor.
9. Checks will not be printed for vendors whose balance is zero or negative. A sample AP check and remittance advice are shown on the following page.
10. After the checks are printed, the screen will display:
"CHECKS FINISHED"
Remove the forms and put regular paper back in the printer.
11. The accounts payable check register is then printed. This check register contains a record of all computer-printed checks. Any checks voided during the check run will be shown as void on the AP check register. The corresponding void checks should be returned to the controller's office for destruction.
12. The program then posts the payments to the AP open item file, the AP distribution file, and the check reconciliation file.
13. Checks and the check register should be burst.
14. Deliver the checks, check register, and voided checks to the controller's office.

FIGURE 13.11 (Continued)

Sample AP Check

VENDOR: 000002				CHECK NO. _____	
OUR INV. NO.	YOUR REF. NO.	INVOICE DATE	INVOICE AMOUNT	AMOUNT PAID	DISCOUNT TAKEN	NET CHECK AMOUNT
001013 001014	HJ	04/25/84 05/23/84	150.00 263.00	150.00 200.00	.00 .00 CHECK TOTAL	150.00 200.00 350.00

FLEET FEET

CHECK NO.	CHECK DATE	VENDOR NO.
000301	06/25/84	000002

PAY
TO THE
ORDER OF RED LINE FREIGHT
350 COMMERCIAL ROAD
BIGTOWN, TEXAS 99996

CHECK NO. _____

CHECK AMOUNT
$*******350.00

NON-NEGOTIABLE

FIGURE 13.11 (Continued)

CASE STUDY

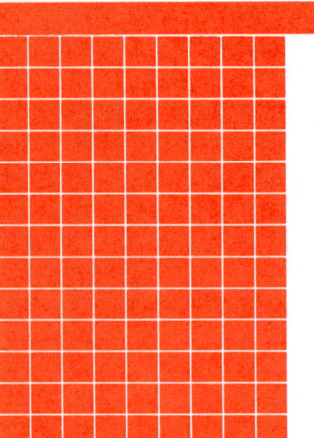

Frank adds these manuals to all the other materials he developed during earlier phases of the systems process, thus creating complete systems documentation. He collects all documents in a three-ring binder with separators dividing the documents from each other.

Copies of the manuals go to Mark Stensaas so his staff can use them to prepare for next Thursday's training seminar. Early that next Thursday, Frank arrives at Mark's office with coffee and doughnuts for everyone. Frank has learned from past training experiences that such thoughtfulness helps break the ice. Instead of sitting at the terminal and entering the data himself, the analyst asks each user to log onto the HP 3000 computer, call up the AP010 program, and run it. Each person enters identical sample data, so everyone can compare results. Looking over their shoulders, Frank offers advice whenever necessary.

Because everyone catches on quickly, the training session takes only 4 hours. "Nice system," says Mark, and his people agree.

Frank next contacts Mark's boss and schedules a lunch for Monday. During the meal Frank summarizes the project, showing some of the reports that will appear.

All the time Frank has devoted to planning and training has really paid off.

WORKING WITH PEOPLE Writing with Style

Analyst Paul Brewer dreaded writing system documentation. When he chose a career in computer science, he had not realized that he would have to write reference manuals. If he had, he would have tried to stay awake in his one college English composition class. FORTRAN was so much simpler and more elegant. Paul forced himself to sit down at his word processor.

Four days later he put the finishing touches on the data-entry user manual. It had been a tedious and frustrating task, but he felt good about the final product. At least his word processor had made revisions a snap. Finally he hit the Print button.

Data-entry supervisor Chris Sprenger thanked Paul for the manual and began to read it, but it didn't take her long to heave a deep sigh. Having majored in English in college, Chris was appalled by Paul's writing style. Unnecessary multisyllabic words, passive constructions, vague technical concepts without adequate concrete examples, and mystifying computer jargon made it all but incomprehensible.

She began circling offensive words, sentences, and even whole paragraphs:

1. "Ascertain whether there are sufficient paper resources for the report to be printed."
2. "Does the payroll package have a check printing capacity?"
3. "Calculation of the expected profit and payback is accomplished in the TRND module of the program."
4. "Care should be exercised by operating personnel to adequately ensure that they not press CLEAR DISPLAY when INSERT CHARACTER is meant to be pressed."
5. "CONTROL and BK should be pressed in the event that operating personnel desire to initiate backspace without erasing."
6. "It is a requirement that user department managers functionally authorize the F-Spec to show their approval."
7. "The project engineer possesses the requisite data verification responsibility."*

Chris rewrote each, using an active writing style, and eliminating all unnecessarily long words and jargon:

1. "Check the paper supply before printing a report."
2. "Can this payroll package print checks?"
3. "The TRND module calculates expected profit and payback."
4. "When you press INSERT CHARACTER, be careful not to press CLEAR DISPLAY."
5. "To backspace without erasing, press CONTROL and BK."
6. "F-Spec approval requires the user department manager's initials."
7. "The project engineer verifies the data."*

Chris debated whether or not to show Paul what she'd done, and finally decided it would be best to give him her honest reaction to the manual. Surprisingly, he loved her editing. "I wish I could have done that myself," he confessed.

"I can teach you some simple tricks," Chris offered.

"Great! And I can translate the jargon for you."

Concerned that her editing might hamper Paul's schedule, Chris volunteered to work overtime to help him, provided he taught her how to use his word processor. They struck a deal whereby she'd edit the whole manual and devise a list of general writing rules if Paul would let her use the word processor to write a magazine article she was working on. After she edited the report, Chris made the following suggestions:

1. Avoid unnecessarily long words, and replace most Latinate words with Anglo-Saxon equivalents. For example, change "The quadriped effected its descent from the arboreal habitat in order to ingest sustenance" to "The squirrel climbed down from the tree to eat."
2. Try to write actively rather than passively. Change "The ball was hit by Bill" to "Bill hit the ball."
3. Try to illustrate all principles and concepts with concrete analogies, examples, anecdotes, and cases. People have a hard time picturing abstractions. For example, compare sequential with random access by saying, "Readers access novels sequentially, whereas they access dictionaries randomly."
4. Eliminate as much computer jargon as possible. Choose common words and analogies to daily experience that can closely approximate computer concepts.

Paul, eager to continue improving his writing skills, asked Chris to recommend a book about clear writing. The next day she brought him a gift: William Strunk and E. B. White's *Elements of Style*. After only 4 weeks, Paul actually found himself looking forward to writing.

*Adopted from *Computerworld*, Page in Depth 16, Jan. 10, 1983. © 1983 by CW Communications, Inc., Framingham, MA 01701. Reprinted from COMPUTERWORLD.

SUMMARY

During the two previous phases of the systems process, the analyst functioned as an architect, designing the system. Now the analyst works as a contractor, coordinating and managing the actual building of the system.

Program, system, and acceptance testing commence when programs are complete. While some testing can occur during coding, complete system testing cannot start until all programs have been written. The analyst must test the whole package of programs by linking them together. Testing can involve artificial or real data input by users and computer services staff. The aim of testing is to locate errors.

Users, managers, and computer operations staff must learn the new system. Each group demands tailored training: managers want to know about the information the system will provide, users must learn to use the system, and operators must be able to keep the system running efficiently.

All three groups need different types of documentation. A complete set of all materials makes up the documentation, which resides in the computer services department for reference.

Since even good systems change, maintenance documentation allows for changes or error correction.

NEW TERMS

Acceptance test
Documentation
Link test
Maintenance
Management documentation
Operations documentation
Program documentation
Program integration test
Run manual
String test
System test
Test data
Testing
Training
User documentation

QUESTIONS FOR REVIEW AND DISCUSSION

Questions appear in three categories. You can find the answers to the A group in this chapter. The B group requires you to apply the material presented here, whereas the C group necessitates investigation or research.

A.1. What three factors must be tested on a new system?
A.2. List three sources of data that are usually available for testing.
A.3. Who needs training when the new system is developed?
A.4. List sources of training.
A.5. List two advantages of documentation.
A.6. List two disadvantages of documentation.
A.7. What are the two types of maintenance activities that can be expected of every system?
A.8. List three sources of training materials.
A.9. List three elements of a run manual.
A.10. List the elements found in a user's manual.

B.1. Match users (a–e) with the appropriate level of training (1, 2, 3).
   a. Computer operators
   b. President of the company
   c. Data-entry operator
   d. Manager of data entry
   e. Manager of computing services

   1. Low level
   2. Medium level
   3. In-depth level

B.2. Write a data dictionary definition of system documentation.

B.3. Write a data dictionary definition of program documentation.

C.1. What percentage of time has been spent maintaining nonstructured systems?

REFERENCES

Robert T. Grauer, *Structured Methods Through COBOL*, Prentice-Hall, Englewood Cliffs, N.J., 1983.

G. J. Meyers, *Software Reliability: Principles & Practices*, Wiley, New York, 1976.

Wayne P. Stevens, *Using Structured Design*, Wiley, New York, 1981.

Robert W. Zmud and James F. Cox, "The Implementation Process: A Change Approach," *Management Information Systems Quarterly*, Vol. 3, 1979.

CHAPTER 14 Management of the Systems Process

- Goals and Preview
- Management Activities: An Overview
- System Operations
- Theories X, Y, and Z
- System Audit or Review
- System Documentation
- Keeping Current
- Case Study: Acceptance of Fleet Feet's New Accounts Payable System
- Working with People: Analysts as Instant Experts
- Summary
- New Terms
- Questions for Review and Discussion
- References

GOALS AND PREVIEW

After reading this chapter you should be able to:

1. Define management's responsibilities within the systems process.
2. Explain the three types of system changes users may want.
3. Describe how a Gantt chart helps manage the systems process.
4. List the components of the final report.
5. Name two professional societies, two books, and two training groups.

The systems process now enters its final phase, during which the system begins full operation. In many respects this phase parallels a family moving into a new home: the family (user) expects the new living quarters (system) to be safe and warm, but only after living in the house for a while will the family fully appreciate its conveniences and flaws. After some time the family may need to undertake repairs or alterations. Appliances break, faucets leak, children strain the living quarters. So it is with a new computer system. To continue fulfilling user needs, a system must accommodate change. A static system eventually dies.

During the systems process, the analyst orchestrated events while management evaluated and approved them. The managing of the systems process requires talent and insight concerning people and organizational structures. In this chapter we focus on the tools that can help management control the three system phases, from initial user request to installation and operation.

MANAGEMENT ACTIVITIES: AN OVERVIEW

All business management activities, whether in profit-centered, governmental, or educational settings, try to control personnel, funds, inventory, relations with customers or vendors, costs, market share, patents, trade secrets, and a host of other factors to achieve an organization's goals. Although management may often feel intimidated by the technical aspects of computers, it must accept responsibility for their management. Consulting firms advocate that management use the same effective management approaches to solve computer-related problems as it has used in more traditional areas. A good manager does not have to become a computer scientist to manage computer scientists.

Keane Inc., a New England consulting firm, has discovered during its consulting to over 350 Fortune 500 companies that six rules govern the successful management of a computer project*:

1. *Define the job in detail*. Determine exactly what work must be done and what products must be delivered. Explicitly evaluate the environment and address all assumptions.
2. *Get the right people involved*. Involve the appropriate users throughout the project, particularly during planning. Involve the appropriate data processing people. Ensure that each member of the project team participates in defining his or her own goals.
3. *Estimate time and costs*. Develop a detailed estimate of each phase of the development process before undertaking that phase. Estimate the components of a job separately to increase accuracy. Do not estimate what you do not know.
4. *Use the 80-hour rule*. Break the job down into "tasks" that require no more than 80 hours to complete. Ensure that each task results in a tangible product. The 80-hour rule provides the framework for scheduling, assigning tasks, identifying problems early, and confirming time and cost estimates, and for evaluation of project progress and individual performance.
5. *Establish a change procedure*. Recognize that change is an inherent part of systems development. Establish a formal procedure for dealing with these changes and ensure that all parties agree to it in advance.
6. *Establish acceptance criteria*. Determine in advance what will constitute an acceptable system. Obtain written user acceptances of products throughout the project so that acceptance is a gradual process, rather than a one-time event at the end.

Since most good managers apply similar rules to other areas, why should they not apply them to management of the data processing function?

Management control Scheduling personnel, equipment, or other resources to achieve maximum results.

Management control is essential throughout the system process. Control involves scheduling personnel and budgeting funds to achieve maximum results at minimum costs.

*From the book *Productivity Management in the Development of Computer Applications* by John F. Keane, Marilyn Keane, Mark Teagan of Keane, Inc. © 1984 Prentice-Hall, Inc. (published by Prentice-Hall, Inc., Englewood Cliffs, NJ 07632).

Schedule Overruns

One managerial tool for monitoring the schedule is the familiar Gantt chart, a visual tool that quickly indicates the status of a given system. When using a Gantt chart to monitor the system process, a manager or supervising analyst records actual events as they occur. The Gantt chart in Figure 14.1 shows that management expected the implementation plan to take a week to complete, whereas it actually took 2 weeks. Program AR010 was scheduled for 3 weeks but only took 2; AR020 should have taken 3 weeks but took 4; AR030 took 1 week less than anticipated; and AR040 finished on time. Such constant awareness of reality helps managers make timely decisions about allocating people and money to solve inevitable problems.

People Problems

Managers routinely spend much of their time monitoring and solving problems concerning **interpersonal relationships** among staff. Such relationships strongly affect the systems process, and problems with them may surface for a variety of reasons: philosophical differences, territorial pride, or personality conflicts. Regardless of their origin, such people problems can imperil a new system just as much as hardware or software failures. In one large retail business, the lead analyst, who had 10 years of experience designing accounting systems, could not communicate effectively with the business manager. Because the two held such radical **philosophical differences** about how the system should function, they spent more time quarreling and playing politics than they did working together to get the system running. Eventually the president transferred the analyst to a task the business manager was not part of, but only after the schedule had slipped 3 weeks.

In another instance the data processing manager of a wholesale beverage firm suffered from **territorial pride**. A subordinate had purchased his own personal computer, learned how to use an electronic spreadsheet, then brought his computer to the office and wanted the data processing manager to provide him with data.

Interpersonal relationships Problems dealing with differences between people that an analyst must solve.

Philosophical differences Interpersonal problems stemming from differences in ideology or theory.

Territorial pride Interpersonal problem related to one person invading another's area.

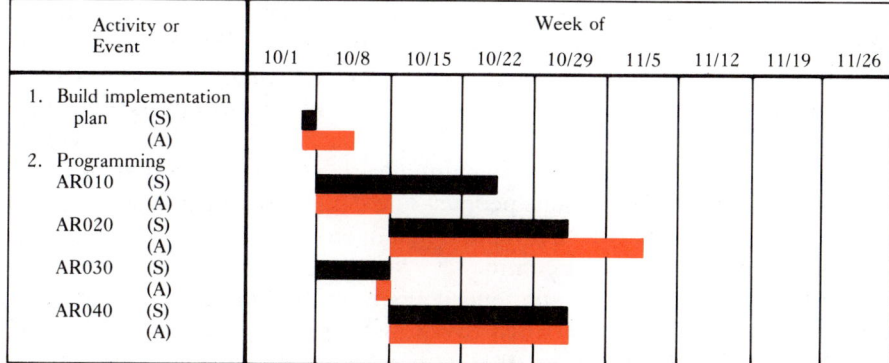

FIGURE 14.1. *Gantt charts help management monitor schedules. Black bars show the anticipated time required for each event/activity and color bars depict what actually happened.*

Theories X, Y, and Z

Though most professional football coaches played the game themselves, their success depends on getting the job done through others by motivating and organizing, directing the team's effort. To motivate people to achieve an organization's goals, a good manager must develop a philosophy about why people behave the way they do. Do people work primarily for money, to get ahead, or just for the satisfaction of seeing an organization achieve its objectives?

In the mid-1950s, Douglas McGregor described the conventional approach to management, which at the time held a certain concept of human behavior. He called it "Theory X":

1. The average human being inherently dislikes work and will avoid it if possible.
2. Due to this fact, most people must be persuaded, rewarded, punished, threatened, and controlled to get them to divert their efforts toward achievement of organizational objectives.
3. People lack ambition, dislike responsibility, prefer to be directed, and want security above all.

A manager who agreed with Theory X would, of course, rather dictatorially direct and control people in order to accomplish organizational objectives.

McGregor went on to insist that such a philosophy was inconsistent with Abraham Maslow's theory of the hierarchy of human needs, which held that people can, and often do, work for higher purposes. The Theory X approach couldn't account for, or allow for, behavior directed toward the satisfaction of personal and social needs. Therefore, McGregor proposed what he considered a more effective philosophy of human behavior, "Theory Y":

1. People do not inherently dislike work, and are not, by nature, passive or resistant to organizational needs.
2. The capacity for assuming responsibility and readiness to direct behavior toward organizational goals is present in everyone.

The data processing manager fancied himself the organization's expert on data processing and resented a subordinate invading his turf. Though he privately endorsed microcomputers, he began publicly to attack them. The problem became so acute that the president of the beverage supply firm had to intercede. Rather than discipline either party, the president sent the data processing manager to a 2-day seminar on creative corporate communication, which taught strong-willed managers how to handle ambitious people. The data processing manager returned with fresh interpersonal skills that helped him overcome his territorial pride without relinquishing management control.

A supplier of bottled water experienced a **personality conflict** when an analyst and a user began to argue over such trivial matters as who would fill the coffee pot in the morning. Their power struggle had less to do with the new computer system than it did with their efforts to impress an attractive female co-worker. The two men had quite different personalities that constantly conflicted. Man-

Personality conflict Interpersonal problems stemming from differences in individuality or disposition.

3. People do not need to be threatened, punished, and controlled in order to get them to be productive.

If one accepts Theory Y, management should arrange conditions and methods of operation which provide opportunities for employees to satisfy their personal and social needs while pursuing organizational objectives.

In 1981, William Ouchi followed up on McGregor's work with what he dubbed Theory Z. In his book he tried to show how American business can profit from certain Japanese approaches to business. Unlike either Theory X or Theory Y, Ouchi argued that the key to increased productivity is involving workers deeply in the organization's goals and giving them strong incentives to do so. He pointed out that such a co-operative approach had made Japanese business extremely productive. Unlike American managers, who seem preoccupied with next year's bottom line, the Japanese took a longer range view of their workers' needs and their organizations' accomplishments. In short, Theory Z recommended:

1. A holistic concern for people, their families, and children, and their intellectual and moral development.
2. Collective decision-making with responsibility still residing with the individual.
3. Trust in individuals to work autonomously without close supervision.

No academic theory can solve all the problems that inevitably occur when people must work together. Any good manager knows that subtlety and intimacy play a major role in pinpointing the right people to assign to the task at hand.

Does your philosphy of human behavior agree with Theory X, Y, or Z? If you are to become a successful manager of people in the post-computer revolution, you will need to evolve an approach that pays attention to more than bottom-line results.

Source: Ross Hallberg, "Smoke and Mirrors," *dataBASE*, August 1983. Used by permission.

agement interceded, threatening to fire both men if they would not cooperate, but the two found that so difficult that one eventually had to go.

Cost Overruns

Since systems frequently cost more than anticipated, management often faces **cost overruns**. However, formulas exist for forecasting costs, and these vary from crudely tripling the analyst's expectations to determining expected costs scientifically.

Two tools for controlling costs are the program evaluation and review technique (PERT) and the critical path method (CPM); see Figure 14.2. The critical path method shows relative priorities among tasks, and PERT extends CPM by linking optimistic, expected, and pessimistic costs with each task. Optimistic costs assume all tasks will go exactly as planned. Expected costs indicate what will probably

Cost overrun A figure that exceeds anticipated expenses.

	Activity or Event	Optimistic Cost	Expected Cost	Pessimistic Cost
1.	Build implementation plan	500	1000	1500
2.	Programming			
	AR010	750	900	1300
	AR020	1200	1400	1900
	AR030	2200	2500	3000
	AR040	680	900	1150
	TOTAL	5330	6700	8850

FIGURE 14.2. PERT provides a tool for controlling costs.

happen, and pessimistic costs provide for the worst possible case in which everything goes wrong.

Mountain Motor's billing system used a PERT/CPM standardized formula to find its average cost:

$$\text{AVERAGE COST} = (1/6) [\text{OPTIMISTIC COST} + 4 (\text{EXPECTED COST}) + \text{PESSISMISTIC COST}]$$
$$= (1/6) [5300 + 4 (6700) + 8850]$$
$$= (1/6) [5300 + 26800 + 8850]$$
$$= (1/6) [40150]$$
$$= 6691$$

According to this method, Mountain Motors should expect to spend $6691 on design and development. Other formulas consider other factors: number of modules, number of programs, system complexity, and number of people involved.

Regardless of the cost forecasting approach, management eventually wants to compare anticipated and actual costs. The closer the two, the better, of course, because accurate forecasting builds confidence not only in budgeting ability, but in the resulting system itself.

SYSTEM OPERATIONS

Management should try to put the system into operation at the most opportune time. For example, it makes sense to switch to a new payroll system at the beginning of a new year or quarter when all prior federal and state government reports have been filed. However, the business manager of a power and water company decided to adopt the new online payroll system for December's pay period because it could deliver year-end reports much more quickly than the old manual system. While this was risky, considering that a new system might experience problems, W-2 forms went to employees by January 4, a new record for the company. Later it discovered only two errors for its 265 employees! Needless to say, the new system's early success won it a lot of confidence and support.

Organizations usually adopt accounts receivable systems at the end of a month before a new billing cycle begins. A flower shop, for example, decided to switch service bureaus from batch to online processing before its May billing cycle. The batch service bureau had required the florist to close books on the 20th day of the month and sent bills on the seventh working day of the next month, but the new online system allowed the florist to post charges up to the last working day of the month, and printed bills only 2 hours after the last data entry was made. During April the store owner ran the two systems in parallel, going fully to the new one only after verifying results. During the parallel operation, the store owner found a rounding error of a penny, which was easily repaired.

No matter when an organization's management chooses to begin relying on a new system, it can expect user questions, problems, and complaints to start immediately, then decrease as users learn the system and analysts solve the problems (Figure 14.3). As time passes these needs for change drop dramatically until the system approaches the end of its life, at which point it experiences a resurgence of change requests because it can no longer meet the needs of users.

SYSTEM AUDIT OR REVIEW

After the system has had a chance to generate regularly desired reports, the analyst conducts a final review of the entire system process, tracking it from preliminary analysis all the way through development. We call this last check a **system audit** or review (Figure 14.4), during which the analyst scrutinizes four aspects of the new system:

1. *Objectives*: Does the new system fulfill the organization's needs?
2. *Costs*: What did the new system cost?
3. *Time*: Was the system delivered on time?
4. *Results*: Does the system produce all intended reports?

To research these four questions, the analyst weighs results of the completed programs, training and operations manuals, and expertise of the computer services staff, and evaluates users' reactions and management comments.

System audit The last check or review of a system to assure it meets objectives and goals.

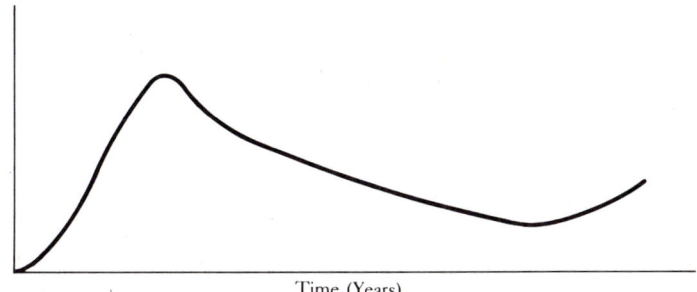

FIGURE 14.3. This graph illustrates the rate of user requests for assistance after system installation.

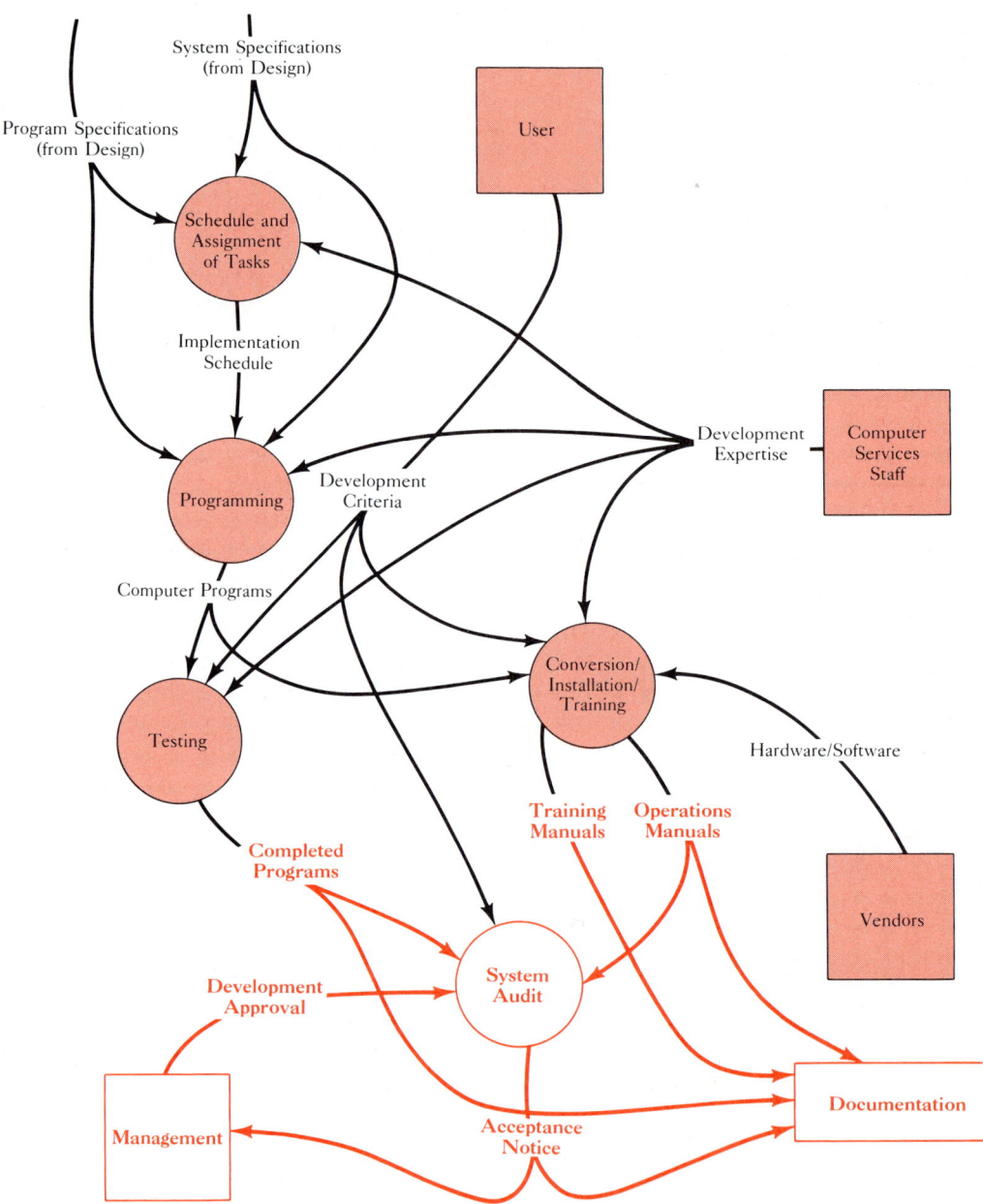

FIGURE 14.4. *System audit or review is the last step in the entire systems process.*

Acceptance notice
Memorandum sent to all parties to notify them of a systems approval.

If all objectives, costs, schedules, and results fall within acceptable limits, the manager responsible for the system and the analyst issue an **acceptance notice** to all concerned parties (Figure 14.5).

The systems process does not always run smoothly. In some situations schedules fall badly behind or actual costs greatly exceed estimates, either of which requires explanation in the system audit. For example, a large midwestern utility company's

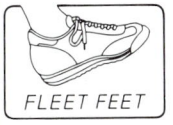

MEMORANDUM   Nov. 19, 1984

TO: All Department Managers

FROM: Walt Carson, Sales Manager, Future Ford

SUBJECT: Acceptance of Sales Analysis System

  For the past 32 weeks, we have been developing a new sales analysis system. Yesterday the system was accepted as complete and is now in full operation. During the last 3 weeks, we have been using the new system and have found all the reports to meet or exceed our original needs.
  The sales analysis by territory has revealed some very interesting and valuable data that until now we could not generate: sales of our new digital/optical record players are very strong in medium-sized metropolitan areas but weaker than expected in very small and very large cities. We are studying this phenomenon to find out why it is occurring.

FIGURE 14.5. *An acceptance memorandum notifies all involved that the organization has formally adopted the system.*

energy-monitoring system cost twice as much as anticipated. In the eventual system and audit, the analyst blamed the overrun on new state regulations that had forced him to redesign the system completely. Fortunately management was aware of the new regulations and had prepared itself for additional costs.

Rarely does an organization completely reject a system at this late stage because reviews and walkthroughs at earlier stages should have revealed any major system deficiencies.

The acceptance memo joins all the other materials to conclude system documentation.

SYSTEM DOCUMENTATION

After notifying all parties of acceptance, the analyst undertakes a final task, assembling all the materials produced during analysis, design, and development (Figure 14.6). If the analyst has done the job thoroughly, completing system documentation should be a breeze. All necessary materials should already exist, so the analyst need only arrange them in order, write a table of contents, and bind them together.

In the days of unstructured analysis and design, analysts dreaded this final task. Materials were scattered and incomplete, and the organization would assign the analyst to more "productive" tasks as quickly as possible. However, structured methodology streamlines paperwork by having it completed at appropriate points throughout the systems process rather than at the end. The widespread use of

```
A. Systems analysis
   1. User's request
   2. Preliminary report
   3. Feasibility study
   4. Budget and schedule
   5. Management action

B. Systems design
   1. Assignment roster
   2. Report formats
   3. Data dictionary
   4. File or data-base layout
   5. Screen formats or file layouts
   6. System specifications
   7. Control factors
   8. Program definitions
   9. Module descriptions
  10. Program specifications
  11. Management action

C. System development
   1. Implementation plan
   2. Computer programs
   3. Test plan
   4. Conversion plan
   5. Training manuals
   6. Operations manuals
   7. System acceptance notice
```

FIGURE 14.6. *The paperwork generated during the systems process forms the complete system documentation.*

text editors (word processing) has also aided system documentation by easing editing and production.

Nevertheless, professionals continue to debate the value of extensive documentation. Opponents cite the redundancy of retaining such items as printer spacing chart report layouts (Figure 14.7a), once the report begins appearing in its actual form (Figure 14.7b). They contend documentation should reflect reality, not intentions. However, proponents of careful documentation argue that it provides a historical record that subsequent users and analysts can find quite helpful, especially when replacing the new system with an even newer one. Regardless of one's persuasion, system documentation does deserve attention.

KEEPING CURRENT

Analysts must constantly strive to keep current with the state of the art. To do this, they enjoy a variety of resources that help sharpen skills: professional societies, periodicals, books, seminars, college classes, extension courses, and training groups.

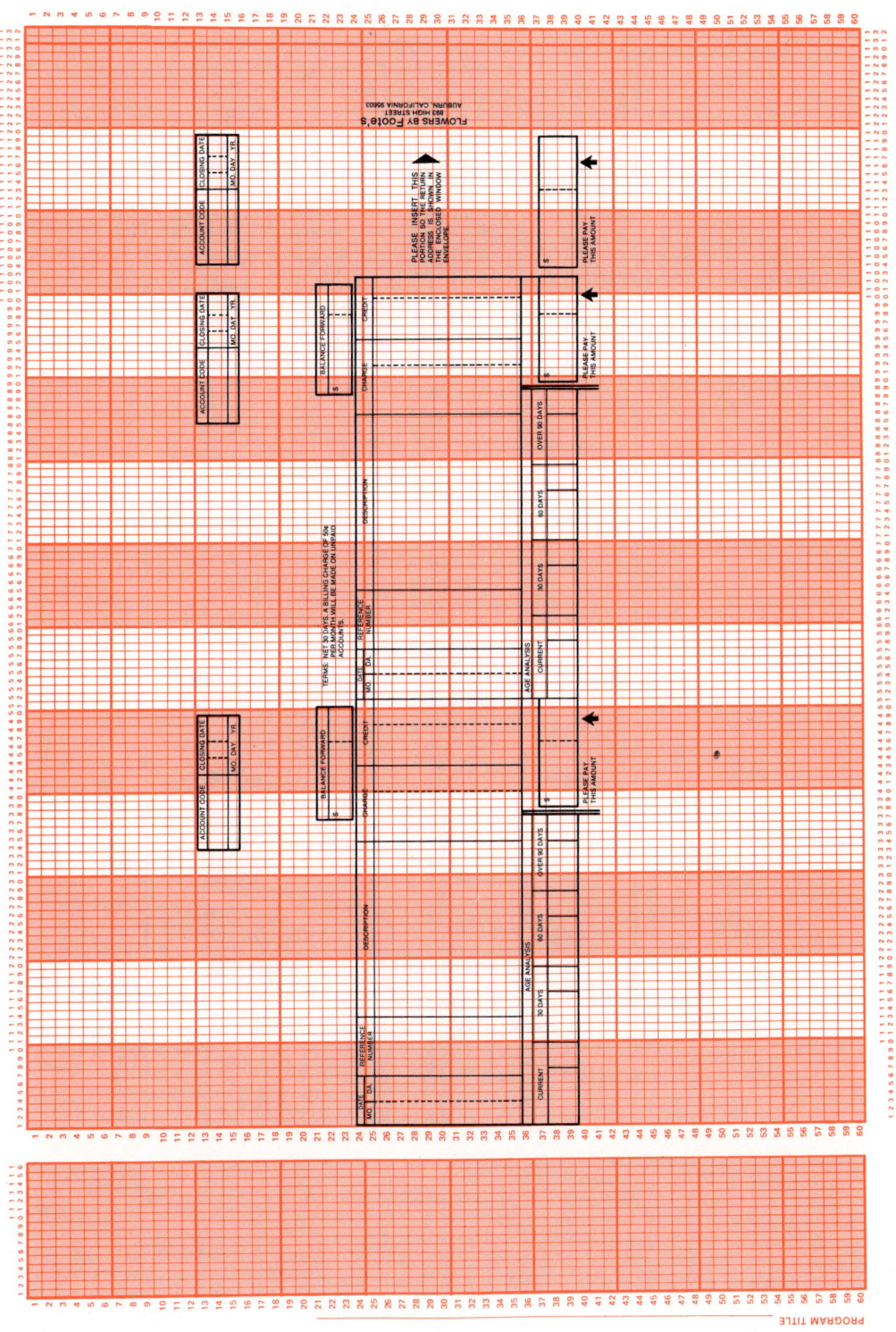

(a) Analyst's report design.

FIGURE 14.7. System documentation includes copies of report designs created during output design. Once a system begins operation, the actual reports supersede the original design.

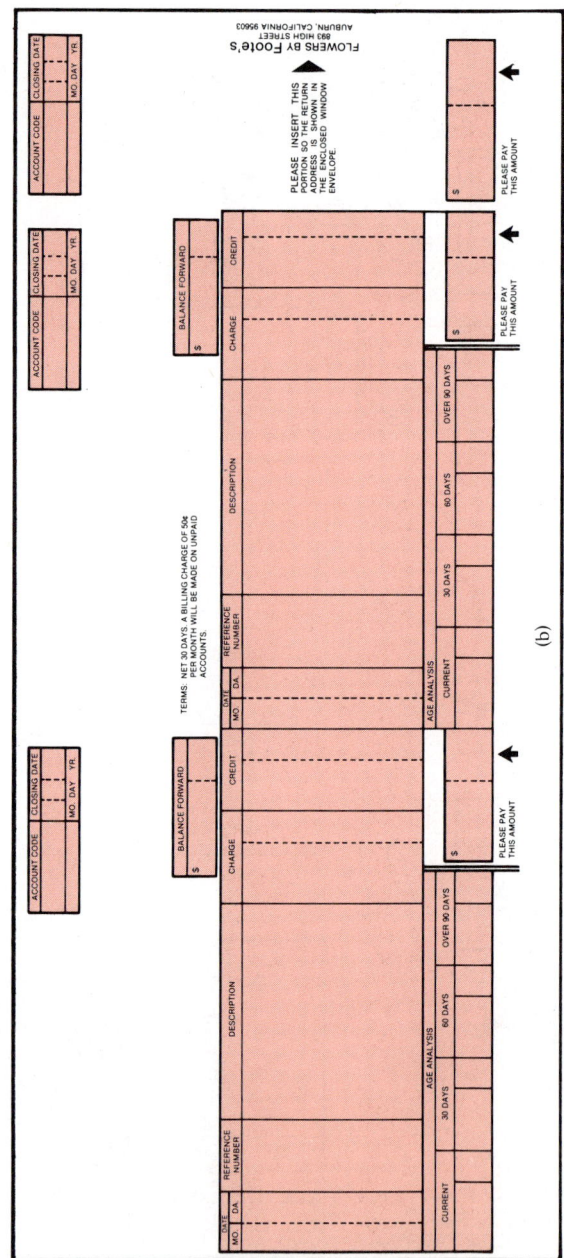

(b) *Final printed report.*

FIGURE 14.7 (Continued)

Professional Societies

Professional societies can provide a rich source of new materials, tools, and contacts. Most of them offer seminars as well as regular regional and national meetings. The four major ones are:

Data Processing Management Association (DPMA). (505 Busse Highway, Park Ridge, IL 60068). DPMA promotes and develops inquiry into the fields of data processing and data processing management and presents a "Computer Sciences Man of the Year" award. The organization includes numerous local chapters that provide monthly meetings with guest speakers.

Association for Computing Machinery (ACM). (1133 Avenue of the Americas, New York, NY 10036). ACM is dedicated to the advancement of the science and art of information processing, the exchange of ideas, and the development of integrity and competence. ACM produces about 30 different periodicals in all areas of computer activities and it presents the Turing Award to distinguished persons in the profession. Over 100 local chapters offer monthly meetings and guest speakers.

Association for Systems Management (ASM). (20 West 33rd Street, New York, NY 10001). This society keeps its members abreast of changes in the field of systems management. ASM has five departments: Data Communications, Data Processing, Management Information Systems, Organization Planning, and Written Communications.

American Federation of Information Processing Societies (AFIPS). This federation of professional societies advocates and disseminates knowledge of constituent societies, some of which are ACM, DPMA, ASM, and the Institute of Electrical and Electronics Engineers (IEEE).

Journals and Periodicals

A multitude of journals and periodicals keep the computer specialist informed. Each of the professional societies (DPMA, ACM, and ASM) publishes journals to which both members and nonmembers can subscribe. In addition, one may subscribe to the following periodicals:

Computerworld (Box 800, 375 Cochituate Road, Framingham, MA 01701). This weekly newspaper covers a wide range of current events. Special or extra editions are devoted to single topics of special interest and appear six times a year. The large number of advertisements from hardware and software suppliers can provide an ongoing awareness of new products.

Wall Street Journal (22 Cortland Street, New York, NY 10007). This daily newspaper is not specifically written for computer professionals but it contains business news relevant to the field. A weekly column in the second section covers computer-related issues.

Byte (70 Main Street, Petersborough, NH 03458). A monthly magazine for the microcomputer fan, this advertisement-laden periodical contains articles relating to hardware and software issues.

Datamation (1301 South Grove Avenue, Barrington, IL 60010). One of the oldest monthly periodicals, it annually surveys industry salaries and summarizes accomplishments of the top 100 computer-related firms.

Infosystems (Hitchock Publishing Company, Hitchcock Building, Wheaton, IL 60187). This monthly magazine presents articles on hardware and software issues.

Modern Office Technology (P.O. Box 95795, Cleveland, OH 95975). This magazine deals with issues related to office workers: word processing, spreadsheet software, microfilm, electronic mail, productivity of clerical workers, and printing.

Special magazines and newspapers have sprung up to serve special interest groups: Apples (*Macworld*), IBM PCs (*PC World*), and other microcomputers, office automation, data communications, and data-base managers.

Books

Books should form part of the analyst's growing reference library. Consult the chapter bibliographies in this book for a wide range of current titles. In addition to these, you may want to collect some on the history and philosophy of computing and on general business topics.

Structured Analysis and System Specification by Tom DeMarco (Yourdon Inc., 1133 Avenue of the Americas, New York, NY 10036; ISBN 0-917072-07-3). This easily readable book teaches the construction and evaluation of data-flow diagrams, data dictionaries, and system modeling.

Structured Requirements Definition by Ken Orr (Ken Orr and Associates, Inc., 715 East Eighth Street, Topeka, KS 66607). Written by one of the developers of the popular Warnier-Orr diagram, this text expands on diagrams and introduces a new tool, the entity diagram.

On Line Business Computer Applications by Alan L. Eliason (Science Research Associates, Inc., 155 North Wacker Drive, Chicago, IL 60606; ISBN 0-574-21405-4). This practical guide to accounting-based computer applications includes sample reports and file layouts.

Database Processing by David Kroenke (Science Research Associates, Inc., Chicago, IL; ISBN 0-574-21100-4). One of the easiest to read books on this topic, its first half deals with the theoretical aspects of data base and the second compares popular data-base systems.

The Computer from Pascal to von Neuman by Herman Goldstein (Princeton University Press, Princeton, NJ 08540; ISBN 0-691-08104-2). This older book traces the developments and individuals involved in the computer industry. It is full of interesting anecdotes and facts.

Theory Z by William G. Ouchi (Addison-Wesley Publishing Company, Reading, MA. 01867; ISBN 0-201-05524-4). It examines the three management systems, explaining why the Japanese approach achieves such high productivity.

In Search of Excellence: Lessons from America's Best-Run Companies by Thomas J. Peters and Robert H. Waterman Jr. (Harper & Row, 10 East 53rd St., New

York, NY 10022). Well over a million copies of this excellent book have been sold. It focuses on corporate management, analyzing the successful management styles of pace-setting companies. The authors illustrate Ouchi's *Theory Z* concept in action.

Megatrends by John Nesbitt (Warner Books, Inc., 666 Fifth Avenue, New York, NY 10103). This work is a synthesis of the 10 major factors influencing contemporary society. Calling our age "the time between parentheses," Nesbitt recommends that everyone prepare to work in a high-information society.

The Third Wave by Alvin Toffler (William Morrow and Company, 105 Madison Avenue, New York, NY 10016; ISBN 0-688-03597-3). Following the author's famous *Future Shock*, this book examines the impact of the technological revolution on jobs, life-styles, work ethic, sexual attitudes, economic and political structures, and philosophies of life.

The Naked Computer by Jack B. Rochester and John Gantz (William Morrow and Company, New York, NY; ISBN 0-688-02450-5). A layperson's almanac of computer lore, wizardry, personalities, memorabilia, world records, mind blowers, and tomfoolery, this book tells you everything you ever wanted to know about computers but did not know how to ask. It is a large entertaining collection of data on computers and computer people.

Training Groups

A booming training industry aimed at upgrading and advancing the skills of managers, analysts, and programmers sprouted in the early 1970s. Among the major ones are the following:

Yourdon Inc. (1133 Avenue of the Americas, New York, NY 10036). Provides seminars across the country (Europe and Asia as well) on structured analysis, design programming, COBOL, Pascal and Ada.

McCormack and Dodge (560 Hillside Avenue, Needham Heights, MA 02194). Trains individuals on their own software as well as on operating systems provided by other vendors. Enrollment grew from fewer than 300 in 1977 to over 3000 in 1981.

Ken Orr and Associates (715 East Eighth Street, Topeka, KS 66607). Orr and Associates provides seminars on Warnier-Orr and entity diagram tools across the country and at customers' sites. Other seminars focus on structured analysis and design.

DMW Group and Database Design Inc. (Ann Arbor, MI 48106): Led by James Martin, this training group specializes in issues relating to the correct design of large and small databases.

Kapur and Associates (776 El Cerro Boulevard, Danville, CA 94526). Specializes in seminars on developing structured systems, emphasizing standardized solutions for classical problems.

Many vendors also provide training for those buying or leasing their equipment. Colleges (2- and 4-year) and universities offer degree, certificate, and extension programs that can often pack a lot of education into a limited period of time.

CASE STUDY

Acceptance of Fleet Feet's New Accounts Payable System

Buoyed by his successful training program for Mark Stensaas' staff, Frank Pisciotta decides to run the new system parallel to the old one. Continued testing for a complete month uncovers a few minor errors, which Frank quickly corrects. Shortly before the end of the month of parallel operation, Frank calls a final acceptance meeting, sending notices to all concerned after first checking by telephone to determine an agreeable time.

After calling the meeting to order, Vice-President Sally Edwards briefly reviews the objectives and goals of the new accounts payable system. She reminds everyone of anticipated costs ($6702 for the detailed analysis, $23,400 for design and development) and schedule (6 months, beginning in mid-September). Then Frank takes charge, presenting the finished system to his colleagues.

As the feasibility study dated September 12, 1984 (Figure 3.9), listed seven objectives, six recommendations, and five intangible benefits, Frank discusses these points. The new system will fulfill all seven objectives:

1. Reduce processing time.
2. Achieve discount savings of $650 per month.
3. Attain better discount rates from vendors.
4. Reduce paperwork.
5. Produce timely reports on user request.
6. Control personnel costs.
7. Easily identify vendors.

Frank's review of the six recommendations reveals:

1. The software cost $22,900.
2. The new AP system solves many old manual AP system problems.
3. Lower costs than anticipated.
4. The AP system links directly to the general ledger.
5. Yearly costs should run about $1000.
6. Payback should be 10–12 months.

Frank also reports on intangible (nonmonetary) benefits:

1. Better service to vendors. Answers to inquiries arrive in a few minutes rather than days.
2. Management receives useful information as to which vouchers to pay and which to hold.
3. Fleet Feet's new AP system outperforms its competitors' systems.
4. The bookkeeping department is enthusiastic about the new system.
5. Fleet Feet utilizes the HP 3000 computer 68 percent of the time compared with 47 percent before the AP system went online.

CASE STUDY

Finally Frank mentions that it took 7 instead of 6 months to deliver the software. Why? Frank was ill for almost 3 weeks in January.

Sally resumes control of the meeting, asking for individual comments. All report a high degree of satisfaction with the new system. Commenting last, Mark Stensaas says he wishes he had had the system a year ago, "It's simple to use and extremely accurate." He thanks Frank for the considerate way in which he treated the people involved.

In conclusion, Sally asks for an acceptance vote. When all vote yes, Sally directs Mark to write a memo, which he and she will sign (Figure 14.8). She concludes the meeting by congratulating Frank on a job well done.

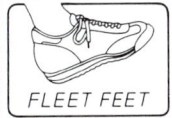
FLEET FEET

MEMORANDUM April 5, 1985

TO: All Vice-Presidents, Managers, and Franchise Owners.

FROM: Sally Edwards, Vice-President, Operations
            Mark Stensaas, Bookkeeper

SUBJECT: Notice of Completion of Accounts Payable System

We have accepted our new online accounts payable system, which is now in full operation. At yesterday's meeting, Frank Pisciotta reviewed the system and it was approved and accepted. We are happy to report that it was completed within the budgeted amount but was a month late (Frank was ill for 3 weeks).

We have been using the system on a trial basis for the past month, and all users are finding that it exceeds their early expectations. Check produced by the new system will be sent to suppliers beginning April 12. If you note any problems or encounter any comments, both negative and positive, please pass them on to us.

We wish to commend all of you for the excellent cooperation you gave Frank these past 7 months.

SE: jks

FIGURE 14.8. Fleet Feet's notice of acceptance for its new AP system.

WORKING WITH PEOPLE   Analysts as Instant Experts

Credentials officer Debbie Dean put down her telephone after scheduling an interview with Ed McMillan, analyst for the State Department of Education. She wanted Ed to explain the Department's planned online credentialing system that would replace the manual one the Department used to track over 37,500 teachers' credentials. Since Ed had taught 10th grade mathematics for 10 years before beginning a new career as a programmer, and eventually an anlayst, Debbie assumed he knew the old credentialing process from experience. As it turned out, he did not.

When it came time for the meeting in Ed's office, Debbie brought along samples of three credential application forms and blank copies of the credentials themselves. Instead of discussing the old system, Ed, eager to display his computer knowledge, launched into a futuristic lecture on new online systems. Whenever Debbie tried to interrupt with a question, Ed put her off, saying "I know, I know, but the new CRT 2000. . . ."

"I can't communicate with the deaf," snapped Debbie the fifth time Ed interrupted her.

She slammed into her office. Should she write Ed a nasty memo about the art of listening? Or should she complain to Ed's boss? This new system would involve a lot of interaction with Ed, and she couldn't stand the thought of spending even 10 minutes with the pompous jerk. When her phone rang, she was surprised to hear a sheepish Ed McMillan on the other end of the line.

"Oh, it's you," she said, unable to keep irritation out of her voice.

"I'm sorry, Debbie, No matter how hard I try, I can't help coming across like an arrogant know-it-all. I do want to work smoothly with you. Can we talk?"

He sounded so sincere, Debbie said, "I have a short fuse myself. Let's do it tomorrow. I'll have everything ready."

The next day Debbie did most of the talking and soon convinced Ed that the credentialing process had changed considerably since he last went through it. She explained that applicants now, for example, were fingerprinted for clearance by local and federal law enforcement agencies. About the only similarity was the size of the credential, still 8½ by 11 inches.

"I'm afraid I assumed too much," Ed said. "You were right to blow up. Don't be afraid to yell at me; sometimes it takes that to get my attention."

SUMMARY

Management must retain responsibility for managing the data processing function. Technical computer expertise is not needed to control people, schedules, and funds. Gantt charts help monitor staff progress, and PERT charts assist in tracking schedules. People problems demand sensitivity to human psychology, not computer technicalities.

System review or audit concludes the system process. All systems begin with a user request and end with the user, management, and computer services staff reviewing the final system, closely examining it to ensure that it achieves the goals, costs, results, and schedules established by the organization. If the system does so, a notice of acceptance goes to all relevant parties.

NEW TERMS
Acceptance notice
Cost overrun
Interpersonal relationships
Management control
Personality conflict
Philosophical differences
System audit
Territorial pride

QUESTIONS FOR REVIEW AND DISCUSSION

Questions appear in three categories. You can find the answers to the A group in this chapter. The B group requires you to apply the material presented here, whereas the C group necessitates investigation or research.

- A.1. What two activities does management supervise?
- A.2. What are the four factors that need to be examined during the system review?
- A.3. List the components of system documentation and indicate at what stage in the systems process they were generated.
- A.4. List two tools managers can use to monitor the progress of a system.
- A.5. Who are the participants in the system review?
- A.6. List the inputs to the system review.
- A.7. What are the outputs from the system review?
- A.8. List three people-oriented problems that may occur for an analyst.
- A.9. When is a new system placed into operation?
- A.10. Who decides when to place a new system into operation?

- B.1. Why are users included in the system review?
- B.2. Why should a user sign the notice of acceptance and not a member of the computer services staff?

- C.1. What do you suppose happens when a system is not accepted?
- C.2. List three examples of what might cause a system to be rejected.
- C.3. Find a copy of Nesbitt's book *Megatrends* and list the corporate goals of two different businesses.
- C.4. Peters and Waterman list factors affecting the future. List six of these factors.

REFERENCES

Keane Inc., *Principles of Productivity Management in the Development of Computer Applications*, Prentice-Hall, Englewood Cliffs, N.J., 1985.

T. Kidder, *The Soul of a New Machine*, Avon, New York, 1983.

K. Orr, *Structured Systems Development*, Yourdon Press, New York, 1977.

William Ouchi, *Theory Z*, Addison-Wesley, Reading, Mass., 1981.

T. J. Peters and R. H. Waterman, Jr., *In Search of Excellence: Lessons from America's Best-Run Companies*, Harper & Row, New York, 1983.

E. Yourdon, *How to Manage Structured Programming*, Yourdon Press, New York, 1976.

APPENDIX A
Business Systems Today

- Goals and Preview
- Information and Decision Making
- The General Ledger System
- The Accounts Receivable System
- The Accounts Payable System
- The Payroll System
- The Electronic Spreadsheet and VisiCalc
- Other Systems
- Summary
- New Terms
- Questions for Review and Discussion
- References

GOALS AND PREVIEW

After reading this appendix you should be able to:

1. List the four common accounting applications of the computer.
2. Draw a diagram of the information flows between the primary accounting applications.
3. Define the primary outputs from various application systems.
4. Define the data requirements for four application systems.
5. Describe an electronic spreadsheet.

In Chapter 1 we defined the systems cycle (analysis, design, and development) and we watched a typical small business complete the first step, preliminary and detailed analysis.

Besides studying about computers, analysts also should be learning about the most commonly computerized business systems. Since most businesses first employ the computer in financial applications, analysts need to understand the basic accounting needs of organizations. Fortunately you do not need a heavy accounting background to understand or appreciate these needs.

This appendix will familiarize you with important terms and systems you will undoubtedly encounter as a systems analyst. After briefly examining each system, we look at the reports it produces, including the data requirements. Then we evaluate each system's impact on an organization.

INFORMATION AND DECISION MAKING

Today's business environment is becoming increasingly complex. Not only do state, federal, and city regulations and laws influence business, but unions and employee bargaining groups have developed tremendous power. Businesses face stiff competition from within the United States and from foreign firms. Inflation and recessions cause a whole host of problems of their own. Proliferation of problems and decreasing productivity have sent management scrambling for new ideas and more efficient methods of operating the companies.

In such an environment, managers demand more and more information to make critical decisions. To guide their companies successfully, they must make important decisions on the basis of reliable, accurate, and timely financial data. Decisions based on out-of-date, inaccurate, or inappropriate data can seriously damage current and future business. For example, if the author of a book on systems analysis based his presentation on computer equipment available before 1980, the book would fail to attract orders and thus lose money for its publisher.

In an effort to combat these problems and to improve sound decision making, many business organizations eagerly applied computers in the 1970s, hoping to improve profits and increase productivity. Managers appreciate the computer's speed and accuracy in financial applications because it can:

Accounting A control system that maintains the financial records of an organization.

Account An individual record for each asset, liability, or owners' capital.

1. Increase managerial effectiveness.
2. Reduce operating costs.
3. Improve the effectiveness of clerical workers.
4. Facilitate management of cash receipts and disbursements.
5. Add confidence in the reliability of data.

General ledger The complete record of assets, liabilities, equity, revenue, and expenses.

Data processed into meaningful information help managers direct their companies profitably and productively.

Accounting is a control system that maintains the financial records of an organization. It involves establishing methods and procedures for collecting and recording financial transactions, organizing the data, and preparing and interpreting summary reports.

Accounts receivable Records of debts owed to a business by its customers; often abbreviated as AR or A/R.

Fundamental to the accounting process is an **account**, a record in which the results of similar transactions are accumulated. For example, a small organization may establish an account entitled "Utility Expense" and give it account number 10152. All telephone, water, gas, and electricity expenses would be put into this account. A large organization, on the other hand, may set up a separate account for each of the various types of utility expenses and assign them separate account numbers: 10022 for telephone; 10049 for gas; 10053 for water.

Accounts payable Records of amounts owed to creditors; often abbreviated as AP or A/P.

Accounting systems often link four subsystems to a central system called the **general ledger** (GL) system, which constitutes the entire collection of an organization's accounts (Figure A.1). The **accounts receivable** (AR) system tracks money customers or clients owe the organization. The **accounts payable** (AP) system records monies the organization owes others. In many respects accounts payable is the opposite of accounts receivable.

Payroll System calculating employee wages, withholdings, and benefits.

Most of us recognize **payroll**. A person works for hourly pay or wages, commissions, tips, or a salary. The payroll system converts labor and tips into earnings

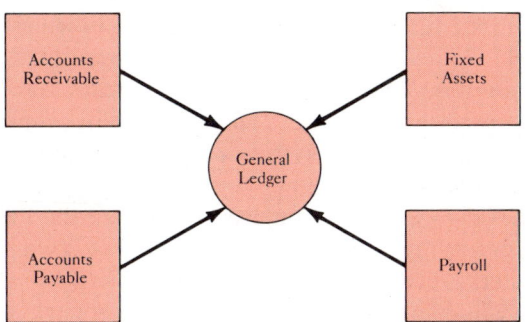

FIGURE A.1. Common business systems. This figure shows that accounts receivable, accounts payable, fixed assets, and payroll all feed data to general ledger. (Courtesy MCBA®. MCBA is a registered trademark of MCBA, Inc.)

and determines how much federal, state, and city tax to deduct. Payroll systems also keep track of other types of deductions (such as those for life and health insurance premiums), vacation time earned, and accumulated sick leave.

Last are **fixed assets**. A fixed asset is anything acquired for long-term use in the business. This includes desks, computers, a building, and property. The fixed asset system records all such assets, and may even track each asset's maintenance history, depreciated value (for tax purposes), physical location, and insurance cost.

These five interlinking systems were the first accounting systems to be computerized, though many other systems appeared over the years, some of them specific to an industry. For instance, large mail order businesses, such as Sears Roebuck or L. L. Bean, need an order-entry system to reduce the time between receipt of a customer order, the processing and filling of the order, and the shipment of goods to the customer. Similarly, a certified public accountant needs a client write-up system to keep tax records for individuals and smaller businesses. Hospitals need a system for tracking the services used by patients (X-ray, emergency, drugs, bandages, food) and the fees for each.

In the remainder of this chapter, we focus on the four most common systems: the general ledger, accounts receivable, accounts payable, and payroll. We also examine the electronic spreadsheet and order-entry, inventory, and fixed asset systems. Spreadsheet systems enable users to transform data into a grid of rows and columns that can be combined from summary statistics (totals, averages, maximums, minimums, standard deviations, or counts). Order entry reduces the time between the receipt of a customer's order and the shipment of goods. Inventory tracks the products a firm has on hand to reduce the number of stockouts and losses, and maximize the return on investment capital. Fixed assets keep depreciation records for properties a firm owns that meet requirements specified by federal law.

Fixed assets Records of all items of value owned by an organization.

THE GENERAL LEDGER SYSTEM

As you may recall from Chapter 1, Mountain Motors, Inc., is an auto-care chain that performs preventive auto care for customers—batteries, tires, front-end alignment, and air conditioning service. In its day-to-day operation, Mountain Motors buys supplies, pays its employees, rents office space, bills customers, and incurs other expenses (telephone, utilities, office supplies) and, of course, it receives payments from customers. The controller needs to know how the firm is doing: Is it losing money or is it profitable? To provide this important information, Mountain has to set up accounting systems, with the general ledger as the focal point for determining profitability.

Keep in mind that the general ledger accumulates all the financial transactions of an organization according to its chart of accounts (Figure A.2). A large conglomerate may collect, process, and maintain transactions for each unit or subsidiary but eventually will combine all the results in one report. The purpose of the general ledger is to organize the data to produce a company's financial statement and other reports used by management.

Before entering transactions in the general ledger, a firm lists them in a tem-

Current Assets (100–121)		Sales (400–499)	
100	Petty Cash	400	Sales
101	Cash	404	Sales Discounts
103	Notes Receivable	453	Freight
104	Inventory	460	Purchases
108	Prepaid Rent	465	Purchase Returns and Allowances
109	Prepaid Taxes	467	Cost of Goods Sold
110	Long-Term Assets		
111	Plant	Operating Expenses (500–599)	
113	Equipment	501	Sales Salaries
		505	Sales Commissions
Intangible Assets (161–180)		522	Fuel Expense
161	Patents	523	Postage Expense
162	Franchises	528	Advertising Expense
163	Trademarks	551	Officers' Salaries
		570	Payroll Taxes
Liabilities (200–219)		579	Legal Fees
201	Notes Payable	583	Bad Debt Expense
202	Accounts Payable	590	Building Repair
203	Salaries Payable		
204	Rent Payable	Other Expenses (600–699)	
210	Bonds Payable	601	Interest on Notes
		602	Interest on Bonds
Stockholder's Equity (300–399)		603	Interest on Mortgage
301	Common Stock		
303	Preferred Stock	Other Income (700–799)	
310	Treasury Stock	701	Interest Revenue
330	Retained Earnings	703	Rental Income
		704	Investment Income
		705	Miscellaneous Income

FIGURE A.2. Chart of accounts for a typical business. Every business establishes a chart similar to this one when it commences operation. The chart shows the account number and description but does not identify amounts of money in the accounts. This list is not a complete list but only shows accounts that are typical for most businesses.

porary, chronological file called the **general journal**, or just the journal. It shows the date of the transaction, the amounts involved, the particular accounts affected by the transaction, and a brief explanation of the transaction. It is similar to your personal checkbook where you record the dates, check numbers, amounts, and the parties to whom you write checks. You probably keep it up to date, maintaining it chronologically.

Periodically (daily, weekly, monthly), or when a certain number of transactions accumulate in a general journal, organizations transfer all transactions to their respective accounts in the general ledger. We call this process of transferring transactions **posting**.

General ledger systems utilize double-entry bookkeeping procedures, which are based on the **accounting equation**:

$$\text{Assets} = \text{Liabilities} + \text{Owners' equity} + \text{Revenue} - \text{Expenses} - \text{Withdrawals} - \text{Drawing}$$

The equation must always be in balance. If a transaction increases an asset account, it must also either increase a liability or owners' equity account or decrease another asset account. Increases and decreases are accomplished through **debits** and **credits**. For asset and expense accounts, debits represent increases and credits represent decreases. For liability, owners' equity, and revenue costs, debits represent decreases and credits represent increases. Revenue and expense accounts are subsidiaries to owners' equity because at the end of an accounting period, they are closed to retained earnings, an owners' equity account.

To illustrate, suppose a customer uses a charge account to buy a $100 alternator from Mountain Motors. When the customer charges the $100, Mountain Motors makes two entries: (1) a debit of $100 to the customer's Accounts Receivable account and (2) a credit of $100 to the Sales account. These two entries offset each other, that is the debits equal, or offset, the credits.

Before computers, businesses posted to the general ledger by hand, a laborious and time-consuming process that required extraordinary efforts to ensure correctness. When a business installs a computerized general ledger, transactions are entered into the computer and eventually stored on a reel of tape or a magnetic disk.

Computerization of the general ledger does not guarantee 100 percent accuracy because the person entering the data can post data to an improper account. The machine does, however, perform rapid additions and subtractions extremely accurately, and a simple program can increase a general ledger system's accuracy by forcing it to reject all transactions in which debits do not offset credits. This higher degree of accuracy is one of the primary goals of computerizing accounting systems.

General ledger systems generate several reports used by management and outsiders to evaluate a company's financial position and operating status (Figures A.3*a* and A.3*b*). Two important reports are the profit and loss statements (P&L), usually called the income statement, and the balance sheet, or statement of financial position (Figure A.3*c*). The income statement shows an organization's revenue, expenses, and net profit over a certain time period. It usually precedes the balance sheet and indicates the organization's profitability since the last income statement.

The balance sheet recapitulates all totals carried in the general ledger and reflects

General journal The chronological collection of many different types of transactions.

Posting Transferring transactions from the journal to a ledger.

Accounting equation The rules governing how financial transactions are recorded.

Debit The left-hand side of an account recording indebtedness.

Credit The right-hand side of an account recording payments or other values received.

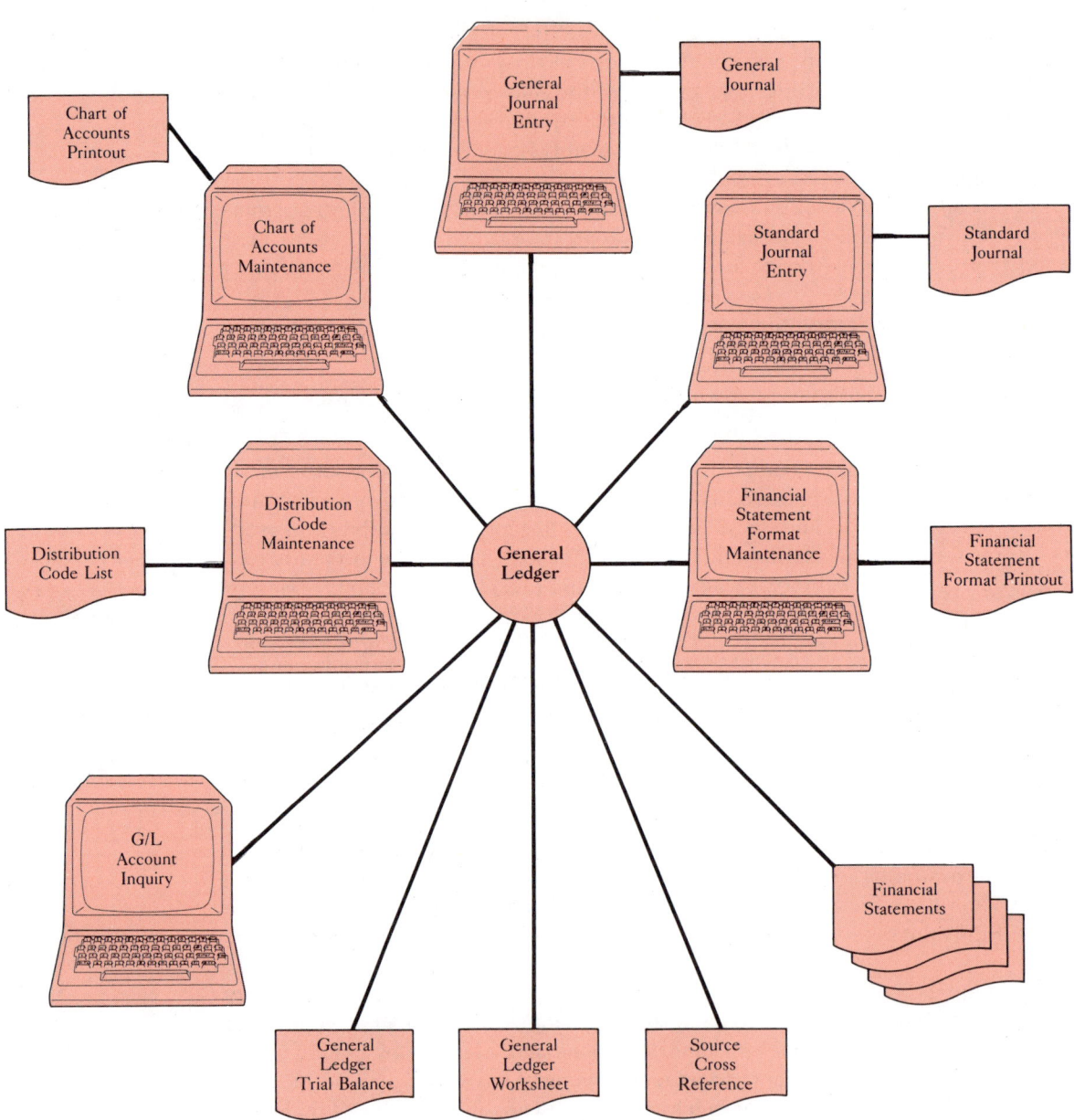

(a) Pictorial view of the general ledger.

FIGURE A.3. Sample inputs and outputs from a general ledger system. (Courtesy MCBA®. MCBA is a registered trademark of MCBA, Inc.)

Report Name	Purpose of Report
Chart of accounts printout	Lists all the accounts used by the organization including the account number, account name, and type of account (asset or liability).
General journal	Chronological list of all transactions. It shows dates, amounts, accounts affected, and an explanation of the transaction.
Standard journal	List of recurring general journal entries. May be fixed amounts posted each period or a variable amount may be entered each period and posted.
Financial statement layout maintenance	Permits user to format and content financial statements.
Financial statements	Balance sheet, Income statement, Statement of retained earnings, and Notes to financial statements.
Source cross reference	Shows the GL transactions in order by date within account number.
General ledger worksheet	Columnar schedule used to conveniently summarize accounting data.
General ledger trial balance	List of each account with its credit or debit balance. Used to check the accuracy of the posting operations because the total of all debits must equal the total of all credits.
GL account inquiry	Allows all transactions for a particular account to be displayed on the terminal.
Distribution code list	Automatic spread by predetermined percentages of general and standard journal transactions.

(b) Typical reports produced by the general ledger system.

FIGURE A.3 *(Continued)*

FLEET FEET
PROFIT AND LOSS STATEMENT
Prepared by Moore and Smith, CPAs
and without opinion expressed or implied by
this financial statement is unaudited

For Month Ended June 30, 1984

Revenue from Sales			
Sales		$205,000	
Less: Sales Returns and Allowances	$ 3,100		
Sales Discounts	3,900	− 7,000	
Net Sales			$198,000
Cost of Goods Sold			
Merchandise Inventory, June, 1, 1984		92,000	
Purchases	106,000		
Freight In	4,000		
Delivered Cost of Purchases	110,000		
Less: Purchases Returns and Allowances	$ 620		
Purchases Discounts	5,400	−6,020	
Net Purchases		103,980	
Cost of Goods Available for Sale		195,980	
Less: Merchandise Inventory, June 30, 1984		−90,200	
Cost of Goods Sold			105,780
Total Revenue			$ 92,220
Operating Expenses			
Rent Expense		7,000	
Utilities Expense		1,500	
Wages Expense		11,800	
Advertising Expense		8,000	
Supplies Used Expense		1,280	
Insurance Expense		200	
Depreciation Expense		7,000	
Total Operating Expense			36,780
Net Profit (Loss)			$ 55,440

(c) The profit and loss statement summarizes the company's performance over a specific period of time.

FIGURE A.3 (Continued)

ASSETS	1982	1981
Current assets:		
Cash	$ 4,852,900	$ 80,100
Short-term investments, at cost which approximates market (includes accrued interest of $3,108,500 in 1982 and $1,627,700 in 1981)	90,426,900	110,570,800
Earned research and development revenues receivable:		
Related parties (Note 1)	1,620,200	961,900
Others	1,871,500	2,303,300
Other current assets	1,207,200	277,900
Total current assets	99,978,700	114,194,000
Property, plant, and equipment, at cost (Notes 1 and 2):		
Land	2,427,600	612,100
Buildings and building improvements	6,156,500	3,808,900
Machinery and equipment	9,775,400	4,252,500
Leasehold improvements	9,652,900	1,090,600
Furniture, fixtures, and equipment	770,200	603,400
Construction-in-progress	7,009,500	2,678,500
	35,792,100	13,046,000
Less accumulated depreciation	4,024,300	2,430,800
Net property, plant, and equipment	31,767,800	10,615,200
Other assets	817,800	424,000
	$132,564,300	$125,233,200

	June 30,	
LIABILITIES AND STOCKHOLDERS' EQUITY	1982	1981
Current liabilities:		
Accounts payable	$ 558,000	$ 1,659,100
Accrued liabilities (Note 8)	684,200	529,100
Income taxes payable	550,400	—
Unearned research and development advances	1,592,600	—
Other current liabilities (Note 8)	503,800	30,800
Current portion of debt obligations	7,500	191,100
Total current liabilities	3,896,500	2,410,100
Debt obligations	354,000	287,600
Deferred income taxes	—	84,000
Commitments and contingencies (Note 2)		
Stockholders' equity (Notes 3, 4, and 6):		
Common stock, $.01 par value; 75,000,000 shares authorized; issued and outstanding 21,975,610 for 1982 and 21,948,610 shares for 1981	219,800	219,500
Capital in excess of par value	132,644,800	132,123,300
Deficit	(246,600)	(4,751,400)
Less:		
Notes receivable secured by common stock	(1,892,200)	(2,212,500)
Unearned compensation expense	(2,412,000)	(2,927,400)
Total stockholders' equity	128,313,800	122,451,500
	$132,564,300	$125,233,200

(d) The balance sheet displays the company's overall performance. It is usually presented on a comparative basis, showing the current period versus the prior period.

FIGURE A.3 *(Continued)*

the overall position of the company at a specific time. The statement lists the assets of the business in decreasing order of liquidity; cash, accounts receivable, inventory, and long-term assets. After the list of assets, the liabilities appear (accounts payable, long-term debts, mortgages), followed by the owners' equity (stock and retained earnings). As you can see from Figure A.3*d*, a balance sheet should always balance, with total assets equaling total liabilities and owners' equity.

Managers and accountants calculate various ratios from these two reports to determine trends (for instance, how much sales are increasing in relation to operating expenses) or to measure overall business health. Income statements and balance sheets assist personnel in planning, budgeting, reporting to stockholders or owners, securing loans, controlling operations, and measuring performance.

THE ACCOUNTS RECEIVABLE SYSTEM

An important source of data to the general ledger is accounts receivable (AR), which records money owed an organization for goods it has sold or services it has rendered but for which it has not yet been paid; see Figures A.4a and A.4b.

To illustrate, let us watch a transaction flow through the Fleet Feet shoe company's AR system. Katie Baker enters the store to buy a pair of running shoes for $45.60. Instead of paying by check, cash, or charge card, Katie puts her purchase on her Fleet Feet charge account by signing an order form (Figure A.4c). Since the order form contains all data relevant to a sale, it becomes the primary source of data for accounts receivable. On Fleet Feet's books, Katie's purchase is recorded as a debit to accounts receivable and credit to sales.

A month later Fleet Feet sends Katie a statement, or bill, showing the month's purchases, payments received during the month, any unpaid balances from previous months, finance charges, and the total balance due (Figure A.4d). When Katie sends a check for $45.60 to Fleet Feet, the company reduces the balance in her account by that amount. Fleet Feet's cash on hand increases by the same amount (remember that accounting requires double entries). From an accounting perspective, Katie's payment would cause a credit to the accounts receivable account and a debit to Fleet Feet's cash account. The customer's payment becomes an additional source of input data to the AR system.

Balancing Methods

Balance forward A type of AR system where customer payments apply to the oldest transaction.

Open item A type of AR system in which payments are applied to specific charges.

An important accounts receivable system concept involves balancing customer accounts. Two balancing methods prevail: **balance forward** and **open item**. In using the balance forward method, the company applies a customer's payment to the oldest recorded transactions in the account. Suppose, for example, Katie already owes $73.00; with her purchase for $45.60, she now owes a total of $118.60. If she pays $45.60 this month, Fleet Feet reduces her balance to $73.00 ($118.00 − $45.60). If the balance forward method is used, Katie's payment is applied to her initial balance due ($73.00) and she still owes $27.40 for last month's debt ($73.00 − $45.60) plus $45.60 for this month's purchase.

If Fleet Feet were using the open item method, the company would consider each transaction individually, applying each payment to a specific charge incurred by the customer. In this case Katie owes Fleet Feet $73.00 on past bills and $45.60 on the current one. She can stipulate that her check for $45.60 pay the latter rather than the former. In either case she still owes $73.00, but for different purchases.

Both AR systems are popular. Since the balance forward method does not require a record of past transactions, it is simple to use. The open item method, on the

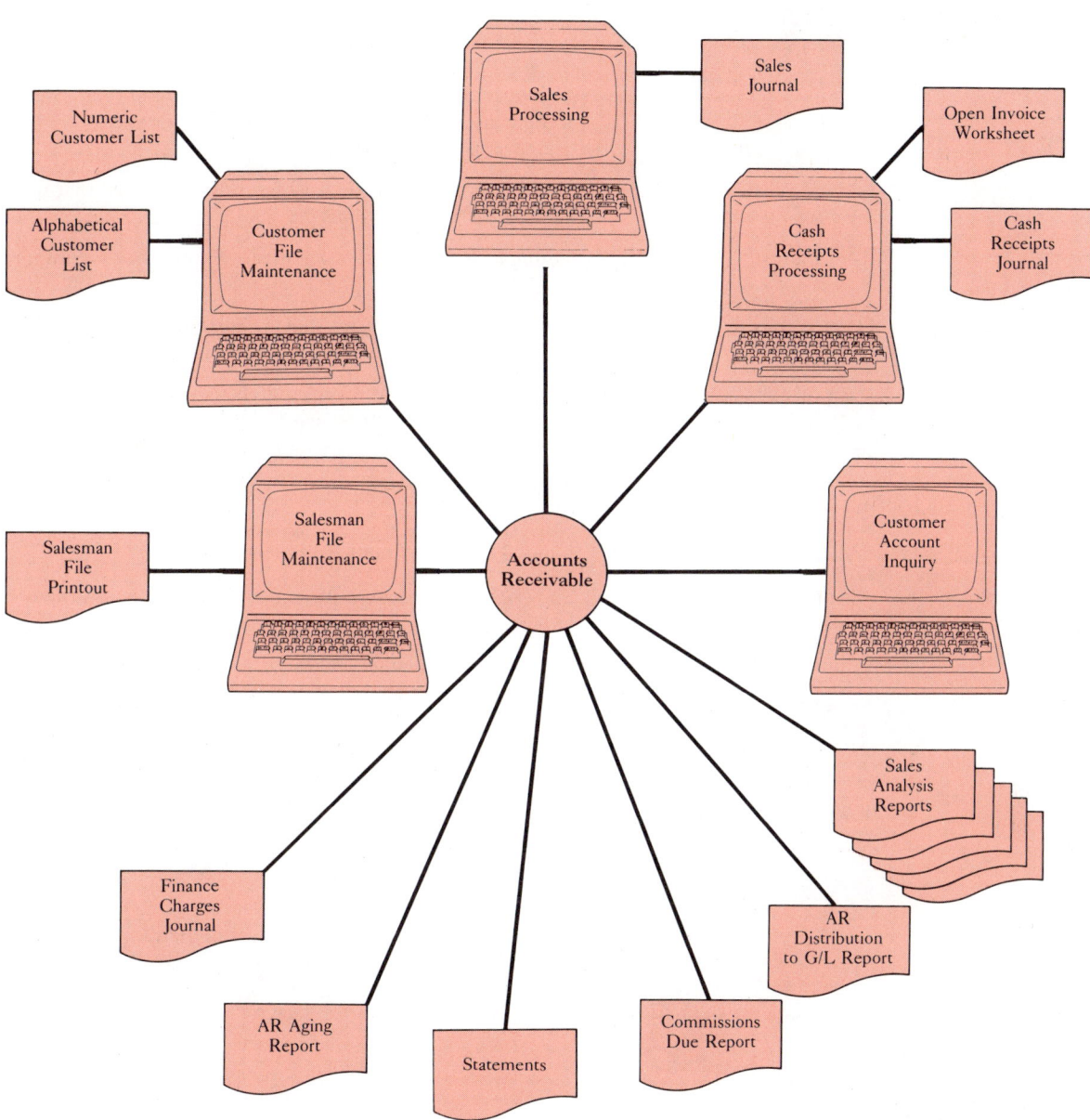

(a) Pictorial view of accounts receivable.

FIGURE A.4. Sample inputs into and outputs from the AR system. (Courtesy MCBA®. MCBA is a registered trademark of MCBA, Inc.)

Report Name	Purpose of Report
Alphabetic list of customers	A cross-referenced listing of all customers and their account numbers. Sometimes also lists addresses, balances, and telephone numbers.
Numeric customer list	Similar to the alphabetic list of customers except that the list is ordered by customer number.
Sales journal	Chronological list of all sales made. Shows the amount, date, customer, and item(s) ordered.
Open invoice worksheet	List of items not sent for specific customer orders. Only applicable for an open item accounts receivable system.
Cash receipts journal	Shows each payment made, the check number, date of receipt, and invoice to which the payment was applied.
Customer account inquiry	Shows all open items after a specified date for a selected customer on the terminal.
Sales analysis reports	Sales statistics for each salesperson, including customer number, amount, and total for each salesperson.
AR distribution to GL report	Shows the total amount of money posted to each general ledger account.
Commissions due report	List of all salespeople and their sales for this time period.
Statements	See Figure A.6c for a sample of this report.
AR aging report	Lists each customer and the amounts owed in each aging category with totals for each aging category.
Finance charges journal	List of finance charges made to customers for unpaid amounts.
Salesperson file printout	Lists the pertinent data about each salesperson: name, address, SSN, sales this year, sales last year, sales this period, sales last year during the same period, commissions earned, returned sales.

(b) A variety of useful AR outputs.

FIGURE A.4 (Continued)

FLEET FEET		
SACRAMENTO, CALIFORNIA 95815		
442-3338		

SOLD BY	DATE	
SE	9/12/84	19

NAME	KATIE BAKER
ADDRESS	888 MAIN STREET
CITY	SACRAMENTO, CA 95816

☐ CASH ☒ CHARGE ☐ MDSE. RET'D
☐ C.O.D. ☐ PAID OUT ☐ PD. ON ACCT.

QUAN.	DESCRIPTION	AMOUNT	
1	RUNNING SHOES	44	50
	TAX	1	10

No cash refunds. Credit or exchanges only.

RECEIVED BY *Katie Baker*

TOTAL $45 | 60

NO. **93374**   Thank You

(c) A sales clerk fills out Fleet Feet's order form, which goes to the business office for posting to the customer's account.

FIGURE A.4 *(Continued)*

```
                          STATEMENT
  FLEET FEET SPORTS, INC.                                    [FLEET FEET logo]
       SACRAMENTO, CA


  Katie Baker                                  DATE      10/15/84
  888 Main Street
  Sacramento, CA 95816
                                               CUSTOMER NO.   872

                                               $ _____
                                                 AMOUNT ENCLOSED
           DETACH THIS PORTION AND RETURN WITH YOUR REMITTANCE
```

INVOICE DATE	INVOICE NUMBER	REFERENCE	CHARGES	CREDITS	BALANCE
8/6/84	#93002	Track suit	$73.00		$73.00
9/12/84	#93374	Running shoes	$45.60		

PREVIOUS BALANCES DUE			FLEET FEET SPORTS, INC. SACRAMENTO, CA PHONE 442–FEET	TOTAL BALANCE DUE
30-60 DAYS	60-90 DAYS	OVER 90 DAYS		
$73.00				$118.60

(d) Fleet Feet expects payment within 30 days and sends bills on the first of the month to each customer. If customers do not pay on time, Fleet Feet charges 2 percent interest per month on the unpaid balance.

FIGURE A.4 *(Continued)*

other hand, does require such a record, but it provides the customer with more data for decision making. Perhaps the customer wishes to dispute a charge or needs a record of past transactions for income tax purposes. Major banks apply the balance forward system to their credit cards, whereas the open item method is more common among suppliers of building materials, wholesalers, and some retailers.

Aging

Aging Classifying the amount of a debt relative to time.

Aging is another important AR concept. Suppose you visit your dentist for a tooth extraction, after which you receive a bill asking for payment in 30 days. If you fail to pay, the dentist will remind you that you are late. To your dentist your account has aged because time has elapsed between the original and the most recent billing dates. Most companies demand payment in 30 days or less and they keep track of unpaid accounts. They usually classify them as from 31 to 60 days late, 61 to 90 days late, and over 90 days late. Aging data, displayed in a report called the aged trial balance, may help determine a customer's reliability, indicate debt-collection procedures, or stimulate discounts to promote prompt payment. Both balance forward and open item methods keep track of the total amounts owed by each customer in each aging category.

The AR system can generate numerous reports depending on the needs of an organization. See Figure A.4 for examples.

THE ACCOUNTS PAYABLE SYSTEM

Accounts payable (AP) also provides inputs to the general ledger, detailing transactions concerning an organization's unpaid bills; see Figures A.5*a* and A.5*b*. These debts arise from such sources as bills from suppliers, goods purchased, services used, and loans. In tracing a transaction through Fleet Feet's AP system, we begin with the fact that the store has run out of a running shoe. The store manager phones the vendor and orders 15 pairs in various sizes and colors for a total of $629. To the vendor this order becomes an account receivable of $629; to Fleet Feet it represents a debt, or an account payable of $629. The vendor ships the order and Fleet Feet accepts delivery with the understanding that it will pay for the shoes when it receives the bill.

The vendor then sends the buyer an invoice listing charges for the shoes. This **invoice** or voucher, states that the buyer owes the vendor a certain amount of money. Fleet Feet creates a voucher record that includes the invoice number, vendor number, amount of the invoice, and the date payment is due. The transaction's own voucher number uniquely identifies it. From an accounting perspective, Fleet Feet records the transaction as a credit to accounts payable and a debit to purchases.

Invoice An itemized statement of merchandise sent to purchasers.

During the month Fleet Feet examines its vouchers and considers making payments. A vendor often offers a discount to customers for prompt payment. For example, a vendor may give a 2 percent discount if payment arrives within 10 days of the invoice date. If the invoice for shoes totals $690.00, a 2 percent discount reduces the cost of the shoes by $13.80. If Fleet Feet takes advantage of the discount by paying the bill within 30 days, it debits accounts payable for $690.00 and credits purchases for $576.20 and purchase discounts for $13.80. To both vendor and buyer, the age of an AP voucher is significant because it can influence the amount of money that changes hands and the credit status of the buyer.

Accounts payable systems can automatically track the age of vouchers, enabling organizations to take advantage of discounts, or, on the other hand, indicating which unpaid vouchers will increase with interest charges. Obviously a company cannot successfully manage its finances without such data.

Vouchers constitute the primary data inputs to an accounts payable system. The primary output from an AP system is a check (Figure A.5*c*) to the vendor.

480 APPENDIX A BUSINESS SYSTEMS TODAY

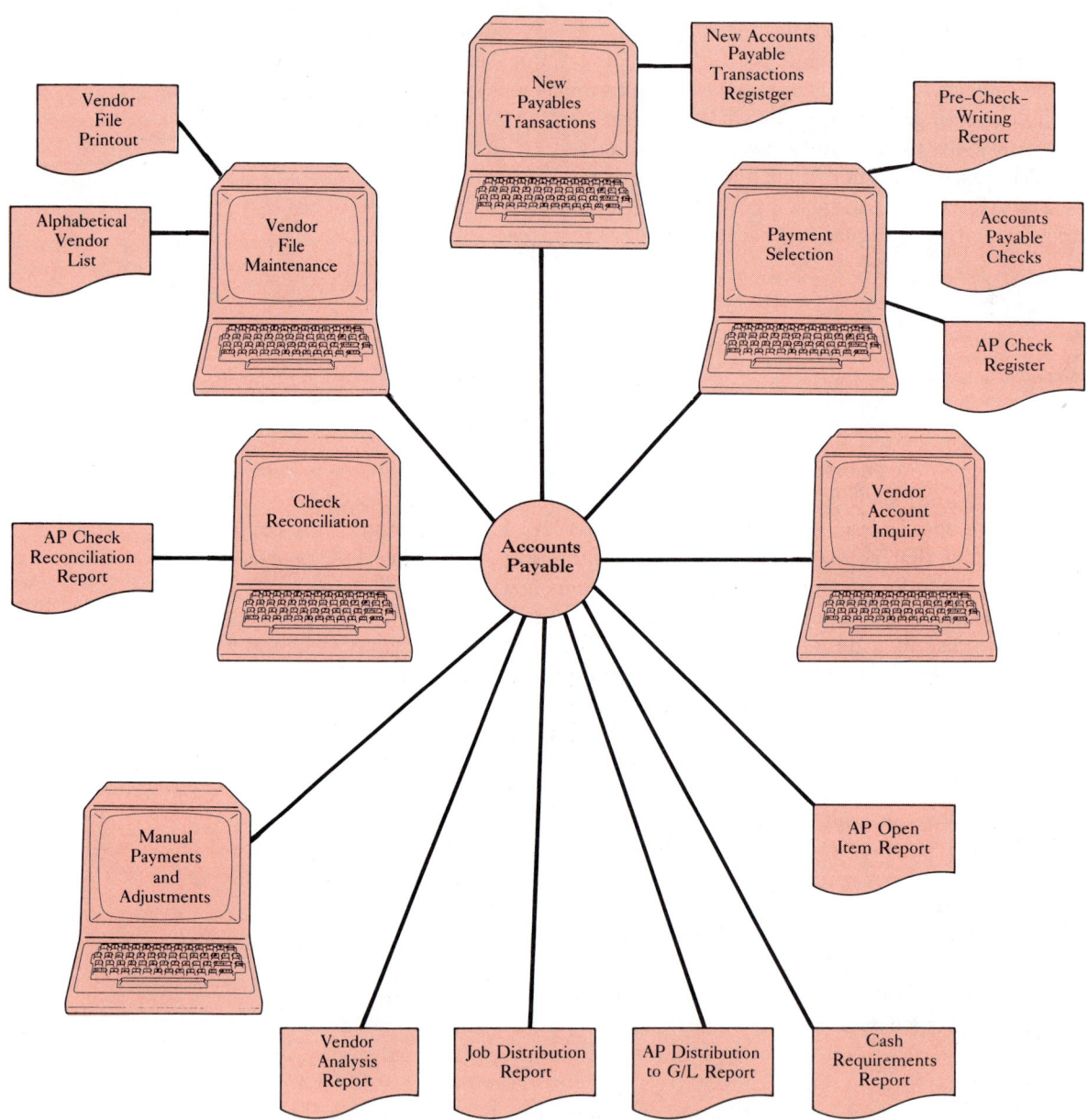

(a) Pictorial view of accounts payable.

FIGURE A.5. Sample inputs into and outputs from the AP system. (Courtesy MCBA®. MCBA is a registered trademark of MCBA, Inc.)

Report Name	Purpose of Report
Numeric vendor list	List of vendors, including their addresses, telephone numbers, and sometimes the contact person for the vendor. The list is in order by vendor number.
Alphabetic vendor number	Similar to the numeric vendor list except that the list is in order by vendor name.
New accounts payable transaction register	List of all new charges made to the AP system. The list is chronological, showing the date, amount, vendor, GL accounts affected, and a description of the transaction.
Pre-check-writing report	List of all checks to be written before they are printed on check stock. Permits checks to be proofread before they are printed.
AP check	See Figure A.7c for example of the check.
Check register	Record of each check produced by the AP system, to which vendor it was written, and the vouchers it is paying.
Vendor account inquiry	Displays all vouchers for a selected vendor. Items are shown in voucher number order.
AP open item report	Lists the vouchers due for payment along with the associated discount terms.
Cash requirements report	Shows all items past due, current, and optionally payable plus any lost or valid discounts for any range of vendors.
AP distribution to GL report	Lists of vouchers, how each was applied to the general ledger, and account totals.
Job distribution report	List of charges applied to specific jobs.
Vendor analysis report	Analysis of this year's activities compared with last year's, showing total purchases and discounts taken.
Manual payments	Adjusts due dates, discount amounts, and discount dates for individual vouchers.
AP check reconciliation	List of all checks and their status: voided, paid to a vendor, or cashed by a vendor.

(b) A variety of useful AP outputs.

FIGURE A.5 *(Continued)*

VENDOR: 000002						CHECK NO. _____
OUR INV. NO.	YOUR REF. NO.	INVOICE DATE	INVOICE AMOUNT	AMOUNT PAID	DISCOUNT TAKEN	NET CHECK AMOUNT
001013		04/25/84	150.00	150.00	.00	150.00
001014	HJ	05/23/84	263.00	200.00	.00	200.00
					CHECK TOTAL	350.00

CHECK NO.	CHECK DATE	VENDOR NO.
000301	06/25/84	000002

FLEET FEET

CHECK NO. _____

CHECK AMOUNT
$*******350.00

PAY TO THE ORDER OF
RED LINE FREIGHT
350 COMMERCIAL ROAD
BIGTOWN, TEXAS 99996

NON-NEGOTIABLE

(c) Accounts payable check sent to vendors.

FIGURE A.5 *(Continued)*

THE PAYROLL SYSTEM

Payroll represents one of the most complex business systems: Figure A.6a. Tax rates fluctuate, taxing procedures vary, union contracts change. Some workers require hourly wages, some receive salaries, and others earn commissions. Some employees submit expenses for meals, gas, and lodging for reimbursement. As with the other interlinking systems of accounting we have examined, payroll also has useful outputs beyond the paycheck. Management can obtain numerous reports from a payroll system, including job distribution and union deduction reports and W-2 and 941A forms. See Figure A.6b for other items produced by the payroll system.

Let us suppose Fleet Feet hires a new person. To initiate payroll procedures, the company collects certain pertinent data about the new employee and enters those facts into its computerized payroll system. Pertinent data may include:

1. Employee number.
2. Employee name, address, and social security number.
3. Date of birth, date of hire.
4. Gross pay.

APPENDIX A BUSINESS SYSTEMS TODAY **483**

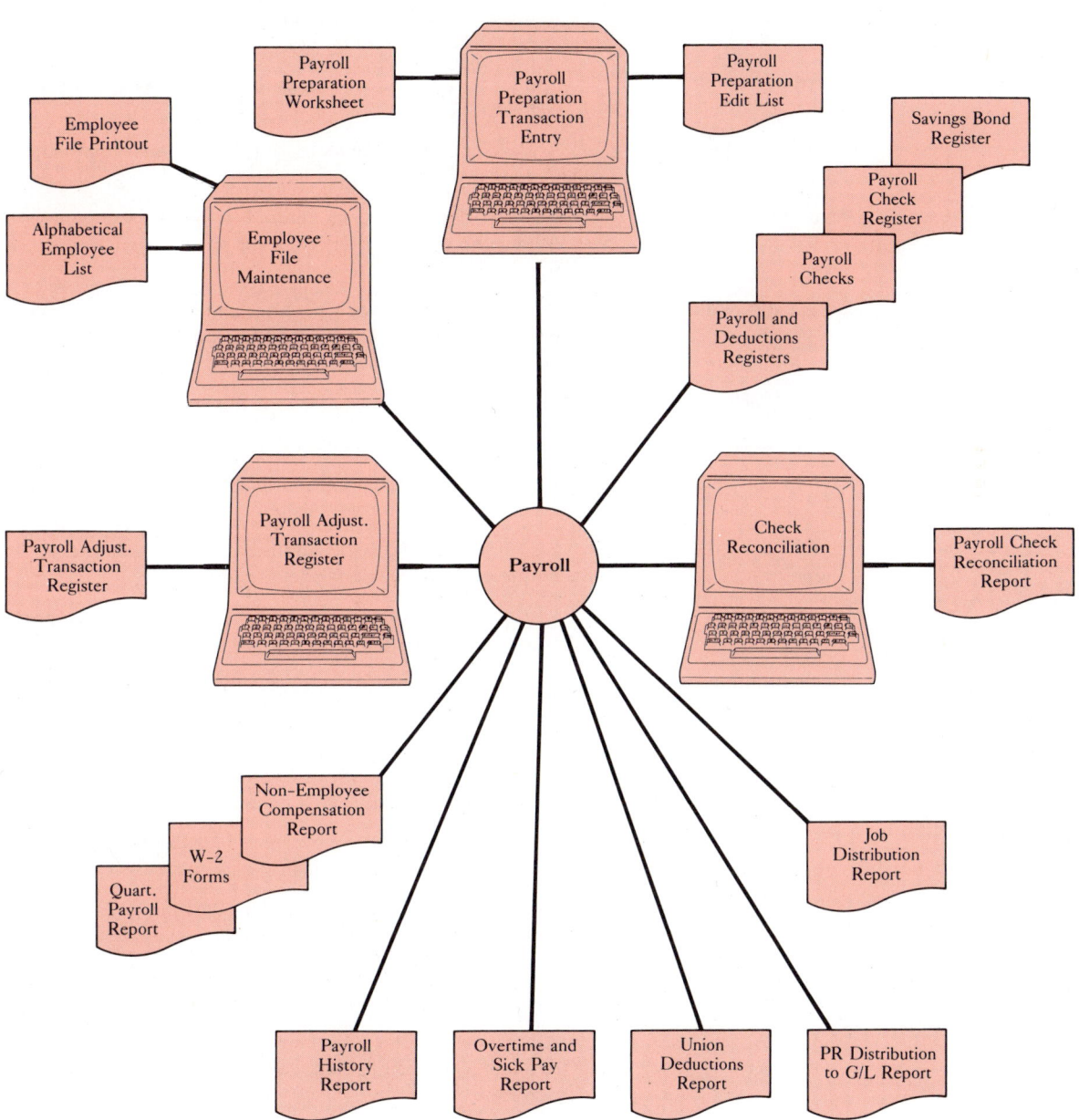

(a) Pictorial view of payroll.

FIGURE A.6. Sample inputs into and outputs from the payroll system. (Courtesy MCBA®. MCBA is a registered trademark of MCBA, Inc.)

Report Name	Purpose of Report
Numeric employee file printout	Lists all employees, giving their addresses, social security numbers, and all their earnings, deductions, quarter-to-date, year-to-date, and other pertinent data.
Alphabetic employee list	Similar to numeric list except the order is by employee last name.
Payroll preparation worksheet	A worksheet with blanks for employee hours, temporary deductions, earnings, and expenses.
Payroll preparation edit list	List of all pertinent data entered for this payroll period. Allows the data to be checked for accuracy before the payroll cycle is begun.
Payroll and deductions register	List of each employee's earnings and deductions for this pay period before the paycheck is printed. The data can be verified and corrected.
Payroll checks	See Figure A.8d for a sample of this report.
Payroll check register	List of payroll checks and all vital data on those checks.
Savings bond register	List of savings bond contributions by employees and matching money by the business.
Payroll check reconciliation report	List of all payroll checks and their status: voided, cashed by the employee, or issued and uncashed.
Job distribution report	List of each employee's cost to specific jobs. This report is used to track labor costs on jobs.
PR distribution to GL report	Shows the amounts to be applied to the various accounts in the general ledger. The report is in order by GL account number and by check date.
Union deductions report	List of each employee's contributions as well as the business's contribution to a union(s).
Overtime and sick pay report	List of each employee's overtime hours and cost plus sick time and pay accumulations.
Payroll history report	List of each employee's pay history for this year as well as last. Used for comparison purposes.
Nonemployee compensation report	List of money paid to employees in nonsalaried areas: mileage, commissions, expenses, food, etc.
W-2 forms	A separate page for each employee's earnings and deductions, which is filed with the Internal Revenue Service.
Quarterly payroll report	List of each employee's earnings during the quarter. The report also shows the taxes withheld as well as the totals for all employees.
Payroll adjustment transaction register	List of all correcting entries to be made to the payroll system: changes in names, addresses, withholdings, etc.

(b) Reports produced by the payroll system. This chart does not include the many reports that can show the labor cost associated with a particular department, job, or unit within a multicompany organization.

FIGURE A.6 *(Continued)*

(c) For workers paid on an hourly basis, a time card is used to collect data for input to payroll. The time card shows the employee's name, days and hours worked, time off, gross income, deductions, and net amounts. The time card is filled out by the employee but the computer adds the totals and monetary computations.

FIGURE A.6 *(Continued)*

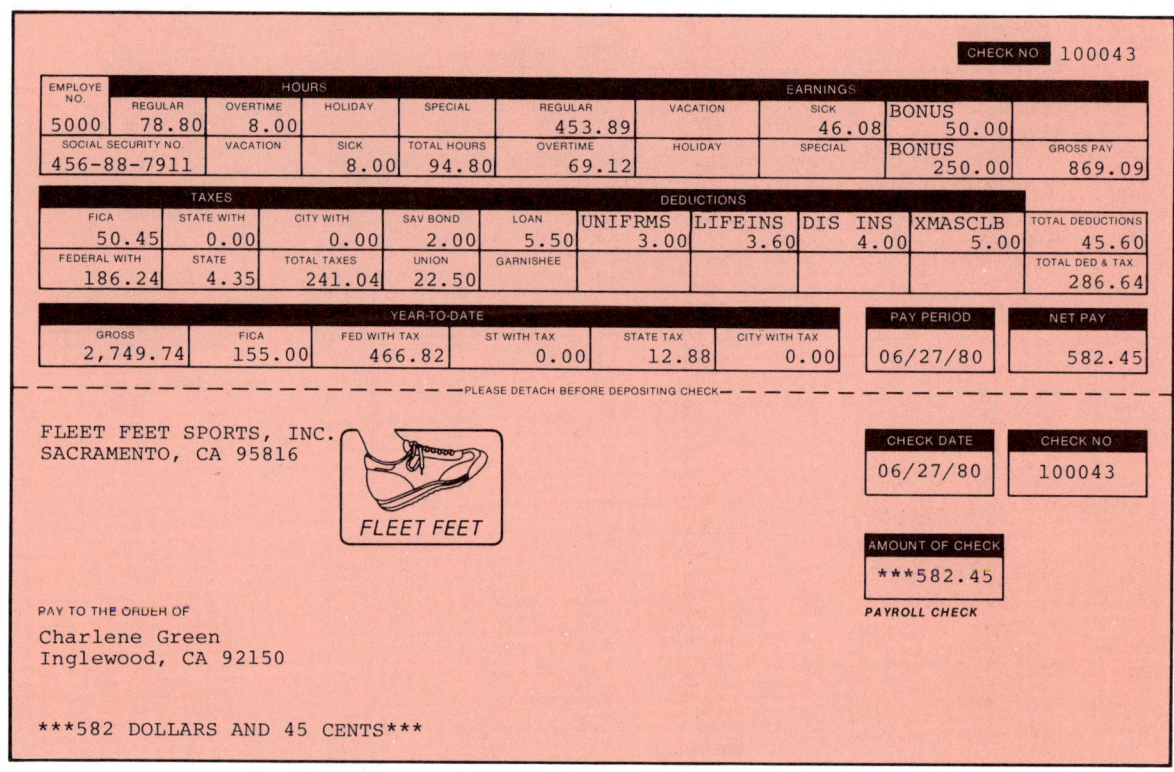

(d) The payroll check is the primary goal of the payroll system. All the data from the time card are included on the top portion of the check so the employee can verify that the amounts are correct.

FIGURE A.6 (Continued)

5. Pay rate: per hour, week, or month.
6. Information for calculating taxes, such as number of dependents.
7. Miscellaneous deductions, such as United Way and union dues.

Upon completing a payroll period, the new employee reports data about his or her work on a time card (Figure A.6c). On the basis of the time card and the employee's personal data file, Fleet Feet's payroll system determines the amount the employee deserves and calculates deductions (social security, federal, state, and local taxes, etc.) to be withheld from the total. The system then prints a check in the amount to be given to the employee (Figure A.6d). In most cases the check itself shows total earnings and deductions, resulting in the employee's net earnings.

THE ELECTRONIC SPREADSHEET AND VISICALC

So far we have examined the five systems most businesses need for accounting purposes. A company can computerize any of these or other systems. Another system, developed fairly recently, is the **electronic spreadsheet** (sometimes also called an electronic worksheet), which has revolutionized the way we work with columns and rows of data. A spreadsheet is a grid of columns and rows of data that facilitates the manipulation of data; for example, you change one number and the program changes all related numbers. We label columns (vertical) with letters or names (A, B, C, . . .) and rows (horizontal) with numbers (1, 2, 3,. . .). Thus B4 represents the intersection between the second column and the fourth row. The intersection of any column with any row displays a relationship between them (Figure A.7). Familiar symbols such as + and − represent arithmetic operations we wish to perform on the data. We can use key words such as "sum," "average," and "sqrt" (square root) to allow the electronic spreadsheet to perform routine operations.

Electronic spreadsheet Software using rows and columns to permit automatic recalculation of data.

The user "scrolls" (moves the cursor on a terminal or personal computer) to the left, right, up, or down to bring different parts of the worksheet into view. The rows and columns the spreadsheet user sees on the screen of the CRT are the window. The intersection of each row and column on the screen corresponds to a cell. Each cell can be assigned a label or a title, a formula, or an item of data; as the user scrolls, the screen displays the assigned label, title, formula, item of data, or the result of applying the formula to the data item.

Suppose the user decides to title location A1 (column A, row 1) as "Student Name," B1 as "Exam 1," C1 as "Exam 2," D1 as "Term Paper," and E1 as "Total." The user then enters Donna Leak in position A2, 90 in B2, 80 in C2, and 85 in D2; the formula E2 (+B2+C2+D2) calculates 255 and places it in E2. Later the user enters the scores for the other students and the spreadsheet calculates all the totals and averages, instantly and automatically.

	A	B	C	D	E
1	Student Name	Exam 1	Exam 2	Term Paper	Total
2					+B2+C2+D2
3					+B3+C3+D3
4					+B4+C4+D4
5					+B5+C5+D5
6	Averages	@AVERAGE (B2..B5)	@AVERAGE (C2..C5)	@AVERAGE (D2..D5)	@AVERAGE (E2..E5)

(a) On a spreadsheet rows are numbered and vertical columns are named with letters. The last column, E, and the last row, 6 (in this case), indicate desired calculations.

	A	B	C	D	E
1	Student Name	Exam 1	Exam 2	Term Paper	Total
2	Donna Leak	90	80	85	255
3	David Fenollio	80	75	90	245
4	Nancy Bernman	60	70	80	210
5	Don Price	90	80	70	240
6	Averages	80	76.25	81.25	237.5

(b) The user enters the data in predetermined rows and columns and the spreadsheet software calculates the averages and totals as specified in that row.

FIGURE A.7. The spreadsheet user keys data into rows and columns, allowing the spreadsheet to display relationships.

Many individuals and organizations use spreadsheets to process data into information. Building contractors use spreadsheets to track anticipated and actual expenses and then calculate differences in dollars or percentages. An architect might find the electronic spreadsheet valuable in calculating sizes of supporting beams for floors and roofs under varying weight conditions. You can, of course, design a spreadsheet with pencil and paper, but the electronic spreadsheet makes the tool more powerful by automatically performing desired calculations.

One of the first and most successful electronic spreadsheets, VisiCalc (the Visible Calculator), hit the market in 1979. Since its introduction competing systems such as Lotus 1-2-3, Symphony, and Multiplan, some with more features than the original VisiCalc, have appeared. The VisiCalc system itself has grown since 1979 and now includes a VisiSeries composed of:

1. *VisiFile:* a data base or file manager.
2. *VisiTerm:* software to make the microcomputer act as a terminal and permit it to communicate with other microcomputers or mainframe computers.
3. *VisiDex:* a time manager and personal data organizer file and retriever.
4. *VisiTrend/Plot:* facility to draw graphs and charts of data.
5. *VisiSchedule:* capability to assist in scheduling projects that have several elements, each due at a certain time.
6. *VisiOn:* an operating system permitting concurrent processing of two or more applications.

To illustrate how electronic spreadsheets work, assume that Sally Edwards, Fleet Feet's vice-president, has collected sales statistics by category covering the past 3 years (Figure A.8*a*). She wants to use VisiCalc to help her analyze those sales, showing total sales for each year, growth of sales in each category, and projections for the next 2 years. Using the electronic spreadsheet, Sally designs a grid specifying the necessary data (Figure A.8*b*).

While reviewing the analysis, Sally discovers an error in the 1980 running shoes sales: $36,123 should read $44,836. To make this correction manually and then recalculate the totals would be a tedious task. With the electronic spreadsheet, Sally can enter the revised data and let VisiCalc do the work (Figure A.8*c*).

The software for spreadsheet systems such as VisiCalc costs about $200, operates

Sales Revenues	1980	1981	1982
Running shoes	36,123	45,642	56,812
Tennis shoes	45,826	54,982	60,376
Clothing	10,564	11,986	13,978
Miscellaneous	5,625	6,148	8,915
Sporting goods	15,875	14,987	16,865

(a) Sales statistics John has for the years 1980 through 1982.

FIGURE A.8. Fleet Feet sales statistics analyzed by a spreadsheet. (Courtesy Timberline Systems, Inc., Portland, OR 97223.)

APPENDIX A BUSINESS SYSTEMS TODAY **489**

	1980	1981	1982	Growth Rate	Projected 1983	1984
Sales Revenues						
Running shoes	36,123	45,642	56,812	25	71,249	89,355
Tennis shoes	45,826	54,982	60,376	15	69,369	79,702
Clothing	10,564	11,986	13,978	15	16,080	18,499
Miscellaneous	5,625	6,148	8,915	27	11,336	14,413
Sporting goods	15,875	14,987	16,865	3	17,450	18,055
Totals	114,013	133,745	156,946	17	185,484	220,025
Sales Percentage						
Running shoes	32	34	36	7	38	41
Tennis shoes	40	41	38	−2	37	36

(b) John Schug's original spreadsheet.

	1980	1981	1982	Growth Rate	Projected 1983	1984
Sales Revenues						
Running shoes	44,636	45,642	56,812	13	64,274	72,717
Tennis shoes	45,826	54,982	60,376	15	69,369	79,702
Clothing	10,564	11,986	13,978	15	16,080	18,499
Miscellaneous	5,625	6,148	8,915	27	11,336	14,413
Sporting goods	15,875	14,987	16,865	3	17,450	18,055
Totals	122,726	133,745	156,946	13	178,509	203,386
Sales Percentage						
Running shoes	37	34	36	0	36	36
Tennis shoes	37	41	38	2	39	39

(c) Sally's table after making the change in the 1980 running shoes.

FIGURE A.8 *(Continued)*

on many different microcomputers, and can store over 250 columns and 250 rows of data. Once users establish relationships and enter data, they can store everything on a disk, thus creating an electronic file for later retrieval and use. Some spreadsheets include security provisions, such as passwords, to prevent unauthorized use or tampering. Finally, most spreadsheets contain predefined functions such as logarithms, square roots, variances, and means to permit sophisticated calculations. Many microcomputer users buy their machines just to use VisiCalc.

OTHER SYSTEMS

While general ledger accounting and spreadsheet systems are common to all kinds of organizations, users have other systems to serve unique purposes. Some examples are order-entry, inventory, and fixed asset systems.

Order Entry

Order entry Part of AR system providing automatic labels, packing slips, and billing data.

Often part of an accounts receivable system, **order entry** allows a company to process customer orders rapidly by providing warehouse personnel with picking labels that they use to locate stock in the warehouse. The labels give the location (slot or bin number), number, and description of the item, the customer's name, and the date. The order-entry system also prints the packing slips that workers place in shipping cartons for the customer (Figure A.9). Packing slips show what is shipped to the customer versus what was ordered.

An order-entry system can make a dramatic difference to a firm that ships a variety of different products. The picking label (Figure A.10) lists items according to location so a worker spends less time hunting for them. Faster response time reduces labor costs and makes customers happy. Furthermore, the vendor can begin the billing process immediately, thus improving positive cash flow.

Inventory Records of the merchandise or other items on hand.

An order-entry system is closely related to an **inventory** system. As a company fills and ships orders using picking labels and packing slips, it automatically creates data about dwindling stock.

Manufacturers traditionally maintain surplus stock in warehouses, which they must tightly control—too few items may lead to shortages and lost sales, and too much stock ties up valuable financial resources. Programs can determine the proper inventory reorder schedule and annual requirements.

An inventory system stores the following data about each item in stock.

1. Description, price, unit of measure, cost, vendor, and location in warehouse.
2. Reorder schedule, safety stock level, and seasonal variation.
3. Quantities on hand, on order, and on reorder.
4. Historical activities such as quantity sold last month, last year, year-to-date, and month-to-date.

These data facilitate an inventory status report (Figure A.11) and a reorder report (Figure A.12).

A computerized inventory system provides management with a valuable tool, reducing the likelihood of overstock and improving cash flow. Cash resources become available for expansion, research and development, and other undertakings. In addition, the inventory system supplies management with information about stock turnover (what is selling and at what rate, and what is not selling), which helps decision makers improve the organization's profitability.

SLOT 09-175	1 OF 1	PACK/SIZE 24 TO 52
New Balance		
ITEM NO. 14512	CUST. NO 793	DATE 10/18/84
Fleet Feet		
653	18	S 3/100

FIGURE A.9. *Picking label sent to warehouse personnel to help them.*

```
          DELIVERY MANIFEST
hudson
house
TRUCK NO   TRAILER NO   DOOR NO   LOAD NO   CUBICS   WEIGHT   INVOICE NO.   ROUTE   DATE
                                             5.0      202     F39933                5/17/
                                            (FEET)   (POUNDS)
SPECIAL
INSTRUCTIONS _____ D

     Fleet Feet           CUST. 1071            Pieces
STORE _____ NO. _____
     Sacramento, California          COMPUTER COUNTS    DELIVERY COUNTS
ADDRESS _____
                                    GROCERIES _____
                          *         REPACKS     0    _____
METHOD OF                           CIG. (CTNS) 0    _____
UNLOADING _____ LABELS? YES
                           (YES OR NO)
                                    DELI.       0    _____
         SPEEDOMETER READING        MEAT             _____
 START              STOP            FROZEN     16    _____
                                                     _____
DRIVER SIGN.         DATE                            _____
UNLOADING     AM  UNLOADING   AM    TOTALS           _____
START TIME    PM  STOP TIME   PM

RECEIVED BY                         REMARKS
          (MUST BE SIGNED)          (USE REVERSE SIDE FOR ADDITIONAL COMMENTS)
```

FIGURE A.10. *A typical packing slip. The actual shipment may not be identical to the order, thus reflecting stock shortages in the warehouse.*

```
Fleet Feet                          Stock Status Report for 08/13/84
Page:    1


ITEM      DESCRIPTION          LOC     QTY      QTY      REORDER    AVG
NUMBER                                 ON       ON       LEVEL      COST
                                       HAND     ORDER

A100      SHOESTRINGS           SAC      7        20        12       0.49

A200      RACQUET               SAC      3        10         3      29.75
                                LA       5         0         3      29.75
                                DAVIS   12         0         4      29.95

A300      STEREO                DAVIS    4         0         1      65.47

A402      NIKE BLUE 8.5C WOMAN  CHICO    3         2         2      19.89
```

FIGURE A.11. *An inventory report is a detailed list of the stock a company owns, its location, and its value. If only one warehouse is involved, then the location will refer to an area in that building; if there are several warehouses, the location identifies which have stock of an item and how much. In our example, item A200 is in stock in Sacramento, Los Angeles, and Davis.*

```
Fleet Feet
Reorder Report for 08/13/84
Page:    1


MANUFACT'R     PART NUMBER     CUR-QTY     ORDR-QTY     UNIT COST     ORDER-COST
CINZANO        87D3R45            2            2          132.43         264.86
NIKE           2400BLUE           2            5           16.79          83.95
RAFTCRAFTS     1392A9             6            8            6.95          55.60
PANASONIC      26A2PAD            2            5          154.95         774.75
SUPERSOUND     52CASSETTE         2            3          275.00         825.00
```

FIGURE A.12. *The computer is programmed to calculate the ideal reorder point for all items. Whenever that point is reached for an item, the computer automatically prints an order form. The reorder report shows what has been ordered and in what quantity. (Westware Systems, 2455 S.W. 4th Ave. Ontario, OR 97914.)*

Many inventory systems automatically trigger reordering when inventory reaches a predetermined amount. When that point is reached, the computer places an order to replenish stock. This works well under normal economic conditions, but as economic fluctuations occur, management must guide the system to respond to changes in supply and demand.

Fixed Assets

Another system tied to the general ledger is fixed assets, which tracks a firm's long-term operational assets (land, buildings, equipment, furnishings, and machines). The fixed asset system records the value, maintenance history, and depreciation (if appropriate) of each asset. To accomplish this the system requires the following data.

1. Purchase price, date, and vendor.
2. Identification number, description, location, maintenance schedule.
3. Depreciation method, salvage value, expected life, current depreciated value.

A fixed asset system enables an organization to maintain tighter control over these assets. It will, for example, keep track of when an asset needs or receives repair. If management can evaluate an individual asset's expense history, it can determine the optimal replacement time. Also, with accurate and accessible data about each asset on hand, management can decide on insurance needs more easily. Figure A.13 shows a typical fixed asset report listing each asset, its description, date of purchase, cost, reserve, year-to-date depreciation, and salvage value.

```
FLEET FEET                                          FIXED ASSETS RESERVE LEDGER
ANNUAL REPORT AS OF 06/29/84
PAGE 2

ASSET NO.   DESCRIPTION          DOP     COST    RESERVE    YTD-DPER    SALVAGE

A60         FURNITURE            01/79   1000    243        65          100

A62         TYPEWRITER           09/79   1200    283        72          100

A65         TYPEWRITER           08/79   1200    260        75          100

A77         PHOTOCOPIER          01/82    900    450        150          50

A99         PERSONAL COMPUTER    02/82   4500     75        75          500
```

FIGURE A.13. A fixed asset report lists pertinent information about all assets of a given type. This report lists the assets owned by Fleet Feet, including asset number, description, date of purchase, cost, and depreciation data.

SUMMARY

One of the major business systems in use today is the general ledger, which summarizes results from all other systems into two reports—the balance sheet and the profit and loss statement. It has four important subsystems.

The accounts receivable (AR) system records an organization's sales to others, so the company can collect money owed by its customers. The primary reports produced by the AR system include the customer statement and the aged trial balance, or aging schedule.

The accounts payable (AP) system compiles the debts of an organization. The primary outputs of this system are checks to suppliers.

Every business requires a payroll system. The primary output from the system is an employee's paycheck, though the payroll system also can produce many other useful reports, including tax forms for federal, state, and local government agencies.

VisiCalc, an example of the electronic worksheet, allows users to manipulate two-dimensional tables with ease. Many people view the worksheet system as one of the most powerful systems ever developed.

Other widely used systems include order entry, inventory, and fixed assets.

NEW TERMS

Account	Credit	Inventory
Accounting	Debit	Invoice
Accounting equation	Electronic spreadsheet	Open item
Accounts payable	Fixed assets	Order entry
Accounts receivable	General journal	Payroll
Aging	General ledger	Posting
Balance forward		

QUESTIONS FOR REVIEW AND DISCUSSION

Questions appear in three categories. You can find the answers to the A group in this chapter. The B group requires you to apply the material presented here, whereas the C group necessitates investigation or research.

A.1. In the double-entry system of accounting, the general ledger requires that each transaction be recorded with balancing entries. Why is this so?
A.2. List the four primary accounting systems.
A.3. List the primary outputs from the AP, AR, and GL systems.
A.4. List the primary outputs from the order-entry, fixed asset, and inventory systems.
A.5. Make a list of four reasons why managers appreciate computers.
A.6. What are the two types of balancing methods found in accounts receivable systems?
A.7. What three time intervals do AR systems use for tracking the age of accounts?
A.8. What discount might a supplier offer a customer to encourage prompt payment?

B.1. Redraw Figure A.1 to include the inventory system.
B.2. Redraw Figure A.1 to include the order-entry system.
B.3. Why is order entry a part of the accounts receivable system?
B.4. Suppose an error were made in the AP system and the wrong vendor number was associated with a voucher. What would happen?
B.5. Suppose an error were made in the AR system and a wrong customer number was associated with an invoice. What would happen?
B.6. Suppose the wrong employee number were used on an employee's time card in the payroll system. What would happen?
B.7. In an inventory system, the unit of measure is kept for each item. Give three examples of units of measure.
B.8. In a payroll system, if an employee quits, is fired, or retires, when can the employee's data be removed from the system?
B.9. Does the bill in Figure A.4*d* show a balance forward or open item accounts receivable system?
B.10. Why does an AR system provide customer lists in order by customer number and alphabetical by customer name?

C.1. Some general ledger systems use a 13th month for year-end closing. Why?
C.2. There is a third type of balancing method in accounts receivable called balance only. How would you find out about this method? Do so and describe it.

REFERENCES

Alan L. Eliason, *Online Business Computer Applications*, Science Research Associates, Chicago, 1983.
Dan Fylstra and Bill Kling, *Advanced VisiCalc User's Guide*, Personal Software, 1984.
Nanci Lee, *Elementary Accounting*, Dryden Press, Hinsdale, Ill., 1984.

Glossary

The boldface number in parentheses at the end of each definition indicates the text page on which the term is first mentioned. Definitions are purposefully short, to reflect common usage and to aid in easy learning.

Acceptance criteria Rules determined in advance describing what constitutes an acceptable system. **(446)**

Acceptance notice Memorandum sent to all parties to notify them of a systems approval. **(452)**

Acceptance test Verification that the system meets its original goals and objectives. **(422)**

Access security Control procedure restricting access to a DBMS to approved users. **(278)**

Account An individual record for each asset, liability, or owners' capital. **(466)**

Accounting A control system that maintains the financial records of an organization. **(466)**

Accounting equation The rules governing how financial transactions are recorded. **(469)**

Accounts payable Records of amounts owed to creditors; often abbreviated as AP or A/P. **(466)**

Accounts receivable Records of debts owed to a business by its customers; often abbreviated as AR or A/R. **(466)**

Action entry Indicates via dot or X whether something should happen in a decision table. **(130)**

Action stub A list of all the processes involved in a decision table. **(130)**

Ada A programming language developed by the U.S. Department of Defense. Named for Ada August Lovelace, generally credited with being the first programmer. **(16)**

Aging Classifying the amount of a debt relative to time. **(478)**

Alternatives A list of potential solutions to a problem. **(83)**

Analysis documentation All the written reports produced during the first phase of the systems process. **(89)**

Analysis walkthrough System review at the end of analysis conducted to uncover analysis errors. **(138)**

ASCII The 7- or 8-bit binary coding system often found on micro- and minicomputers. An acronym for American Standard Code for Information Interchange. **(239)**

Asset An item of value owned by an organization: cash, equipment, land, patents, and so on. **(467)**

Auditing Tracing a transaction through a system to verify accurate processing. **(196)**

Backup Control procedure whereby files are copied on a routine or special basis. **(255)**

Backup file A copy of another file created to protect against loss of the original file. **(238)**

Balance forward A type of AR system where customer payments apply to the oldest transaction. **(474)**

Balance sheet The primary output of GL; it shows assets, liabilities, and stockholders' equity. **(469)**

495

BASIC A programming language developed at Dartmouth College in the 1960s. An acronym for Beginners All-Purpose Symbolic Instruction Code. (**256**)

Batch processing A technique that processes a number of similar transactions in groups. (**327**)

Batch total Control technique whereby the computer compares a filed sum with a predetermined one. (**296**)

Block mode Sending data to the computer in groups of characters. (**182**)

Blocking Grouping logical records together to form a single physical record. (**242**)

Blocking factor The number of logical records per physical record. (**242**)

Bohm and Jacopini Authors of a 1964 paper defining the three fundamental control structures. (**19**)

Bond A measure of the fiber content in paper. (**179**)

Bottom-up approach Testing approach that checks lowest level routines first, and then each module. (**394**)

Bus network A long line with computers or peripherals attached at various points. (**337**)

Byte A configuration of 8 bits representing a single character of data. (**241**)

Capacity A measure of the number of users or devices a network can support. (**337**)

Case Control structure supplementing IF-THEN-ELSE. (**359**)

Change procedure A formal process detailing how to alter or modify a system. (**446**)

Chart of accounts A formal list of all a firm's accounts and their identifying numbers. (**468**)

Check digit Control technique using a formula or pattern to check a field's correctness. (**296**)

Circle Represents the conversion of inputs to outputs in a data-flow diagram. (**20**)

Class test Validation technique for determining whether data are numeric or alphabetic. (**295**)

Closure The conclusion of an interview. (**47**)

COBOL A programming language developed for business data processing applications. An acronym for Common Business-Oriented Language. (**16**)

Code test Validation technique for matching data with predetermined values. (**295**)

Coding Translating module descriptions into a set of computer instructions. (**393**)

Cohesion A measure of a module's strength to accomplish a single task. (**110**)

Collision When two or more record keys calculate to the same location in the file. (**250**)

Column headings Captions appearing over vertical units of data. (**183**)

Combination test Validation technique used when groups of data are all present at the same time. (**295**)

Completely connected network A network providing a path or link between all stations. (**333**)

Condition entry A list of all the yes/no permutations in a decision table. (**130**)

Condition stub A list of all the necessary tests in a decision table. (**130**)

Context data-flow diagram Diagram showing the system in its most general form. (**77**)

Control footing The final part data item that shows the total for a specific group of data. (**184**)

Control headings Captions or titles separating one group of data from another. (**183**)

Control structure A pattern for building the logic of a computer program. (**121**)

Control system Set of procedures that help ensure proper and approved system operation. (**195**)

Conversion The changeover from one system to another. (**403**)

Cost overrun A figure that exceeds anticipated expenses. (**449**)

Coupling The measure of control and interdependence among modules. (**110**)

CPM Critical path method—planning and scheduling tool showing relationships and schedules for a project. (**161**)

Credit The right-hand side of an account recording payments or other values received. (**469**)

Critical path Pathway through a network that consumes the longest time from start to finish. (**161**)

Cross-footing Totaling all columns and rows of data to assure accuracy. (**196**)

Data-base design Design stage where the analyst decides on the system's file storage needs. (**154**)

Data dictionary Definition of each term, data item, or file used during analysis or design. (**128**)

GLOSSARY

Data entry Collecting data for future processing. (**291**)

Data-flow diagram Graphical description of a system's data and how the processes transform the data. (**125**)

DBMS Data-base management system. Software package permitting many applications to share information. (**154**)

DDL Language used to define a data base. An acronym for data definition language. (**272**)

Debit The left-hand side of an account recording indebtedness. (**469**)

Decision Control structure where the next activity depends on a conditional test. (**358**)

Decision table A chart with four sections listing all the logical conditions and actions. (**130**)

Decision tree A graph representation of a decision table. (**135**)

Decomposition The continual breaking down of a task into its elementary components. (**38**)

Design review A decision-making meeting to determine whether the system meets organizational needs. (**372**)

Design walkthrough System review at the end of design conducted to uncover design errors. (**138**)

Detail flowchart A type of flowchart depicting the steps necessary to write a program. (**115**)

Detail line Displays data for a single transaction. (**183**)

Detail report Displays all the pertinent facts about a group of transactions. (**185**)

Detailed analysis Expands beyond preliminary analysis to an elaborate explanation of the solution. (**21**)

Development Final phase of systems cycle: programs are written, tested, and installed. (**109**)

Development walkthrough System review of the completed system to determine if it complies with original plan. (**139**)

Direct conversion The immediate changeover from one system to another. (**407**)

Direct entry Data collection whereby a machine reads data directly into the computer. (**293**)

Direct file A file storage method that uses the record key to locate a record in the file. (**249**)

Discount An incentive reduction in price given to a buyer for early payment. (**85**)

Distributed processing Sharing of computerized systems among geographically dispersed computers. (**331**)

DML Commands written to retrieve or store data in a DBMS. An acronym for Data Manipulation Language. (**274**)

DMS-II A network DBMS operating on Burroughs computers. An acronym for Data Management System number two. (**272**)

Documentation Written materials describing a new system or changes made to an existing system. (**426**)

Double entry A bookkeeping approach requiring an equal debit entry for each credit entry. (**469**)

Dual intensity A terminal capable of displaying data in two levels of brightness. (**182**)

Dumb terminal A data-entry device that cannot check data entered by the operator. (**300**)

EBCDIC The 8-bit binary coding system most often found on IBM computers. An acronym for Extended Binary Coded Decimal Interchange Code. (**239**)

Electronic spreadsheet Software using rows and columns to permit automatic recalculation of data. (**487**)

Estimating time and cost A detailed list of each task, depicting its duration and expense. (**446**)

Exception report Displays data about a single or selected group of transactions. (**186**)

Expense The cost of obtaining revenue. (**466**)

Extended entry Type of decision table displaying values to be tested in the condition entry. (**134**)

External reports Reports sent to customers, clients, or others outside of the organization. (**173**)

Fact finding Learning as much as possible about the present system. (**71**)

Feasibility study A report including costs, alternatives, schedules, and background information. (**21**)

Feedback Comparing actual output with desired output for system adjustment. (**5**)

Fixed assets Records of all items of value owned by an organization. (**467**)

Flat file Data structure lacking the parent–child or any other record relationship. (**467**)

Flexibility A measure of the ability to expand or shrink a network without difficulty. (**337**)

Flowchart Graphic representation of the method of a problem's solution. (**111**)

Follow-up Summary of findings sent to the interviewee. (**47**)

FORTRAN A programming language used for mathematical, scientific, and engineering applications. An acronym for FORmula TRANslator. (**256**)

Gantt chart Scheduling tool that uses horizontal bars to depict project schedule and progress. (**67**)

General journal The chronological collection of many different types of transactions. (**469**)

General ledger The complete record of assets, liabilities, equity, revenue, and expenses. (**466**)

Hardware order A contract with a vendor to supply equipment. (**152**)

Hashing algorithm A procedure for converting a record key to a location in a file. (**249**)

Hierarchical relationship Parent–child data structure in which each child belongs to one parent. (**269**)

HIPO A set of charts emphasizing the functions of a system or computer program. An acronym for Hierarchy plus Input Process Output. (**115**)

Host language A DBMS that permits access using high-level-language constructs. (**275**)

IMAGE A network DBMS operating on HP computers. (**272**)

Implementation plan Outline showing activities, times, and events of development. (**388**)

IMS A hierarchical DBMS operating on IBM computers. Stands for information management system. (**275**)

Input design Design stage where the analyst decides how the system will collect data. (**155**)

Intelligent terminal Data-entry device that can check data without support from main computer. (**300**)

Internal reports Reports used by managers and others within an organization. (**173**)

Interpersonal relationships Problems an analyst must solve that deal with differences between people. (**447**)

Interview A meeting between analyst and relevant personnel for fact finding. (**45**)

Inventory Records of the merchandise or other items on hand. (**490**)

Invoice An itemized statement of merchandise sent to purchasers. (**479**)

IPO A detail chart listing inputs, processing steps, and outputs of a module. An acronym for Input Process Output. (**115**)

ISAM A file access method permitting both sequential and direct access to records. An acronym for Indexed-Sequential Access Method. (**248**)

Iteration A control structure that allows repetition as long as a condition remains true. (**122**)

Job definition Determining what work must be done or what products delivered. (**446**)

Key entry Data collection whereby a user enters data on a terminal or other similar device. (**293**)

Key-to-disk device Data-collection device that allows data to be recorded on a hard or floppy disk. (**300**)

Key-to-tape device Data-collection device that allows data to be recorded on a reel of tape. (**300**)

KSAM A file access method similar to ISAM. An acronym for Keyed Sequential Access Method. (**249**)

LAN (Local Area Network) A medium- to high-speed data communication system designed for intrabuilding use. (**337**)

Leveling Expanding a data-flow diagram to more specific details. (**38**)

Liability Debts owed by a business under an agreement with a creditor. (**469**)

Limited entry A type of decision table listing a y or n response for each condition. (**134**)

Line structure Structure wherein overall responsibility and authority are assigned to top-level management. (**7**)

Link test Verifying that programs work as intended. (**418**)

Looping A control structure that allows a set of tasks to be repeated. (**360**)

Loosely coupled Program modules that are entirely independent of each other. (**361**)

Maintenance Changing or modifying an existing system to fix errors or provide enhancements. (**433**)

Management control Scheduling personnel, equipment, or other resources to achieve maximum results. (**446**)

Management documentation Written materials

for supervisors, including overview, goals, costs, and schedules. (**427**)

Master file A file holding permanent or semipermanent records. (**238**)

Mixed entry A type of decision table mixing values in the condition and action entries. (**134**)

Module A discrete or identifiable single-function program unit. (**109**)

Module design Determining the nature and purpose of each module a program requires. (**157**)

Module integration Validating the coupling of modules in a program. (**402**)

Module testing Validating the correctness of each module. (**402**)

Multiple choice Questionnaire for responses to a predetermined set of possible answers. (**74**)

Multiple Record Layout Form A document used by an analyst to describe the format and fields in a file. (**308**)

Nassi-Shneidermann chart A tool that divides rectangles into halves to depict logical structures. (**137**)

NCR A type of paper that uses special chemical coatings to produce copies. An acronym for no carbon required. (**180**)

Network Parent–child data structure wherein child records can have more than one parent. (**270**)

Observation Monitoring an existing system to verify understanding and its operation. (**74**)

Online processing A technique that allows user access to data but does not permit updating. (**329**)

Open ended (1) A type of decision table that permits access to another decision table. (2) Questionnaire items that respondents must answer in their own words. (**134**)

Open item A type of AR system in which payments are applied to specific charges. (**474**)

Operations documentation Written materials for the operations staff explaining how to run a system. (**430**)

Optical reader Data-collection device that scans preprinted forms to collect data. (**304**)

Order entry Part of AR system providing automatic labels, packing slips, and billing data. (**490**)

Organization chart Flowchart identifying responsibilities and lines of authority. (**7**)

Output design Design stage where the analyst determines the content and format of reports. (**153**)

Owners' equity The portion of assets to which an owner has a direct claim. (**469**)

Package design Bringing together all design specifics. (**157**)

Packed decimal Numeric storage method that places two digits in each 8-bit segment. (**241**)

Page footing Appears at bottom of each page and may display page numbers, dates, or totals. (**184**)

Page heading Appears at top of each page and may display dates, page numbers, or program name. (**183**)

Parallel conversion Simultaneous operation of two systems cross-checking the results of each. (**403**)

Pascal A structured high-level programming language. Named for Blaise Pascal. (**16**)

Paycheck The primary output of a payroll system showing gross earnings and net amount. (**486**)

Payroll System calculating employee wages, withholdings, and benefits. (**466**)

Performance A measure of how quickly a network can respond to a user request. (**337**)

Periodic report Results printed at specific intervals, for example, every month. (**187**)

Personality conflict Interpersonal problems stemming from differences in individuality or disposition. (**448**)

Personnel selection Involving the right people throughout a project. (**446**)

PERT chart A graphic tool to assist task scheduling and estimate completion time. An acronym for Program Evaluation and Review Technique. (**160**)

Phased conversion Gradual replacement of one system with another. (**405**)

Philosophical differences Interpersonal problems stemming from differences in ideology or theory. (**447**)

Posting Transferring transactions from the journal to a ledger. (**469**)

Preliminary analysis Consists of evaluation of request, analysis of request, and management action. (**21**)

Preliminary report Contains the problem review, findings, recommendations, cost, and schedule. (**21**)

Presence test Validation technique for ensuring that all data have been collected. (**295**)

Present value Amount of a future cost measured in today's dollars. (**85**)

Presentation of findings Final step in detailed analysis where management makes a go/no-go decision. (**68**)

Process design Design stage during which the analyst decides how the system will function. (**156**)

Profit The return received by a business after all costs have been met. (**469**)

Program definition A detailed description of each individual program in the system. (**157**)

Program design A detailed description of every task a program performs. (**354**)

Program documentation Written materials explaining the details of each module in a system. (**427**)

Program file A file holding COBOL, Pascal, or other source-language statements. (**238**)

Program integration Verification that individual programs work together as expected. (**418**)

Program specifications Functional descriptions of each program the system needs. (**152**)

Program testing Checking the correctness of an entire program. (**388**)

Program walkthrough Peer review to find errors, omissions, faulty logic, or improper language use. (**395**)

Program writing The conversion of the program design into a computer language form. (**393**)

Pseudocode A concise English-like description of what users want the computer to do. (**120**)

Query language Data manipulation language designed for use by noncomputer-trained personnel. (**276**)

Query report Shows the details about one transaction, usually on a CRT. (**185**)

Questioning The fact-collecting component of an interview. (**45**)

Questionnaire Document designed to gather information from those involved with the system. (**72**)

Range test Validation technique for determining if data are within upper and lower bounds. (**295**)

Rank Questionnaire where respondents prioritize responses on a high to low ranking. (**74**)

Rating Questionnaire where respondents rank satisfaction on a 1 to 5, A to E, scale. (**74**)

Real-time processing A technique whereby data processing occurs rapidly enough to control an activity. (**330**)

Reasonableness test Validation technique that rejects data that should never occur. (**295**)

Record count A control procedure that calculates the total number of records in a file. (**254**)

Rectangle Represents a file in a data-flow diagram. (**125**)

Refinement Dividing a task into more specific details. (**360**)

Relational Data structure using two-dimensional tables to store data. (**270**)

Reliability A measure of the likelihood of a network failure. (**337**)

Remote job-entry device (RJE) A device transmitting data to another location for further processing. (**327**)

Repetition Control structure permitting looping or iteration. (**358**)

Report footing Overall and final values that appear at the end of a report. (**184**)

Report heading Titles that appear at the beginning of a printed document that shows titles. (**183**)

Revenue The inflow of cash or other assets for goods sold or services performed. (**469**)

Review and assignment Stage where the analyst appoints people to tasks and sets goals and schedules. (**68**)

RFP A written document on which the analyst specifies hardware needs. An acronym for Request for Proposal. (**341**)

RFQ A written document on which analysts write a description of functional needs. An acronym for Request for Quotation. (**341**)

Ring network A network providing links between computers but not to a central computer. (**335**)

Run manual Operator manual showing how to handle errors, special forms, files, and security. (**430**)

Sample documents A collection of printed forms or reports for input or output by a system. (**81**)

Schedule overrun A task taking longer than anticipated. (**447**)

Schema The definition or description of a data base, including fields and records. (**155**)

Scratch, or temporary, file A temporary file created to hold data between processing actions. (**238**)

Security A measure of the protection of a network against unauthorized use. (**198**)

Selection A control structure that allows tests to be made for events or conditions. (**121**)

Self-contained A DBMS with its own special language for retrieving and storing data (**274**)

Sequence A control structure where each action follows the next in a linear fashion. (**121**)

Sequence test Validation technique that assures data fall in a correct order. (**295**)

Sequential file A file that stores records according to a key field. (**244**)

Sign test Validation technique that compares a value with zero. (**295**)

Slide Control technique that detects an improperly placed decimal point. (**297**)

Software order A contract with a vendor to supply programs. (**152**)

Soundex A procedure for converting alphabetic keys to numeric values. (**250**)

Source document The paper containing the original data (**298**)

Span of control Refers to the number of modules subservient to a parent module. (**111**)

SQL Relational DBMS operating on IBM computers. An acronym for Structured Query Language. (**272**)

Square Represents a source or destination of data in a data-flow diagram. (**36**)

Staff structure Departments in an organization serving other departments. (**8**)

Standards A collection of rules analysts and programmers must follow. (**393**)

Star network A network providing a path or link to one central computer. (**333**)

Statement A bill sent to customers advising them of their debts to an organization. (**478**)

Straight binary Numeric storage method using the base 2 number system. (**241**)

String test Evaluation of programs where output from one program is input to the next. (**418**)

Structure chart A flowchart-like diagram showing the hierarchy of program modules. (**119**)

Structured English A synonym for pseudocode with limited vocabulary and limited syntax. (**120**)

Structured methodology Set of rules, guidelines, and tools facilitating the systems process. (**16**)

Structured walkthrough A peer review of a system conducted to uncover errors. (**138**)

Stub An abbreviated version of a program module written to facilitate programming. (**393**)

Summary report Provides overall totals for specific groups of detail records. (**185**)

System Combination of resources working together in converting inputs to outputs. (**4**)

System audit The last check or review of a system to assure it meets objectives and goals. (**23**)

System documentation The materials assembled from analysis, design, and development. (**453**)

System flowchart A flowchart showing the interaction among programs. (**112**)

System specifications The description of the input, output, control, and file designs for a system. (**22**)

System test Users verify that a system performs as expected. (**418**)

Systems analysis The first phase in the systems process, identifying problems. (**6**)

Systems analyst Person performing the systems study. (**6**)

Systems design The second phase in the systems process, specifying output, input, and data base. (**6**)

Systems development The last phase in the systems process, including writing, testing, and installation. (**6**)

Systems life cycle Expected time a system will operate, from time of conception to removal. (**20**)

Territorial pride Interpersonal problem related to one person invading another's area. (**447**)

Test data The collection of data the analyst plans to use in proving the system's accuracy. (**418**)

Test plan The method of assuring the system's correctness. (**158**)

Testing (1) Verifying that a system performs as expected. (2) Checking a program to see that it performs as expected. (**399**)

Theory X Human behavior thesis based on dislike of work, and lack of ambition and control. (**448**)

Theory Y Human behavior thesis that insists managers satisfy high-level needs. (**448**)

Theory Z Human behavior thesis that recommends worker involvement, trust, and concern. (**449**)

Tightly coupled Program modules that are intertwined to a great degree. (**361**)

Top down (1) Testing approach that checks highest level modules first, and then lower level ones. (2) Dividing a system's major modules into increasingly detailed modules. (**108**)

Training Explaining to all parties how a system operates. (**423**)

Transaction, or detail, file A file describing the details of a business activity. (**238**)

Transaction logging Control procedure that automatically produces copies of old and new records. (**278**)

Transposition Control procedure that checks to ensure correct order of the entries. (**297**)

Turnaround document Computer outputs that serve as input for a subsequent activity. (**173**)

User Person or group initiating the systems study. (**10**)

User documentation Written materials that teach people how to use a system. (**427**)

User request Usually initiates the systems process. (**39**)

Utility Software to permit backup, boosting capacity, purging, or transferring a DBMS. (**277**)

Utilization A measure of the ability of a network to keep track of usage statistics. (**337**)

Validation Data-collection control procedure using the intelligence of the computer itself. (**295**)

Verification Data-collection control procedure wherein the operator rekeys data. (**295**)

Visual verification Control technique where the computer displays data for the operator to check. (**295**)

VSAM A file access method that uses a table to locate a record in the file. An acronym for Virtual Sequential Access Method. (**248**)

Visual Table of Contents (VTOC) Method of diagramming program modules in order of priority. (**115**)

Walkthrough A peer review of a system conducted to uncover errors. (**138**)

Warm-up Establishing rapport during an interview. (**45**)

Warnier-Orr diagram A design tool that uses brackets to group related items or operations. (**135**)

Windowing Splitting of a CRT screen to show multiple activities or options. (**182**)

Index

Acceptance criteria, 446
Acceptance notice, 452
Acceptance test, 422
Access security, 278, 280
Access versatility, 267
Account, 466
Accounting, 102, 466
  equation, 469
Accounts Payable, 466, 479–482
Accounts Receivable, 466, 474–479
Action entry, 130
Action stub, 130
Ada, 16, 369
Adabase, 198, 277
Aging, 478
Alias, 129
Alternatives, 83
American Federation of Information Processing, 457
Analysis, 109
  detailed, 21
  documentation, 37, 89
  preliminary, 21
  request, 42
ANSI, 111
AP, 24, 40, 113, 117, 126, 127
  check, 80, 82
  context DFD, 78, 79
Apple, 12
  Macintosh, 15
AR, 5
ASCII, 239
Asset, 467
Association for Computing Machinery, 457

Association for Systems Management, 457
Attribute, 271
Audit, system, 23, 451
Auditing, 196
Average cost, 450

Backup, 255
Backup file, 238
Balance forward, 474
Balance sheet, 469
BASIC, 256
Batch processing, 327
Batch total, 296
Baud, 181
BCD, 240
Binary, straight, 241
Blinking video, 182
Blocking, 242
  factor, 242
Block mode, 182
Bohm, Corrado, 19, 358
Bond, 179
Bottom-up approach, 394
BPI, 242
Brace, 128
Brackets, 135
Bricklin, Dan, 123
Budget and schedule, 37
Burroughs, 16, 198
Burster, 180
Bus network, 337
Byte, 241
Byte, 457

Capacity, 337
Case, 359
Chart
  of accounts, 468
  Gantt, 67
  organization, 7–9
  PERT, 160–163
Change procedure, 446
Check digit, 296
Child data set, 269
Circle, in data-flow diagram, 20
Class test, 295
Closure, 47
COBOL, 16, 17, 19, 102, 123, 157, 164, 256, 367
Codd, Dr. Edgar, 276
Code test, 295
Coding, 393
Cohesion, 110
Collision, 250
Column heading, 183
Combination test, 295
Communication skills, 30
Completely connected network, 333
Computerworld, 457
Condition entry, 130
Condition stub, 130
Context Data-Flow Diagram, 77
Controls, 155, 230
  backup, 255
  check digit, 296
  data base, 277, 278
  data-entry, 294–298
  footing, 184

heading, 183
management, 446
record count, 254
slide, 297
structure, 121, 136, 358
system, 195
total, 296
transportation, 297
Conversion, 403
direct, 407
facilities, 409
phased, 405
procedure, 410
program, 408
Cost overrun, 449
Count, record, 254
Coupling, 110, 363
CPM (Critical Path Method), 161–163, 449
Credit, 469
Critical Path Method, 161–163, 449
Crossfooting, 196
CRT, 155, 181, 190, 191

DASD, 244
DASDL, 272
Data, 4
external, 291
internal, 291
security, 267
set, 269, 272
test, 418
Database, 371
controls, 278
design, 154
management, 197
management systems, 266
type, 269
Data entry, 291
controlling, 294–298
screen design, 299, 305
Data Definition Language (DDL), 272
Data dictionary, 128–130, 154, 199
system, 129
Data element, 128
Data-Flow Diagram (DFD), 20, 36, 125, 252, 354
analysis, 37
arrows, 36, 125
box, 36
circle, 20, 125

context, 77
open-ended box, 125
rectangle, 36
square, 36, 125
Data Manipulation Language (DML), 273
Datamation, 458
Data names, 395
Data processing department, 8
Data Processing Management Association (DPMA), 457
dBase II, 272, 277
DBMS, 154, 157, 266
advantages, 267, 285
capabilities, 277
disadvantages, 285
hierarchical, 269
network, 270
relational, 270
DDL, 272
Debit, 469
DEC, 260
Decimal, packed, 241
Decision, 358
Decision table, 130
Decision tree, 135
Decision support system, 76
Decollator, 180
Decompose, 109
Decomposition, 38, 360
DeMarco, Tom, 19
Design, 109
data base, 154
file, 154, 251, 253
input, 155
module, 157, 354
output, 153
overview, 151–159
package, 157
process, 156
program, 157, 354
review, 158, 372
review and assignment of tasks, 152
screen, 190–191
walkthrough, 138
Detail file, 238
Detail flowchart, 115
Detail line, 183
Detail report, 185
Detailed analysis, 21, 67
Development, 109, 386

Development walkthrough, 138
DFD, 36
analysis, 37
arrows, 36, 125
box, 36
circle, 20, 125
context, 77
open-ended box, 125
rectangle, 36
square, 36, 125
Diagram, Warnier-Orr, 135–137, 157
Dijkstra, E., 18
Direct access, 244
files, 248
Direct conversion, 407
Direct entry, 293
Direct files
advantages, 250
disadvantages, 250
Discount, 85
Disk
drives, 237
floppy, 244
hard, 244
Distributed processing, 331
Distribution, 200
DML, 274
DMS-II, 24, 272, 277, 371
Document, source, 298
Documentation, 426
analysis, 89
system, 453
Domain, 271
Double entry, 469
Dual intensity, 182
Dumb terminal, 300

EBCDIC, 239
Electronic Funds Transfer Act of 1979, 198
Electronic spreadsheet, 486–489
Ergonomics, 425
Estimating time and cost, 446
Ethernet, 337
Evaluation of user's request, 39
Exception report, 186
Expense, 466
Extended entry, 134
External data, 291
External report, 173, 175
Eye strain, 425

INDEX **505**

Facilities conversion, 409
Fact finding, 68, 71
Feasibility study, 21, 37
Feedback, 5
File
    consolidation, 267
    design, 154, 251
    direct access, 248
    independence, 267
    layout sheet, 251
    names, 395
    sequential, 242
    types, 238
Fixed assets, 467, 492
Flexibility, 337
Floppy disk, 244
Flowchart, 111
    detail, 115
    program, 115
    system, 112
Follow-up, 47
Footing
    control, 184
    page, 184
    report, 184
Forms mode, 182
FORTRAN, 256, 369, 440
Frankston, Bob, 124
Frequency, 200

Gantt chart, 67, 90, 91, 388, 447
General journal, 469
General ledger, 25, 27–29, 466, 467–474
GO TO, 19, 357, 399
Guidelines
    data dictionary, 128
    data-entry screens, 299
    DFD, 126, 127
    file design, 253
    flowcharts, 111
    IF, 399
    modules, 366–367
    pseudocode, 122, 123
    screen design, 190–191
    terminal users, 425
    walkthrough, 138

Hard copy, 182
Hard disk, 244
Hardware, 11
    order, 152
Heading, 174

column, 183
control, 183
page, 183
report, 183
Hexadecimal, 240
Hierarchical, 269
HIPO, 115
Hopper, Grace, 17
Host language, 275
HP, 25, 140, 198, 256

IBM, 276
    PC, 11, 260
    3080, 11
    3380 disks, 237
IF, 399
IF-THEN-ELSE, 121, 358
Image, 198, 272, 277
Implementation plan, 388
IMS, 277
Information, 4
Infosystems, 458
Ingres, 272
Input design, 155
Inquiry, 198
Intelligent terminal, 300
Interblock gap, 243
Internal data, 291
Internal report, 173
Interpersonal relationships, 447
Interview, 43, 45
    preparation for, 45
Interviewing, 72
Inventory, 25, 490
Inverse video, 182
Invoice, 82, 479
IPO, 115, 119, 120, 371
ISAM, 248
Iteration, 122, 360

Jacopini, G., 19, 358
Job definition, 446
Job description, 43, 44
Journals and periodicals, 457

Keane Inc., 446
Key entry, 293
Key-to-disk, 300
Key-to-tape, 300
KSAM, 249

LAN, 337–339
Language
    host, 274

query, 276–277
self-contained, 275
Layout, multiple record, 308
Ledger card, 81
Letter of authorization, 43
Leveling, 38
Liability, 469
Life cycle, system, 20
Limited entry, 134
Line
    detail, 183
    structure, 7
Link test, 418
Local Area Network (LAN), 337
Logging, transaction, 278
Logical system, 77–81
Looping, 360
Loosely coupled, 361
Lotus 1-2-3, 260, 277, 488

Machine language, 16
Maintenance, 433
Management
    control, 446
    documentation, 427
    information systems, 51
    training, 424
Manual entry, 293
Master data set, 269
Master file, 238, 256
MB, 248
McGregor, Douglas, 448
Meeting, conducting a, 61
Methods of data entry, 291
Mixed entry, 134
Modern Office Technology, 458
Modula-II, 369
Module, 109, 356
    design, 157
    guidelines, 366–367
    integration, 402
    length, 357
    rules, 395
    testing, 402
Moore Business Forms, 230
Multiple-choice questionnaire, 74
Multiple record layout, 308

NASA, 330
Nassi, I., 137
Nassi-Shneidermann chart, 137, 138, 157
NCIC, 329

NCR paper, 179, 180
Network, 270
  bus, 337
  completely connected, 333
  local area, 337
  ring, 335
  star, 333
New York Times project, 19

Observing the system, 74
OCR, 294
Online processing, 329
Open-ended decision table, 134
Open item, 474
Operations
  documentation, 430
  training, 424
Optical character reader, 294, 304
Optimistic cost, 450
Order, 200
Order entry, 490–492
Organization
  chart, 7, 43
  line, 7
  staff, 7
  structures, 6
  types, 6
Orr, Kenneth, 135
Ouchi, William, 448
Output design, 153, 173
Owners' equity, 469

Package design, 157, 369
Packed decimal, 241
Packing slip, 82, 491
Page
  footing, 184
  heading, 183
  numbers, 174, 195
Paper
  NCR, 179, 180
  types, 179, 200
Parallel conversion, 403
Parent data set, 269
Pascal, 16, 256, 363
Passwords, 195
Payroll, 25, 466, 482–486
People problems, 447
Performance, 337
Periodic report, 187
Personality conflict, 449
Personnel selection, 446
PERT, 160–163, 388, 391, 449

Pessimistic cost, 450
Phased conversion, 405
Philosophical differences, 447
Picking label, 490
Plan, implementation, 388
Plus sign, 128, 136
Posting, 469
Preliminary
  analysis, 21, 39, 67
  report, 47, 49
Presence test, 295
Presentation of findings, 68
Presentation to management, 68, 86–90
Present value analysis, 85
Prime number, 249
Printer, 176
  character-at-a-time, 177
  cost, 177
  impact, 178
  layout chart, 188
  line-at-a-time, 177
  nonimpact, 178
  speed, 178
Procedures, conversion, 410
Process design, 156
Processing
  batch, 327
  distributed, 331
  on-line, 329
  real-time, 330
Professional society, 457
Profit and loss statement, 469
Program, 176
  building master file, 154
  conversion, 164, 408
  definition, 157
  design, 354
  development, 267
  documentation, 427
  file, 238
  flowchart, 115
  independence, 267
  integration test, 418
  maintenance, 267
  sorting a transaction file, 150
  source, 91
  specifications, 24, 152, 158, 369–371
  structures, 18
  testing, 388
  utility, 149

  walkthrough, 395
  writing, 393
Pseudocode, 120, 140, 157

Query language, 276–277
Query report, 185
Questioning, 45
Questionnaires, 72
  multiple-choice, 74
  open-ended, 74
  rank, 74
  rating, 74

Random access, 249
Range test, 295
Rank questionnaire, 74
Rapport, 72
Rating questionnaire, 74
Reader, optical, 304
Real-time processing, 330
Reasonableness test, 295
Record count, 254
Record layout sheet, 252
Record names, 395
Refinement, 360
Relational data sets, 270, 277
Relational DBMS, 274–275, 284
Reliability, 337
Remote job entry (RJE), 327
Repetition, 135, 358
Report
  design, 187–199
  detail, 185
  detailed findings, 98
  exception, 188
  external, 173
  footing, 184
  heading, 183
  internal, 173, 174
  periodic, 187
  query, 185
  summary, 185
Request
  for proposal, 341
  for quotation, 341
  for systems services, 41
Resource utilization, 337
Review
  assignment of tasks, 68, 152
  design, 372
RFP, 341
RFQ, 341
RJE, 327

RPG, 256
Run manual, 430

Sample documents, 81
Savings, 83
Schedule overrun, 447
Schema, 155, 272
Scratch file, 238
Screen
    design guidelines, 190–191
    layout chart, 192
    splitting, 182
Scrolling, 182
Secured video, 182
Security, 198, 200, 337
    access, 278, 280
Selection, 121
Self-contained, 274
Sequence, 121, 358
    test, 295
Sequential file, 242
Sheet, file layout, 251
Shneidermann, B., 137
Sign test, 295
Slave data set, 269
Slide, 297
Software, 16
    order, 152
    supplier, 76
Soundex, 250
Source document, 293, 298
Span of control, 111
Special information, 267
Specifications
    program, 24, 152, 370
    systems, 22, 152
Spreadsheets, 123, 124, 423, 486–489
SQL/DS, 24, 272, 277, 284
Square, in data-flow diagram, 36, 125
Staff structure, 8
Stale date, 230
Standards, 393, 395
Star network, 333
Statements, 395, 478
State-of-the-art, 167
Stock tab, 179
Straight binary, 241
String test, 418
Structure, control, 358

Structured English, 120
Structured methodology, 16, 108
    cohesion, 110
    coupling, 110
    iteration, 122
    module, 109
    pseudocode, 120
    selection, 121
    sequence, 121
    span of control, 111
    top down, 108
    walkthrough, 138
Structured programming, 18
Structured walkthrough, 138
Stub, 393, 397
Summary report, 185
Symphony, 488
System, 4
    analysis, 6, 10
    analyst, 6, 10
    audit, 23, 451
    components, 5
    computer-based, 4
    control, 195
    design, 6, 22
    development, 6, 22
    documentation, 453, 454
    flowchart, 112
    life cycle, 20, 23
    observation, 74
    process, 109
    program, 152
    specifications, 22, 152, 157
    study, 21, 46
    test, 418
System R, 274

Tax Reform Act of 1976, 198
Temporary file, 238
Terminals, 14, 181–183, 190
    advantages, 182
    dumb, 300
    guidelines, 425
    intelligent, 300
Territorial pride, 449
Test
    acceptance, 422
    data, 418, 429
    plan, 158
Testing, 399, 418
    bottom-up, 394

    link, 418
    program, 388
    program, integration, 418
    string, 418
    system, 418
Theory X, Y, Z, 448
Tightly coupled modules, 110, 361
Time stamp, 195
Timecard, 485
Top-down approach, 108
Total, 174, 200
    batch, 296
Traditional files, 266
Training, 423
Transaction file, 238, 256
Transaction logging, 278
Transposition, 297
Trial balance, 174
Tuple, 270
Turnaround document, 173, 175

Unit testing, 402
UPC, 293
Upgrade, 77
User, 10, 39
    documentation, 427
    training, 424
Utility routines, 277

Validation, 195, 295
    tests, 295
VDT, 181
Verification, 195, 295
VisiCalc, 124, 486, 488
VisiPlan, 260, 488
VisiSchedule, 390, 488
VisiTerm, 260, 488
Visual Table of Contents (VTOC), 115–119, 157, 159, 393
Volume, 200
VSAM, 248

Walkthrough, 138, 139, 371
    program, 395
Wall Street Journal, 457
Warm-up, 45
Warnier, Jean-Dominique, 135
Warnier-Orr diagram, 135–137, 143, 371
WHILE, 121–123
Windowing, 182
WordStar, 260
Writing style, 440